"ENLIGHTENED ZEAL"

The Hudson's Bay Company and Scientific Networks,
1670–1870

"Enlightened Zeal" examines the history of the Hudson's Bay Company's involvement in scientific networks during the company's two-hundred-year chartered monopoly. Working from the company's voluminous records, Ted Binnema demonstrates the significance of science in the company's corporate strategies.

Initially highly secretive about all of its activities, the HBC was by 1870 an exceptionally generous patron of science. Aware of the ways in which a commitment to scientific research could burnish its corporate reputation, the company participated in intricate symbiotic networks that linked the HBC as a corporation with individuals and scientific organizations in England, Scotland, and the United States. The pursuit of scientific knowledge could bring wealth and influence, along with tribute, fame, and renown, but science also brought less tangible benefits: adventure, health, happiness, male companionship, self-improvement, or a sense of meaning.

The first study of scientific research in any chartered company over the entire course of its monopoly, *"Enlightened Zeal"* expands our understanding of social networks in science, establishes the vast scope of the HBC's contribution to public knowledge, and will inspire new research into the history of science in other chartered monopolies.

TED BINNEMA is a professor in the Department of History at the University of Northern British Columbia.

"Enlightened Zeal"

The Hudson's Bay Company and Scientific Networks, 1670–1870

TED BINNEMA

UNIVERSITY OF TORONTO PRESS
Toronto Buffalo London

© University of Toronto Press 2014
Toronto Buffalo London
www.utppublishing.com
Printed in Canada

ISBN 978-1-4426-4697-1 (cloth)
ISBN 978-1-4426-1475-8 (paper)

∞

Printed on acid-free, 100% post-consumer recycled paper with vegetable-based inks

---

Library and Archives Canada Cataloguing in Publication

Binnema, Ted, 1963–, author
Enlightened zeal : the Hudson's Bay Company and scientific networks, 1670–1870 / Ted Binnema.

Includes bibliographical references and index.
ISBN 978-1-4426-4697-1 (bound). – ISBN 978-1-4426-1475-8 (pbk.)

1. Hudson's Bay Company – History.   2. Hudson's Bay Company – Research – History.   3. Science and industry – Northwest, Canadian – History.   4. Science – Social aspects – Northwest, Canadian.   I. Title.

FC3209.S35B55 2014   971.2'01   C2014-900134-7

University of Toronto Press acknowledges the financial assistance to its publishing program of the Canada Council for the Arts and the Ontario Arts Council.

University of Toronto Press acknowledges the financial support of the Government of Canada through the Canada Book Fund for its publishing activities and the assistance of the University of Northern British Columbia Office of Research through a publication grant.

This book has been published with the help of a grant from the Canadian Federation for the Humanities and Social Sciences, through the Awards to Scholarly Publications Program, using funds provided by the Social Sciences and Humanities Research Council of Canada.

To the memory of Harry J. Groenewold

Contents

List of Illustrations ix

Acknowledgments xi

Preface xv

1 Introduction 3

Part I: The Hudson's Bay Company and Science, 1670–1821

2 "A Profound Secret": The Adventurers and the Fellows from the 1660s to 1768 47

3 "Desirous to Encourage Science": The Transit of Venus of 1769 and the Hudson's Bay Company's Collaboration with the Royal Society, 1768–1774 75

4 "Amends for the Narrow Prejudices": The Hudson's Bay Company and Science in an Era of Competitive Expansion, 1774–1821 95

Part II: The Hudson's Bay Company and Science, 1821–1870

5 "Benevolent Intentions": The Hudson's Bay Company, the Royal Navy, and the Search for the Northwest Passage, 1818–1855 129

6 "The Liberal Spirit": David Douglas, Edinburgh, and the Douglas Legacy, 1823–1870 169

7 "Disinterested Kindness": The Hudson's Bay Company and North American–Based Science, 1821–1870 199

8 "Knowing the Liberal Disposition": The Hudson's Bay Company and the Smithsonian Institution, 1855–1868 238

Epilogue 290

Conclusion 294

Notes 299

Bibliography 405

Index 439

Illustrations

Some of the black-and-white illustrations listed below are also reproduced in the colour plate section, which follows page 224.

1.1 Hudson's Bay Company Territories at Their Maximum Extent, 1821–1838   4
1.2 *Hudson's Bay House – Fenchurch Street (London)*, 1854 [also colour plate 1]   5
1.3 *Sir George Simpson* [also colour plate 2]   16
2.1 Rupert's Land and the HBC's Transatlantic Transportation Route, 1670–1768   59
2.2 "The Effects of Cold"   63
2.3 "Ring-Tail'd Hawk" [also colour plate 3]   64
2.4 Hudson's Bay Company Explorations to 1772   69
3.1 The Transit of Venus   76
3.2 J.H. Lambert's Proposed Meteorological Network, 1771   84
3.3 "Account of Several Quadrupeds"   87
4.1 Hudson's Bay Company Territories, 1774–1821   96
4.2 Thomas Hutchins on the Freezing Point of Mercury   101
4.3 "A Plan of Part of Hudson's Bay and Rivers, Communicating with York Fort and Severn," by Andrew Graham, ca. 1774   107
4.4 "A Map exhibiting all the New Discoveries in the Interior Parts of North America" by Aaron Arrowsmith, 1795, with additions to 1802 [also colour plate 4]   115
5.0 *Hudson's Bay Company's House, Fenchurch Street, 1843*   128
5.1 Arctic Explorations Sponsored or Assisted by the HBC, 1821–1838   136
5.2 Map of the Arctic Coast of America, 1837   149
5.3 Arctic Explorations Sponsored or Assisted by the HBC, 1845–1854   154

5.4 *John Rae*, by Stephen Pearce [also colour plate 5]   162
6.1 The Columbia District, 1821–1838   175
6.2 Hudson's Bay Company Territories, 1838–1870   182
7.1 Paul Kane's Portrait of John Henry Lefroy (1845–1846)   214
7.2 Paul Kane, Field Sketch: *Kee-a-kee-ka-sa-coo-way: Head Chief of the Crees*, 1848 [also colour plate 6]   217
7.3 Paul Kane, *Kee-A-Kee-Ka-Sa-Coo-Way, Man Who Gives the War Whoop* [also colour plate 7]   219
7.4 Distribution of Forests   230
8.1 Smithsonian Institution Building from the Northeast with Flowers along Mall (1860s)   239
8.2 Spencer Fullerton Baird, 1867   247
8.3 Robert Kennicott Shortly Before Going to the HBC Territories   251
8.4 Robert Kennicott's Travels in the Mackenzie River District   254
8.5 "Robert Kennicott, Explorer, in Field Outfit"   257
8.6 Bernard Rogan Ross   267
8.7 Chief Factor Roderick McFarlane   270

Acknowledgments

It is a pleasure to pay tribute to those people who have assisted me as I researched for and wrote this book. I must begin by thanking my colleagues (fellow faculty members and graduate students) in the history department at the University of Northern British Columbia. It is difficult to place a value on the collegiality that has existed for as long as I have been there. I thank Jon Swainger (who proofread the whole manuscript), Jacqueline Holler, and Dana Lightfoot in particular. I shudder to think of how often I have bored them with stories about what I was finding and writing. Marianne (Marika) Ainley encouraged me during the early part of this project. Although Marika has since died, her encouragement and advice was important when I was uncertain whether I wanted to attempt a foray into the history of science. While I was researching Mark Sarrazin and Yvan Prkachin gave me valuable assistance.

Several people read and commented on earlier versions of the entire manuscript. For generously taking so much time to provide their valuable feedback, I owe a huge debt of gratitude to Maria Van Harten, David Vogt, David Holm, and Bill Waiser. Bill even went so far as to accompany me on home-and-home hiking trips in the Babine Mountains and Grasslands National Park. Jack Nisbet and Peter Broughton read and commented on portions of the manuscript. Jack fed me bits of evidence even as I was making late revisions. Peter and his wife Marilyn also showed me their hospitality on several of my trips to Toronto. The opportunity to chat about mutual interests in their living room was a highlight. I also wish to thank David and Rosemary Malaher, and Andy Korsos for helpful suggestions and support. David and Rosemary have been as gracious hosts to me in Vancouver as Peter and Marilyn have been in Toronto. Who knew that the historical guild could be so congenial? Debra Lindsay, having learned that I was busy with this project, also contacted me on her own initiative to offer me advice.

At UNBC I wish to thank Mary Bertulli and Liz Tait in the interlibrary loan office for their tireless and cheerful efforts to meet my continual requests for material. The staff at the Hudson's Bay Company Archives, Provincial Archives of Manitoba, also readily and willingly went beyond their call of duty to assist me. I also want to thank the National Portrait Gallery, the Art Gallery of Ontario, the Royal Ontario Museum, the Library and Archives Canada, and the Hudson's Bay Company Archives for generously permitting me to reproduce some of the artwork in their collections at such low cost.

The Social Sciences and Humanities Research Council of Canada awarded me a grant which, despite the fact that the final product evolved significantly from my original proposal, supported much of the research and writing of this book. I wish to thank the council for its support. I also wish to acknowledge Barbara Belyea, with whom I shared the SSHRC award. Our mutual interest in Peter Fidler brought us together. Although our actual research agendas diverged, Barbara has continued to be an encouragement.

Even when the researching and the writing of this study did not take me away from home, I must admit that I was often absorbed and preoccupied with the project. I thank my wife, Helen, and my children, Derek, Kathryn, and Josiah, for their patient tolerance of my absences, literal and figurative, and my idiosyncrasies. The families of academic writers may well have the worst of all possible worlds: they must live daily with distracted and engrossed writers, unable to entertain realistic dreams of sharing in the windfalls of a bestseller. But my family accepts me all the same.

The citations in this book pay insufficient tribute to the scholars who have preceded me in the study of the history of science in the Hudson's Bay Company. Many very fine books and articles have already been written on several topics and periods examined in this book. Writing a history of science that spans two hundred years of history would have been impossible without the work done by others. That the present study sometimes defends interpretations that differ from the literature should not obscure the fact that I continue to be impressed by the literature that exists. A project such as this one, geared towards a book that seems to be a mile wide and an inch deep, has been daunting for me. Deeper research would certainly have revealed errors and distortions. I also acknowledge that if the history of science in chartered monopolies receives the scholarly attention it deserves, future studies will certainly reveal many shortcomings of this one. None of the shortcomings should be attributed to those who I have acknowledged here. If however, this book is judged to have made a worthwhile contribution in its time, a substantial measure of credit goes to the scholars who have preceded me, and to those to whom I have paid specific tribute here.

I don't know if it is part of the process of aging, but as I have gotten older, I have reflected more on those who helped nurture my fascination with the past. I have been fortunate enough to have had the opportunity to co-edit *festschriften* published to honour my graduate supervisors. This book is dedicated to the memory of my first undergraduate history professor, Harry J. Groenewold, at The King's University College. He deepened my interest in history and helped me understand its enduring relevance. I recall "rewarding" him with awfully indifferent work when I was an undergraduate student, but somehow he still instilled in me a passion for inquiry that drove me on. He is missed.

Preface

When I began the research that produced this book, I did not intend to write a history of the Hudson's Bay Company's contributions to science, and I did so only because it seemed necessary. Longer ago than I am willing to admit, I undertook to understand better the HBC trader, Peter Fidler, whose remarkably erudite journals contributed so much to my understanding of the human and environmental history of the northwestern plains. Knowing that we shared an interest in Fidler, Barbara Belyea (of the English Department at the University of Calgary) and I secured a grant from the Social Sciences and Humanities Research Council of Canada in 2002 to study this enigmatic, little-known figure. Barbara and I soon realized that our questions, interests, and approaches differed significantly, so we agreed to pursue separate studies.[1]

As my research progressed, discoveries kept changing the direction of my inquiries. The first discovery was that, although I had been intrigued by Fidler because of his observations about the aboriginal peoples and the environment of the northern plains, Fidler's own obsessions lay primarily in astronomy, mathematics, and meteorology. I soon admitted to myself that I was not going to understand Fidler satisfactorily unless I better understood his interests in those fields of knowledge. Even when Fidler joined the HBC as a young man in 1788, he already knew a lot about astronomy, the history of astronomy, and even about the history of the HBC's contributions to scientific knowledge. I did not. Furthermore, although I could not discover explicit evidence about why Fidler joined the HBC, it struck me as entirely possible that his knowledge of the history of science within the company, and his own curiosity might have influenced his decision. Furthermore, I learned that during his entire career with the HBC, Fidler kept abreast of intellectual trends in Great Britain. How was I to understand Fidler and his journals without also understanding what Fidler knew, or thought he knew? Facing a lacuna in the literature, I reluctantly stepped into what turned out to be a quicksand of fascinating evidence about the history of science in the HBC.

That is when I began to realize that the HBC's contributions to public knowledge were far larger than the literature had led me to expect. Why were these contributions so large? Eventually I concluded – and this is a central argument of this study – that the HBC's large contributions to science were made possible by the development of extensive networks that linked metropolitan and elite scientists, company directors, and HBC officers in North America (and to a lesser extent HBC labourers and aboriginal people) in mutually beneficial and satisfying relationships.

Even well into my research, I still contemplated producing just a few articles that would set the context for an eventual study of Fidler, but then one more discovery motivated me to carry on with this study: that there is no general interpretive survey of the history of science in any chartered monopoly anywhere in the world, nor any history of science in a non-scientific institution. The lack of general studies struck me as unfortunate not only because the literature on any given company, including the HBC, seemed contradictory and incomplete, but also because corporations were so important to scientific work in British colonies before the mid-nineteenth century, and because corporations continue to be important sponsors of scientific activity, even to the present day. Historians of science have also recently grown very interested in the networks that tied metropolitan and elite scientists with lay practitioners in the hinterlands. When I realized all of these things, I understood that social historians of science would welcome a study of the HBC's contributions to knowledge throughout its history as a chartered overseas monopoly, and that some might even be inspired to compile broad histories of science in other chartered monopolies.

Now, years after beginning a project I never intended to undertake, I hope that I will still have the opportunity to publish another study placing Peter Fidler and his journals in their social and intellectual contexts, but in the meantime this study represents the results of "background" research that has expanded into a book-length study in which Fidler, because much of his knowledge was not shared, plays only a minor part.[2]

"ENLIGHTENED ZEAL"

The Hudson's Bay Company and Scientific Networks, 1670–1870

1 Introduction

Science is never just about science, knowledge never just about knowledge. This is especially true of knowledge shared publically. That the public has often viewed science as a disinterested, unselfish, and even altruistic search for truths reflects the fact that scientists and promoters of science have often found it useful, and sometimes imperative, to convince others – and even themselves – that *their* science at least, is unbiased and apolitical.[1] So widely accepted is this notion of science pursued for its own sake that not only the prestige but also the trustworthiness of science has long been connected with the myth of its objectivity.[2] The history of the contributions of the Hudson's Bay Company (HBC) to public knowledge during the time it enjoyed a chartered monopoly (i.e., exclusive trading licence) in vast regions of North America well illustrates these points. In fact, a case study of the HBC between 1670 and 1870 reveals much about the history of knowledge networks in the British Empire.

Most chartered monopolies were long-distance corporations, and many, such as the HBC, were permitted actually to rule their overseas territories (see figure 1.1).[3] They owed their privileges to governing authorities – the Crown, parliament, or, as was the case with the HBC, to both. But what obligations did their charters carry? The answer to that question was vague and ever changing. The HBC's charter, for example, was an award that did not explicitly place any obligations upon the company. Whatever their duties, their exclusive privileges made monopolies the targets of many critics and rivals. The destiny of any chartered monopoly depended upon the good judgment of its directors. Monopolists could not behave as non-chartered private corporations did. But neither could they act as departments of government;

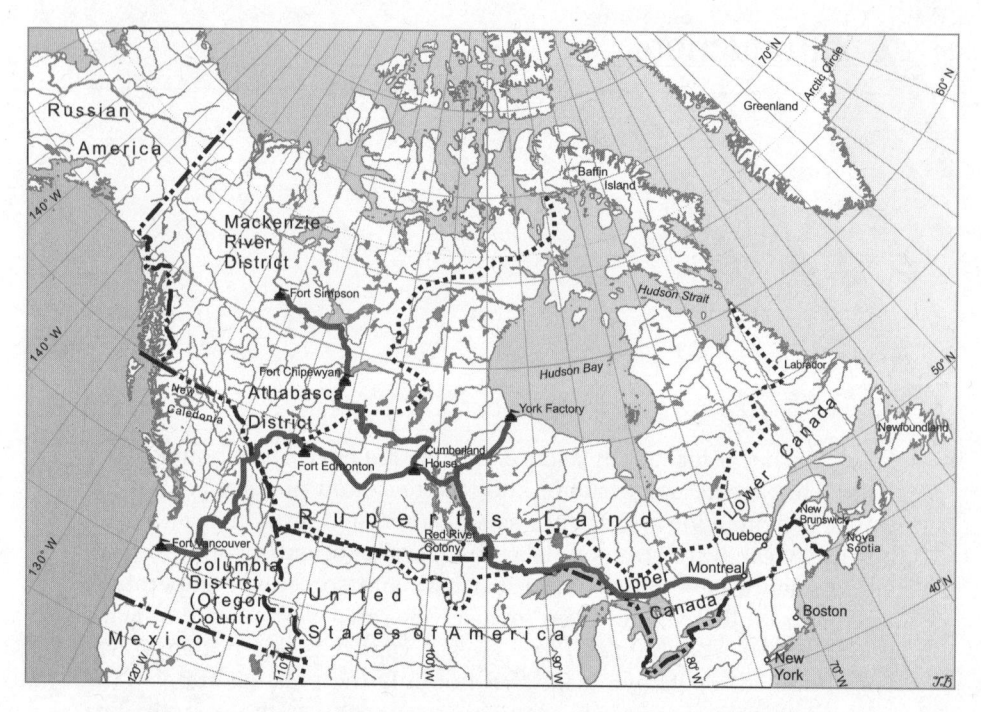

1.1 Hudson's Bay Company Territories at Their Maximum Extent, 1821–1838. The HBC's Royal Charter applied to Rupert's Land only (the border of the HBC claims is shown by dotted line). The border between Rupert's Land and the United States was adjusted to follow the 49th parallel in 1818. The British Parliament granted the HBC an exclusive trading licence to additional territories in 1821, including the Columbia District (Oregon Territory) and the vast Northwest Territory that included the Mackenzie District and the Athabasca District. The border between the British territories and Russian America was determined by treaty in 1825. The heavy grey lines show the main transportation routes of the HBC after 1821. Source: Ted Binnema.

they were private corporations with shareholders' interests to defend. For much of its history before 1870, the HBC was headquartered on Fenchurch Street in England's commercial core, the old City of London (see figure 1.2), not in Westminster, the home of the British government.

Chartered monopolies have been characterized as arms of the state, as commercial enterprises interested only in activities that could contribute directly to their bottom line, or as both.[4] Two Dutch historians

1.2 *Hudson's Bay House – Fenchurch Street (London), 1854.* Watercolour by Thomas Colman Dibdin. For much of its history as a chartered monopoly, the HBC was headquartered on Fenchurch Street, in London City's commercial district. Hudson's Bay House is the four-storey building on the left. Also see colour plate 1. Source: HBCA P-220.

have argued in regards to the Verenigde Oostindische Compagnie (Dutch East India Company [VOC]) that science "never played a significant role in the policy of the VOC."[5] Another historian recently summed up his assessment of that same company by writing that "the Dutch East India Company was certainly no scientific society" – a statement presumably meant to convey more than its literal meaning.[6] The English East India Company (EIC) also appears to have been unenthusiastic about cooperation with savants. John Gascoigne, after generalizing that "keen monopolists" such as the EIC were "unlikely patrons of scientific entrepreneurship," concluded that the EIC's "rearguard action in maintaining its privileges and autonomy was one of the reasons why [Joseph] Banks's attempts to mobilise the East India Company to embrace scientifically-based new products and enterprises had only limited success."[7] In sum, the literature characterizes chartered monopolies as indifferent to science at best, hostile at worst.

The historian of science Mark Harrison noted that, whatever the presumed attitude of corporations towards science, in the British Empire "most scientific work before the mid-nineteenth century was not conducted in formal colonies but under the auspices of corporations such as the English East India Company or individuals working independently of the state."[8] But he further explained that despite the prominence of corporate science in the British Empire, and "although many historians have examined scientific work conducted under the auspices of the English East India Company, its activities have still to be considered in their totality. Other 'long-distance corporations,' such as the Levant Company, have scarcely been examined."[9] The inherently complex and ambiguous position of chartered monopolies vis-à-vis the governments that chartered them and the nations in which they were based also makes them excellent vehicles for better understanding the complex interactions among metropolitan, elite scientists and scientific organizations, chartered corporations, their directors, and colonial employees, and the indigenous people that inhabited chartered corporations' overseas territories.[10]

The HBC is a particularly good candidate for study because it enjoyed officially sanctioned exclusive trading and governing privileges during two centuries that were crucial to the development of modern science. Its founding roughly coincided with the emergence of Baconian science. The HBC was established fifty years after Francis Bacon published *Novum Organum* (1620), eight years after the Royal Society of London was chartered, five years after the Royal Society published its first issue of the *Philosophical Transactions*, and just as the first public museums were emerging in Europe and Great Britain.[11] It was chartered even as the idea of curiosity was being transformed from something prideful, dangerous, and feminine to something commendable, noble, quintessentially European, and male (when appropriately harnessed).[12] The company retained its charter until 1870, just as science was professionalizing, as the generalist scientific traveller was vanishing, as scientists were increasingly becoming based at universities, as governments were assuming a larger role in supporting scientific research, and as the participation of the general public in most branches of scientific activity was declining dramatically. The British government never had any official representative on the HBC's London Committee; nor did it interfere with the HBC's actual rule in its territories. In economic terms, the HBC was small compared with most of the other British-chartered monopolies: its capital stock never exceeded much more than £500,000.[13] The human population of its territories was small. But at its height the HBC had jurisdiction over three million square miles – comprising a quarter of North America, and almost a quarter of the British Empire at its height.[14] For most of the time between 1670 and 1870, Europeans travelled and researched in the HBC's territories only with the assistance of the company. Those realities make the HBC and its territories particularly worthy of historical study.

Nonetheless, the scholarship on the HBC is comparable to Mark Harrison's characterization of the literature on the East India Company. Although the literature on the history of science in the HBC is not small, the many studies are focused on specific topics and particular periods, and each addresses distinct research questions.[15] No one has yet attempted a broad history of science in the company. This is unfortunate because, as John Pickstone has argued, "when we work on a particular time and place, we can easily lose sight of the 'big picture' and the fact that we only know what something is when we know what it is *not*. ... 'Periods' illuminate each other."[16] In short, the literature makes it as difficult to discern the broader significance of the history of science in the HBC as it is to recognize coherence in the history of science in any other chartered monopoly. By presenting arguments about the overall significance, broad themes, general trends, and important milestones, this book provides a "big picture" of the history of science in the HBC that Pickstone endorsed and for which Harrison seems to have called. But it is not a comprehensive history of science in the company. A tremendous amount of scientific data and knowledge preserved in unpublished HBC documents – some of it evidently gathered in self-consciously scientific ways – was not shared publically until after 1870. This book focuses on knowledge that was shared with the learned community before 1870. Furthermore, it is oriented primarily towards the social dimensions of science in the HBC in an effort to discover when, how, and under what conditions scientific enterprise in the HBC flourished or failed to flourish. It is intended to contribute particularly to the growing literature in the history of scientific networks.[17]

This book argues that after an initial century of secrecy, the HBC's contributions to public knowledge grew significantly between 1769 and 1870, by which time the company had a well-established reputation as a generous patron of science. These contributions cannot be properly understood if we assume that the company must have acted as a mere representative of the British state or the British Empire, that the company acted as a disinterested and apolitical sponsor of science, or that the company supported only that science that might contribute directly to its bottom line. The history of science in the HBC can be understood only as the product of intricate symbiotic networks that linked the HBC as a corporation with metropolitan virtuosi and scientific organizations, some of the HBC's directors as individuals, and the company's North American officers and aboriginal trading partners in North America. Those networks functioned because they were nurtured by myriad motives – corporate and individual – including wealth, influence, possessions, various forms of recognition including tribute, fame, and renown, and less tangible benefits: adventure, health, happiness, male companionship, self-improvement, or a sense of meaning.

Historians have noted that scientific activity did not flourish in North America, even in the United States, before the middle of the nineteenth century, primarily because North America had comparatively few wealthy leisured people. The British American colonies and the United States were not scientific wastelands, but compared with Europe relatively little scientific study occurred in any of the formal British settlement colonies before the 1820s – particularly compared to French colonies, in which the state sponsored scientific study as early as the eighteenth century.[18] Most elite science in the British settlement colonies was conducted by European visitors. A historian of science in Jeffersonian America, in seeking to explain the relatively small contributions made by residents of North America even into the nineteenth century, has argued that "it was inevitable that the garnering of the first harvest in American natural history should have been done by outsiders, who, unhampered by the daily demands of a practical calling, were free to indulge their passion for exploring the varied productions of nature in a new world."[19]

If the historical literature on the history of science in North America, the British Empire, and chartered companies is accurate, then compared with the far more populous British North American colonies, and compared with most chartered monopolies, the contributions of the HBC to science, particularly after 1769, were extraordinarily large considering the small number of European employees and subjects who lived in the HBC territories.[20] Indeed, had the residents of other British colonial territories (or the employees of other British overseas companies) submitted scientific data and specimens to Great Britain on the scale that HBC employees did during the eighteenth and nineteenth centuries, the British scientific community would have been overwhelmed. Various geographical, environmental, demographic, and logistical factors only partly explain this reality.

Despite the obstacles, scientific research in the HBC was much less difficult than it was in many other companies. The tropical climates of the territories in which most chartered companies operated – the Dutch East Indies for example – were regarded not only as "the enemy of a concentrated mind" and destructive to scientific instruments and specimens, but also as unpleasant and unhealthy enough to cause the hasty death or departure of many educated and inquisitive Europeans.[21] The pervasiveness of disease and ill health, which were intended to be the focus of the attention of surgeons, also helps explain a letter on the part of the Court of the EIC that explained that they "generally disapproved" of having any of their medical officers "prosecute enquiries into matters connected with natural history."[22] Although the winters were cold in most of the territories in which the HBC operated, many HBC officers spent many healthy and productive years – even entire careers – working for the HBC in North America.[23] Thus the HBC's surgeons usually had ample time to pursue research in natural history – research for which their studies had prepared them. The presence of a stable corps of long-time residents in the periphery greatly aided those who wanted to establish and maintain effective long-distance scientific networks.

Scientists who actually travelled to the HBC territories also faced relatively few perils. There was always some danger of accidents along the long transportation routes, but travellers faced virtually no danger of contracting any of the dangerous diseases that stalked travellers in the tropics. The temperate climate of the HBC's territories was also relatively kind to sextants, telescopes, compasses, watches, and other scientific instruments, specimens, and books, even along the damp and humid northwestern coast of North America.

Distance was less tyrannical for the HBC than for most European-chartered companies. The great distance between Europe and the theatre of operations of most chartered companies interfered greatly with their contributions to science. For example, the distance from Europe to Batavia (Dutch East Indies) resulted in intellectual isolation and significant delay and difficulty in acquiring, repairing, and replacing scientific instruments and books.[24] If the VOC had actively encouraged and supported scientific research in the East Indies, these difficulties may well have been mitigated, but, as one colonial governor rued, "in the East Indies the aim of making money is the principal object of study."[25] Two historians concluded that "intellectually, Batavia was indeed positioned on the outskirts of the literary and scientific world."[26] The HBC territories were less peripheral to that world. The HBC transportation routes, long as they were, were short and safe compared with those of the various East India companies.[27] Rarely did passengers on HBC ships die. It took far less time to transport people, instruments, and books between London and Hudson Bay than between most European colonies.[28] Furthermore, because the company directors actually supported much of the research carried out in its territories, it was far easier for HBC officers to acquire books and instruments and to have them repaired.

If environmental, geographical, logistical, and demographic realities made it possible for the HBC to contribute knowledge, they did not make it inevitable, or even probable, that it would. Other factors came into play. The HBC eventually did contribute so much to knowledge production because the interests of the corporation as a whole, of learned institutions, and of many individuals – including elite savants, scientific travellers, HBC directors and officers, and aboriginal people – encouraged the formation of fertile knowledge networks.

From the perspective of the HBC as a corporation, decisions about whether or not to contribute to the production and sharing of knowledge, and whether or not to be *seen* to contribute, were complex but important ones that extended far beyond considerations of the interests of the state, or the company's own narrow economic or territorial interests. In fact, the company supported much inquiry from which it could expect to gain no direct commercial advantage. But this does not mean that the company did not gain from sponsoring science. It benefitted in important ways. Probably the most important was that such patronage established and maintained an invaluable corporate image in Great Britain. The company's support of science earned it an image as liberal, generous, enlightened, disinterested, zealous, and open – useful to combat the stereotype of chartered monopolies as "mean and selfish," conservative, benighted, apathetic, and secretive.[29] Enjoying that reputation grew increasingly important as Great Britain modernized and liberalized. There is no doubt that this positive image served the ongoing political and economic interests of the company, but perhaps ironically, the science that best served to achieve the company's aims was that which could most convincingly be portrayed as having no potential to advance the company's economic and political interests.

A company such as the HBC that enjoyed exclusive (but revocable) privileges had to respond judiciously in the first place to not only the growing influence of the public in Britain, but also to evolving attitudes towards public and ostensibly disinterested science. The directors needed to be alert to the changing expectations of chartered companies. They needed to respond to requests by governments, institutions, and individuals in light of the immediate and long-term interests of the company. Chartered monopolies never lacked opponents and critics (found among courtiers, parliamentarians, rival merchants, customers, seaports, tradespersons, missionaries, humanitarians, inhabitants of the company's territories, and the British public), but if the HBC is any indication, they did not suffer much on account of a secretive policy towards science until the Enlightenment. Then as the eighteenth and nineteenth centuries wore on, they needed to be ever more mindful of their relationships with governing authorities, scientific institutions and individuals, and the public. Increasingly seen in a liberalizing world as archaic relics of the past, chartered monopolies ignored evolving public and political expectations at their peril.

The most important and obvious benefit that the HBC enjoyed because of its support of science came in a form still in general use in the scholarly world: public tribute.[30] The public tribute is a published statement of indebtedness written by scholars – often tediously repetitive and formulaic, but prominently positioned and effusively thankful – that compensates those who support research. The value of public tribute grew, as did the influence of the myth of disinterestedness. The sociologist of science Warren O. Hagstrom described the dynamics of this myth as he observed it in the 1970s. He argued that science operates around "an exchange of social recognition for information," and yet "scientists usually deny that they are strongly motivated by a desire for recognition" because, "the expectation of return gifts (of recognition) cannot be publically acknowledged as the motive for making the gift."[31] Scholarship is ostensibly an exchange of gifts – research "grants" in exchange for scholarly "contributions." This is merely a variation on a pattern that is centuries old.

Tribute, particularly when paid by the scientific elite – who were supposedly as apolitical and disinterested as their science was – provided the company with promotional material that no advertising or lobbying could ever replicate. Tribute, paid by respected people and institutions who supposedly had no vested interest in the company, that repeatedly described a chartered monopoly as "enlightened," "disinterested," "liberal," and "benevolent" – words not normally associated with chartered monopolies – helped create and maintain a positive image of the company instrumental to the long-term interests of those on Fenchurch Street. And politicians and government officials cared what those learned men who were not dependent on the company said, as testimony before the British Select Committee on Aboriginal Peoples

(1836), and the Select Committee on the Hudson's Bay Company (1857) shows.

The science that the HBC supported did not need to have the potential to provide any direct economic benefit to the company – indeed, the purer the science seemed to be, the greater its cachet.[32] It was useful if the science was popular or prominent. Tributes published in popular books reached a broad audience, whose sympathies the HBC also needed to cultivate. But not all of the science had to have popular appeal. Neither did it matter much if most readers skimmed over the tributes. The company could ensure that the right eyes read them. As the myth of science as disinterested and apolitical grew more influential, its potential political utility increased. By the nineteenth century, the HBC's promotion of science contributed to an image that was likely to generate sympathy among decision-makers, particularly those in the Colonial Office, which mattered most to the company.

The company's assistance to scientists also frequently garnered published endorsements of its treatment of aboriginal peoples and of its monopoly. Again, such portrayals of the company as humanitarian, written by respected, learned men and appearing in influential publications independent of the company's control carried far greater weight than any material the company might produce on its own. Such depictions were most valuable in the 1830s, when humanitarians concerned about the circumstances of indigenous peoples in the empire were particularly influential in British government and society. When such testimonies appeared just as Parliament was considering the renewal of the company's licence, so much the better.

The cultivation of learning served the corporate interests of the HBC in other ways. Recruitment and retention was among them. By encouraging their officers to "improve" themselves by, among other things, reading books and contributing to science, the London Governor and Committee ensured that they could attract and retain educated, intelligent, and ambitious men. An officer in the HBC was much more likely than a supervisor in most chartered monopolies to have to manage a small isolated trading establishment without the companionship of other officers. The HBC needed a core of intelligent, trustworthy, self-confident, and experienced leaders who could calmly make judicious decisions pertaining to a wide range of complex and rapidly changing circumstances, often without being able to confer with other officers – to say nothing of consulting with the London directors, from whom a response might take three years. Aside from having to keep account books and detailed journals, officers had to decide how to comply with instructions from distant superiors that proved impractical under local circumstances, how to adjust to the non-arrival, shortage, or surplus of trade goods, how to supervise subordinate posts and outposts, how to plan, build, and maintain buildings, how to maintain discipline, order, and morale among labourers and tradesmen (and their families), how (unless there was a surgeon present) to treat medical issues among the inhabitants of the post and aboriginal people, how to deal with food shortages, and how to interact appropriately and profitably with indigenous people. On such matters, not only the success of a trading post but also the lives of the post's residents might depend.

The company's support of science did occasionally produce direct economic and territorial rewards (by identifying the commercial value of resources found in its territories, facilitating the geographical expansion of the company's operations, or strengthening British or HBC land claims against foreign or domestic rivals), and were occasionally tied to Great Britain's and the company's territorial ambitions. But just as often, HBC-assisted scientific activity actually undermined the company's hold on its territories. Expectations of direct economic benefit or territorial expansion were far less important than the less tangible but very important considerations related to the company's "brand," and its recruitment and retention of talented men.

The HBC rarely, if ever, initiated partnerships with savants or learned organizations. It did not have to. The benefits of HBC assistance to such men and organizations were obvious. At a time when there was little public funding for knowledge production, the company's sponsorship offered natural historians, travellers, explorers, institutions, museums, and horticultural societies golden opportunities for success, prominence, and recognition. Science needed the company at least as much as the company needed science, but it was the promise of mutual benefits that really nurtured flourishing knowledge networks. So, it was natural that, as new scientific societies emerged in Great Britain, the United States, and Canada in the nineteenth century, cooperation between the HBC and savants increased accordingly.

As late as 1781, the Royal Society of London was the only society in England dedicated to scientific study.[33] Thereafter, and especially after 1820, scientific organizations proliferated. Many of them, including the British Museum, various horticultural organizations, and the Royal Geographical Society, were joined by the Admiralty, the Artillery, and North American scientific societies in seeking the assistance of the HBC.

Fortunately, their requests came during the halcyon years of the HBC. The HBC was remarkably likely to respond favourably to requests from scientific organizations and individual scientists after 1821.

The HBC's economic and political influence in Britain may have been small compared with more famous chartered monopolies, but the HBC exercised significant control over vast expanses of northern North America, including subarctic and arctic regions that were at times of great interest to scientists, scientific travellers, and explorers. Before the Enlightenment, scientific travel was difficult and expensive everywhere because scientists generally had to negotiate their way onto commercial ships.[34] During the entire period between 1670 and 1870, it was either impossible or extremely inconvenient and expensive for anyone to travel through most of the HBC territories without the assistance of fur traders, and between 1821 and 1870 visitors needed the help of the HBC specifically. But the company developed an efficient system for moving people and goods over long distances. Without incurring great costs itself, it could render a visitor's trip through its territories – one that would otherwise be impractical – not only relatively safe, convenient, and comfortable, but cheaper than travel through formal British settlement colonies. And as one late nineteenth-century account shows, the men at HBC posts did not treat such visitors as interlopers, but as highly favoured guests whose attentions were eagerly sought:

> The arrival of a traveller from the outer world is the greatest episode in the monotonous every-day life of the post. The community find in him an inexhaustible fount of enjoyment; and if he be of a communicative disposition, his store of news and narrative will do service in payment of his weekly board-bill for an indefinite period. To such a one, much more than to a passing officer from another fort, the hospitalities of the fort are extended in the most liberal manner. An apartment is assigned him for his sole occupancy during the period of his sojourn. He is free to come and go when and where he listeth, means of locomotion being furnished on demand. His companionship is eagerly sought by all; and the fortunate individual who secures his preferred acquaintance excites at once the envy of less favoured ones. Nothing is left undone to prolong his stay, and when he finally departs, he is sent upon his journey freighted with the good wishes of the isolated post, and is certain of the same cordial treatment at his next stopping-place.[35]

If scientists were unable or unwilling to travel to the territories themselves, the HBC could easily arrange the free delivery of enough natural history specimens or scientific data to overwhelm sizable institutions. But those who wished to benefit from the company's largesse were well advised to understand how to conduct themselves. The HBC was not a charity.

The virtuosi had to repay the company. Sometimes they could do so by providing advice, recommendations for employees, and occasionally advice about the marketability of resources. But the prudent understood that the most important currency by which the company's assistance could be compensated was the kind of positive publicity that the company otherwise could not buy. Most scientists successful at attracting the HBC's patronage grasped the obligatory nature of public tribute. In 1860, when Joseph Henry, head of the Smithsonian Institution, wrote to thank the HBC's London Governor and Committee for the "support and cooperation of officers" of the company, he assured them that "it will always be a pleasure independent of duty, to bear testimony to this countenance on all possible occasions."[36] Henry understood what corporate directors wanted to hear.

A host of individuals connected with the HBC were also crucial to science. As individuals, the governors of the HBC had reason to support science. Often, public tribute paid to the corporation also mentioned governors by name – ensuring that they too shared in the prestige and recognition enjoyed by the company. Governors were also lavished with private letters of thanks from influential scientists, and with gifts (especially in the form of books and stuffed natural history specimens), rewards and awards (ranging from monetary awards to knighthoods and baronetcies), and other forms of recognition (including having species named after them). Some HBC directors sought public recognition. Over the course of his career, George Simpson, the powerful North American governor of the HBC from 1822 to 1860 – John Galbraith aptly described him as the "viceroy" of the HBC territories – increasingly sought to be known and recognized by powerful and influential men, including scientists (see figure 1.3).[37] According to one historian, his frequent interaction with powerful politicians and diplomats between 1839 and 1845 "developed in him an appetite for association with great men and great causes."[38] After negotiating an agreement with the Russian American Company in St. Petersburg, Russia in 1839, he set off on what he argued was the first overland trip around the world in 1841–42 (via the HBC territories, Siberia, and European Russia). His narrative of that trip was published in Britain and the United States in 1847. Actually partially written by other personnel in the HBC, it was more impressive as a travel book than as a scientific work, but according to one historian,

16 "Enlightened Zeal"

1.3 *Sir George Simpson.* This mezzotint portrait of Governor George Simpson by James Scott, after Stephen Pearce in 1857 shows Simpson briefly interrupted from his work. The portrait reveals how important it was to Simpson to be perceived as an important and influential decision-maker, an older yet vigorous man of action, and a man of books. For Stephen Pearce's painting of Simpson, see colour plate 2. Source: © National Portrait Gallery, London, England (NPG D31633).

it was intended to enhance Simpson's reputation as a learned man.[39] Directors in London also wished to be thought of as learned. John H. Pelly, London Governor of the company from 1822 to 1852, and Nicholas Garry, deputy governor from 1822 to 1835, were two of 535 founding members of the Royal Geographical Society. When the society's members were listed in its journal, Pelly and Garry made sure that not only their names, but also their positions in the HBC were indicated.[40]

Social historians of knowledge networks pay careful attention to the complex relationships among elite and metropolitan scientists and the lay practitioners in the hinterlands. In order to understand those relationships well, it is important to consider who those people were. While elite scientists generally worked in one of the major metropolitan centers and lay practitioners were often marginal in more than one sense, it is nonetheless crucial to distinguish among metropolitan, elite, hinterland, and lay sciences. Roy MacLeod's influential definitions of metropolitan and colonial science emphasize *how* they are done.[41] But they thus conflate metropolitan with elite science and science carried out in the geographical margins with lay science. In this book, metropolitan and hinterland sciences are defined according to *where* they are done.[42] Specifically, metropolitan science was science done in the major centers of Western Europe (or later in the major urban centers of North America), because that setting made its science possible or enhanced its perceived value. Hinterland science was science done in locations geographically peripheral to these centers precisely because the research was feasible only outside of those centers, or because the perceived value of the specimens, data, or evidence was tied to the location of their origins.

An important point is that although hinterland science may not have been as prominent as metropolitan science, the prestige of many scientific activities was dependent on the geographical locations in which they were carried out. A wide range of scientific activities might apply. Many astronomical or geomagnetic phenomena were observable only outside Europe, or had to be observed in remote locations as well as in Europe in order to be useful. The frigid winter temperatures in the lands surrounding Hudson Bay meant that certain experiments that could not be carried out in Western Europe might be conducted in Rupert's Land. Meteorological registers might be kept anywhere, and natural history specimens gathered anywhere, but London scientists placed different significance upon a meteorological diary kept at York Factory than one kept at Bristol, and valued a specimen of an American coot (*Fulica americana*) collected at Lake Athabasca differently than that of a Common coot (*Fulica atra*) found on the Thames. The material culture of the Inuit or Chipewyan was assumed to have significance that analogous artefacts from Europe or eastern North America did not have. Possessing specimens is one thing, but actual observation in situ, increasingly valued among European intellectuals as empiricism gained sway, gave the collector in the hinterlands a standing that metropolitan scientists had to respect.[43] Indeed, metropolitan intellectuals often frankly acknowledged the fact. In short, the value of evidence and data gathered in the peripheries was linked to location and to the trustworthiness of the collector in the hinterlands. Its value was often also enhanced because it pertained to intellectual questions and scientific theories that evidence gathered in the metropolis, or at least exclusively in the metropolis, could not.

On the one hand we know that virtuosi in the major centers of Europe often disparaged scientific travellers and amateur collectors.[44] On the other hand such travellers and collectors had unique opportunities to earn recognition for their contributions. The geographical distance between collectors and British and American scientists, which might have made communication more difficult, was compensated for by the fact that the exotic evidence that people in the geographical margins could supply had particular potential value to those savants. Evidence from outside Europe was far more likely to be unfamiliar to metropolitan scientists than analogous evidence gathered in Britain, and it usually spoke to questions about universality, diversity, and uniformity. This helps account for the fact that British and American scientists took great interest in the HBC territories even before the HBC facilitated communication between their officers and these intellectuals. Since metropolitan savants understood that evidence gathered outside of Western Europe was a means towards their own recognition, profit, and fame, many were highly motivated to encourage, flatter, and reward any potential collectors and informers in the hinterlands who might supply them with specimens, data, and other information. Although the present study is focused on the HBC, it sheds light on the relationships between metropolitan scientists and hinterland collectors and informants generally.

It is useful to base a definition of elite and lay science on Roy MacLeod's definition of metropolitan and colonial science.[45] Elite scientists were those who defined scientific orthodoxy. Their science was science done by a relatively few people – wherever they were – whose scientific contributions had earned them sufficient standing among their peers to shape the priorities, conventions, theories, beliefs, and approaches favoured within the learned societies of Western Europe, and among those who worked within the orbit of those societies. They were well read and highly skilled (although not necessarily formally educated), and some earned at least a significant portion of their income from their scientific work. Lay science was conducted by people outside the "priesthood" who had insufficient standing to influence natural philosophers, natural historians, and scientists the way elite scientists did. They generally had less education than elite scientists, and generally did not earn a significant portion of their income from their contributions to science. They may have been more or less aware of and influenced by the priorities, debates, and conventions of the elite, and some may have addressed the theoretical debates of elite scientists insofar as their knowledge and inclinations led them, but they had little ability to change the directions, priorities, or beliefs of those elite scientists, regardless of how insightful their work might have been. Put negatively, their science was accepted as authentic science only when validated by the elite.[46] Stated positively (as many HBC men would have been inclined to), their activities, when authenticated by the elite, offered amateurs the thrill of attaining a level of scientific standing.

During the late eighteenth century and early nineteenth century, people of humble birth and little formal training or education – indeed of unexceptional intelligence – had more opportunity to contribute to knowledge production than they did in later years. For much of the period under study, it was still possible for a single person to comprehend a great range of Western scientific knowledge. There were few specialists and many generalists. The boundaries between fields of knowledge were indistinct and fluid, and the realms of natural philosophers and other savants, were not as separate as they later became. An HBC officer at an isolated post and without specialized schooling might still publish in learned journals. In the late nineteenth century, the disciplines became more distinct and professionalized, and scientific literature became more specialized, abstract, and expensive. Scientists were increasingly hosted at universities, and funding increasingly came from governments and their large granting agencies. As a result, the link between elite scientists and lay practitioners became more attenuated. A history of science in the HBC offers glimpses into lay science, the motivations of its practitioners, and the intricate networks that connected elite scientists and lay practitioners in the eighteenth and nineteenth centuries before those transformations really took hold.

Most lay scientists did not aspire to great influence or fame. Most were happy to toil in relative obscurity. Many were enthusiasts. They were amateurs (lovers) in the original sense of the word. Generally speaking, depending on their own knowledge and priorities, they gathered facts either according to the conventions and expectations of elite scientists, or in naive and unsystematic ways. In many respects, studying lay practitioners of science in the HBC may teach us much about lay practitioners of science in Europe, (especially Great Britain), British America, and the United States, for they shared much in common. In sum, this book sheds light on the significance of science carried out in the hinterlands, whether practiced in the HBC or other overseas corporations, and on lay science, whether practiced in the HBC territories or in Great Britain, Europe, or the United States.

As most of the literature on the history of science published before the 1990s shows, historians of the "big ideas" and "great men" in science might be able to ignore lay and hinterland practitioners.[47] In 1959, Bernard Cohen argued that, although "there were individuals whose contributions to science had a very considerable merit, ... the fact remains that throughout most of the nineteenth century the major ideas in science ... were produced in Europe rather than in America."[48] Cohen's argument still stands. Referring to the Dutch East Indies, two historians argued that "the scientific effort ... was an initiative taken by Europeans, modeled after European examples, and leaning on European-made instruments. Apart from the localized subjects of study, especially in natural history, the colonial setting had little influence on the science undertaken."[49] And in 2008, two other historians noted that "rather than using their travels to construct novel theories of nature as a whole, most Atlantic travelers acted on behalf of metropolitan patrons who were their social and institutional superiors."[50]

But if intellectual historians of science can ignore lay and hinterland science, social and cultural historians of pre-twentieth-century science must take its many practitioners seriously because many elite and metropolitan scientists themselves did so. Many scientists benefitted not from the reluctant compliance, but from the enthusiastic devotion of the collectors, correspondents, assistants, and donors who supplied them. Elite and metropolitan scientists in several fields, meteorology and natural history especially, depended on intricate networks of collectors and correspondents. Susan Scott Parrish argued that "colonial subjects in America were not mere collectors for the knowledge makers of the metropole. European correspondents depended on locals for their kinds of expertise: identifying a novel specimen, understanding its properties or behavior, reporting on or depicting the specimen in its live and natural context, or seeing the interdependence of plants and animals."[51] In their introduction to an excellent recent collection of articles, James Delbourgo and Nicholas Dew likewise noted that "gradually but surely, scholars are moving away from histories focused on seemingly isolated metropolitan knowers, ... toward a social history of the interconnections between the radically different peoples that made and circulated early modern knowledge."[52] Delbourgo and Dew explained that, "the fundamental project" to which historians of science have turned in recent years "is a profound rethinking of traditional metropolitan narratives of center-and-periphery, involving the recovery of a variety of 'peripheral' actors' agencies, and the reimagination of knowledge production from their perspectives. The straight lines of communication that scholars of scientific 'diffusion' once traced from center out to periphery, ... are now being replaced by an intricate latticework of intersecting itineraries and competing agencies."[53]

They added that "the American correspondents of Europeans were not servile drones but shrewd self-fashioners who sought to turn European recognition to local advantage in cultivating their status as cosmopolitan knowledge makers in the provinces and serving provincial agendas as well as metropolitan ones."[54]

Astute scientists, whether they travelled to the HBC territories or corresponded from a distance, understood the difference between networks peopled by grudgingly compliant subordinates, and those populated by men who thought of themselves as ardent and valued partners in research. But it took significant social intelligence and empathy for scientists and scientific organizations to generate and nurture networks of lay supporters and suppliers who eagerly devoted their time and energy to their programs. Those who used such intelligence were well rewarded. Scientists with access to HBC officers passionate enough to collect and preserve specimens according to instructions, to keep detailed and careful scientific journals, to answer specific questions, and to enlist the help of aboriginal people as collectors and informants, could succeed beyond anything they might have hoped to achieve on their own. As valuable as it was to have company directors urge their officers to assist them, scientists also needed to understand how to cultivate and maintain the enthusiasm of the officers themselves. A visiting scientist could accomplish much by working diligently during a visit to the HBC territories, but he could accomplish far more by also infecting HBC men with the same zeal for his research that had driven him to travel such great distances to the HBC territories himself. This meant more than conveying a sense of the importance of his scientific work. It meant cultivating the friendship, admiration, and esteem of HBC men whose backgrounds and perspectives might be very different from his own. By the same token, a metropolitan scientist might be happy with the results of a circular sent by the governors of the company urging its officers to respond to his request for assistance, but he could reap far more if he also convinced lay collectors that they could be partners and beneficiaries, not mere servants, of science. The history of science in the HBC reveals how important affability, generosity, and infectious enthusiasm could be, and how a scientist's willingness and ability to befriend, honour, reward, flatter, cajole, and encourage his suppliers affected his own success.

The HBC never had two thousand permanent employees; and only a handful of men trained in science travelled to the company's territories before 1870. There were no permanent European colonies there until after 1812, and the first short-lived scientific society was established at Red River only in the 1860s. The population was too small and dispersed to support such an institution. But none of these obstacles prevented significant contributions to public knowledge.

Metropolitan scientists were lucky; officers in the HBC were well positioned to become lay contributors to knowledge production. For one thing, they had at least as wide a range of reasons for pursuing inquiry as the directors of the HBC and metropolitan scientists did, and these motivations extended well beyond economic ones. Carl Berger has argued that "natural history was born of wonder and nurtured by greed."[55] Margaret C. Jacob has argued more pointedly that though "explorers and travelers had many motives, to be sure none exceeded in intensity the drive for wealth and profit."[56] Whether we accept those arguments or not, we fail to understand the dynamics of knowledge production – even in a commercial enterprise such as the HBC – if we do not consider non-pecuniary motives. Relatively few officers of the HBC were ever paid directly or rewarded monetarily for their contributions to science, and most of the few who were could not have anticipated these payments or awards before their contributions were made.

HBC employees always included a small but significant number of literate and fairly well-educated officers and other employees hoping for promotion. At all times, but particularly after 1821, when the company's hold on the trade of much of its territories was more secure, its directors particularly valued well-educated men with managerial skills (as opposed to the energetic and courageous men that were favoured especially between 1810 and 1821).[57] Overwhelmingly, these were the men who contributed to science. And HBC men could contribute to science with far less effort than most European settlers in North America. They were sometimes encouraged and occasionally ordered to assist savants, but even without encouragement HBC officers posted along the Hudson Bay who wished to assist scientists needed only to pack specimens in crates or write descriptions and deliver them to the company's ship to have them sent, free of charge, to governors in central London who could then convey them to scientists, some of whom were also colleagues and fellow members of learned societies, only a few city blocks away. Officers were also free of worries over how to pay the expenses of transporting specimens, sometimes thousands of kilometres overland, to the coast. When the Governor and Committee encouraged communication, the lack of any natural history society or other scientific institutions in the HBC territories did not prevent the formation of links between the HBC territories and scientists in the metropolis. Furloughs and retirement in Great Britain or Canada[58] also permitted HBC traders with years of experience in North America to meet directly with interested scientists in London, Edinburgh, Washington, and Montreal.

HBC officers were also able to contribute so much to science because they were salaried men who lived in facilities and ate food paid for by the company. Most did not intend to spend the rest of their lives in the fur trading territories. Free from the difficulties of establishing farms or businesses or finding work in a settled colony, and spared the costs and anxieties of providing shelter and food for European wives and families, these men were better positioned to pursue interests in science than were the vast majority of immigrants to North America.

As stressful and dangerous as an officer's job might occasionally be, the day-to-day life of an officer – indeed, any HBC employee – tended to be dull, lonely, and stultifying. While an officer might be very busy in spring and fall, at other times he might be able to complete his work in only a few hours a day. Furthermore, during evening hours at any time of the year, employees had few options to fill their leisure time. In 1820, Daniel Williams Harmon noted that "no other people, perhaps, who pursue business to obtain a livelihood, have so much leisure, as we do. Few of us are employed more, and many of us much less, than one fifth of our time, in transacting the business of the Company. The remaining four-fifths are at our own disposal."[59] Harmon was referring to the life of a trader in the North West Company (NWC), but much the same could be said for HBC officers. In August 1827, James Hargrave, a HBC trader stationed at York Factory wrote that "my time for half of the year is at my own disposal, spent in exercise, hunting, fishing, or reading – the season of active employment is spent in the Counting House or the Store and tho' duty often presses so heavy as to tire the staunchest industry, – when the busy season is over quiet and ease again are ours."[60] More than thirty years later, during his visit to the HBC territories, Robert Kennicott, a naturalist with the Smithsonian Institution, noted that the HBC officer's "duty is almost nothing beyond his actual presence. A little less than two months in the year is sufficient for all the writing. No wonder then they become lazy."[61] Leisure was the prerogative of the HBC gentleman, but as Kennicott implied, boredom, indolence, and homesickness stalked many HBC officers. Those who read, studied, and researched, those who kept their minds active and alert, identified more closely and communicated more often with their fellow officers, tended to make better decisions on behalf of the company, and found their jobs more satisfying. Furthermore, their serious and studious examples helped them keep discipline among the employees at a post. It was better that a trading post manager tutored his men in mathematics or writing, led them in religious services, or read Shakespeare's plays to them (as John Rae did during his arctic explorations), or that he studied metamorphosing beetles, than that he succumbed to idleness, apathy, and alcohol abuse, as some officers did.[62]

In 1844, during his visit to the Columbia District (Oregon Country), the German naturalist Karl Andreas Geyer was obviously impressed with the intellectual life at HBC posts. He had botanized in the upper Mississippi River region in the late 1830s and had travelled to the Columbia District overland from St. Louis, so he was able to compare HBC posts with the posts of the American Fur Company. After noting that the HBC trader, Peter Skene Ogden, showed him a book manuscript he was writing, he added that "other gentlemen of the Company likewise busy themselves in their leisure time with literary efforts. They have a library at [Fort] Vancouver and constantly circulate books from one fort to another. Here one finds Lyell's Geology and the Asiatic Journal. Another type of life here from that in the American Fur fort!"[63] He also noted that Archibald McDonald, Chief Factor at Fort Colville, "is a contributor to the British Museum and one of [William Jackson] Hooker's correspondents, also an honorary member of the Botanical Society in London. Indeed, the officers of the Hudson's Bay Company are a group of rather substantial, educated men."[64] Even when he was at Fort Colville, Geyer noticed that "Hooker writes to Mr. McDonald about a new species [of cactus] which [David] Douglas found at Fort Walla Walla and of which he left only a description."[65] At about the same time, Jean-Baptiste-Zacharie Bolduc, a French-Canadian missionary with significant scientific interests of his own, was similarly impressed upon his arrival at Fort Vancouver. "Who could believe," he wrote,

> that even in this country there are men who are well educated even in natural sciences. Nothing is truer, their number, it is true, is not very great but there are enough of them to make mention of it. Several times, I have had occasion to speak of natural history, physics, chemistry, astronomy, etc. At Vancouver there are some instruments such as globes of land and sky, an electrical machine, a voltaic pile, a galvanic trough, etc., etc. As for the last two, they do not know how to use them, so Governor [John] McLoughlin is waiting for me to come and put everything in order and to make some experiments. I, myself, have a little electrical machine that I built. The only thing that fails me is some tin which I replaced with a sheet of lead that was used to wrap some tea in boxes. Lots of people opened their eyes with fright when they saw for the first time the phenomenon produced by the machine in motion, and even more commotion was caused by the experiment of the Leyden jar. The Indians are convinced that I have a tamanwas, a very powerful protecting spirit, and even that I have supernatural powers.[66]

The intellectual life that Geyer found in the Columbia District in the 1840s was not unusual for the HBC. Scholars have already discussed the many impressive libraries at HBC posts, both private libraries and subscription libraries.[67] Their studies show that no later than the 1790s, and probably earlier, many HBC men had easy access to learned books on a wide variety of topics. Because the company did not charge for shipping and delivery, an HBC officer in a remote corner of the HBC territories did not have to pay more for a book than did a person in London. And because some officers received publisher's catalogues, they could order books before they were published. Books were expensive, but officers often lent books, magazines, and newspapers to one another. They even developed a few circulating libraries. Indeed, according to an 1879 account, "the Company has established extensive libraries for the use of the officers and servants in many of the larger stations in the north, from which supplies for the adjacent smaller posts may be drawn, so that the diligent reader may command new supplies from time to time."[68] In short, the impediments to the development of a learned culture among HBC officers in North America were not insurmountable.

"Knowledge is now become a fashionable thing," reads the preface of one of Benjamin Martin's books of lectures published in England the mid-1700s, and Martin was in a position to know. His inexpensive popular science books, scientific instruments, and entertaining public demonstrations and lectures catered to a rapidly growing population in Great Britain (and America) with an interest in science and a desire for scientific instruments.[69] The history of science in the HBC sheds light upon the cult of knowledge and self-improvement among lay practitioners at a time when people outside the guild were particularly important to British science and to science in the HBC. In short, science was a hobby for many.

Many of the HBC's officers were drawn from the ranks of educated and literate people to whom Benjamin Martin referred. Like many people in Great Britain, they read, studied, and researched more for personal pleasure, curiosity, wonder, fulfilment, self-improvement – even worship – than to achieve fame, wealth, or influence. Perhaps no one in the HBC hinted better at the ordinary rewards of study than the HBC surgeon and Copley Medal winner Thomas Hutchins when, in August 1772, at the head of a long document describing the natural history of the Hudson's Bay region, he quoted this passage from Adam Ferguson's *Essay on the History of Civil Society* (1767): "In the recess of better employment the time which is bestowed on Study, if even attended with no other advantage, serves to occupy with Innocence the hours of leisure, and set bounds to the pursuit of ruinous and frivolous amusements."[70]

Some HBC men must certainly have studied nature to help themselves feel more at home in what was at first an unfamiliar environment. Research could stave off homesickness, depression, or loneliness. Should this seem odd, consider that, as one historian of science has pointed out, "resources of time, money, and labour were willingly spent in pursuit of an aim outsiders deemed trivial or even mad."[71] It may indeed seem odd even to us that some fur traders recorded weather statistics three or more times a day for years on end, but these men obviously found comfort in doing so. Some of them, after all, diligently kept meteorological journals that they evidently never shared with scientists.

That many HBC officers were inclined towards intellectual pursuits was one thing; metropolitan scientists needed to convince them to contribute to *their* research agendas. Most hinterland and lay collectors were pleased to do so if given the opportunity. Relationships within knowledge networks could not flourish if they were competitive or distrustful. They had to be mutually satisfying. Certainly elite metropolitan scientists enjoyed greater wealth and renown and took on fewer hardships and risks than their peripheral counterparts, but that bothered few lay scientists in the geographical margins. Most were delighted when scientists praised them and repaid them with public tribute, when scientists sent them gifts that were difficult to acquire locally, when scientists named species after them, and when the company recognized and rewarded their efforts. But for officers in the HBC, internal motives go a long way to explaining why they assisted scientists. This was probably true of many lay scientists regardless of location.

During the first sixty years of the HBC's history, some employees actually risked their careers with the company by contributing to science. In later years, they were more likely to be rewarded. Between the late 1760s and 1821, the spirited rivalry between the HBC and the NWC influenced the HBC's contributions to science. After the merger of the HBC and the NWC in 1821, opportunities for promotion within the HBC diminished significantly.[72] During the same period, the company's directors showed greater interest in supporting science. When the company's directors encouraged contributions, many men must have responded at least in part to curry favour with Governor George Simpson or the London Governor and Committee. Often requests came in the form of circulars which an individual officer could ignore. But astute men understood how to interpret a letter from the Governor that opened with "you are well aware of the desire of the Company to promote the interests of science by all the legitimate means in its power."[73] Few could have missed the implication that a positive response to such requests would be greeted with approval by the directors – and would influence decisions relating to promotion. And the fact that many men were promoted who were known to be active contributors to science provided clear evidence that the London Governor and Committee did not resent its officers devoting considerable time and energy (and it is difficult to believe that the directors did not assume that officers devoted at least some of the company's goods) to scientific activities.

Encouragement like that helped to cultivate the development of a scientific culture among its officers.

For some, the decision to share scientific knowledge and specimens was about earning recognition in the world of science. Lay scientists were unlikely to realize great fame, although lay practitioners in the hinterlands had far greater potential to achieve recognition or remembrance than those in the metropolis. Contributing to the work of prestigious scientific organizations such as the Royal Society, British Museum, or the Smithsonian Institution, or to the work of renowned scientists was a significant inducement in and of itself. Many HBC men were clearly flattered by the tributes paid by well-known scientists and institutions, often in print, to their contributions. Some also experienced great delight at seeing their own knowledge published in scientific journals, or having species named after them. Such forms of recognition addressed what must be nearly universal human yearnings to feel that one has made a difference, to leave legacies, and to be remembered after one's death.

The practice of science also acted as a social marker. Science was linked to class. Bernard Cohen once explained that the lack of a significant scientific tradition in North America "did not surprise Alexis de Tocqueville. ... To him, it seemed clear that science was an aristocratic pursuit."[74] Tocqueville was not mistaken. The pursuit of science everywhere was dominated by the social elite – men who had ample leisure time. One historian has noted of England that meteorological journals (which were commonly kept by HBC men) were "closely linked with privilege and leisure, ... kept almost exclusively by people of independent means and stationary residence."[75] The same could be said of other branches of scientific research throughout the Western world. By the end of the nineteenth century a growing number of scientists were employed by governments and universities, but generally speaking, the pursuit of science was the arena of wealthier leisured people throughout the period examined in this book.

During the eighteenth century, the growth in British society of a comfortable class of people desiring respectability saw the practice of lay science broaden. To be sure, the "priesthood" still acted as the intermediaries between the truth and the toilers, but the portion of the population that had the money and time to pursue learned activities expanded. The day when professional scholars could count on large grants to carry out research that lay people were not intended to contribute to or to even understand, or treat paid research assistants with indifference or contempt was still in the future. In the meantime, the connection between the practice of science and the life of the gentry meant that being known to have contributed to science carried with it considerable social prestige. By the mid-eighteenth century curiosity was regarded as a marker of the social elite.[76] HBC gentlemen – and HBC officers were addressed as gentlemen – cannot have missed this fact.[77] The gentlemanly protocols that marked the correspondence between HBC officers and metropolitan virtuosi suggest that HBC officers could find that their relationships with those savants usually reinforced rather than undermined their sense of place in a social hierarchy, and reassured them that they lived in remote confines of civilization, rather than in savage wilderness.

Overwhelmingly, the lay practitioners of science in the HBC were indeed from among the officers – ship captains at first, but also the surgeons, some of whom were hired partly to serve as naturalists, and the officers in charge of major posts or districts. But they also included some non-officers, such as clerks, apprentice clerks, postmasters, and assistant traders. Their practice of science signified the class pretensions of HBC "gentlemen." It served to distinguish them from tradesmen and labourers within the company and to enhance their social position in the outside world – not an irrelevant concern for HBC officers, many of whom anticipated leaving fur-trade country when they retired from the trade. Social standing was not less of a concern for those who had aboriginal wives and mixed-blood children. The typical HBC officer appears to have been very conscious of his reputation. During the late nineteenth century, when HBC officers posed for photographs, they generally did so with clothing and hair that conformed to the fashions of polite urban society. Long-haired, leather-clad, fur-hatted traders were a rarity among HBC officers.[78] When Sir Edward Poore visited the HBC territories in 1849 he himself noted that he arrived with "long hair, earrings, leather trousers fringed & all the other fixings belonging to a half breed."[79] Few HBC traders were impressed. Chief Factor Peter Skene Ogden noted that "although I am not overfond of dress, still I am of opinion that rules of propriety should in all things be strictly observed which I must say Sir Edward lost sight of."[80] HBC "gentlemen" played the part. From their attire to their association with influential men of science, ambitious HBC men wished to be set apart from the labourers and less ambitious men in the company.

As inconsequential it might seem, the very fact that knowledge networks operated on exchanges of letters was itself a great encourager of industry. Correspondence bound participants together in bonds of emotionally satisfying male companionship. HBC officers themselves rarely got together, so the exchange of letters (both personal and work-related) was a large part of their lives. But there were never enough letters, especially from those outside the company. One account describes the "red-letter day" that arrived, "once or twice during the winter season upon which the mail arrives":

> bringing a great budget of letters to be answered and periodicals from the outer world. In the answering of letters considerable difficulty is experienced from the absence of anything new to write about. To obviate this and produce the requisite novelty, the writer generally succeeds in composing a single letter having the desired degree of spiciness. This he copies, and sends to all those friends whom he is desirous of placing under the obligation of an answer. Thus, for many days after the arrival of a mail, occupation for the long evenings is easily found, until the returning dog-train bears his correspondence away, and with it that method of passing time.[81]

This account helps explain why regular correspondence with a scientist-friend in London, Edinburgh, or Washington was so highly treasured among officers. With such friends, there was always something to write about. And from such unrelated learned correspondents, HBC men clearly experienced much satisfaction. When letters were accompanied by newspapers, magazines, or books, so much the better.

Relatively few non-officers contributed to science. Among them were clerks whose participation in knowledge networks was probably motivated by desire for promotion, but others were probably induced, more or less willingly, by the instructions of their superiors. For example, in 1859, Bernard Rogan Ross, unable to keep meteorological records himself, instructed the assistant trader in his post, Andrew Flett, to keep the journal.[82] If tradesmen and labourers within the HBC contributed significantly to science they were not recognized for doing so. It appears that they generally preferred wiling away leisure hours by playing games (cribbage, whist, and other card games, as well as backgammon, dominoes, chess, and quoits being favourites in the late nineteenth century), engaging in athletic contests (canoe, dog, and horse racing), playing music and dancing (with violin and flute being favourite instruments), story-telling and conversation, or by hunting or fishing recreationally.[83] If those men did act as collectors they must have expected to be paid to collect. Their contributions must then have been channelled through the officers. Few had the ability to write letters with the elegance and flair normally exhibited by learned correspondents. Neither did many likely aspire to the same kinds of external validation that the officer class sought.

The role of aboriginal people in knowledge networks is poorly documented, but it must have been significant.[84] The predominant relationship between HBC traders and aboriginal people was one of economic partnership. Furthermore, social mixing, including intermarriage, between Europeans and indigenous people occurred in the fur trading regions of northern North America at a scale unknown anywhere else in British America.[85] Evidence from the second half of the nineteenth century, the period for which the documentary evidence is most abundant, reinforces the fragmentary evidence from earlier periods that aboriginal people contributed proportionately more significantly to scientific knowledge of the HBC territories than in any other region of Anglo-America. Aboriginal people routinely served not only as trappers, but also as guides, couriers, and hunters for traders throughout the HBC territories. That meant that explorers and scientific travellers easily hired them in those same roles. But aboriginal people's contributions to scientific collecting were more direct than that. They knowingly collected many specimens and shared much knowledge for scientific purposes. When, in the late 1840s, a Cree or Ojibwa person somewhere south of Hudson Bay carefully processed a fox according to a special request from John James Audubon "with the skull bones, legs &c complete," he, like many other aboriginal people who collected natural history specimens over the years, had likely been told of the reason for the unusual request.[86] An 1808 letter from the Governor and Committee to James Bird, the trader in charge at Edmonton House, hints intriguingly at how aboriginal people might have provided ethnological specimens. In that year the London Committee sent Bird "an excellent Gun of superior workmanship," with an explanation that "the Committee has thought proper to make some return to the Indian Chief who sent them last year an Indian dress."[87] There is enough evidence to conclude that much of the natural history knowledge, and, in at least some contexts, most of the natural history specimens that officers and scientists acquired were purchased from aboriginal trading partners, friends, or spouses, even if the incompleteness of the evidence prevents us from ascertaining the true scale of the transfer of knowledge and material.

The HBC contributed meaningfully and intentionally to some of the more highly respected and prominent fields of study to which it could naturally be expected to contribute at the time, and in ways most likely to earn the company and its employees recognition and respect. Given that the HBC was headquartered in London, it would have been possible for the company to adopt a policy of assisting the most famous elite natural philosophers in London and Edinburgh. But that would have made little sense. If chartered overseas companies were going to contribute to science, it was only natural that they would contribute to scientific activity in their overseas territories. Thus the HBC's support came overwhelmingly in exploration, surveying and cartography, and the observational sciences: astronomy, meteorology, natural history, and ethnology. A contextualist social history of science, the kind to which many historians have turned in recent years, should acknowledge what the historical actors themselves thought about "science," and what kinds of science they admired: it should not focus merely on that science that contributed most to today's scientific thought.[88] That includes, but is not limited to, recognizing that the very meaning of terms like *science*, *natural history*, and *natural philosophy* were fluid, and that the boundaries between fields of knowledge moved between 1670 and 1870. In 1670, *science* could refer broadly to any kind of systematic knowledge. Only gradually did it assume a definition similar to today's.[89] More importantly, approaches to many sciences now regarded as quaint and unsophisticated, or even pseudoscientific and disreputable, were formerly de rigueur. This was the case with the HBC's involvement in exploration, cartography, and surveying, as well as natural history, meteorology, and ethnology.

Until relatively recently, many historians considered geographical exploration as an activity quite distinct from science. But in the 1970s, the prominent historian of exploration William H. Goetzmann argued that exploration "should be thought of as integral to any history of science."[90] Goetzmann's argument is widely accepted today. One historian of science has recently noted that "disciplines like cartography, mapmaking, and natural history are no longer marginal to the narratives of the origins of scientific modernity."[91]

For several reasons, it is essential that this book include aspects of the history of exploration, surveying, and cartography in its examination of science in the HBC. Many historical actors did not draw a distinction between knowledge gained by geographical exploration and other kinds of knowledge. In Francis Bacon's utopian novel, *New Atlantis*, Solomon's House has a gallery to celebrate "principal inventors," featuring, most prominently, Christopher Columbus.[92] Into the eighteenth century the *Philosophical Transactions of the Royal Society* published articles related to exploration. Furthermore, during the seventeenth century exploration became "scientized" in a process influenced by Bacon's philosophy of science.[93] James Cook's voyages in the eighteenth century represent another major milestone in the process of the scientization of travel.[94] Cook's was not the first European exploration motivated primarily by scientific aims, but beginning in 1769, the scientific objectives of many voyages of exploration were often central, and even paramount.[95] Geographical exploration and surveying by the HBC was sometimes geared as much to answering questions of interest to British thinkers outside the company as they were to meeting the company's immediate business needs. Even more important for present purposes, scientists often accompanied explorers in ways that bolstered arguments that lands were not viewed primarily by covetous "imperial eyes," but by "liberal," "enlightened," or even humanitarian eyes.[96] In other words, the prestige of exploration, which was almost instinctively assumed to be carried out for personal, corporate, and national gain, could be greatly enhanced by any evidence that costly expeditions were, in fact, selfless contributions to humanity. Thus, although scientific research was often only a secondary aim of exploration, the scientific activities were instrumental in bolstering claims for the disinterested or even humanitarian nature of exploration. This is certainly the case with the search for the Northwest Passage between 1818 and 1854. This study examines the history of geographical exploration and surveying when the results of those activities were shared with prominent geographers and cartographers, and particularly when the instructions given to the explorers, the actual activities of the expeditions, or the published accounts of the explorations emphasized the scientific dimensions of exploratory expeditions.

Natural history is not in the scientific mainstream today, but throughout the period under study, natural history was a much-respected field of study.[97] It, like geographical exploration, was especially important as a hinterland science, and the natural history of North America had particular significance for European intellectuals. Bernard Cohen explained that "from the sixteenth century to the end of the nineteenth, America remained a naturalist's paradise. ... from Oviedo and Acosta in the sixteenth century to Darwin and Asa Gray in the nineteenth, these data not only filled the gaps in the whole description of nature on this globe, but were the source, or perhaps the occasion, of challenging ideas that upset general preconceptions."[98] This book is not geared towards examining how the specimens, data, and evidence gathered in the HBC territories did or did not influence scientific thought at the time; the point is that they were gathered in response to scientific trends at the time, according to the expectations of scientists, and in order to be useful to scientists. This understanding is important to a social history of science because it was upon those bases that the scientific activity derived its stature in the scientific and non-scientific world.

Almost immediately after the European discovery of the Americas, European thinkers were intrigued by the marked difference between the climate of Western Europe and eastern North America in the same latitudes, and that difference sparked first confusion and then some grand theories about the North American environment.[99] The development and improvement of meteorological instruments (especially thermometers and barometers) in the seventeenth century permitted scientists to compare climate and weather. Moreover, for almost the entire time between 1670 and 1870, many scientists believed that the systematic keeping, collecting, and studying of meteorological journals from many places offered the best hope of enabling them to predict weather. Today we know that they were wrong. But, the tremendous efforts taken by many HBC officers to keep and submit meteorological registers that conformed to the expectations of scientists in Europe were rooted in a hope that those weather records would contribute to what was regarded as a significant scientific project, one that – had it succeeded – would even now be regarded as momentously important.

In short, the HBC did not drive scientific trends, but it was not oblivious to them. The company's directors and officers wanted to be perceived as contributing effectively to the kinds of sciences that were prominent at the time. HBC officers were not trend-setters to be sure, but neither would metropolitan and elite scientists have regarded the scientific questions to which they directed their attention, the approaches that they took, or the methods they used to preserve specimens or gather data as trivial, esoteric, unscientific, or outdated. Compared with the Galapagos Islands, the HBC territories were the locus of few scientific discoveries that are renowned today. Still the conduct of science there reflected trends in Europe remarkably well. It had to; its contemporary social and cultural significance required it.

Aside from defending the central arguments already presented above, this book provides a critical narrative history of science in the HBC. Since readers are not assumed to have a thorough prior knowledge of the history of the HBC, the contents and arguments of each chapter are summarized below, alongside an outline of the corporate history of the HBC.[100] The historical context helps explain many aspects of the history of science in the company.

The three chapters of part 1 of this book explore the period between 1670 and 1821 when the company often faced significant commercial competition and rivalry. Although the English Crown granted the HBC a charter to Rupert's Land in 1670, for most of the period before 1821 the company's monopoly was a mere chimera. For long periods between 1686 and 1714 the French occupied some of the trading posts on Hudson Bay. Even as late as 1782 the French attacked and destroyed some of the company's major coastal trading posts. Moreover, for most of the time before 1760, competitors operating out of New France drew off many of the furs from the Hudson Bay drainage basin. Beginning in the 1740s vehement critics in Great Britain also began questioning the legitimacy of the company's charter. This was one of many such attacks on the company's exclusive trading privileges that punctuated the history of the company.

Chapter 2 explores scientific activity in the HBC during the first century of the company's existence, a period during which scientific networks did *not* flourish.[101] For nearly its entire first century, the HBC did not support scientific research in its territories because it regarded the communication of any information to outsiders as a threat to the company's interests. The company's policy was not unreasonable. Chapter 2 illustrates why the directors of overseas companies easily perceived that the risks of contributing to science outweighed the potential benefits. In short, as late as 1768 scientific activity in Rupert's Land had probably done more to undermine the company's interests than to bolster them. Furthermore, the experiences of the company during its first hundred years (as well as in its second century) showed that scientific activity in the company's realm could backfire in such unpredictable ways that it would always be impossible to eliminate the inherent risks associated with supporting science in its territories. That fact goes a long way to explaining why the company's directors remained cautious throughout the corporation's history about what visitors to their posts might learn.

But it is not as if the HBC rebuffed many overtures from scientists during its first century of existence. The Royal Society – the only scientific organization with which the company might have collaborated in that century – did not show much interest in cooperation with the HBC before the 1760s. Given that the company's secretiveness evidently mirrored the policies of most chartered companies and some European governments towards their overseas territories at the time, the company did not suffer for its secretive policy until attitudes towards science changed.

The second chapter also shows that although the reluctance of company directors to support science did prevent the formation of substantial networks tying metropolitan intellectuals with potential informants in the hinterlands, individuals within a company such as the HBC might still contribute to science even when the company's directors were uncooperative and scientific organizations were not actively seeking cooperation. The HBC's small contributions to science were made primarily by ship captains who regularly travelled between North America and London and by surgeons who visited with metropolitan scientists while on furlough.

The single most important milestone in the history of the HBC's cooperation with science occurred when the Royal Society of London approached the company to request its assistance with its planned observations of the Transit of Venus in 1769. Only five years later, the HBC's cooperation with the Royal Society had expanded to produce a network that connected the Royal Society with several lay collectors and observers in Rupert's Land. The scope of their work included several fields of knowledge, particularly astronomy, natural history, and meteorology. Chapter 3 examines those crucial years between 1768 and 1774. It argues that two men, Samuel Wegg (treasurer of the Royal Society and member of the London Committee of the HBC) and William Wales (astronomer), were key to the success of this cooperation, from which the Royal Society gained valuable data at little cost and the company earned tribute from the foremost scientific institution of the time. In other words, attitudes in both the Royal Society and the HBC were crucial to the change.

More broadly, the third chapter suggests that, although historians of science have already acknowledged that the Transit of Venus was an important milestone in the history of the expansion of European science, the significance of that event may yet be underestimated. Some British astronomers at the time argued that the level of Great Britain's support of observations of the Transit of Venus was a gauge of the nation's honour. The diffusion of European scientists, scientific instruments, scientific books, and European scientific values to far corners of the globe affected the practice of science outside Europe in ways that we do not yet fully appreciate. The HBC at least, never returned to its secretive policy towards science.

When British forces conquered New France in 1760, the directors of the HBC might have hoped that they would soon enjoy the benefits of their monopoly, but instead competition from Montreal-based traders only stiffened. Although British subjects did not compete with the HBC from posts established along Hudson Bay itself, between 1763 and 1821 firms based in Montreal captured much of the trade of the northern interior of North America, and after 1778 most of that trade was consolidated in the hands of one partnership, the North West Company. By the early 1770s Montreal-based competitors were threatening the profitability of the HBC.[102]

Until 1774, the HBC's North American activities were limited almost exclusively to seven factories at the mouths of rivers along the coasts of Hudson and James Bays. But between 1774 and 1821 the HBC responded to competitors operating out of Montreal with a dramatic inland expansion of its operations. The rapid westward expansion began with the establishment of Cumberland House in 1774. By 1799, the company had established Acton House (Rocky Mountain House) in the shadows of the Rocky Mountains (see figure 4.1, p. 96).[103] The HBC also made a feeble attempt to gain a foothold in the Athabasca country (outside Rupert's Land) between 1799 and 1806, but after the NWC resorted to brutal methods to defeat the HBC there the HBC abandoned that initiative temporarily.[104] However, the HBC grew more aggressive again in the 1810s. In 1812 it permitted the establishment of the Red River Colony (Selkirk Colony) astride important supply routes for the NWC, and in 1814, that colony's governor, Miles Macdonell, issued a proclamation forbidding the exportation of pemmican (a crucial foodstuff for the NWC) from the colony. Two years later, the HBC assertively moved into the Athabasca country.[105]

The years between 1816 and 1821 were the worst in the history of the HBC. In 1816, as the competition between the HBC and NWC became financially ruinous to both companies, the increasingly bitter rivalry between the companies descended into virtual private warfare. Hostilities centred particularly around the Red River colony and in the Athabasca country. Between 1816 and 1821, conflict consumed the attention and revenues of both companies.[106]

Chapter 4 examines the HBC's contributions to science from the beginning of its inland expansion to its merger with its last surviving

Canadian competitor, the North West Company, in 1821. Although these were very difficult years for the company, during that time the HBC contributed significantly to the mapping of the interior of the continent. The company's contributions to the work of British cartographers during those years were a by-product of its business activities. The company had little knowledge of the North American interior in 1774 – a fact that its critics and competitors had emphasized since the 1740s. But when the company expanded inland after 1774, its London directors needed to know where its operations were located. That is why it hired its first formally trained surveyor in 1778. Rather than keep its knowledge secret, the HBC began almost immediately to share the results of its surveys with British cartographers. It even allowed people outside the company to influence its surveying agenda. The fourth chapter argues that, thanks in large part to the company's London Governor, Samuel Wegg, the HBC contributed significantly to British public knowledge of northern North America in this period. Contrary to the impression conveyed by much of the historical literature, despite the fact that the NWC contributed more to exploration than the HBC, the HBC contributed far more to British mapping of the northern half of North America than did the NWC. The maps helped the company in its rivalry with the NWC. The company also began sharing some its documents with savants. The most important of these documents were the journals of Samuel Hearne. The HBC was in a position to share so much of its geographical knowledge because that knowledge gave it virtually no direct advantage (because the NWC already possessed greater knowledge). Its cooperative stance explains why by 1784 the company's directors were credited with having "made amends for the narrow prejudices of their predecessors."[107]

The four chapters of part 2 explore the period between 1821 and 1870. The circumstances facing the HBC help explain why, although it was easier than ever for the HBC to contribute to science after 1821, it was also more important that it do so. The merger of the HBC and the NWC in 1821 represents a major turning point in the corporate history of the HBC. The violent rivalry between the NWC and HBC convinced the British Colonial Secretary Lord Bathurst to urge representatives of the two companies to negotiate a settlement.[108] Those negotiations produced that June a merged company that retained the name of the Hudson's Bay Company headquartered in London. The British Parliament rewarded the merged company by acknowledging the company's claim that Rupert's Land encompassed the entire Hudson Bay drainage basin, and adding to those territories a twenty-one-year exclusive licence to the fur trade of the regions in which the NWC had dominated: the Athabasca and Mackenzie River drainage basin, New Caledonia, and the Columbia District (see figure 1.1, p. 4).[109]

By 1825, the HBC had returned to profitability. The next thirty-five years were the company's halcyon period. Between 1825 and 1860, it never paid dividends of less than 10 percent.[110] According to one of the most prominent historians of the company, the directors of the company in that period "were probably as able a group of directors as presided over any British business of that day."[111]

Profitability and stability made it easier for the HBC to contribute to science. So did the altered post-1821 corporate structure. The merger brought many NWC men into the HBC, some of who – Peter W. Dease, John McLoughlin, Edward Smith, Colin Campbell, and George Barnston among them – were talented men in a good position to help the HBC contribute to science and exploration in the ensuing years. Subsequent to the merger, the HBC also sought different attributes in its officer class. After 1821, the directors' emphasis on careful, efficient management favoured well-educated men – especially Scottish men, who typically had a better education than the English. The officer class of the HBC became increasingly likely, for various reasons, to be willing and eager to contribute to science.

Despite prosperity, the HBC did face new dangers after 1821. Although its commercial competitors were few after 1821, its critics and adversaries in Great Britain multiplied. In a liberalizing Great Britain chartered monopolies were increasingly tempting targets for politicians seeking to win easy political points. "Attacks upon its privileges," wrote John Galbraith, "won easy applause in Parliament."[112] Beginning in 1836, an increasingly organized humanitarian movement criticized the company's treatment of aboriginal people. By the end of the 1840s, with free trade widely accepted, few politicians and officials were apt to defend the company publically (although officials in the Colonial Office were remarkably supportive of the company privately). But the company repeatedly needed the assistance of British politicians, diplomats, and officials. The HBC's exclusive licence to its territories outside Rupert's Land was subject to parliamentary review and renewal every twenty-one years, and the legitimacy of its charter was repeatedly under attack.[113] Meanwhile, no one could predict the fate of the HBC's 1670 charter should the British government decide to subject it to judicial review.

The company's hold on some of its territories – especially the Columbia District west of the Rocky Mountains – was also rendered uncertain by foreign rivals. In the 1820s and 1830s, growing Russian claims to and activity in the north Pacific and far northwest required the British and Russian governments to define the boundary between Russian and British North America in an 1825 treaty that still did not prevent conflict between the Russian American Company and the HBC nearly to the end of the 1830s.

Threats also included the uncertain status of the Columbia District (Oregon Country) – an issue that became a crisis by late 1844. Even after the British government ceded a portion of the Columbia District to the United States in 1846, the HBC continued to require the assistance of the British government in defending its interests internationally.[114]

Dominating the HBC's operations for most of this period was the imperious and indefatigable George Simpson.[115] First sent to North America by the London Committee in 1820, Simpson was appointed governor of the HBC's Northern Department after the merger. His authority soon spread over all of the HBC's North American operations. The London Governor and Committee exercised ultimate authority in the Company, but by retaining their confidence and respect, Simpson was able to rule the company's North American officers almost dictatorially. Simpson and his officers understood that the company needed to be seen to sponsor science, but jealous of the needs of the company, and conscious of his own social standing, Simpson ensured that the pursuit of science promoted the interests of the company and of himself personally. From 1821 until Simpson's death in 1860, those who would undertake exploration and science in the HBC territories had to deal with this towering figure.

Part 2 of this book, then, deals with a period in which (although it was easier for the company to contribute to science) it was more important than ever that the HBC demonstrate its generosity. The actual contributions of the company to science between 1821 and 1870 are so large and diverse that they are examined in four chapters. Together, they show how scientific activity, once established, gained momentum. Various metropolitan scientists and scientific organizations, realizing that the HBC's directors were willing supporters of science, were encouraged to appeal to the company for assistance.

Chapter 5 examines the HBC's involvement with arctic science between 1818 and 1855. In the arctic – actually outside the HBC's territories – the HBC began by assisting overland Royal Navy expeditions to search for a Northwest Passage, a search that the Admiralty portrayed as scientific and disinterested rather than strategic. The company assisted the British Navy's overland expeditions beginning in 1818, but it could do so far more effectively after the merger in 1821. Then, in the late 1830s, as the deadline of the company's first parliamentary licence approached (due to expire in 1842), it launched its own arctic expeditions. When it did, it mimicked the Royal Navy by ensuring that the ostensibly neutral and humanitarian exploratory expeditions also included scientific mandates that were more plausibly "pure." This chapter also presents some of the most direct evidence that influential men within the company believed the company's contributions to science were crucial to its efforts to maintain a positive corporate image in Great Britain. The search for the Northwest Passage also illustrates the resurgence of wonder and science in a Romantic age. Samuel Taylor Coleridge first coined the term "second scientific revolution," one marked particularly by "Romantic science."[116] Romantic science reconciled high emotion, including awe, wonder, and terror, with science. In that world the perilous – even the reckless – pursuit of knowledge was particularly admired, especially among the public. These sensibilities provided an avenue to fame for HBC arctic explorers beginning with Samuel Hearne and extending at least to John Rae, who provided perhaps a spectacular example of the solitary scientific explorer, utterly committed to his mission, and oblivious to all danger and discomfort.[117]

One of the most remarkable scientists to travel to the HBC territories was the Scottish naturalist, David Douglas. His first opportunity to visit the HBC territories in 1824 arose because of the great growth of interest in Great Britain in exotic plants for their ornamental, botanical, and economic value. Plants from the northwest coast of North America, particularly its conifers, were of great interest to horticulturalists because the similarities of the climate of that region and the British Isles made those plants suitable for British outdoor gardens (and commercial forests). Scientists and horticulturalists were also interested in that region because of the diversity of beautiful plants available there, although they were interested in the flora and fauna from around the world, including the subarctic and arctic regions of North America. David Douglas first arrived as a collector on behalf of the London Horticultural Society in 1824. He died tragically in 1834 at the age of 35, but as chapter 6 explains, the HBC's cooperation with David Douglas inspired and influenced many others to contribute to science for decades after his death. Thus, Douglas helped to establish scientific networks that long outlived him. The evidence also reveals something of the depth of feeling that accompanied the male companionship forged by shared fascination with natural history. The chapter also explores how men hired or supported by the HBC sought to parlay their experience collecting outside of Europe to careers and to prominence as naturalists or gardeners in Great Britain or Europe.

Until the 1830s, the HBC's scientific assistance was granted almost exclusively to people and organizations based in Great Britain, but beginning in the 1830s, HBC personnel assisted some American scientists, including John K. Townsend, Thomas Nuttall, and John James Audubon, and in 1842 it also began cooperating with Canadian-based travellers. Chapter 7 explores this scientific cooperation with particular attention to the unanticipated consequences for the HBC of the company's assistance to those scientists and travellers. Specifically, the HBC's contribution to American science probably contributed to American interest and settlement in the Oregon Country in the late 1830s and early 1840s, and to the American move to annex the region in the mid-1840s.

The HBC's contributions to Canadian-based science followed a similar pattern in that they appear to have resulted in encouraging the development of an expansionist movement in Canada. The HBC's contributions to Canadian-based science were small until the company was presented with the opportunity to assist an international survey of terrestrial magnetism promoted by some of its advocates as "The Magnetic Crusade."[118] Dwarfing the impressive international efforts to observe the Transit of Venus in 1769, the geomagnetic survey of the 1830s and 1840s inaugurated and became the model for "big science."[119] This book explores several aspects of the survey's legacy. Part of the project saw the British Artillery establish a geomagnetic and meteorological observatory in the small city of Toronto in the colony of Canada. With the assistance of the HBC, John Henry Lefroy, the officer in charge of the Toronto observatory, spent eighteen months in the HBC territories in 1843 and 1844, conducting geomagnetic research in some of its most remote regions. Soon after his return to Toronto, Lefroy facilitated an extended trip by the young Toronto-based artist, Paul Kane, to the HBC territories.

It is no mere coincidence that Toronto-centred Canadian interest in the HBC territories developed in the late 1840s shortly after Lefroy's and Kane's return to Toronto. The standing, reputation, and connections that those two men acquired in Toronto society by the mid-1850s allowed them to participate in and contribute to the emergence of the Canadian Institute as the first influential public scientific organization in the city. Furthermore, during the year or two before the Canadian expansionist movement burst onto the scene (and the years thereafter), the meetings of the Canadian Institute put Toronto scientists, colonial politicians and officials, and Canadian expansionists together in the same room, sometimes to attend to Kane's papers and exhibitions about the HBC territories. Indirectly then, the HBC's assistance to Canadian-based travellers in the 1840s contributed to a Toronto-centred Canadian expansionist movement that adopted some stridently anti-HBC rhetoric beginning in 1856. This chapter also presents some of the most explicit evidence that HBC directors sought to reward influential men who had defended the company.

Beginning in the mid-1850s, the HBC faced a kind of criticism rarely seen since the 1770s. Allegations that the company clung unreasonably to an illegal charter, and that it treated inhabitants of its territories unjustly arose again, but critics in Britain and Canada also began to accuse the company of deliberately concealing the potential of Rupert's Land to support permanent British agricultural settlement. The company needed to respond to such criticism carefully. It had to face a British Parliamentary Committee struck to investigate it in 1857, and it had little choice but to cooperate with British and Canadian expeditions sent to Rupert's Land between 1857 and 1860 to investigate the possibility that southern portions of its territories territory might be suitable for agricultural settlement. Findings that Rupert's Land was suitable for agricultural settlement were not necessarily threatening to the company's interests, but accompanied as they were by the anti-HBC rhetoric of Canadian annexationists (rhetoric used to defend an argument that the HBC should not be compensated for the loss of its charter), the company's officials might have been excused if they feared Canadian expansionism as much as they did American assertions of manifest destiny.

The central focus of chapter 8 is on a period of cooperation between the HBC and the Smithsonian that was not simply a natural expansion of older connections between the HBC and American scientists. HBC employees began contributing to the Smithsonian Institution by the mid-1850s, but between 1857 and 1868, the Smithsonian became the primary beneficiary of the HBC's assistance to science. The Smithsonian was fortunate that a significant scientific culture had already developed in the HBC by the 1850s, but chapter 8 argues that the social acumen and awareness of the directors and scientists at the Smithsonian Institution ensured that the scientific efforts and sympathies of the entire company, from its London Governor and Committee, and its North American governor, to its officers in the field, were reoriented towards contributing to its scientific program. Thus, this book opens with an examination of a period in which scientific networks did not flourish, and closes with an examination of the most effectively managed scientific network in the history of the HBC.

Chapter 8 also examines many aspects of the history of science that were probably important during much of the history of the HBC but have been rendered invisible in earlier periods by the lack of evidence. For example, aboriginal people were probably paid for providing plant and animal specimens, geological samples, and information destined for scientists for much of the history of the HBC, but the documentary evidence from later years is far more abundant than it is for the early years. The ability of HBC traders to tap aboriginal communities as effectively as they did was probably rooted in the fact that the relationship between aboriginal peoples and HBC traders was generally cooperative, quite different from the unequal and coercive relationship that colonized peoples experienced in what Mary Louise Pratt has defined as the "contact zone."[120] In the HBC territories, aboriginal peoples often willingly collected, processed, and delivered specimens and provided information to traders on terms not much different from the way they delivered furs, provisions, and services to the same men: as part of well-established systems of freely negotiated exchange. Whether scientists actually visited the HBC territories or stayed in the metropolis, they had access to this exchange system through the auspices of the company and its employees.

This book shows that mutually beneficial and satisfying relationships were crucial to science in the HBC. Scientific activity in the company's territories languished or flourished depending on the degree to which the HBC's Governor and Committee, scientific organizations, and a host of individuals, ranging from company directors and savants in the metropolis to lay practitioners and aboriginal peoples in the hinterlands, perceived and seized opportunities to participate in symbiotic relationships with others whose backgrounds and interests might have been very different from their own. In so doing it contributes to our understanding of a wide range of themes in the history of science and the history of the British World before 1870.

PART I

The Hudson's Bay Company and Science, 1670–1821

2 "A Profound Secret": The Adventurers and the Fellows from the 1660s to 1768

In some ways the circumstances in London during the 1660s and 1670s were favourable for the development of fruitful cooperation among Britain's political, economic, and scientific elite. An informal "Invisible College" of natural philosophers dedicated to the accumulation of knowledge through experimentation and observation had begun meeting in 1645. Thanks to the Royalist sentiments of its "fellows," in July 1662 King Charles II – only fifteen months after he was crowned in Westminster Abbey – granted this group a formal charter. Thus was born the Royal Society of London.[1]

Less than a decade after the formal establishment of the Royal Society, several Fellows of the society – including Sir Christopher Wren, Sir Paul Neile, Sir Philip Carteret, and Sir James Hayes – invested in an expedition sent in 1668–9 to explore (and trade in) Hudson Bay in hopes of finding a passage to the Orient. Seeking to protect their investments, some of these "Adventurers" petitioned the king for a charter. And so, on 2 May 1670, King Charles II granted a charter to eighteen men, forming the "Company of Adventurers of England trading into Hudson's Bay," better known as the Hudson's Bay Company.[2] The prominent Royalists, so closely associated with both the Royal Society and the HBC, had the potential to form a firm foundation for a three-way alliance among natural philosophers, merchants, and the Crown. But that alliance failed to form during the first century of the HBC and Royal Society, apparently because neither the company nor the society sought or embraced opportunities to cooperate. The modest contributions to scientific knowledge about the HBC territories came as the result of the efforts of remarkable individuals who worked without the official support of either the HBC or the Royal Society.

Sir Francis Bacon planted the philosophical seeds for an alliance of merchants and natural historians half a century before the HBC was chartered. Bacon's *Novum Organum* (1620) rejected Aristotle's *Organum* – a reliance on logic and reason (deduction) – and championed instead a *novum organum* (new instrument): the accumulation of knowledge and theory through observation and experimentation (induction). Furthermore, Bacon believed that exploration outside Europe had an important role to play in the accumulation of knowledge. However, given the small role that England had played in exploration to that time, it was natural for Bacon to draw upon Iberian examples. A highly symbolic title page in *Instauratio Magna* ("Great Renewal" [1620]), in which the first edition of *Novum Organum* appeared, portrayed a ship sailing beyond the pillars of Heracles. The now famous illustration was not original; it was inspired by Spanish illustrations from 1606.[3] The illustrations symbolize the eagerness of Europeans to extend their geographical and intellectual reach beyond the *nec plus ultra* of the ancient world. Bacon explained that "if anyone attempts to renew and extend the power and empire of the human race itself over the universe of things, his ambition (if it should so be called) is without a doubt both more sensible and more majestic" than those who aimed to extend their power over their own countrymen (which he described as "common and base"), or "to extend the power and empire of their country among the human race" (which he believed had "more dignity, but no less greed"), for "the empire of man over things lies solely in the arts and sciences. For one does not have empire over nature except by obeying her."[4]

In what was probably his most influential work, *New Atlantis*, published posthumously in 1627, Bacon explored these ideas further. Inspired again by the Spanish Empire (the Solomon Islands having been discovered and named in 1568–9), the narrator of that short utopian novel was part of an expedition that encountered a land west of Peru, called "Bensalem."[5] The Christian inhabitants of Bensalem supported a scientific organization called "Solomon's House" with thirty-six "fellows." The purpose of Solomon's House was the acquisition of "the knowledge of Causes, and secret motions of things; and the enlarging of the bounds of Human Empire, to the effecting of all things possible." The means to accomplish these purposes included "the Preparations and Instruments" that made empirical study possible, and the "Employment and Offices" of "fellows" of Solomon's House. The division of labour among the fellows implied a network that tied metropolitan natural philosophers to scientific travellers. Fully a third of the fellows were "Merchants of Light" who "sail into foreign countries under the names of other nations (for our own we conceal); who bring us the books, and abstracts, and patterns of experiments of all other parts."[6] Not only his most famous aphorism, "knowledge is power," but his broader approach to science shows that Bacon believed that science and exploration were closely linked to power. He extolled the philanthropic use of such power, but was obviously alive to the potential of humans to use their knowledge to dominate other humans. The early Fellows of the Royal Society and the early Adventurers of the HBC appear to have well understood the social, economic, and geopolitical value of scientific knowledge, whether or not they used that knowledge in ways that Bacon would have admired.

The charters of the Royal Society and the HBC imposed no specific obligations upon the organizations, but the preambles of the first and second charters of the Royal Society (1662 and 1663) made the connection between exploration and knowledge explicit. They noted that "we have long and fully resolved Ourself to extend not only the boundaries of the Empire, but also the very arts and sciences."[7] From their beginnings then, the Royal Society and the HBC were well placed to pursue scientific and geographic knowledge and, for that matter, for the HBC to deploy some of the Royal Society's "Merchants of Light."

Despite these apparently auspicious beginnings for the HBC's potential contributions to British science, the company contributed little to public knowledge before 1768. There is no evidence that the HBC's London Governor and Committee ever shared any of their knowledge with British natural philosophers during that period. On the contrary, the HBC was an obstacle to the flow of information about the Hudson Bay region to savants. Maps of the world show that European knowledge of northern North America lagged well behind that of most areas of the world from the 1500s to the 1700s. It seems as though most (perhaps all) of the communication between Bay men and intellectuals before 1768 occurred when HBC employees circumvented the company's London-based Governor and Committee. These were primarily ship captains, relatively well-educated men whom the Governor and Committee could not easily control because they travelled regularly between Hudson Bay and the metropolis, but also included some officers based in North America who communicated directly with scholars during year-long furloughs or after retirement. Their contributions before 1768 came in two waves. During the first – perhaps more a ripple than a wave – information flowed mostly (perhaps solely) through ship captains to the Fellows of the Royal Society, with modest significance. During a second wave, between 1725 and 1750, several employees contributed more significantly to navigation, natural history, and meteorology. But the Governor and Committee, by 1750, could very legitimately conclude that the flow of knowledge since 1670 had served the interests of some of its employees and many of its critics, competitors, and rivals, but not those of the company itself. On the other hand, they could also understand that the company's secretiveness, which appeared to serve the company well for many years, had more recently become the company's Achilles heel. Still, the company moved only cautiously to share its knowledge before the late 1760s.

That the HBC's contributions to public knowledge and its communications with the Royal Society were not particularly plentiful in its early years illustrates common attitudes towards knowledge in the period. The historian Alison Sandman has recently reminded us that although secrecy in science is often perceived as abnormal – we "tend to assume that knowledge was expected to circulate" – the borders between what was considered rightfully kept secret and what ought to be shared have moved over time.[8] Iberians, for example, were secretive with their knowledge of their empires for centuries.[9] Hapsburg Spain's policy of *arcane imperii* extended not only to cartography and navigation but also to natural history.[10]

If the Portuguese and Spanish guarded their knowledge carefully, the Dutch were not much more forthcoming. The *Eerste Schipvaart* (first voyage) of the Dutch to the East Indies from 1595 to 1597 marks the beginning of Dutch influence in Indonesia. To be sure, the first voyage (made possible by charts and maps leaked to the Dutch) had scientific objectives: to measure magnetic deviation of the compass needle, and to map the stars of the southern hemisphere.[11] One historian of science has concluded that the Dutch effort to map the southern sky "served first and foremost a purely scientific purpose, namely to fill the long-existing gap in our description of the complete heavens," rather than navigational purposes.[12] The pureness of the effort is open to question. The voyage's investors insisted that the savants turn over their evidence to them, and much of it was never published. Some observations from that voyage, and some ethnological and astronomical observations made during another expedition were published in 1603.[13] But "Dutch navigators did not obtain new data on the southern sky after 1603, as globes published after that date seem to confirm."[14]

The historical context then, suggests that there is little reason to believe that Bacon (or anyone else at the time) believed that scientific knowledge ought to be widely disseminated. The *New Atlantis* does not imply that he did. The very fact that he was obviously inspired by the secretive Spanish Empire suggests that Bacon was untroubled by secrecy. As Delbourgo and Dew have noted of Solomon's House, "the true political objectives of its members ... are shrouded in secrecy."[15] Although Bacon is justly famous for his philosophy of science today, in his lifetime Bacon communicated remarkably little with other natural historians. A child of privilege, parliamentarian, courtier, attorney general, and Lord Chancellor, Bacon was first a man of politics, statecraft, and the court. For Bacon, knowledge was important because of its utility; its strategic value meant that it might not be shared widely. Thus, the fact that the HBC, a product of courtly privilege, gathered knowledge and decided to share or hoard it strategically and cautiously, made it in its own way as Baconian as the Royal Society. The Pyensons have noted that until the mid-seventeenth century, "thinkers tended to guard and keep secret what they knew, fearing that good ideas might be stolen by a rival." It was only gradually with the scientific revolution, they explain, that people began to see scientific knowledge as "shared and communal."[16] If communal culture was relatively new even within scientific circles in the early years of the HBC, it is easy to understand why it might not occur to intellectuals to seek valuable evidence from commercial corporations at that time.

Fellows of the Royal Society had the potential to be Baconian "Merchants of Light" even before the HBC was chartered. No later than June 1662, the Fellows were "considering with themselves, how much they may increase their *Philosophical* stock by the advantage, which *England* injoyes of making Voyages into all parts of the World ... appointed Master [Laurence] *Rooke* ... to think upon and set down some *Directions* for *Sea-men* ... of which the said Sea-men should be desired to keep an exact *Diary*."[17] Rooke's directions, published in the first volume of the society's *Philosophical Transactions*, focused particularly on questions relating to compass variation, inclination of the dipping needle, the dynamics of tides, maps of coastlines, and descriptions of "meteors." The Royal Society proposed to collect seamen's diaries which would be made available to the Fellows and the Admiralty. A second article in the same volume of the *Philosophical Transactions* – this one written by Robert Boyle – was directed more towards overland travellers. It asked for information relating to the "Heavens," and "the Air, the Water, and the Earth" (including latitude, longitude, weather, flora, fauna, and human inhabitants) of any country, "that the Inquisitive and Curious, might by such an Assistance, be invited not to delay their searches of matters, that are so highly conducive to the improvement of *True Philosophy*, and the welfare of *Mankind*."[18] In keeping with these efforts, in early 1664 Dr. John Beale presented a letter to the society "concerning Capt. [Thomas] James's Voyages and wintering in charleton island in Hudson's Bay," and in 1665, the society showed interest in information provided by Pierre Esprit Radisson and Médard Chouart, Sieur des Groseilliers.[19] Only seventeen days after the HBC was chartered, Henry Oldenburg, secretary of the society, presented information gathered by New Englander Zachariah Gillam, who had captained the *Nonsuch*'s voyage to Hudson Bay in 1668–9.[20]

The "Adventurers of the Bay" also had the potential to forge an alliance between science and commerce. E. E. Rich, one of the most prominent historians of the HBC, wrote that "it is odd to note how many of the early adventurers were Fellows of that [Royal] Society, that is to say – as things then were – men known, or wishing to be known, for their enlightened curiosity."[21] Indeed, of the eighteen original Adventurers identified in the HBC charter, at least four were also Fellows of the Royal Society. Furthermore, Christopher Wren, who invested heavily in the HBC, was president of the Royal Society from 1680 to 1682.[22] But the HBC charter itself hints that, although these men may have wished to be known for their "enlightened curiosity," their economic interests, rather than their scientific curiosity, explains why they were "Adventurers." According to the HBC charter, the eighteen original adventurers had "humbly besought Us to Incorporate them and grant unto them and theire successors the sole Trade and Commerce" of the waters of Hudson Strait, Hudson Bay, and the lands surrounding them, after these men had "at theire owne great cost and charge undertaken an Expedicion for Hudsons Bay in the North west part of America for the discovery of a new Passage into the South Sea and for the finding some Trade for Furrs Mineralls and other considerable Commodityes," and had "made such discoveryes as doe encourage them to proceed further in pursuance of theire said designe by meanes whereof there may probably arise very great advantage to us and our Kingdome."[23]

During the 1670s, the Fellows of the Royal Society continued to gather information about Hudson Bay, but not in ways that suggest the cooperation of the Governor and Committee of the nascent HBC. Furthermore, the information gathered by the Society shows that the

Fellows of the Royal Society continued to be interested in Hudson Bay for the commercial and strategic value that the knowledge might give. The minutes of a Royal Society meeting of 18 April 1672 note that "there were read some observations concerning the voyage lately made [in 1670–1] to East Hudson's Bay and the state of that country and its inhabitants, communicated to Mr. [Henry] OLDENBURG [the Secretary], upon his inquiries, by captain GUILLIAUME and Mr. BAILEY, two of the chief persons employed in that voyage, who had wintered there."[24] The interview, which must have taken place after October 1671, when Zachariah Gillam and Charles Bayly had returned to London, was shaped at least as much by the interests of investors who had financed the voyage as by disinterested curiosity.[25] The interviewers sought highly practical information and observations pertaining to the navigation of Hudson Strait and Hudson Bay, the variation of the compass needle and the tides there, and the aboriginal people, soil, plants, and animals of the Hudson Bay region.

Oldenburg's interview of Zachariah Gillam and Charles Bayly was not published until the 1750s, but in 1675, John Seller, Hydrographer to the King, published a brief report of Gillam's 1668–9 voyage to Hudson Bay.[26] The Royal Society did publish, in 1674, a translation of a Dutch paper about attempts to discover a northeast passage to the Orient. Appended to this paper was a two-page discussion – of obscure authorship – of the possibility that a practical passage from the Atlantic to Pacific Oceans might exist via Hudson's Bay, but this discussion referred to the explorations of Henry Hudson and the theories of José de Acosta, not to any information gained from the HBC.[27] Still, the fact that the very expeditions that led to the creation of the HBC were geared towards the discovery of a "new Passage into the South Sea," and the fact that the Fellows of the Royal Society continued to show interest in such a passage during the 1670s, foreshadowed the long and ambiguous relationship that the HBC was destined to have with the search for the Northwest Passage.[28] For most of the first 150 years of the company's existence, its Governor and Committee cooperated only reluctantly with efforts to discover a passage to Asia via Hudson Strait because they perceived, with some good reason, that the presence of outsiders in Hudson Bay, and, for that matter, the discovery of a convenient passage to the Orient via Hudson Strait, threatened the interests of the company.

During the 1680s and 1690s, the Royal Society and its members continued to get their information about the Hudson Bay region either primarily or exclusively from HBC ship captains. At the 26 June 1681 meeting of the Royal Society, Christopher Wren, then the president of the Royal Society, explained – during a meandering discussion that touched upon such diverse topics as the natural history of Ireland and slave nutrition in Jamaica – his beliefs concerning Indians along Hudson Bay, including his understanding that they lived robustly until they were 130 or 140 years old.[29] Wren, however, gave no indication of the source of his information. His snippets of (mis)information, and Oldenburg's 1672 interview were noteworthy in 1757 when they were first published, because at that time they contained virtually the only ethnological information that the HBC willingly shared with the Royal Society, or any other English natural philosophers up to that time.

In 1690, Edmond Halley presented "a white Hudson-bay Partridge whose feet were all overgrown with a thick Down to preserve them from the Cold."[30] And in 1691, Halley spoke of icebergs reported to him by a Captain of a HBC ship. In 1699, the society received "a girdle from Hudson Bay & also a lower jaw of a Beaver" and in 1701 Sir Robert Southwell presented information on the natives along the western shore of Hudson Bay that he had received from the HBC ship captain James Young, who himself likely acquired much of the information from Henry Kelsey, one of the first HBC men to travel a significant distance inland from the bay.[31] A report written in 1682 by John Nixon, the HBC's North American governor from 1679 to 1683, suggests how some of the specimens trickled to the Fellows of the Royal Society. Nixon wrote that "I have sent a small Indian popuce [papoose] skin with a girdle marked AC with Isac Rede [Isaac Read] for a present to the Earle of Shafsberie, I hope yor. Honours will let it pass, I have given to one of Captain [Nehemiah] Walkers men, John Pye two Mouse [moose] skins for ane of the Cape Indian lances that I have sent home to yor. Honours by Captain Cobbie and I desire that it may be presented, with my service to Sr. James Hayes."[32] The only surviving version of Nixon's report was found, curiously, in the papers of Robert Boyle, another Fellow of the Royal Society who was also an Adventurer.[33] That report, which assesses the company's affairs in North America is the oldest surviving "detailed and substantial account of conditions by the Bay," but it contained no information of scientific interest. It must have been of purely business interest to Boyle.[34] Of greater scientific interest were a few pages found in Boyle's papers relating to ice conditions in Hudson Bay and the magnetic variation at Charlton Island.[35]

The HBC's contributions to science during the late seventeenth century were evidently small compared with those of the English East India Company (EIC), but the EIC was not an enthusiastic contributor to science either. During the 1670s the company's directors did little to support science, leaving it to a few curious and motivated servants to make contributions on their own initiative.[36] In late 1676, the directors allowed the young Edmond Halley passage aboard the EIC ship *Unity* to map the stars of the southern hemisphere from St. Helena.[37] Then, in the late 1690s, the EIC contributed natural history specimens to the Royal Society. On 23 November 1698, Hans Sloane, secretary of the Royal Society, addressed a letter to the EIC expressing the "most humble and hearty thanks" of the society's assembly for the "plants, drugs, &c., of the East Indies, with the account of the uses of them." He promised that the Royal Society would "take care that your most Honorable Company, whose prosperity they truly wish, shall have all publick acknowledgements due for such a favour."[38] The Fellows of the Royal Society had good reason to wish for the EIC's prosperity: the society owned stock in the company.[39]

Aside from the fact that Western Europeans did not yet assume that scientific knowledge should be shared, several factors may explain why the HBC's (and to some extent other companies') contributions to science were so small before 1701. First, during the 1660s and 1670s surprisingly few Fellows of the Royal Society were eminent natural philosophers. Fellowship in the Royal Society was conferred upon quite a number of people primarily for their connections to the king. Indeed, the links between the early Adventurers and the government and imperial policy were at least as important as their links with the Royal Society. For example, Sir Paul Neile, probably the most politically influential person with membership in both the HBC and the Royal Society, evidently had a great love of the sciences, particularly astronomy, but had no significant scientific accomplishments. His position with the Royal Society was probably rooted in his ability to mediate between the king and the society, a role he apparently filled with alacrity.[40] Neile was in a similar position to mediate between the HBC and the society. He attended HBC General Council meetings, sometimes held at the lodgings of Prince Rupert at Whitehall, during the very same years he was "always ready to help his colleagues [in the Royal Society] and ever to be remembered for his devoted services in its early years to the Society."[41] Neile may have been well placed to facilitate scientific communication between the HBC and the Royal Society, but there is no evidence that he ever did. The rest of the Adventurers were also wealthy men with close connections to the English Crown, and with other English overseas endeavours.[42] For example, the Earl of Shaftesbury and Peter Colleton were both Lord Proprietors of Carolina – a fact that explains how John Nixon, a former deputy to Colleton in Carolina, came to the HBC.[43] So, the web of connections that tied the Adventurers of the HBC to the Crown, the Royal Society, and to other organizations was important, but it apparently did relatively little to ensure that English overseas travellers shared their knowledge of foreign parts, or even to ensure that English natural philosophers sought that knowledge very assiduously.

Not only were few Adventurers prominent intellectuals, few men were active in both the HBC and the Royal Society for very long. Philip Carteret died in 1672 before attending his first HBC council meeting. James Hayes remained active in the HBC but became inactive in the Royal Society after 1670, as did Paul Neile by 1680.[44] Peter Colleton and the Earl of Shaftesbury sold their HBC stock in 1678 and 1679 respectively, and all of the courtiers with HBC stock were gone by 1680.[45] Meanwhile, the Royal Society stagnated in the 1670s. Attendance at meetings was poor and few new recruits with scientific credentials were elected to the Fellowship.[46] During the 1680s, the Royal Society was rejuvenated, but just as it was, crises hit the HBC.

That the HBC's contributions to science in the early years were small is also probably attributable to the fact that the fraternity of savants was small enough that there was not a large enough population to devote much interest to Rupert's Land. Until thirteen of Britain's North American colonies declared independence in 1776, scientific networks among metropolitan virtuosi and collectors in those colonies were probably enough to satisfy British savants.[47] The Fellows of the Royal Society seem rarely to have actually thought of the HBC as potential "Merchants of Light." Even the prominent men of science with connections to the HBC did not seem to think often of the HBC as a source of scientific knowledge. Robert Boyle illustrates this fact. Boyle was an eminent scholar, an active Fellow of the Royal Society until 1674, and the author of one of the Royal Society's appeals for information from travellers. He was in a position to forge ties between overseas travellers and the scientific community. He was an investor in the HBC, although not a member of the HBC's Committee, and a director of the EIC from the early Restoration until 1677.[48] Boyle's primary interest in the EIC however, was evidently rooted in his fervent desire to proselytize Native peoples.[49] Christopher Wren, another respected, if somewhat credulous, man of science with connections to both the Royal Society and HBC between the 1660s and the 1680s, was more famous as an architect than as a natural philosopher. As a Fellow in the days of the Invisible College, and an Adventurer before 1670, he was probably the most important link between the HBC and English savants between the late 1660s and the early 1680s.[50] Wren was active and influential in the HBC's London Committee during the early 1680s, bringing to the committee "not only a readiness to advance money at difficult times but a shrewd and active judgment, assiduous attention as a committee-man, and willingness to serve in the committee for trade as Vice-Deputy Governor."[51] Wren did present some information from the Hudson Bay region to his colleagues in the Royal Society, but it was insignificant. Furthermore, during the 1680s his involvement with both organizations diminished as his architectural commitments grew.

If anything, the HBC became increasingly secretive early in its history, and although the London Committee was not concerned primarily with preventing its members from cooperating with intellectuals, the company's policies affected all communication between its members and outsiders. A fledgling company with risky and costly operations, the HBC's knowledge quickly became its most important strategic asset. The prominent historian of the HBC Glyndwr Williams has explained that "one of the Company's main defences against potential rivals was secrecy, an obsessive guarding of the knowledge and expertise accumulated by its servants over the years of living, trading and navigating in the sub-Arctic conditions of the Bay region."[52] If, as Bacon averred, knowledge was power, secrecy was crucial. As early as 1679, London Committee members were required to take an oath promising not to divulge any of the company's knowledge to people outside the committee, and not to remove books from the company's offices without the approval of the committee. The company's ship captains and North American servants faced similar rules, not only to prevent the unauthorized flow of information outside the company but also to prevent employees from engaging in illicit private trade. The company, for example, ordered that all journals, charts, and logbooks be handed over the committee as soon as ships arrived in England.[53] It also ordered its servants "not to presume to bring any chest, boxe or bundle on shoare until they have been searched under penallty of looseing all their wages," and insisted that "none of our servants do send any intelligence to, or carry on any correspondence with any person whatsoever in London or elsewhere relating to the affairs of the Company."[54] The dangers of knowledge falling into the hands of domestic and foreign interlopers or foreign governments were not imaginary, as the company's committee and investors soon learned. The secrecy exhibited by the HBC was important to the management of information in other chartered companies as well.[55]

During the 1680s, the HBC began to face crises that must have convinced the Governor and Committee of the wisdom of guarding its knowledge carefully. Gillam had been dismissed by the Governor and Committee during the 1670s because of evidence of misconduct that included engaging in contraband private trade in Hudson Bay. But owing to his knowledge and expertise, the Governor and Committee took a chance on him in 1682, and rehired him to establish a post (Port Nelson) at the mouth of the Nelson and Hayes River (see figure 2.1). He arrived there, however, to find that he had been preceded by two other parties, one from Boston licensed by the Governor of Massachusetts and led by his own son Benjamin – raising obvious suspicions that the father had shared too much information with his son – and the other a French party led by Radisson and Groseilliers. When the HBC Governor and Committee learned about these developments they angrily revoked Gillam's command and ordered him to return to London, but Gillam, having died in a storm in October, was beyond their reach.[56]

Gillam's death did not solve the HBC's problems. In the spring of 1683, Radisson seized the two English posts and captured their occupants. The HBC built a new post (York Fort) to replace Port Nelson in 1684, but in 1686, the French captured the HBC posts at the mouths of the Moose River (Moose Fort), Rupert River (Charles Fort), and Albany River (Albany House). The HBC lost and recovered most of its posts over the next decade, but for several years it held only Albany Fort, and for the entire time between 1697 and 1714, the French held the most valuable location at the mouths of the Hayes and Nelson Rivers (where Port Nelson and York Fort were located).[57] Thus, during the 1680s and 1690s, the HBC had little opportunity to cooperate with the Royal Society. Furthermore, it had ample reason to believe that the flow of information to its domestic and foreign rivals had contributed significantly to the costly crises it faced.

The HBC had hardly been generous with its knowledge of the Hudson Bay region during the first three decades of its existence, but it

2.1 Rupert's Land and the HBC's Transatlantic Transportation Route, 1670–1768. For its first hundred years the HBC's posts were limited almost exclusively to the coast of Hudson Bay (Henley House was established 200 kilometres inland to forestall French encroachments, not to expand HBC operations inland). The HBC's ships traveled from London to the Orkney Islands, and then almost due west along the 60th parallel to Hudson Strait, then to its forts established at the mouths of major rivers along Hudson and James Bays (dotted line). London to York Factory was a trip of about 7800 kilometres. French fur trade supply lines (dashed line) followed rivers and lakes from Montreal into the interior, reaching as far west as the hinterland of York Factory by 1750. Source: T. Binnema.

became all the more secretive during the governorship of Bibye Lake from 1712 to 1743. As Glyndwr Williams noted, "by their very nature monopolistic companies were wedded to commercial secretiveness as a defence against the twin threats of foreign rivals and domestic interlopers. Under Lake the Hudson's Bay Company pursued this policy with obsessive intent, carefully guarding the knowledge accumulated by its servants."[58] As had been the case earlier, the company was worried about domestic and foreign interlopers and foreign powers, but the example of the Royal African Company offered more reasons to be secretive. That company had been chartered two years after the HBC, but interlopers undermined its potentially lucrative monopoly on the British slave trade during the 1680s. Then, in 1698, following a campaign on the part of English merchants, the Royal African Company lost its monopoly.[59] The EIC faced similar attacks on its monopoly following the Glorious Revolution of 1688, an event that also damaged the position of the HBC.[60] Understandably, the HBC Committee preferred as little visibility as possible. The Governor and Committee were not concerned primarily with preventing their men from contributing to science, but the company's secretive policy did mean that between 1701 and the mid-1720s, neither the Governor and Committee nor the company's employees contributed anything of significance to public knowledge. Then, between the 1720s and 1740s, HBC employees once again contributed to science, but evidently without the encouragement of the Governor and Committee.

Lake's strategy worked. During most of his governorship, the HBC enjoyed prosperous obscurity. Its bayside posts faced competition only from traders from New France who dealt with Indians from inland posts. At least as important, the company's secrecy seems to have ensured that the HBC avoided the kinds of controversies that the Royal African Company and EIC faced. Public knowledge of the Hudson Bay region, small in 1700, was largely unchanged forty years later. John Oldmixon's *British Empire in America* (1708) was the first book to deal in any depth with the activities of the HBC. Oldmixon explained that much of the information for his chapter entitled "History of Hudson's-Bay" came from "original Papers, he having had in his Possession the Journal of a Secretary of the Factory."[61] These were the journals that Thomas Gorst kept in 1670–1 and 1672–5, primarily at Charles Fort (Rupert House). It is not clear how Oldmixon acquired these journals, but his silence on the subject suggests that he did so without the assistance of the company.[62] When Oldmixon published a second edition of his book in 1741, he explained that he had left the chapter from the 1708 edition unchanged because, although "application was made to Persons concerned in the Affairs of the Company ... it being not come to Hand before the Book was printed," he was unable to update it.[63] For several decades, company officials seemed to believe that outsiders' ignorance was the company's bliss. However, by 1741, the company's secretive policies were becoming controversial – at least in part because of the scientific contributions of one of its ship captains, Christopher Middleton.

Ironically, the first really noteworthy surge of contributions by HBC employees to the world of science came between the 1720s and 1740s, decades when the Governor and Committee maintained its guarded stance. Given that ship captains had hitherto been the most important HBC men to contribute to science, it is not surprising that a ship captain, Christopher Middleton, became the first HBC employee to achieve a scientific reputation in his own right – a reputation significant enough to earn him the Royal Society's Copley Medal.[64] But two officers in North America, Alexander Light and James Isham, also communicated with men of science during the 1730s and 1740s. Their efforts contributed significantly to another Copley Medal – one won by George Edwards in 1750. The efforts of Middleton, Isham, and Light must have been noteworthy for the Governor and Committee of the company because they highlighted the potential risks and benefits to the company of its employees' scientific efforts.

After serving aboard privateers during Queen Anne's War (1701–13), Christopher Middleton made the first of his sixteen voyages to Hudson Bay in 1721 onboard the HBC's *Hannah*. Already an accomplished and inquisitive navigator, he immediately began taking notes on the magnetic variation in the Hudson Strait and Hudson Bay. His first four seasons with the HBC gave him the material for a table presented to the Royal Society by Edmond Halley in February 1726, and published in the Royal Society's *Philosophical Transactions* later that year.[65] In 1731, 1732, and 1735 he published other papers relating to weather and compass variation in Hudson Bay, and in 1735 he published measurements of latitude determined by four new quadrants, including "Mr. Smith's Prismatic," "Mr. Hadley's," and those made by "Mr. John Elton," and "Mr. Caleb Smith and Mr. William Ward."[66] Although Middleton is best known today for having been one of the first (perhaps the very first) to test John Hadley's quadrant (an octant, really), which had been introduced to the Royal Society in 1731, Middleton was particularly impressed with Caleb Smith's quadrant.[67] On the strength of his contributions Middleton became, in 1737, the first HBC employee elected Fellow of the Royal Society.[68] During that same year, he published two articles. One explained that he found that in some cold and foggy places he needed to warm his compass in order to keep it functioning, and the other more substantial piece reported on his testing of a new azimuth compass.[69] Middleton then published a paper explaining that when sea ice was melted, it was virtually free of salt.[70] In 1741–2 he led an expedition that determined that there was no passage to the Orient via the Hudson Bay south of the sixty-fifth parallel – an expedition that provided him with the evidence he needed to publish the paper in *Philosophical Transactions* that won him the Copley Medal in 1742 (see figure 2.2).[71] In 1743 Middleton published the first detailed map of the western coast of Hudson Bay as far north as Repulse Bay (66° N).[72]

In 1737 – the year Middleton was elected to the Royal Society – another HBC officer, Alexander Light, originally hired in 1733, conveyed the first known collection of natural history specimens from Hudson Bay to London's natural historians when he travelled to England on furlough.[73] How Light, a shipwright, came to the company, is not known, but given that Light had evidently earlier provided the prominent British naturalist, George Edwards, with natural history specimens from Maryland and South Carolina, John Richardson may have been at least partially correct when he suggested in the 1820s that the HBC had hired Light "on account of his Knowledge of Natural History."[74]

Light appears to have initiated a period of fruitful interaction between HBC servants and British naturalists. For example, when he was in London on furlough in 1737–8, Light showed his specimens to George Edwards, who included illustrations of them in the first part of his *Natural History of Uncommon Birds* (1743–51), a work which, more than eighty years later, John Richardson still described as "the most original and valuable work of the kind in the English language."[75] In two places in that volume Edwards paid tribute to Light, "a friend of mine residing at Hudson's Bay," for information on birds and mammals. Elsewhere he mentioned a "Gentleman employ'd in the *Hudson's-Bay* Company's Service" who brought a live "White Tailed Eagle" and a live "Spotted Hawk or Falcon" to England and presented them to Dr. R. M. Massey, who kept live birds.[76] This "gentleman" may be Light, but might also be Captain Middleton. In reference to a "Spotted Hawk or Falcon," Edwards wrote: "It pitched on a Ship belonging to the *Hudson's Bay* Company in *August* 1739, as the Ship was returning Home."[77]

[157]

II. *The Effects of* Cold; *together with Observations of the* Longitude, Latitude, *and* Declination *of the* Magnetic Needle, *at* Prince of Wales's Fort, *upon* Churchill-River *in* Hudson's Bay, North America; *by Capt.* Christopher Middleton, F. R. S. Commander *of His* MAJESTY'S *Ship* Furnace, 1741-2.

Read Oct. 28. 1742.

I Observed, that the *Hares, Rabbets, Foxes* and *Partridges*, in *September*, and the Beginning of *October*, changed their native Colour to a snowy White; and that for Six Months, in the severest Part of the Winter, I never saw any but what were all white, except some *Foxes* of a different Sort, which were grizzled, and some half red, half white.

That Lakes and standing Waters, which are not above 10 or 12 Feet deep, are frozen to the Ground in Winter, and the Fishes therein all perish.

Yet in Rivers near the Sea, and Lakes of a greater Depth than 10 or 12 Feet, Fishes are caught all the Winter, by cutting Holes through the Ice down to the Water, and therein putting Lines and Hooks. But if they are to be taken with Nets, they cut several Holes in a strait Line the Length of the Net, and pass the Net, with a Stick fastened to the Head-line, from Hole to Hole, till it reaches the utmost Extent; and what Fishes come to these Holes for Air, are thereby entangled in the Net; and these Fish, as soon

X 2 as

2.2 "The Effects of Cold." This article in the *Philosophical Transactions of the Royal Society* earned Christopher Middleton the Royal Society's Copley Medal in 1742. It is derived from a paper Middleton sent to Hans Sloane. Source: *Philosophical Transactions of the Royal Society*, 42 (1742): 157.

(107)

The RING-TAIL'D HAWK.

THIS Bird seemed to me to be of the Size of a Common Crow. The Wing when closed is fourteen Inches long; the Leg from the Foot to the Knee is three Inches long. It differs from other Hawks of its Size, in having a smaller Bill, and longer and slenderer Legs, in Proportion, than I have observed in any of the greater Kinds of Hawks. It resembles our Sparrow-Hawk in the Slenderness of its Legs, and small Bill, and is something like it in Colour, tho' at least four Times its Magnitude.

Its Bill is of a dark Horn-Colour, or blackish, the Nostrils cover'd with a yellow Skin, which encompasses the upper and lower Chaps, and extends from the Angles of the Mouth as far back as the Eyes; this Skin is beset thinly with black stiff Feathers; the Bill is hooked, as in all of this Kind, but hath no Angle on the Edges of the upper Chap, as is common, but only a little Bend or Wave; the Head is of a dusky or blackish Colour, having a little White on the Forehead joining to the Bill, and a light Mark passing above each Eye; the Feathers beneath the Bill are also light colour'd for a little Space; the Fore-part of the Neck and Hinder-part of the Head are of a Clay-colour, intermixed with a dusky Brown; the upper Part of the Neck, the Back and Wings, are of a dark dusky Brown; the Edges of some of the Middle Quills are Ash-colour; the Inside of the Wing is White, except the Ends of the Quills, which are dusky; the inner Coverts are sprinkled with small brown Spots, and the inner Webs of the Quills are faintly barred a-cross with narrow dusky Lines; the Rump and Covert Feathers of the Tail are white; which Whiteness joining with the white Feathers beneath the Tail, forms a white Ring round the Tail; the middle Feathers of the Tail are dusky, the next on each Side of a blueish Ash-colour, the outermost White, all transversly marked with seven or eight dusky Lines; the Tail beneath is almost White, the Bars hardly appearing through the Feathers; the Breast, Belly, Thighs, and Covert Feathers under the Tail are White, intermixed with some Reddish-brown Spots transversly waved on the Breast and Thighs, in the Form of Hearts on the Belly, and in half-moon-like Spots on the lower Belly and the Coverts under the Tail; the Legs and Feet are of a bright Yellow or Gold-colour, the outer Toe joined the middlemost by a Membrane, the Claws are Black.

Hudson's-Bay in *North America* is the native Place of this Bird, from whence it was brought, with many others, by Mr. *Isham*, who has obliged me extremely by furnishing me with more than thirty different Species of Birds, of which we have hitherto had little or no Knowledge, the far greatest Part of them being Non-descripts. As I shall in the Course of this Work have Occasion frequently to mention the above curious Gentleman's Name, it will be here necessary to let the Reader know, that Mr. *Isham* has been employ'd for many Years in the Service of the *Hudson's-Bay* Company, and has, for some Years past, been Governor under them at different Times, of several of their Forts and Settlements in the most Northern habitable Parts of *America*; where at his leisure Times, his commendable Curiosity led him to make a Collection of all the Beasts, Birds, and Fishes of those Countries, as well as the Habits, Toys, and Utensils of the native *Americans*. The Furs of the Beasts, and the Skins of the Birds were stuffed, and preserved very clean and perfect, and brought to *London* in the Year 1745. Mr. *Isham* is now in *London*, [1749] where he will stay for a short Time, and has favour'd me with the Pleasure of his Conversation.

Nothing exactly agreeing with the above describ'd Bird, can be met with in our natural Historians. What comes nearest its Description, is the *Ring-Tail* described by *Willughby*, P. 72.

Toe

2.3 "Ring-Tail'd Hawk." In his prize-winning *Natural History of Birds*, George Edwards frequently paid tribute to the HBC and its men. In the note on the bottom of this page from volume 3, Edwards paid tribute to James Isham for supplying him with a specimen of the "Ring-Tail'd Hawk" (probably the Sharp-shinned Hawk or the Cooper's Hawk) and many other birds. Also see colour plate 3. Source: University of Wisconsin Digital Collections.

2.3 (Continued)

The fact that neither Light nor Isham was on a ship returning to England in 1739 suggests that Middleton – who was – may have been one of the people who brought live birds back to London. In the second volume of his compilation (1747), Edwards acknowledged Light's assistance several more times.[78] In May 1744 Edwards made a presentation to the Royal Society of "a large white owl from Hudson's Bay, with a description of its way of life."[79] The specimen and Edward's information about the owl must have come from Light.

James Isham, who had arrived in Rupert's Land in 1732, made later but more significant contributions to the work of George Edwards in the 1740s. It is possible that Isham, a Londoner, knew George Edwards before he joined the company, but it is also possible that Alexander Light introduced them. Isham appears to have submitted four boxes of "trees Herbs &c" in 1738, and collected specimens of animals thereafter.[80] Today Isham is best known for his "Observations on Hudson's Bay," a document written in 1742–3, and published in the twentieth century. "Observations" is most valuable for its remarkably perceptive analysis of the natural history of the Hudson Bay lowlands. Ironically however, this document may have gone unread in Isham's lifetime.[81] In 1745, when Isham went to England on furlough he brought specimens and written observations that he probably presented to George Edwards directly. Isham's specimens were the basis of thirty-one of the thirty-three birds pictured in the third volume (1750) of *Natural History of Uncommon Birds*. Edwards gushed with tribute. In a footnote to the first description (of the "Ring-Tail'd Hawk") Edwards explained that Isham

> obliged me extremely by furnishing me with more than thirty different Species of Birds, of which we have hitherto had little or no Knowledge, the far greatest Part of them being Non-descripts. ... at his leisure Times, his commendable Curiosity led him to make a Collection of all the Beasts, Birds, and Fishes of those Countries, as well as the Habits, Toys, and Utensils of the Native Americans. The Furs of the Beasts, and the Skins of the Birds were stuffed, and preserved very clean and perfect, and brought to London in the Year 1745. Mr. Isham is now [in 1749] in London, where he will stay for a short Time, and has favour'd me with the Pleasure of his Conversation.[82]

Then, in a footnote to the last description (of the "Little Brown and White Duck") Edwards explicitly paid tribute to Isham "to whose

Curiosity and good Nature I am beholden for the greatest part of the Subject-Matter of this third Part of my *History of Birds*; and I believe the curious Part of the world will not think themselves less obliged to Mr. Isham than I acknowledge myself to be."[83]

The many specimens supplied to Edwards by HBC traders obviously did much to enhance the reputation of Edwards (who won the Copley Medal in 1750 for his *Natural History of Uncommon Birds*), and to make European naturalists more aware of the natural history of the northern part of North America (and other parts of the globe).[84] Indeed, some of the species that were new to British naturalists had been described decades earlier by others, and evidently forgotten, but Edwards and the HBC officers were instrumental in making them known to European science.[85]

The Governor and Committee of the HBC had little reason to regret the contributions of Light and Isham to natural history However, they did have reason to regret Middleton's contributions to navigation and meteorology, because of the link Middleton forged with Arthur Dobbs, who emerged as the most vociferous critic of the HBC during the mid-eighteenth century. Dobbs, aside from being politically influential, was a man of some scientific standing, having published in the Royal Society's *Philosophical Transactions* as early as 1723.[86] Glyndwr Williams has noted that Middleton's first paper "attracted the attention of Arthur Dobbs, an Ulster landowner and an influential member of the Irish House of Commons, whose long-standing interest in trading matters was after 1731 directed towards the finding of a northwest passage."[87] Meetings in 1733 with Samuel Jones, the Deputy Governor of the HBC, and in 1735 with Governor Bibye Lake convinced Dobbs that the company officials were not showing appropriate interest in the search for the passage.[88] Governor Lake and the Committee may well have understood how not only the discovery of a passage to the Pacific but also merely the search for one threatened the interests of the HBC, but the company's lack of enthusiasm can be explained more simply by the fact that its own expensive and tragic explorations had already convinced the Governor and Committee that no practical strait to the Orient was to be found via Hudson's Bay. In or around 1716, James Knight, the intelligent and ambitious overseas governor of the HBC, gathered maps and information from aboriginal people that convinced him that the Northwest Passage, and copper mines, could be found via Hudson's Bay.[89] He persuaded the Governor and Committee to support an expedition to search for the Northwest Passage from Hudson Bay north of the sixty-fourth parallel. The Knight expedition ended badly: Knight and all of his men died, and the HBC apparently lost between £7000 and £8000.[90] The company sent several more expeditions northward from Churchill in the late 1710s and early 1720s, partly to look for Knight and his men, and partly to search for the fabled strait to the Orient.[91] They seemed to confirm that such explorations were expensive, risky, and futile.

But Dobbs was persistent. Unimpressed with the response of Lake and Jones, Dobbs, having read Middleton's contributions to the *Philosophical Transactions*, decided to approach the captain in 1735. Middleton must have informed the governor about Dobbs, for Lake authorized Middleton to communicate with Dobbs.[92] Lake also relented and, to satisfy Dobbs, sent two sloops northward from Churchill during the summer of 1737 in an expedition that combined exploration and trading. It was unsuccessful on both counts, and Lake must have hoped that he would satisfy Dobbs by assuring him that the expedition failed to find "any the least Appearance of a Passage."[93] But Middleton, who was at Churchill Fort when the sloops came back, wrote Dobbs that the crew, which he thought were unqualified for the task, had travelled no farther north than 62°15 N.[94]

Dobbs became convinced that the Governor and Committee were determined to prevent the discovery of a passage that he believed certainly must exist.[95] Middleton must have agreed, and his cooperation with Dobbs quickly began to undermine the interests of his employer. In 1738, Middleton wrote Dobbs that the company's Governor and Committee "keep every Thing a Secret; and from some Questions I have been lately asked, I found they seem suspicious of my corresponding with you."[96] The committee's suspicions were well founded. If not by that time, then soon thereafter, Middleton injudiciously began taking documents relating to past HBC exploration efforts from the HBC's London headquarters (perhaps, in some cases, the only copies the HBC had), and handing them over to Dobbs.[97] And Middleton even went so far as to suggest to Dobbs that the legality of the HBC's charter might be challenged.[98]

In March 1741, Middleton quit the HBC to join the Royal Navy. Meanwhile, convinced that the HBC was stonewalling, Dobbs persuaded the government of Robert Walpole to support a naval expedition – to be led by Christopher Middleton – to search for the Northwest Passage. After resisting the government for some time, the company finally relented and agreed to cooperate with the expedition.[99] The expedition of 1741–2

2.4 Hudson's Bay Company Explorations to 1772. Henry Kelsey's expeditions of 1690 and 1691 were rare examples of HBC seventeenth-century inland expeditions. The 1740s witnessed a surge of exploration for the Northwest Passage along the west coast of Hudson Bay as far north as Repulse Bay. Anthony Henday was sent inland as an emissary in 1754–5. Hearne accompanied Chipewyan people as far northwest as the mouth of the Coppermine River between 1770 and 1772. Hearne mistakenly believed that the mouth of the Coppermine River was at 71°54′ N and 120°30′ W (marked by an X on the map). Source: Ted Binnema.

consisted of two ships, the *Furnace*, led by Middleton and the *Discovery*, captained by William Moor, another former HBC ship captain and a cousin of Middleton (see figure 2.4).[100] During the winter of 1741–2, Middleton wintered at Churchill, where James Isham was in charge. The news that Middleton and Isham had gotten along reasonably well during the winter subsequently raised further suspicions, both with Dobbs and with the HBC Committee. Middleton's expedition in the summer of 1742 convinced him that there was no passage south of the sixty-fifth parallel (and any passage farther north would be frozen too long each year to be practical), although it provided him with the evidence he needed for the paper that secured him the Copley Medal.[101] Dobbs was having none of it. "A very warm dispute" arose between Dobbs and Middleton when Dobbs accused Middleton of accepting a bribe offered by the HBC to conceal all knowledge of the passage.[102] The two engaged in a heated private pamphlet war during 1744 and 1745 that neither of them won.

By that time Dobbs had abandoned his efforts to induce the HBC to cooperate with efforts to locate the Northwest Passage, but sought, in alliance with others, to convince the government to revoke the company's charter – evidently to acquire concessions for himself and his allies.[103] In 1744 Dobbs published a book as part of his campaign to have the HBC's charter revoked.[104] He could not change the geographic reality. Neither could he convince the government to support another expedition, war with France having broken out in March. In 1745, the British Parliament elected instead to pass "An Act for giving a publick reward to such person, or persons, His Majesty's subjects, as shall discover a North-west Passage through Hudson's Streights to the Western and Southern Ocean of America" (18 Geo. II, c.17). The legislation offered a reward of up to £20,000 for the discovery of the passage.[105] The promised reward enabled Dobbs to persuade enough "Adventurers" to invest in a private expedition to find the Northwest Passage. Thus, in 1746 and 1747 two more ships, the *Dobbs Galley* and the *California*, captained by William Moor (who had sided with Dobbs in his disagreement with Middleton) and Francis Smith, probed the western shores of Hudson Bay in a futile attempt to find the passage.[106] They wintered near York Factory, where James Isham offered them some assistance.[107] While these explorations were underway, Dobbs also attempted to bolster his case with science. He published an article in the *Philosophical Transactions* dealing with Vitus Bering's explorations.[108] But his ships returned in 1748 with bad news. No passage to the Orient had been found.

Unable to collect any reward to recoup their losses, Dobbs and his committee of investors applied, on the strength of their exploration efforts, for a charter granting them a monopoly over the trade in a portion of the Hudson Bay region. By asking for the very same type of exclusive privileges that he had long attacked, Dobbs seriously undermined his case. The application was denied.[109] Thereafter Dobbs and his allies attempted to keep up the heat by publishing two more books by members of the 1746-7 expeditions that cast aspersions on the HBC (and expressed hope that the passage might yet be found via Chesterfield Inlet).[110] One of these, written by hydrographer and mineralogist Henry Ellis, even earned for its author election to the Royal Society of London.[111] The weight accorded to this work may explain why James Isham penned a rebuttal to it.[112]

During this entire time, the HBC elected to answer its critics with silence as often as possible, but it also aimed to make maximum use of its servants' previous contributions to exploration and science.[113] The fact that the HBC Governor and Committee recalled James Isham to London in 1748 to help them defend the company's charter reflects this strategy.[114] Still, the British House of Commons, influenced by the ongoing controversy, and faced with petitions from merchants in port cities throughout the country, ordered an inquiry into the HBC's monopoly. That inquiry sat in March and April 1749. The accusations against the company – and therefore the parliamentary committee's inquiries – were centred on the HBC's efforts to explore the country (especially its efforts to find the Northwest Passage), its efforts to maximize its trade, and the legality of the HBC's charter. Even when faced with this inquiry, however, the company presented relatively little information, and called few witnesses in its defence.[115] The company's charter and monopoly survived, but not unscathed – and certainly no longer invisible. As Glyndwr Williams concluded, "the Company faced a future in which it could no longer rely on obscurity for protection."[116]

His battle against the HBC now decisively lost, Dobbs tried to convey the impression that he had withdrawn from the fray. In the first sentence of a presentation read before the Royal Society on 8 November 1750, Dobbs coyly noted that: "Since my View of doing Good, by making Discoveries of the Great World has been disappointed, upon my Retirement into this little Corner of it [Ireland], amongst other rural Amusements I have been contemplating the Inhabitants of the Little World; particularly that most useful and industrious Society of Bees."[117] But if he had in fact intended to retire from his fight against the HBC, he could not, in the end, resist one veiled parting shot. In 1752, Joseph Robson, a disgruntled former employee of the HBC who had testified against the company at the 1749 inquiry, published a book containing a famous accusation that "the Company have for eighty years slept at the edge of a frozen sea. ... They have kept the language of the natives, and all that might be gained by a familiar and friendly intercourse with them ... and the invaluable treasures of this extensive country a profound secret to Great Britain."[118] Robson's book was the first written by a man with more than one year of experience in the Hudson Bay territories, and it was reviewed positively in the *Monthly Review*, the *Gentleman's Magazine*, and other influential publications.[119] By seeming to offer independent confirmation of many of Dobb's assertions, it restored the reputation of the once discredited Dobbs. In fact, although Robson wrote most of *Account of Six Years Residence*, it was clearly prepared and revised under Dobbs's influence, and portions of the book were actually authored by Dobbs. There is no doubt that Dobbs concealed his involvement with *Account of Six Years Residence* deliberately for strategic reasons.

To be fair to Robson, his book suggests that he was a tenaciously inquisitive man whose curiosity and experiments might have been put to better use. He claimed to have examined on his own initiative the environs of York Factory, where, after cutting a trench to drain land near the fort, the ground "wore quite a new face; the snow did not lie upon it near so long as before, and the grass flourished with a new vigor."[120] He also explained that "I perceived that the garden-ground at York-fort and Churchill-river thawed much sooner and deeper in the space of one month, than the waste that lies contiguous to it; and the same is to be observed in England."[121] But this evidence was used to defend exaggerations about the mildness of the climate, and the "fine improveable lands."[122] He argued that "by the heat ... which the earth here would acquire from a general and careful cultivation, the frost might be so soon overcome, that the people might expect regular returns of feed-time and harvest."[123] Robson's statement is an interesting early articulation of the theory that cultivation of lands ameliorated the climate. Similarly, Robson used evidence from natural history to support his assertion that the Pacific Ocean was near Hudson Bay. He argued that geese arrived at Churchill in the spring before they arrived at the more southerly York Factory because, at the latitude of Churchill, the continent was so narrow that the geese easily migrated eastward from the warm Pacific coast to Hudson Bay.[124] On the other hand, although he was unwilling to praise the company for anything, he did note that while he "might here give a particular description of all the animals peculiar to this country," he acknowledged that this was unnecessary because it "has been already done by other writers about Hudson's

Bay."[125] Robson and Dobbs were not inclined to emphasize the contributions of the HBC to the science of the day – and contributions of Alexander Light and James Isham did not help the company greatly – but here was at least an acknowledgment of the contributions of those two men to the work of natural historians.

The degree of influence that the critics had on the public perception of the HBC is illustrated by the exhaustive and influential summary of global exploration published by John Campbell in 1748. The second volume of *Navigantium atque Itinerantium Bibliotheca* includes a long discussion of the company in which Dobbs's critique of the company is accepted and quoted at length, and another section focused on Christopher Middleton, which opens with the opinion that

> one would have imagined, that after the Company was established, ... with a View to the finding a North-west Passage, continual Attempts would have been made for that purpose, agreeable to the Petition upon which that Charter was granted, and to the Preamble of the Charter itself, which has been so often mentioned; but so far has this been from the Practice of the Company, that they have taken all Methods possible to prevent the Notion of a Passage being found that Way ... by preventing their Captains from publishing their Journals; and as by their Charter they have an exclusive Trade, if they do nothing in this Matter, it is impossible any body else should.[126]

The subsequent discussion of Middleton's explorations suggests that Campbell accepted Dobbs' criticism of the conduct of these expeditions.[127]

Stung by its critics, the HBC stepped up its efforts after 1750 to map the Hudson Bay coastline and interior, to search for the Northwest Passage, and to establish some presence inland. In 1760, Moses Norton (in charge at Prince of Wales Fort) drafted a map that was based on aboriginal maps, or at least on detailed information from Cree informants, which gave him hope that Chesterfield Inlet might lead to a Northwest Passage. That information inspired a reconnaissance by William Christopher and Moses Norton in 1761 and 1762, which indicated once again that there was no strait along the western side of Hudson Bay south of the sixty-fifth parallel.[128] Norton also collected what is now the oldest extant map in the HBC collection that is explicitly attributed to named aboriginal cartographers. In 1762 he asked Idotlyazee, a Chipewyan man, to draw him a map of the Hudson Bay coast to the north of Churchill. Then in 1767 he secured a map, drawn on leather, by Idotlyazee and Matonabbee, the second man being the Chipewyan man who later guided the HBC trader Samuel Hearne on his third journey. Norton transcribed this map and submitted it to the company, and it influenced the instructions given to Samuel Hearne.[129]

But the company did not budge from its public silence. Certainly the circumstances had changed. When Oldmixon published the second edition of his history of the British Empire in 1741, the British public and British politicians had very little information about the HBC or Rupert's Land. Maps drawn by European cartographers before 1780 reveal starkly how European knowledge of Hudson Bay and the regions west of the bay lagged far behind knowledge of most parts of the globe, except those areas farthest from Europe. By 1752, the public and politicians had access to many books, plus the report of a parliamentary committee, which presented highly inaccurate impressions of the climate, soils, and aboriginal peoples of Rupert's Land, and of the likelihood of finding a Northwest Passage via the bay. Nevertheless, the HBC's Governor and Committee made little effort to refute the misrepresentations. Neither does it appear that the HBC's Governor and Committee sought to facilitate communication between their employees and intellectuals in the metropolis. Evidently the company saw greater risks in releasing more information than it saw in the public having misinformation. Of course, only two years after Robson's book was published, the Seven Years War (Great War for the Empire [1754–63]) broke out. The very fact that the HBC, like the entire British world, was threatened by France ensured that any domestic criticism would fall on deaf ears. And after New France fell in 1760, the London Committee could hope that interlopers operating from Montreal would no longer be a problem in the interior. Anyone hoping that the company could slip once again into prosperous obscurity would have reason to be optimistic. By the late 1760s, however, interlopers challenged the HBC as much as they ever had. Even as they did, the HBC Governor and Committee began to cooperate with English natural historians so avidly that several of them soon publicly congratulated the company for its support of science.

3 "Desirous to Encourage Science": The Transit of Venus of 1769 and the Hudson's Bay Company's Collaboration with the Royal Society, 1768–1774

It took a rare alignment of planets – literally – for the Governor and Committee of the Hudson's Bay Company to embrace opportunities to contribute to British science. That alignment occurred in 1769 when Venus passed directly between Earth and the Sun. Historians of science already know that the Transits of Venus in 1761 and 1769 were significant milestones in the history of astronomy and in the history of the expansion of Western scientific endeavour outside of Europe. Captain James Cook's first voyage was the most famous expedition inspired by the Transit of Venus. The main purpose of Cook's voyage was to be at Tahiti "to observe the Passage of the Planet Venus over the Disk of the Sun on the 3rd of June 1769."[1] But Cook's voyage only hints at the significance of the Transits of Venus for European, and particularly British, science. The HBC may well provide the best example. The Transit of Venus of 1769 was the most important turning point in the history of science in the HBC. The preparations for the observations of the transit from Fort Prince of Wales (Fort Churchill) marked the beginning of the company's sponsorship and encouragement of a wide range of sciences.[2] Never again did the company revert to the secrecy it had exhibited during the first hundred years of its existence. Two men, William Wales and Samuel Wegg, were behind the dramatic change in policy. William Wales was one of the two astronomers sent to observe the Transit of Venus, and Samuel Wegg was the person perfectly placed to mediate between the Royal Society and the HBC.

Before 1769 only two of Venus's transits had ever been observed, one in 1639 and one in 1761. John Keill's *Introduction to the True Astronomy* (1721) explains the significance of the transit from the perspective of an early eighteenth-century astronomer:

Once *Venus* was seen within the body or disk of the Sun; but there was but one Man who had the happiness to be witness of the Sight, our Country-Man Mr. [Jeremiah] *Horrox*, who in the Year of Christ 1639, observed it with his Telescope to enter upon the Body of the Sun like a black Spot. This is a Sight which can seldom be observed, for it will not be seen again in the Sun's Body, 'till the Year 1761, upon the 26th day of the Month of *May* in the Morning, at which time all our *Astronomers* will no doubt be busie in making their Observations; for by them our Distance from the Sun can be nearly determined, which before that time is not easily to be ascertained.[3]

As Keill's commentary implies, European natural philosophers were particularly interested in the Transit of Venus because it offered them an opportunity to improve their estimates of the distance between Earth and the Sun, and by extension to improve their understanding of the dimensions of the whole solar system.[4] Before 1761, estimates of the distance between Earth and the Sun differed greatly, from 41 to 87 million miles.[5] But in 1716, only five years before Keill published his words, Edmond Halley alerted astronomers to the fact that an accurate estimate of the solar parallax (from which the distance between the Sun and Earth could be calculated) might be made during the Transit of Venus by having many astronomers view the transit from as widely separated places as possible (see figure 3.1).[6] Since observers

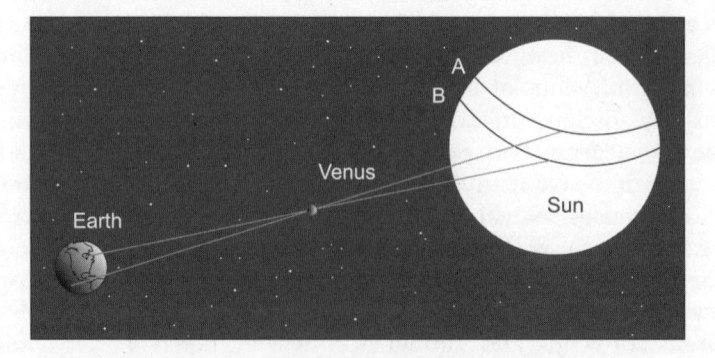

3.1 The Transit of Venus. To two people observing the Transit of Venus from different locations, Venus appears to cross the Sun along different tracks. Using careful observations of the duration of the transit from different vantage points, astronomers hoped to calculate the distance between Earth and the Sun. Source: T. Binnema

in different locations would see Venus cross different parts of the Sun, the duration of the transit would be different in different places. That meant that if the location of each observer could be precisely known, and the duration of the transit in each place could be accurately measured, it would be only a matter of mathematical calculation to determine the distance between the Sun and Earth.

The circumstances under which the Transit of Venus could be useful to astronomers shows why it was so crucial to the spreading of European scientific endeavours around the world. Observations of the transits could contribute to European knowledge of the solar system only if they were observed from those widely separated places on the globe (none of which could be in Western Europe, where neither the 1761 nor 1769 transit was visible from beginning to end), by people who were well educated in European science, using European scientific instruments. These purposes could not be readily accomplished without international cooperation and state sponsorship. The rarity of the transits made this cooperation all the more urgent. If astronomers missed their chances in 1761 and 1769, their next chance to observe the Transit of Venus would not come until 1874. So, with unprecedented international efforts, European men of science, their instruments and books, and their Enlightenment European perspectives travelled abroad, creating to greater or lesser degrees foundations for scientific research in Europe's colonies, and facilitating the conveyance of specimens, data, and knowledge from regions remote from Europe to the European metropolis. Furthermore, although the transits themselves were overwhelmingly of interest to astronomy, the efforts to observe the transits also inspired further efforts in various other observational sciences in the ensuing years.

In his *Venus in the Sun* (1761), Benjamin Martin, the prominent British popularizer of math and science, hyperbolized that "if we make the best Use of each [transit], there is no doubt but Astronomy will, in ten Years Time, attain to its ultimate Perfection."[7] Taking up Edmond Halley's challenge, astronomers from many countries, (but dominated by France) observed the transit from more than 60 places in 1761.[8] As one historian has noted, "the range and intensity of activity directly connected with the eighteenth-century transits of Venus were, by contemporary standards, enormous."[9] George Costard's *History of Astronomy* (1767) explained that after Horrox's observation of 1639, the transit "was observed again *June* the 6th 1761, by great numbers of Astronomers, excited particularly by a dissertation of Dr. *Halley* in the *Philosophical*

Transactions, in which he proposed finding, from that transit, the *Sun's Parallax*, and thence the *distance* of the *Earth* from the *Sun*."[10] That 1761 project was significant in that it represented "the first international scientific enterprise undertaken on a global scale."[11] For various reasons, including the fact that Halley had unrealistically predicted that the distance between Earth and the Sun might be calculated to within an error of 0.2%, the results in 1761 were disappointing.[12] However, Costard optimistically wrote that "there will be another transit of this same planet in the year 1769, June the 3d, at 10^h 10', according to Mr. [Jérôme] *De La Lande*, and consequently invisible at *Paris* or *London*; but by comparing together two observations to be made, one at *Mexico*, and the other to the North of *Petersburgh*, he says, the Sun's Parallax may be determined with double the precision it was in the last."[13]

As this passage betrays, between 1761 and 1769 scientific interest in the transit grew dramatically, but until 1767 interest in Britain lagged behind that in France. In 1761 both the French and the Swedes had supported more observers than the British had, but influential savants in Great Britain argued that the British, with their long maritime tradition and having just won the Seven Years' War, ought not to be outdone again. No one better expressed these sentiments than Thomas Hornsby, the Savilian Professor of Astronomy at Oxford University. In 1765 in the *Philosophical Transactions* he tied efforts to observe the Transit of Venus not only to national prestige, but also to the possibility that an expedition to observe the Transit from the South Pacific might also discover a great southern continent. He concluded his article by warning that

> the several Powers of Europe will again contend which of them shall be most instrumental in contributing to the solution of this grand problem. Posterity must reflect with infinite regret upon their negligence or remissness; because the loss cannot be repaired by the united efforts of industry, genius or power. How far it may be an object of attention to a commercial nation to make a settlement in the great Pacific Ocean, or to send out some ships of force with the glorious and honourable view of discovering lands towards the South pole, is not my business to enquire. Such enterprizes, if speedily undertaken, might fortunately give an advantageous position to the astronomer, and add luster to this nation, already so eminently distinguished both in arts and arms.[14]

For whatever reasons, the British effort, led by the Royal Society, turned out to be "one of the most ambitious scientific projects devised in Britain up to that date."[15] Thanks in large measure to a grant of £4000 from the king, in 1769 Britain supported twice as many observers as the French did.[16] Despite the lingering animosities in the aftermath of the Seven Years' War, the French proved willing to cooperate with these British efforts. Warfare had been one of the factors that interfered with the 1761 observations, so the British sought immunity for Cook and his crew. So convinced were the French that Cook was "out on enterprises that were of service to all mankind," that they ordered that his expedition not be molested.[17]

The HBC had had no opportunity to contribute to the observations of the Transit of Venus in 1761; at no location in Rupert's Land was that transit observable for its entire duration. It was a different matter in 1769. In order to get as accurate as possible an estimate of the solar parallax, British natural philosophers wanted astronomers to observe the transit from as far north, and as far south, as possible. Prompted by Alexander Dalrymple, the Royal Society began its planning for the 1769 transit in June 1767, and at a meeting on 17 November of that year a special committee of that society identified Fort Prince of Wales in the HBC territories as a promising place from which the transit might be observed.[18] Locations in the north were important because the transit would be of longest duration there.[19] The society chose Samuel Wegg, its newly appointed treasurer, to negotiate with the company.[20] Wegg was the obvious choice because, in addition to having been a Fellow of the Royal Society since 1753, he had been a shareholder of the HBC since 1748, and an HBC Committee member since 1760.[21] Fortunately for Wegg and the Royal Society, when the HBC Committee met on 27 January 1768, they decided that "this Committee is very ready to convey the Persons desired with their Baggage and instruments, to and from Fort Churchill, and to provide them with Lodging and Medicines while there, Gratis, they to find their own Bedding: The Company expecting £250 for Diet during their absence from England which will be about 18 Months."[22] Still concerned with guarding its operational secrets, however, the London Committee emphasized that the observers were not "to have the least Insight into the conduct of the Companys Trade or any other of Our Concerns."[23]

Another person instrumental in encouraging scientific research in Rupert's Land was one of the two men chosen to observe the transit: William Wales. Although Wales had expressed a decided preference for observing the transit from a warm place, he and Joseph Dymond were hired, at £200 each, to observe it from Fort Churchill.[24] Wales and

Dymond knew one another, although their temperaments evidently differed dramatically. Both men had worked under Nevil Maskelyne's supervision at the Royal Observatory, Greenwich, in 1765, Wales as one of the more effective computers of the first *Nautical Almanac* and Joseph Dymond as Maskelyne's assistant.[25] Thus Wales and Dymond became the first men with formal scientific training to winter in Rupert's Land. Dymond's conduct at Churchill irked those in charge.[26] But all indications are that Wales was an affable and enthusiastic man who had acquired the social intelligence to interact effectively with men of all social classes. A man of humble birth, without a university education, he had married in 1765 the sister of Charles Green, astronomer at the Royal Observatory, Greenwich. His uncanny ability to navigate ambiguous and ill-defined roles as a man of science in atypical circumstances was crucial to his career.[27] The essayist and poet Leigh Hunt later described Wales as "a good man of plain, simple manners, with a large person and a benign countenance." The essayist Charles Lamb agreed. In Wales he found "a perpetual fund of humour, a constant glee about him, which, heightened by an inveterate provincialism of North country dialect, absolutely took away the sting from his severities."[28]

Wales's easy and pleasant temperament must have stood him in good stead, because in order to observe a seven-hour transit Wales and Dymond had to stay at Churchill for thirteen months! Since HBC ships departed for Hudson Bay only once each year, Wales and Dymond had to leave England aboard the HBC's *Prince Rupert* on 30 May 1768 – two and a half months before James Cook's expedition left for Tahiti, and more than a year before the transit – hoping that the weather would cooperate with their mission to observe the transit. Their long stay in Rupert's Land gave the men ample time to make various astronomical observations at Fort Churchill. They also kept meteorological registers, made a map of the west coast of Hudson Bay, and recorded notes on the natural history and human communities in the vicinity of Fort Churchill.[29]

Fortunately, 3 June 1769, "being the finest day we had for some time," Wales and Dymond's observations of the transit were successful.[30] Although the two men had to wait until 7 September 1769 for the departure of the annual ship from Fort Churchill, they could do so satisfied that they had gathered valuable data.[31] James Cook, who had sharpened his astronomical skills while surveying the coast of Newfoundland between 1762 and 1767, and Charles Green (William Wales's brother-in-law), an astronomer from the Royal Observatory,

Greenwich, were similarly blessed with clear skies at Tahiti. Tahiti was a particularly important location because it was the most southwesterly known land from which the entire transit would be visible, and from which the transit would be of the shortest duration.[32] And so Wales, Dymond, Cook, and Green were among a sizeable number of successful observers in 1769. In total, 151 astronomers observed the transit of 1769 from 77 different locations, producing data for about 600 scientific papers relating to the solar parallax.[33] Wales estimated the distance between Earth and the Sun at 95 million miles.[34] Astronomers were still disappointed with the results of the 1769 observations, but the data did permit them to narrow the range of estimates of the distance between Earth and the Sun to between 92.0 and 96.2 million miles, a considerable improvement over 1761.[35]

The observations of the Transit of Venus had some obvious and immediate effects on branches of sciences apart from astronomy. The first expedition of James Cook illustrates this fact plainly. The primary purpose of Cook's expedition was to observe the Transit of Venus, but Cook's instructions were far broader. The French had ordered immunity for Cook, but the British deceived the French by giving James Cook secret instructions to search for and map "a Continent of Land of great extent" south of Tahiti.[36] The economic and strategic import of these instructions is obvious, but these instructions also related to scientific questions of the day. The Royal Society was interested in proving or disproving a theory that all of the land masses in the well-mapped portion of the globe had to be balanced by a similar-sized land mass in the south Pacific. Evidence of how reasonable Enlightenment thinkers found such a theory is reflected in a passage in Jedidiah Morse's *American Geography* (1792). Christopher Columbus's explorations were cleverly reinvented as if they conformed to the preoccupations of the Enlightenment: "From a long and close application to the study of geography and navigation, for which his genius was naturally inclined, Columbus had obtained knowledge of the true figure of the earth, much superior to the general notions of the age in which he lived. In order that the terraqueous globe might be properly balanced, and the lands and seas proportioned to each other, he was led to conceive that another continent was necessary."[37]

It would be interesting to know how Morse might have tried to explain why Columbus, who had "conceive[d] that another continent was necessary," was actually looking for Asia, and did not believe he found another continent even when he encountered it! At any rate,

Cook's explorations went far towards disproving the theory that there had to be a large enough continent in the south Pacific to counterbalance the continents in the northern hemisphere.

Furthermore, James Cook's patrons expected his expedition to contribute well beyond astronomy and mere geography to natural history and ethnography. His voyage, backed by the most powerful empire in the world and supported by London's Royal Society, was accompanied not only by astronomer Charles Green, but by also naturalist Joseph Banks, botanist Daniel Carl Solander (student of Linnaeus), and artists William Hodges, Alexander Buchan, and Sydney Parkinson (James Cook himself also had respectable scientific credentials).[38] Cook's instructions bring to mind Bacon's "Merchants of Light" and the appeals to travellers published by the Royal Society in 1665 and 1666. Cook was told to study and collect information, specimens, and examples of the natural world:

> You are also carefully to observe the Nature of the Soil, and the Products thereof; the Beasts and Fowls that inhabit or frequent it, the fishes that are to be found in the Rivers or upon the Coast and in what Plenty; and in case you find any Mines, Minerals or valuable stones you are to bring home Specimens of each, as also such Specimens of the Seeds of the Trees, Fruits and Grains as you may be able to collect, and Transmit them to our Secretary that We may cause proper examination and Experiments to be made of them.[39]

Clearly, this whole expedition, not just his orders to search for "a Continent of Land of great extent," was heavily influenced by the priorities of Enlightenment science and the British state.

Just as Cook's efforts in the south Pacific produced results in various branches of science, Wales's visit to Fort Prince of Wales had a significant legacy in the HBC. Wales's ability to contribute to the observations of the Transit of Venus seemed to inspire the HBC's London Committee. On 6 December 1769, shortly after Wales and Dymond returned to London, the company wrote to the Royal Society that "the Committee are and always shall be glad to contribute to the Encouragement of Science and every other National Improvement."[40] And when it came time for the HBC to bill the Royal Society for their support of Wales and Dymond, the governors advised the Royal Society's treasurer, Samuel Wegg, that "being desirous to encourage Science, and other National improvements, they decline being paid anything for freight of the

Instruments thither, and shall be fully satisfied with the £250 stipulated for the accommodation of the Observers."[41]

The contact established between the HBC and the Royal Society in connection with the Transit of Venus in 1769 marks a dramatic turning point in the history of science in the HBC. It marks the beginning of a period in which the HBC Committee explicitly encouraged its officers to contribute to scientific research in its territories and deliberately attempted to act as the conduit through which this research was transmitted to learned men. It also marks a turning point in scientific interest in Rupert's Land, especially within the Royal Society. Something of a transatlantic scientific network was being created. In the realm of astronomy, for example, William Wales soon wrote the HBC surgeon at York Factory, Thomas Hutchins, to request that he observe a partial solar eclipse that would be visible at York Factory. On 6 November 1771, Hutchins noted in the York Factory journals that he and Andrew Graham had successfully observed and timed the eclipse:

> Mr Wales in a Letter to Mr Hutchins having informed us that this Phenomenon would be visible at York Fort a little after the Sun had passed the Meridian, it gave us much uneasiness to see the whole Morning of the preceding Astronomical Day was cloudy so as totally to occlude the Sun, but, contrary to our expectations, about Noon by our Watches the Clouds began to break and the Sun appeared. Mr. [Andrew] Graham attended a meridian Mark in one of the Windows and I stept out with Hadley's Quadrant to take the Sun's Altitude.[42]

Hutchins later also submitted observations of the solar eclipse of 27 October 1780 from Albany Fort.[43] The efforts of William Wales, and some accommodating HBC employees in Rupert's Land, ensured that the cooperation that began with the observations of the Transit of Venus continued afterward.

The branch of knowledge most immediately and directly stimulated by the Transit of Venus was meteorology. Many European savants who studied astronomy and physics were also interested in the weather.[44] This helps explain why some natural philosophers saw in the efforts to observe the Transit of Venus new hope for a systematic study of weather. They were naturally inspired by the very impressive advances in astronomy in the 1600s and early 1700s, and more specifically by the international cooperation exhibited to organize observations of the Transit of Venus in 1769. In 1771, the German astronomer,

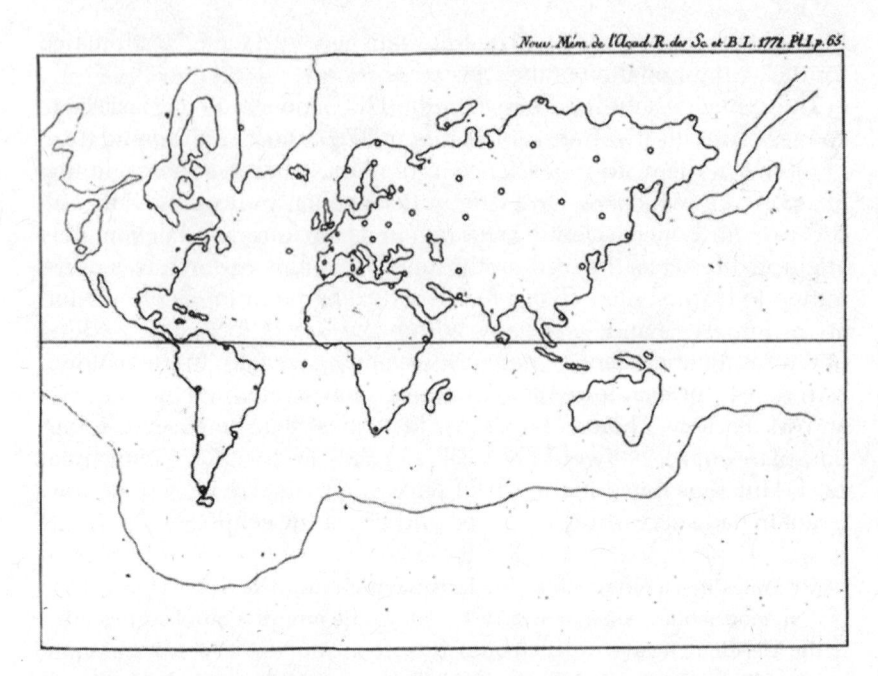

84   The Hudson's Bay Company and Science, 1670–1821

3.2 J.H. Lambert's Proposed Meteorological Network, 1771. In 1771, J.H. Lambert suggested that the Royal Society of London spearhead the formation of a global network of meteorological observatories, evidently including, as this map shows, a station at Churchill. Source: Lambert, "Exposé de quelques Observations."

Johann Heinrich Lambert, published a paper stating that "it seems that to make meteorology more scientific than it is, it is necessary to adopt the practice of astronomers, who, without getting lost in minutia, begin by identifying general laws and regular movements. Thus they are able to recognize anomalies and predict phenomena with a precision that inspires respect for astronomy, even among the most ignorant."[45]

Although Lambert made no explicit reference to the observations of the Transit of Venus, his proposal for accomplishing his goal was obviously influenced by those observations. He proposed that the Royal Society of London should take the lead in establishing a global network of meteorological observation stations. In a sketch map published with the paper (see figure 3.2), Lambert evidently proposed that Fort

Churchill be included among the observatories.[46] The renewed interest in meteorological research in Europe during the 1770s and the growing emphasis on international networks of meteorological stations are attributable to a number of causes, but the international efforts to observe the Transit of Venus seemed to provide the clearest example of the benefits of global efforts to improve scientific knowledge.

The observations of the Transit of Venus may have been a significant encourager of scientific activity in many settings outside Europe. For example, there is reason to believe that efforts in the United States to observe the Transit of Venus in 1769 were crucial to the emergence of the American Philosophical Society for the Promotion of Useful Knowledge as a scientific society of international stature. The roots of the society can be traced back as far as the 1740s, but it is really when two minor societies merged in 1768 as the American scientific community sought to coordinate and publish observations of the Transit of 1769 that the history of the society truly began.[47] Much of the first volume of the society's *Transactions* published in January 1771 were given over to observations of the transit.[48]

The importance of the transit went well beyond the field of astronomy in many locations. It led directly to the beginning of instrumental meteorological research in various European colonies. In Batavia, Johan M. Mohr was a man of learning but of little scientific training before he was taught how to observe the Transit of Venus in 1761. After he observed that transit, he soon began keeping a meteorological journal – probably the first instrumental weather register in the Dutch East Indies, and the scientific society that emerged from his efforts focused heavily on meteorological research in the 1770s.[49] The examples of the HBC, the United States, and the Dutch East Indies suggest that we may yet underestimate the significance of the Transit of Venus in the global expansion of European science.

By the time Lambert published his paper, the observations of the Transit of Venus had already directly produced meteorological research in Rupert's Land. Some of the instruments required by astronomers – including thermometers and barometers – were also useful for meteorological research, so it was natural, when William Wales and Joseph Dymond spent thirteen months at Fort Prince of Wales on behalf of the Royal Society, for them to keep a meteorological journal. Their meteorological journals, published eventually in the *Transactions of the Royal Society of London*, were kept according to contemporary scientific standards, from 10 September 1768 to 26 August 1769. From two to four times per day, the men recorded the air pressure, temperature (on two thermometers placed in different locations), winds, and any atmospheric conditions deemed worthy of note.[50]

Wales and Dymond's meteorological journals appear to have been the first systematic instrumental land-based meteorological journals kept in Rupert's Land, and their visit appears to have acted as a significant boost to subsequent meteorological research. Samuel Wegg, who supported the idea of an international network of weather stations, must also have been a factor.[51] At any rate, when Wales and Dymond left Churchill, they left their thermometers, barometers, and timepieces behind for the use of the company.[52] After the Royal Society asked the HBC to keep meteorological journals, the HBC also asked Wales to send them good-quality thermometers to be used in Rupert's Land.[53] In 1771, William Falconer at Severn House, and Thomas Hutchins at York Factory, began keeping systematic meteorological and phenological journals adhering to scientific standards.[54] The momentum was maintained for several years. At the council meeting of the Royal Society on 23 December 1773,

> it was moved and ordered by ballot, that two barometers, four thermometers and two ambrometers be purchased at the expense of the Society, and sent as an acknowledgement to the Hudson Bay Company, for their considerable and repeated benefactions; with a view that they be conveyed to some of their officers at their settlement to make observations of the state of the weather and send them from time to time to the Society.[55]

The Royal Society's donation was rewarded. Quite a number of meteorological journals kept primarily by Joseph Colen, Thomas Hutchins, John McNab, Philip Turnor, and Ferdinand Jacobs according to guidelines stipulated by metropolitan virtuosi, were submitted to the Royal Society, in whose collections they still remain.[56] Most of these journals were kept in the second half of the 1770s and the second half of the 1780s, although they range between 1774 and 1811.

Another obvious result of the observations of the Transit of Venus was the renewed burst of contributions by HBC traders to the field of natural history. The Royal Society explicitly noted the connection between the observations of the Transit of Venus and the HBC's contributions to natural history (see figure 3.3).

[370]

XXVIII. *Account of several Quadrupeds from* Hudson's Bay*, *by* Mr. John Reinhold Forster, *F. R. S.*

Read May 21, 1772.

1. ARCTIC Fox, Penn. Synopf. of Quadr. p. 155. n. 113. *Canis Lagopus*, Linn.

Severn River.

A most beautiful specimen in its snowy winter furr; this animal seems to be lower on its legs than the common fox, and is prodigiously well secured against the intense cold of the climate, by the thickness and length of its hairs, which are at the same time as soft as silk.

* Among the occasional advantages, which the observations of the last Transit of Venus have procured, that of receiving useful informations from, and settling correspondencies in, several parts of the world, is not the least considerable. From the factory at Hudson's Bay, the Royal Society were favoured with a large collection of uncommon quadrupeds, birds, fishes, &c. together with some account of their names, place of abode, manner of life, uses, by Mr. Graham, a gentleman belonging to the settlement on Severn River; and the governors of the Hudson's Bay Company have most obligingly sent orders, that these communications should be from time to time continued. The descriptions contained in the following papers were prepared and given by Mr. Forster, before his departure on an expedition, which will probably open an ample field to the most important discoveries. M. M.

The

3.3 "Account of Several Quadrupeds." When J.R. Forster's "Account of Several Quadrupeds," was published in the *Philosophical Transactions of the Royal Society* in 1772, the prominent tribute paid to the HBC and its governors and employees at the bottom of the first page must have been very satisfying to Andrew Graham, Samuel Wegg, and all of the directors of the HBC. Source: Forster, "Account of Several Quadrupeds from Hudson's Bay." *Philosophical Transactions of the Royal Society*: 370.

As did the first wave of natural history collecting in the 1730s and 1740s, this rush seems to have been related to a surge of interest among natural historians and the public in England, this time related to Linnaean science.[57] In 1775, Gilbert White wrote to his brother that "anything in the naturalist way now sells well."[58] But William Wales and Samuel Wegg were the most important promoters of science in Rupert's Land. Just as James Cook took naturalists with him when he went to Tahiti to observe the Transit of Venus, Wales and Dymond took advantage of their voyage and long stay in Rupert's Land to keep a journal describing and discussing various phenomena, including icebergs, kayaks, dandelions, insects, fish, soils, and minerals, and the crops grown at Fort Churchill. They also recorded ethnological information regarding the aboriginal inhabitants of the region. All of this work was published in the *Philosophical Transactions*.[59] Probably during their stay in Rupert's Land, Wales and Dymond encouraged traders there to contribute their knowledge to science. Definitely upon their return to London, Wales encouraged the Royal Society to ask the HBC to collect natural history specimens and to keep and submit instrumental meteorological registers.[60]

Several HBC officers immediately complied with the request of the Royal Society. The most enthusiastic of them appears to have been Andrew Graham, the master at Severn House. Graham was on furlough in London in 1769–70, at which time he may have met Thomas Pennant who had recently published his four-volume *British Zoology* (1761–6).[61] Pennant's was the first complete handbook of British birds published in almost a century.[62] Inspired either by Pennant or by the Royal Society's request, Graham submitted sixty-four skins and a manuscript describing the specimens from Severn in 1771. Meanwhile, in the same year, Moses Norton and Ferdinand Jacobs submitted specimens from Churchill and York Factory respectively, and Humphrey Marten did likewise from Albany House.[63] Many of these specimens and a good deal of the information flowing to London must have originated with aboriginal people. For example, Ferdinand Jacobs submitted two specimens – a whooping crane and an American white pelican – from York Factory in 1771.[64] Since those birds are not normally found at the coast, these specimens must have been brought to the coast from the interior by aboriginal people.[65] Evidence from the nineteenth century (discussed in later chapters) suggests that aboriginal people were likely specifically asked and paid for gathering such specimens.

On 22 October 1771, the London Governor and Committee noted that they "have Received by their two ships ... Two Chests containing the Skins of Quadrupeds and Birds which have been sent from Hudsons Bay at the particular desire of some Members of the Royal Society."[66] When all of their ships had arrived in November, the HBC Committee sent eight boxes of specimens to the Royal Society. The Royal Society appears to have been delighted with the response. The society's minutes of 29 November of 1771 note the arrival of

> eight boxes of stuffed and dried skins of Quadrupeds, Birds, etc; and also a collection of stones and Fossils; and there was delivered at the Table a manuscript in Folio entitled "Descriptive and Historical remarks on the several articles sent from Severn River in Hudson's Bay" by Mr. Andrew Graham; and some shorter lists. And it was moved ... that the thanks of the President and Council be given ... for this valueable present, as well as for their readiness upon this and all other occasions to promote science in general, and particularly the knowledge of natural history; and their kind assurances of continuing their communications to the Society.[67]

These boxes included specimens of animal species never before described to science, and information completely new to European natural historians. A week later the HBC Committee received a letter from the Royal Society, "returning the Thanks of a full Council of the Royal Society to this Committee for their very valuable Present ... and expressing the readiness of the said Council to communicate to this Committee the Result of any Tryals respecting those Productions."[68]

In March 1772 the Royal Society decided to establish a special committee on natural history, chaired by Samuel Wegg, to decide how the flood of specimens from Rupert's Land was to be studied.[69] They turned to the German naturalist John Reinhold Forster. Forster was well qualified. He had just published *Travels into North America* (1770–1), an English translation of some of the published accounts of the North American travels of the Swedish naturalist Peter Kalm. Then he published *Catalogue of the Animals of North America* (1771) and *Flora Americana Septentrionalis* (1771), in which he attempted to catalogue all known animal and plant species of North America. Soon Forster and others began to report on the specimens supplied by the HBC. In 1772 and

1773 Forster published several papers in the Royal Society's *Philosophical Transactions*, describing the fish, birds, and "quadrupeds" shipped to London by HBC officers.[70] Daines Barrington also published papers based on ptarmigan and hare specimens submitted to the Royal Society by the HBC in the early 1770s.[71]

When the HBC ships returned to Rupert's Land in May 1772, they carried with them some of the Royal Society's initial reports on the specimens. The London Committee's letters to its officers reflect how quickly the Governor and Committee came not only to encourage but also to direct its officers to continue their collections. To Humphrey Marten they wrote, "we have delivered to Capt. Christopher a Packet of Observations made by the Royal Society of the Birds &c. sent home last year and what further Articles are desired to which we expect Your Due Attention and we now send You under Customary Permit 5 Gallons of proper Spirits to preserve such subjects as may require it."[72] To Andrew Graham, they wrote, with guarded encouragement, that "the History you sent with the Birds &c. was very agreeable, and whatever other Curiosities you may collect, or farther Observations occur to You, will be always acceptable provided they are in the first Instance addressed to the Committee."[73] The thanks and encouragement must have gratified Marten, Graham, and others, but even before they received this feedback they had already gathered more specimens and information. So, the ships of 1772 returned to London with more samples and another remarkable manuscript, this one written by Thomas Hutchins. Hutchins, the HBC's surgeon at York Factory, was better educated than Graham, although he too appears to have been prompted to contribute to science only after the Royal Society request of 1770. In 1772, Hutchins and Graham were both at York Factory, and Hutchins's manuscript was obviously inspired by Graham's earlier "Observations" and benefited from Graham's contributions. Indeed, beginning in 1772 Hutchins and Graham began collaborating on versions of a manuscript with detailed observations of human and natural life along the shores of Hudson Bay – a major contribution for which Hutchins got most of the credit in his lifetime, and for which Graham got most of the credit in the twentieth century.[74] These "Observations" appear to have been written at the request of the British naturalist Thomas Pennant. However, in the preface to his *General Synopsis of Birds* (1781–1801) the eminent ornithologist, John Latham, a friend of Pennant, also mentioned reading these "Observations."[75]

Humphrey Marten was another HBC officer who responded to requests for natural history specimens. He submitted hundreds of specimens of birds and plants to London in the early 1770s, and included "A Short Description of the Birds in a Box," which drew upon his own observations, as well as knowledge gained by other traders, and on interviews with aboriginal people. In his published journals, Samuel Hearne subsequently remarked that "by some mistake," credit for Marten's contributions had been given to Hutchins.[76]

The full extent of the HBC's contributions to science in this period will never be known. Joseph Banks and Daniel Solander introduced plants from Hudson's Bay to the Kew Gardens in 1773 and 1778, but there is no record of how they got those plants.[77] But it is clear that by 1774 the HBC had emerged as one of the most important suppliers of North American natural history specimens in England.[78] Indeed, so impressive were its contributions that they inspired the Royal Society's Committee on Natural History to seek similar cooperation from the East India Company, the Muscovy Company, the Levant Company, and from the British government, but, as one historian of science had recognized, "had the responses been commensurate with the Hudson's Bay Company's contributions, it is difficult to imagine what the Committee on Natural History could have done."[79] Instead the HBC's contributions to natural history trailed off after 1774 as scientific trends in England changed, and as the HBC concentrated on a sudden and rapid expansion of its inland operations. It is somewhat ironic then, that at the moment when the HBC's potential to contribute to natural history suddenly improved because of the expansion of its operations, its actual contributions to that field declined.

When the HBC began to cooperate with the Royal Society in 1769 it quickly began to benefit in several ways from its association with British men of learning. First, it could respond to its critics by pointing to new maps of North America and to thankful tributes published in scientific publications. Indeed, the London Committee's insistence that its employees channel their contributions through them ensured that its secrets were well guarded, but also the committee could better ensure that the HBC's role in sharing information was more explicitly acknowledged than it had been previously. It worked. In contrast to the period before 1771, the accolades were showered much more directly upon the company than upon the actual collectors. For example, in 1772 and 1773 when John Reinhold Forster published three papers in the *Philosophical Transactions* of the Royal Society based on his studies of the birds and mammals and fish sent to the Royal Society by the HBC, his tributes were paid to the company and its directors, not to the officers in North America. His 1773 paper begins with the statement that "the Governor and Committee of the Hudson's Bay Company presented the Royal Society with a choice collection of skins or quadrupeds, many fine birds, and some fish."[80] In his work he also explicitly noted that "the Governors of the Hudson's Bay Company have most obligingly sent orders that these communications should be from time to time continued."[81]

The company benefited not only in public relations, but also more directly from advice and suggestions from the Royal Society. When the company sought to hire a surveyor in 1778, it chose Philip Turnor on the recommendation of William Wales – a recommendation for which the company paid Wales a gratuity of £5.5.0.[82] More significantly, individual naturalists and the Royal Society deliberately took it upon themselves to explore the commercial potential of items it received. For example, in a paper published in the *Philosophical Transactions* in 1772, John Forster reported on experiments conducted with plants used by indigenous North Americans to dye porcupine quills. After expressing his hope that the HBC would import more of these valuable products and learn how Indians used them, he wrote that "the wild inhabitants of North America are certainly possessed of many important arts; which, when thoroughly known, would enable the Europeans to make a better, and more extensive use of many unnoticed plants, and productions of this vast continent, both in physic, and in improving our manufactures, and erecting new branches of commerce."[83] And after they received a bison hide in 1772 the society consulted with tanners, leather dressers, and bookbinders to explore the potential use of bison hides for book bindings. They reported to the company that "the hides of these Animals are found upon Tryal to be as good a Material for Bookbinding as the Hides of Russian Buffaloes," and predicted that each properly preserved skin "may be worth about four shillings."[84] The Royal Society also presented to the London Committee hats and stockings made of buffalo hair.[85] And on 5 May 1773 concerning a swan specimen, the society explained that, "we have put the Skin into the hands of an importer, and we shall perhaps surprise when we inform you that if it had been in a state to be properly dressed, it would have been worth at least a Guinea and an half, so scarce is this commodity at present and so great is the demand for Powder puffs, the best sort of which can only be made from Swan down."[86] The letter continued by explaining exactly how the swans could be prepared in order to maximize their commercial value.

The company wasted little time in informing its officers of the Royal Society's advice. On 12 May 1773, the London Committee wrote Ferdinand Jacobs: "We are informed that the skin of the Wild Swan may probably turn out of some utility in our trade."[87] It took several years before the HBC benefited significantly from the sale of swan skins, but in the longer term, the consequences for the trumpeter swan (*Cygnus buccinator*) were catastrophic. Trumpeter swans were probably never abundant along Hudson Bay, and the population there declined by 1785 as a result of over-hunting, even though the company did not trade many swan skins until 1804.[88] Then as the trade suddenly soared and spread, the population of swans plummeted. By the late 1860s the trade in swan skins, and the swan's population, was a fraction of what it had been. Rumours of their extinction around the 1900 proved to be exaggerated, but the trumpeter swan was nearly wiped out by the millinery trade.

Within a decade of the observations of the Transit of Venus, the HBC Governor and Committee had seen the benefits that could accrue to the company if it willingly contributed to science, and, perhaps more importantly, if it was *seen* and *known* to contribute to science. The assistance of London men of science in recruiting skilled personnel and assessing the commercial potential of the resources of Rupert's Land was probably less important than the political and public-relations benefits of being acknowledged for its assistance to science. Furthermore, the company's cooperation with London savants appears to have incurred little cost to the company. The company's contributions to natural history and meteorology evidently waned after the 1780s, but the Governor and Committee's contributions to other branches of knowledge became even more remarkable thereafter.

The company acted just in time. The publication of Adam Smith's *Wealth of Nations* (1776) marked a milestone in the advance of economic liberalism in Great Britain. In his influential book, Smith argued that joint-stock and chartered companies were rarely competitive or useful to the nation. He conceded that the HBC might be the sole exception. Although he doubted the legality of the HBC's charter, he also dismissed Arthur Dobbs's allegations that the HBC was wildly and irresponsibly profitable, and he doubted that the trade of Hudson Bay could be conducted more effectively than it was by the HBC.[89] Going forward, the HBC's directors could not count on either obscurity or sympathy from politicians or the public. Their experience with the Transit of Venus in 1769 may have been crucial in helping them understand the value of being perceived as patrons of science.

4 "Amends for the Narrow Prejudices": The Hudson's Bay Company and Science in an Era of Competitive Expansion, 1774–1821

Scientific networks unravel easily. And there was reason to expect scientific activity and cooperation to diminish in the HBC territories during the increasingly difficult years between 1774 and 1821. To compete with the North West Company (NWC) the HBC was induced to begin an inland expansion in 1774 (see figure 4.1). But circumstances only deteriorated for both the HBC and NWC as, by the late 1810s, ruinous competition and bitter rivalry threatened to drive both companies into bankruptcy. Only the amalgamation of the companies in 1821 averted disaster. But thanks largely to the efforts of the London Governor and Committee and a number of officers in the company, the HBC had earned, by 1821, a reputation as a generous patron of the sciences.

It might seem incongruous that the company's contributions to surveying and mapping increased during a period of bitter rivalry. There were relatively few risks associated with its contributions to astronomy, meteorology, and natural history between 1769 and 1774, but it might have seemed to be potentially dangerous to contribute to British maps that would inevitably be available to the HBC's competitors. Closer examination reveals that it was entirely natural that the company would share what it knew of North American geography. The HBC's knowledge of the geography of the interior of North America lagged behind that of its competitors between 1774 and 1821. In the unlikely event that maps would be useful to new Montreal-based competitors, those competitors would threaten the NWC more than the HBC. On the other hand, by contributing to British maps of North America, the HBC's directors put the company in a good position to defend itself against its critics (including those in the NWC) who argued that its charter should be revoked because of its failure to accept the obligations that these critics claimed came with its charter.

4.1 Hudson's Bay Company Territories, 1774–1821. The period between 1774 and 1821 witnessed a dramatic geographical expansion of the HBC's operations inland. The HBC's Governor and Committee also employed surveyors Philip Turnor, David Thompson, and Peter Fidler to survey the interior. The results of those surveys were shared with London cartographer, Aaron Arrowsmith. Source: T. Binnema.

Although the HBC's knowledge of the interior was less extensive than that of the NWC, its knowledge conformed more closely to the standards of scientific cartography. Developments in surveying and cartography raised the expectations and prestige of cartography significantly during and after the 1760s.[1] Beginning in 1778, the HBC employed surveyors who were able to supply British cartographers with the kinds of data they needed to produce maps that met European standards of accuracy. The NWC was not in a similar position until 1797. In the meantime, the HBC seized opportunities to contribute to mapping in ways that could embarrass its NWC rivals. Even thereafter, the HBC continued to contribute to British maps in ways that earned it many accolades.

More than anyone else, Samuel Wegg was responsible for changes in the HBC's informal policies towards science in this period. Wegg had been instrumental in orchestrating the HBC's cooperation with the Royal Society's efforts to observe the Transit of Venus from Fort Prince of Wales in 1769. At that time, Wegg was the treasurer of the Royal Society, and a member of the HBC's London Committee. Thereafter, Wegg became increasingly influential in the HBC. In 1774, he was named deputy governor of the company, and from 1782 to 1799, he was simultaneously London Governor of the HBC and the treasurer of the Royal Society. Throughout the years of Wegg's influence, the HBC adopted a remarkably generous distinction between information that rightfully might be kept secret because of its strategic and commercial importance, and information that might be shared with scholars.[2] Indeed, although the HBC's contributions to various branches of knowledge ebbed and flowed, the most remarkable development of the period between 1774 and 1821 was the HBC's growing cooperation in the more obviously strategic and commercially significant realms of surveying and mapping. As a result of the company's generosity, European maps of the northern interior of North America improved dramatically during the period.

Thanks to the efforts of William Wales, meteorological journals conveyed some of the first and most plentiful scientific evidence from Hudson Bay to London in the first five years following the Transit of Venus in 1769. Most of the meteorological journals that were submitted to the Royal Society in the late 1770s and early 1780s had been kept by men—such as Thomas Hutchins and John McNab—who had also contributed to other branches of science.[3] But the flow of meteorological journals appears to have diminished significantly by the 1780s. It also seems that the company no longer facilitated contributions after 1800, probably because interest in meteorology waned both within the company and within the Royal Society at that time. There is good reason to believe that many of the journals that made it to the Royal Society's permanent collection did so because of the efforts of Joseph Colen and Peter Fidler, without the cooperation of the London Committee.

Joseph Colen, from Cirencester, Gloucestershire, joined the HBC in 1785 as a writer for Humphrey Marten, resident chief at York Fort.

When Marten retired the next year, Colen was promoted to replace him – becoming, according to explorer and trader David Thompson, "the most enlightened Gentleman who had filled that station."[4] Little is known about Colen's life prior to joining the company, but he had experience in "mechanic & mercantile Affairs" and was a "perfect master of his Pen in writing and figures."[5] He was obviously well educated and curious: "a man of ideas" according to his biographer, "an innovator and a thinker, rather than a shrewd businessman."[6] Colen directed his interests towards meteorology, medicine, the collecting and husbandry of trees and shrubs (he sent shrubs from Hudson Bay to his employers in London), and foremost towards reading.[7] If his claim to have owned 1400 volumes of books when he left the company was true, his was probably the largest personal library in the history of the HBC's territories.[8] Although the evidence is scanty, Colen also appears to have been a mentor and encourager of sorts to two very different English servants in the HBC, Peter Fidler and Joseph Howse. Peter Fidler, from Bolsover, Derbyshire, joined the company in 1788 and spent his first winter under Colen's supervision at York Factory. He – as will become apparent – developed a friendship with Colen that lasted beyond Colen's departure from Rupert's Land in 1798. Joseph Howse was, like Colen, from Cirencester, and subsequently was described as Colen's "pupil."[9]

Colen submitted several meteorological journals to the Royal Society that he could not have kept himself. Some date from before he joined the company. Furthermore, although Colen's time was a period of great spatial expansion for the HBC, Colen himself never travelled farther inland than a few days' journey from York Factory. Nevertheless, he submitted quite a number of meteorological journals to the Royal Society that were taken from inland locations – ranging from Lake Athabasca (from the time that Philip Turnor and Malchom Ross were at the North West Company post there in 1791-2) to Cumberland House.[10] Furthermore, most of the journals he submitted for York Factory between 1795 and 1797 appear to be in Peter Fidler's hand, suggesting that Fidler and Colen cooperated to transcribe the York Factory journals for submission to the society.[11] These facts also suggest that Colen took the initiative to have meteorological journals from various posts transcribed and submitted to the society, or at least that he became a willing supplier of journals to the society. Colen continued to convey HBC weather records to the Royal Society long after he retired from the company's service. Peter Fidler kept meteorological journals during most of his career with the HBC (which spanned from 1788 to 1821), and some of them (from 1808 to 1811) are in the Royal Society Archives. The evidence suggests that Fidler attempted to submit many more to the Royal Society, but that he sent them via Joseph Colen. In December 1811, more than a decade after he had left the company, Colen sent a letter to Joseph Banks, apologizing "for my not forwarding to you annually the Meteorological Journal as usual, sent me by Mr Peter Fidler from Hudson's Bay – You will herewith receive a Packet – Remarks made three seasons following, being a regular continuation – to then I have had the honor of conveying to you – and which I beg may thro' your means be presented to the Royal Society."[12] The fact that Fidler submitted his journals through Colen rather than the company suggests that he doubted whether the committee would pass them on to the Royal Society. But Banks too may have been unreliable. The evidence that Colen yearly sent Peter Fidler's meteorological journal to Joseph Banks suggests either that Banks failed to forward those journals to the Royal Society, or that the Royal Society did not preserve all of the journals it received. At any rate, the fact that the journals have not survived reflects the fact that interest in meteorology waned in the HBC and in scientific circles between 1790 and 1812.

At least some of the HBC's unpublished meteorological data reached learned circles during the late eighteenth century. The climate of the Hudson Bay region was of some interest to Europeans because Hudson Bay was in the same latitudes as Great Britain, and yet, as the instrumental records now proved, it was obviously far colder in the winter. Naturalists evidently did consult some of the meteorological records of the HBC not long after they were sent to Britain. In his *Arctic Zoology* (1784–5), Thomas Pennant noted that the temperatures along Hudson Bay, according to Thomas Hutchins, were as high as 103° F in the summer and as low as −45° F in the winter.[13]

The HBC had supported enough instrumental meteorological research by the mid-1770s to establish the fact that the coastline of Hudson Bay did not have a climate suitable for the agricultural pursuits that its critics had suggested that it did. Thus, sharing that kind of information with outsiders suited the interests of the company very well. But confirmation that Rupert's Land experienced frigid winter temperatures also quickly opened up a new field of research. Generally speaking, little physical or experimental science took place in European colonies before the mid–nineteenth century. But HBC posts had one physical "asset" – frigid winters – that suited them for such sciences. Thomas Hutchins was the first to seize upon the opportunities of the cold to perform experiments. Hutchins was in England on furlough in 1773–4, during which time he must have communicated with Fellows of the Royal Society. At the behest of the Royal Society and the encouragement of Samuel Wegg, Hutchins had conducted and reported on experiments with the dipping needle at Stromness (Orkney), the Hudson Strait, Moose Factory, and Albany Fort in 1774–5.[14] He then performed some not entirely satisfactory experiments on the freezing point of mercury in 1775 while at Albany.[15] The freezing point of mercury was of considerable scientific interest at that time because only in 1772 did most natural historians accept the fact that mercury could freeze – news that undermined perceptions of its usefulness in thermometers.[16] Hutchins' paper, published in the *Philosophical Transactions of the Royal Society* in 1776, captured the attention of Joseph Black, an Edinburgh chemist, who, in 1779, wrote Andrew Graham – then in retirement in Edinburgh – about how Hutchins' experiments on mercury might be improved. Hutchins completed the refined experiments – the first well-designed research on the freezing point of mercury – at Albany Fort in the winter of 1781–2, and his report was published in 1783 (see figure 4.2).[17] For this research he became in 1783 the second HBC servant to win the Royal Society's prestigious Copley Medal. By that time Hutchins had returned to London to take up the position of the HBC's corresponding secretary, a position from which he may have been able to offer some encouragement for research in Rupert's Land.[18]

Hutchins' experiments encouraged the well-known British chemist and Copley Medal winner Henry Cavendish to organize the use of Rupert's Land as a laboratory to conduct experiments with the freezing point of other liquids.[19] For example, in the mid-1780s, Cavendish convinced John McNab, surgeon and master at Henley House (at the forks of the Albany and Kabinakagami Rivers, see figure 4.1), to conduct a number of experiments with the freezing point of various liquids. Cavendish reported that McNab's "endeavours have also been attended with much success, as he has not only shewn many remarkable circumstances relating to the freezing of nitrous and vitriolic acids, and the phænomena of freezing mixtures; but has also produced degrees of cold greatly superior to any before known."[20] The results of further experiments were published two years later.[21]

The HBC's contributions to the natural history of its territories continued to be especially valuable for the company's public reputation. Although the flow of natural history specimens supplied by the HBC to the Royal Society diminished during the 1780s, the specimens continued

[*303]

*XX. *Experiments for ascertaining the Point of Mercurial Congelation.* By Mr. Thomas Hutchins, *Governor of* Albany Fort, *in* Hudson's Bay †.

Read April 10, 1783.

THE following experiments, to determine the freezing point of quicksilver, were made by the direction of the Royal Society, at Albany Fort in Hudson's Bay, situated in the latitude of 52° 14′ North and 82° West longitude from Greenwich.

The instruments used in these experiments were simply thermometers, except the apparatus F and G, furnished by Mr. CAVENDISH, and of these a more satisfactory idea will be formed from the annexed drawing than could be conveyed by words alone; I have, therefore, only specified a few particulars, so that each instrument may be distinguished from another.

I have compared the instruments with each other for several weeks in the various temperatures, to adjust, with the greater precision, the relative degrees on the scales; which was the more necessary as they differed very much.

The five first experiments were made exactly according to the directions sent to me by the Society, in order to obtain the point of congelation. The two succeeding ones are also made in the manner they directed, to endeavour to ascertain the greatest degree of contraction mercury is capable of; then follow two

† This paper having been for some time mislaid, could not be printed in its turn. This accounts for the double paging and signatures.

VOL. LXXIII. *S f experiments

4.2 Thomas Hutchins on the Freezing Point of Mercury. Thomas Hutchins received the Royal Society's Copley Medal in 1783 in recognition of this report, published in the Royal Society's *Philosophical Transaction*, on experiments undertaken at Fort Albany to determine the freezing point of mercury. Source: Thomas Hutchins and Joseph Black, "Experiments for Ascertaining the Point of Mercurial Congelation." *Philosophical Transactions of the Royal Society of London* 73 (1783): 303.

to make an impact. The HBC shared natural history specimens with other friends of the company. Thomas Pennant, for example, noted having seen natural history specimens from the HBC in the "most magnificent museum" of Sir Ashton Lever.[22] But the company's increasingly generous approach to its employees' records is best exemplified by its use of Samuel Hearne's journals.

Samuel Hearne, who served the HBC from 1766 to 1787, was one of its more intelligent and energetic employees. In the words of one of his biographers, "as an explorer and writer" Hearne embodied "an interesting combination of physical endurance and intellectual curiosity."[23] The editor of one of the editions of his journals has judged that he was "head and shoulders superior to every other North American naturalist who preceded Audubon."[24] He is best known for the explorations he undertook between 1769 and 1772. Inspired by Chipewyan accounts of distant "copper mines" and by the quest for the Northwest Passage, Hearne was sent inland from Churchill. In 1769 and 1770 he was forced to abandon two attempts, but between 1770 and 1772 a Chipewyan man named Matonabbee led him safely from Churchill to the mouth of the Coppermine River along the Arctic Ocean and back.

Hearne had no inkling that intellectuals would ever read his journals. "My ideas and ambition," he later wrote, "extended no farther than to give my employers such an account of my proceedings as might be satisfactory to them."[25] Given the Governor and Committee's secrecy up to that point, Hearne's expectations were reasonable. But beginning sometime in the 1770s, the HBC committee, particularly "through Mr. Wegg's interest," took the unusual move of sharing manuscript copies of Hearne's journals with London naturalists and geographers, including Thomas Pennant and Alexander Dalrymple.[26] While Hearne was in England during the winter of 1782-3 (after the French had seized and destroyed Churchill and Hearne had been forced to return to England), Samuel Wegg introduced Hearne to Thomas Pennant and other savants.[27] Pennant was also given access to "Mr. Graham's MS" and information gained from Thomas Hutchins.[28]

By the mid-1780s, tantalizing summaries of Hearne's explorations were reaching print. When he published his *Arctic Zoology* in 1784-5, Pennant noted that he had "perused [Hearne's] journal and had frequent conversation" with Hearne.[29] In fact, aside from publishing some of the insights he gained from Hearne's zoological knowledge, Pennant summarized Hearne's travels and his descriptions of "Coppermine" and "Dog-ribbed" Indians.[30] Pennant also described an event that was eventually to become one of the most famous and controversial purple patches in Hearne's journals after they were published: the "Massacre at Bloody Fall." Pennant explained that Hearne's Chipewyan hosts attacked a band of Copper Inuit, during which "a young woman made her escape, and embraced Mr. Hearne's feet, but she was pursued by a barbarian, and transfixed to the ground."[31] The fragmentary reports of Hearne's journals produced an appetite for more. When Dr. John Douglas, Bishop of Salisbury, published the first official edition of the journals of James Cook's voyages in 1784, he took the opportunity to quote a section of the manuscript journal and state that "the publication of this [journal] would not be an unacceptable present to the world."[32]

Hearne quit the HBC in 1787. Back in London, he revised his journals for publication, encouraged by several men of science, including the naturalist Thomas Pennant (who may have urged him to add his very perceptive natural history observations to his journals), and William Wales (who, in October 1792, successfully facilitated the sale of Hearne's manuscript to the prominent London publishers Strahan and Cadell), and by editor John Douglas.[33] Fortunately, when Hearne asked the Governor and Committee for permission to see his original journals (or copies of them), they agreed "with the greatest affability and politeness."[34] The subsequent posthumous publication of Hearne's journals in 1795 is arguably more historically significant than Hearne's journeys themselves. The journals' enthusiastic reception in England not only made available valuable evidence about the natural history, geography, and peoples of a vast region of northern North America, but also made Hearne famous well outside scientific circles. He became the first among several hyper-masculine, Romantic adventurer-scientists of North American arctic exploration.[35] His journals quickly captured the attention of the British public, including its Romantic poets. An episode in Hearne's published journal inspired William Wordsworth to pen "The Complaint of a Forsaken Indian Woman," first published in 1798.[36] But considerations other than Hearne's public fame probably motivated the company directors to cooperate with the publishers. Certainly the fact that Hearne's journals, by placing the mouth of the Coppermine River at 71° 54' North, strongly supported the company's assertion that any Northwest Passage to Asia would be obstructed by ice for almost the entire year helps explain why the Governor and Committee welcomed their publication – albeit more than two decades after

Hearne's explorations were made – but the publication of those journals, as we shall soon see, represented a significant milestone in the history of the company's communication with outsiders.

The HBC's contributions to cartography between 1774 and 1821 deserve special attention. That the HBC contributed little to European knowledge of the geography of North America before the 1780s was not merely out of secrecy. The company knew relatively little about what lay inland of the coast until the 1750s. As managers of a commercial enterprise, the Governor and Committee in London always wanted maps, but often reminded its servants that exploration and mapping were "to be made Subservient to promote the Company's Interest and increase Their Trade."[37] To be fair, the company's Governor and Committee had, since the 1680s, asked its officers to send men inland on explorations, and they had supplied bayside posts with quadrants and other surveying instruments by 1715, but officers found few men able and willing to survey the interior.[38] In 1690 they found someone willing, but not very able. Henry Kelsey was sent inland from York Factory in 1690 and 1691, but he had no training in surveying or navigation. His enigmatic journals (portions of the 1690 journal are written in stilted verse) prohibit historians from reconstructing his travels beyond generalities.[39] The only post established inland before 1774 was Henley House, at the confluence of the Albany and Kabinakagami Rivers in 1743 – although that post was a minor post built to defend the bayside posts, not to expand the HBC's commerce into the interior.[40] The Governor and Committee had no desire during this period to share their geographical knowledge, but as late as 1754, the HBC had remarkably little knowledge of the interior to share, and as late as 1775, it had few maps of the interior.[41]

The HBC's attitude towards the exploration of the interior changed between 1754 and 1778 largely because Montreal-based traders infiltrated the hinterlands of most of the HBC's posts, cutting dramatically into the company's fur returns. The company was probably also spurred on by the embarrassing attacks launched by critics of the company in Britain during the 1740s and early 1750s. During many of the winters between 1754 and 1772, the company sent emissaries inland from York Factory to spend winters with Cree bands as they travelled through the interior. These men were instructed to try to convince peoples in the interior to travel to the bay to trade with the HBC rather than to trade with interlopers from Montreal. None of the men sent inland had any training in surveying or mapping, and none had any scientific mandate. Indeed, many of them were illiterate. Only a few, including Anthony Henday (in 1754–5) and William Tomison (in 1767–8 and 1769–70) were literate enough to provide much substantial information to the company.[42] But even in the case of Henday, one of the ablest employees, the London Committee concluded that he was "not very expert in making Drafts with Accuracy or keeping a just Reckoning of distances other than by Guess which may prove Erroneous."[43]

Samuel Hearne's journeys from Churchill mark a watershed in the history of HBC explorations. The circumstances surrounding Samuel Hearne's three journeys between 1769 and 1772 suggest that they were influenced by the presence of William Wales and Joseph Dymond at Churchill in 1768–69. Unlike any of the men sent from York Factory before him, Hearne was sent inland with a quadrant. Hearne may have received some training in the use of surveying instruments during his time in the Royal Navy, but even if he had, he probably also received some training at Churchill during the 1768–9 season, when William Wales was in residence there. Hearne's actions show how significant he considered his quadrant to be. When he broke his quadrant in August 1770 during his second attempt to reach the fabled "copper mines," he elected to return immediately to Churchill even though he was already more than five months' travel and about 600 kilometres from Churchill.[44] On the third journey, Hearne broke his replacement quadrant – a thirty-year-old Elton's quadrant – but this time not until March 1772 when he was already returning to Churchill.[45] Given, however, the fact that he was still north of Great Slave Lake when he broke that quadrant, it is understandable that Hearne's published journals state that "I cannot sufficiently lament the loss of my quadrant, as the want of it must render the course of my journey from Point Lake, where it was broken, very uncertain."[46] Hearne's lament notwithstanding, the potential significance of his journey was further undermined by the infrequency with which he had used the quadrant (if his published journal gives any indication), and the inaccuracy of the observations he made.[47] But it is no coincidence that the first quadrant known to be carried inland by a HBC servant was carried from Fort Churchill shortly after William Wales had been there. And the influence of Wales is further suggested by the fact that Matthew Cocking in 1772 was probably the first HBC servant to carry a quadrant inland from York Factory.

In 1772, when the Governor and Committee had accepted the fact that they needed to build posts inland from York Factory to compete with their Montreal-based competition, Andrew Graham conceded that "the Accounts given us by Men sent Inland were incoherent and unintelligible."[48] And thus Graham turned to Matthew Cocking, a literate, capable, and willing Englishman, second in charge at York Factory. Cocking was given some rudimentary training in surveying and when he returned to York Factory in the spring of 1773, he provided the company with the intelligible journals they sought.[49] Similarly, in 1775 and 1776, the company sent Edward Jarvis, a young surgeon, and John Hodgson, an apprentice acquired from London's Grey Coat Hospital School, on several expeditions inland from Albany Fort.[50]

The year 1774 was another watershed in the HBC's efforts to understand the interior regions of North America, from which most of its furs came. In 1774 Andrew Graham supplied a map – unmistakably influenced by Native information as well as information gathered by its servants – that includes recognizable portrayals of Lakes Winnipeg, Manitoba, and Winnipegosis, and the Saskatchewan River system beyond the forks of the North and South branches of that river (see figure 4.3).[51] More importantly, in 1774 the HBC began an inland expansion that eventually took it across the continent. After contentedly and profitably operating solely from factories at the mouths of rivers on the shores of the Hudson Bay for over a hundred years, the HBC finally responded to its competition by building Cumberland House on the Saskatchewan River system in 1774. In the ensuing twenty-five years, the HBC established posts as far west as the foothills of the Rocky Mountains, and as far northwest as Lake Athabasca. The impetus for this expansion was not scientific, but commercial. Competing partnerships operating out of Montreal had so effectively infiltrated the hinterland behind the bay that the HBC could no longer be content to operate solely from bayside trading posts. In the next several decades, NWC and HBC posts proliferated throughout the western interior of North America and as far as the Pacific Ocean, and explorers operating on behalf of the NWC, and to a far lesser extent those employed by the HBC, explored areas previously unknown to Europeans.

As commercial enterprises, the HBC and its Canadian competitors explored and surveyed only insofar as these efforts supported their trade. Thus, because of their different approaches to the trade, their need for exploration, surveying, and mapping differed. The Canadian partnerships supported remarkable explorations because they depended upon continually discovering and tapping rich new sources of valuable furs.[52] Until 1797, their traders and partners depended upon geographical information and maps that were inexact by the standards

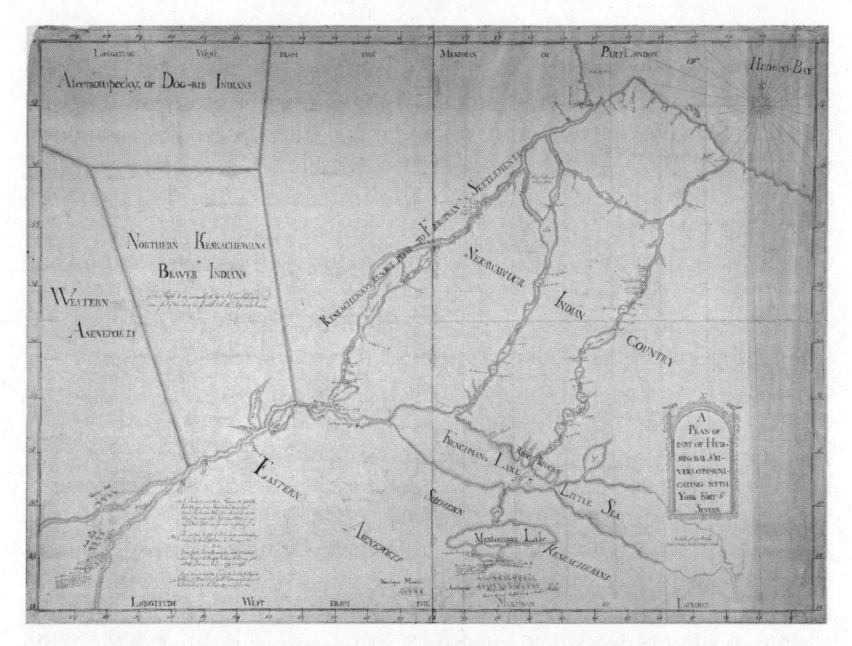

4.3 "A Plan of Part of Hudson's Bay and Rivers, Communicating with York Fort and Severn," by Andrew Graham, ca. 1774. This map submitted by Andrew Graham in 1774 relied on information supplied by Matthew Cocking and William Tomison, but was also influenced by information gathered from aboriginal people. The Eagle Hills of present-day Saskatchewan are depicted at the bottom left. Source: HBCA G2/17.

of Western cartography at the time. But they were partnerships that did not have the deep pockets or long-term investment horizons to justify the expense of systematic surveys of the territories in which they operated. Not until the NWC had eliminated most of its Canadian competitors did it have the stability and size to change its approach. The HBC, by contrast, did not have the personnel, skills, or experience to match the Canadians' probing of far-off *terra incognita*. Instead, the HBC concentrated on acquiring and exploiting more precise information about territories in which the company was already active.[53]

Despite the attacks on HBC secrecy and inaction in the late 1740s and early 1750s, and despite the company's increasing openness during the late 1760s and early 1770s, the company did not immediately begin to share its knowledge of the interior. As far as the results of its employees' surveys of the Hudson Bay region were concerned, Sverker Sörlin's description of the approach of the Dutch East Indies Company until the late 1770s was as true of the HBC; "their discoveries were not treated as discoveries to be shared with the scholarly world, but rather as closely guarded trade secrets."[54] Until the late 1770s, the HBC Governor and Committee kept to itself all of the geographical knowledge that it was collecting. None of the maps made by the prominent British geographers and cartographers of the time reflected any of this knowledge. The company changed its approach during the early 1780s. At some time before 1784, the company allowed selected geographers, cartographers, and naturalists to see Hearne's manuscript journals or extracts from them.

The company's Governor and Committee also sought to survey the interior more systematically. Having launched an inland expansion, the company's directors needed to know exactly where the company's inland establishments and inland rivers were situated. Thus, the company set out to hire "three or more Persons well skilled in the Mathematicks & in making Astronomical Observations ... to travel Inland with the Title of Inland Surveyors."[55] The company understandably turned to William Wales for advice, and on Wales's recommendation it hired Philip Turnor in 1778.[56] In this way the HBC hired its first formally trained surveyor. Turnor began his work as soon as he arrived at York Factory in August 1778. When the ship returned to London that same summer, it carried with it the results of Turnor's quick survey of the environs of York Factory.[57] Turnor, like other HBC surveyors that succeeded him, was rarely put in the position of exploring lands not already known to Europeans, indeed, even to HBC traders. Instead, from 1778 to 1782 he spent most of his time surveying the major established travel routes and pinpointing the coordinates of the HBC's inland trading posts (such as Cumberland House, Hudson House, Henley House, and Gloucester House) in order to provide the London Governor and Committee with the precise geographical information they desired. Thereafter, the London Committee seemed satisfied. Rather than being sent on further surveys or explorations, Turnor was put in charge of trading posts from 1782 to 1788. This use of Turnor reflects another policy of the HBC: even its trained surveyors were traders first and foremost. However, Turnor returned to London in 1788 to draw maps for the company and probably to consult with the company directors.

The HBC must have begun sharing the results of its surveys soon after Wegg's promotion to the Governorship of the HBC in 1782. Preparations were then underway for the publication of the official journals of James Cook. The journals, published in 1784, included a map of the world completed by Lieutenant Henry Roberts, who had sailed with Cook on his second and third voyages. In his explanation of the sources for his map, Roberts offered a long description of the information provided to him by the HBC, which reads in part:

> The whole of Hudson's Bay I took from a chart, compiled by Mr. [John] Marley ... with which I was favoured by Samuel Wegg, Esq; F.R.S. and Governor of that Company, who also politely furnished me with Mr. Hearne's Journals, and the map of his route to the Coppermine River, ... together with the survey of Chesterfield Inlet made by Captain Christopher and Mr. Moses Norton, ... and the discoveries from York Fort to Cumberland, and Hudson Houses ... extending to Lake Winnipeg, from the drafts of Mr. Philip Turnor, made in 1778 and 1779, corrected by astronomical observations. ... [58]

The Governor and Committee of the HBC could not have hoped for a more prominent place in which to have its cooperation with geographers and cartographers publicized than the official journals of James Cook. The committee must also have found the adulatory discussion of the HBC in the introduction to those journals very gratifying. Its public relations value may explain why in the ensuing years, Samuel Wegg and the London Committee made maps, charts, and geographical information available to several of the elite members of the cartographical fraternity in England.

Soon after the HBC began sharing its surveys with London geographers and cartographers, those savants also began shaping the company's surveying agenda. Alexander Dalrymple, the East India Company's hydrographer, doubting some of the maps he had seen, began in the 1780s to urge the HBC to make a careful survey of Lake Athabasca because, if the maps were correct, Lake Athabasca might be a crucial link in a practical river route across the continent. If so, it would advance his aims to combine the operations of the HBC and the EIC.[59] Dalrymple, a Fellow of the Royal Society and friend of Wegg, was known to have "dined week after week" with Samuel Wegg.[60] Therefore, he was in a position to influence Wegg and the London Committee.

Dalrymple's hopes for (and doubts about) Lake Athabasca were founded on maps by Samuel Hearne, and the Montreal-based trader, Peter Pond, whose 1785 maps were based on his travels in the Athabasca country in 1778–9.[61] Those maps suggested that Lake Athabasca might extend as far west as 132° W (it actually extends only to 112° W) and might drain into a river that flowed to Cook's Inlet (at present-day Anchorage, Alaska) on the coast of the north Pacific. Coupled with Dalrymple's correct belief that Hearne's astronomical observations at the mouth of the Coppermine (which placed the mouth of the Coppermine River at about 72° N) were wrong, and the results of the explorations of James Cook in 1778 (which not only provided an accurate map of the outer coast of northwest North America as far north as Icy Cape, but also showed how lucrative sea otter pelts could be if sold in China), this map offered new hope that an inlet or large river might yet be found to link Lake Athabasca with the Pacific Ocean.

Peter Pond's 1785 maps had not been the product of disinterested inquiry. In October 1784, NWC directors Benjamin and Joseph Frobisher submitted to the British Government an application requesting that the NWC be given a ten-year monopoly on the trade of the Athabasca country and trade along the Lake Superior-Lake Winnipeg route, in consideration of their discoveries in those regions. To support the application, the company submitted maps of both regions, including Peter Pond's map. It also advanced a plan for further exploration, and promised "to transmit from time to time ... correct Maps of those Countries, and exact account of their Nature and productions, with remarks upon everything useful or curious, that may be met with in the prosecution of this plan."[62] The tactic backfired. The NWC was soon to be outmanoeuvred by the HBC. The NWC did garner some positive press. For example, in 1790, the *Gentleman's Magazine* published a letter from Quebec jurist Isaac Ogden that described Pond as "a Gentleman of observation and Science." Pond, Ogden incorrectly asserted, was "instructed fully in the Knowledge of Astronomy &c &c. His Lat. is undoubtedly right, and his Long. is near right."[63] Dalrymple thought otherwise. In a memorandum written in February 1790, Dalrymple questioned Pond's skills and his character. After referring to Pond's "*ignorance* or impudence," he added that "Pond is a native of the United States, and cannot therefore be deemed to be attached to this Country. He also pretends to the Sovereignty of the Lands adjacent to the Arathapeskow Lake, so that by encouraging him we may be fostering a viper in our bosom." Far better, Dalrymple wrote, to turn to "the information of Mr. Turnor whom the Hudson's Bay Company have sent into those parts and from whose Astronomical abilities we may reasonably expect competent Information."[64]

Dalrymple's memorandum does much to explain Philip Turnor's return to North America in 1789. He was sent back to prepare for a multi-year survey in the Athabasca region.[65] In his journals of the survey, Peter Fidler noted that "our sole motive for going to the Athapescow is for Mr. Turnor to survey those parts in order to settle some dubious points of Geography, as both Messrs Hearne and Pond fixes those places in their respective maps far more to the westward than there is good reason to think them."[66] (Turnor and Fidler always incorrectly assumed that the lake Samuel Hearne called "Athapuscow Lake" [Great Slave Lake] was Lake Athabasca.[67])

While he was back in North America Turnor was to train promising HBC servants in the art and science of surveying. After rejecting Thomas Stayner and George Hudson as potential students, Turnor trained David Thompson during the winter of 1789–90. David Thompson was an apt choice. An apprentice acquired from the Grey Coat Hospital School in Westminster (London), Thompson had already had some training in surveying. But Turnor feared that it was unwise to take Thompson with him on the Athabasca expedition. One of Thompson's eyes had become inflamed, and Turnor worried that Thompson had not yet recovered sufficiently from a broken femur that he had suffered in 1788. Thus, between June and September of 1790 he also trained Peter Fidler.[68] Fidler, who had joined the HBC in 1788, also appears to have been well schooled. In 1789 Joseph Colen had described him as "a good Scholar and Accountant" and "a sober steady young man."[69] Between September 1790 and July 1792, Turnor, with Fidler and Malchom Ross, surveyed the rivers and lakes between Cumberland House and Great Slave Lake, via Île-à-la-Crosse, the Methye Portage, and Lake Athabasca (see figure 4.1).

Turnor's and Fidler's surveys of Lake Athabasca were connected with George Vancouver's planned explorations to identify and map the navigable inlets of the Pacific Coast – explorations that were being delayed by the Nootka Crisis (1789–90 [during which Spain and Great Britain disputed each other's claims to possession of portions of the northwest coast of North America]) even as Turnor, Fidler, and Ross travelled the Athabasca Country. The surveys of Turnor, Fidler, and Ross confirmed Dalrymple's suspicions that Pond had placed Lake Athabasca too far west. Furthermore, Vancouver found no long inlet or large navigable river penetrating into the interior of North America from the Pacific Ocean (his maps did not even include the Fraser or Skeena Rivers). The results of Vancouver's expedition, Alexander Mackenzie's 1789 exploration of the Mackenzie River, and Mackenzie's 1793 expedition to the Pacific Ocean later offered further evidence that no commercially practical cross-continent route existed in Lake Athabasca's latitudes.

The HBC did not keep the results of its surveys secret. Turnor's observations and some subsequent surveys (including Fidler's observations of the location of Buckingham House) were published in 1794 when Turnor was back in London.[70] The directors of the HBC could have taken great pleasure in the degree to which their surveys must have embarrassed the partners of the NWC. Pond's 1785 map was discredited. Furthermore, notwithstanding the "sole motive" of Turnor and Fidler's 1790–2 expedition purportedly being related to "dubious points of geography," and notwithstanding the NWC's lack of resistance to the expedition on those grounds, the knowledge gained by Turnor and Fidler during this survey was subsequently crucial to Peter Fidler's unsuccessful efforts between 1802 and 1806 to establish the HBC's permanent presence in the Athabasca country.

Turnor left Rupert's Land for the last time in the fall of 1792, but the company's contributions to the British cartography of North America picked up only after he left. The HBC retained Turnor for at least three years as he compiled his information and drew maps for the company, including a particularly valuable map depicting "Hudson's Bay and the Rivers and Lakes between the Atlantick and Pacifick Oceans" for which the Governor and Committee decided to pay him £100 on 14 January 1795.[71] After that, Turnor appears to have left the company, but during a transitional period he may have worked with Aaron Arrowsmith, who became, beginning in 1795, the company's unofficial cartographer. Arrowsmith, who emerged as the pre-eminent cartographer in London, produced frequently updated and oft-copied maps of North America – and the most respected maps of North America – for about fifty years beginning in 1795.[72] Turnor's surveys from between 1778 and 1792, and other HBC maps, were incorporated into Arrowsmith's first great map of northern North America published in January 1795 – a map that Arrowsmith dedicated to the Governor and Committee of the HBC.[73] These facts suggest that Arrowsmith's cooperation with the company began before 1795.[74] An updated map, issued later in 1795, added material based on Hearne's journey to the Arctic Ocean.[75] Richard Ruggles was correct, then, when he argued that beginning in 1795 "geographical information was regularly transferred from company sources to the Arrowsmith drafting tables and hence to other cartographers and the public."[76] After the publication of Hearne's journals and Arrowsmith's map in 1795, European explorations for the Northwest Passage ceased until after the end of the War of 1812. This must have been highly satisfactory for the Governor and Committee of the HBC.

With Turnor's final departure from Rupert's Land in August 1792, the surveyor's mantle was cast upon his two young protégés, David Thompson and Peter Fidler. Having been left behind when Turnor and Fidler headed north in 1790, David Thompson worked hard to capture the positive notice of the London Committee. Thompson began by surveying the route between Cumberland House and York Factory during his trip to the bay in the summer of 1790, and even before Turnor and Fidler had left Cumberland House, he wrote to the committee from York Factory, asking for surveying equipment in lieu of a set of clothes and offering to make surveys of the coast of Hudson Bay.[77] By taking the initiative, Thompson had attracted the positive notice of the committee. So, in the fall of 1792, despite Turnor's return to England, the HBC had two surveyors in Rupert's Land. That fall, William Tomison, inland chief, assigned Peter Fidler to survey the entire route from York Factory to Buckingham House, and to accompany some Peigan as they travelled southward from Buckingham House. Fidler made surveys as far southwest as the upper reaches of the Oldman River, near the present-day Alberta-Montana border. Given that it had become clear that no practical overland transportation route across North America existed north of the North Saskatchewan River, this assignment suited the desires of the HBC and London geographers. Joseph Colen, the resident chief at York Factory, was more interested in finding a shorter route to the Athabasca. So, during the same fall, he sent David Thompson northward to discover whether there was a more direct practical river route from York Factory to Lake Athabasca than the one used by the Montreal-based Canadian traders (the one Turnor and Fidler had surveyed).[78]

The following several years appear to have been frustrating for David Thompson. Thanks to various distractions and delays, he was not able until the summer of 1796 to confirm that no feasible trade route existed through the so-called Muskrat Country directly west of Hudson Bay between York Factory and Churchill. The HBC also made only modest use of Fidler's surveying skills in the same period, although there is no evidence that Fidler was bothered by the fact. After returning from his winter with the Peigan in 1792–3, he surveyed the North Saskatchewan River upstream from Buckingham House to the site where Edmonton House was to be built. Then, for two years, although he took lunar distances to determine the longitude of York Factory, he did not go on any surveying expeditions. Instead, he was assigned to York Factory until the fall of 1795. Then in September 1795 Fidler surveyed the Swan River district, just west of northern Lake Winnipegosis, where the HBC was just beginning to establish new posts.[79] After the spring of 1796 Fidler went on no surveying expeditions until 1799, although he did take lunar distances to determine the locations of Buckingham House (1796–7) and Edmonton House (1797) during the course of his travels.[80] By that time, Fidler was the only surveyor the company still had.

The HBC's commitment to surveying and exploration may have been meagre, but its efforts to pass on the results of its surveys and explorations were impressive. Fidler conveyed the information he gathered during his surveys of 1792–3 and 1795–6 to the Governor and Committee in the form of maps and journals, and the company obviously submitted them to Aaron Arrowsmith quickly thereafter. Although Fidler's original maps have since disappeared, Arrowsmith used them to update his map of North America, and, in an odd insertion, Arrowsmith explicitly included Fidler's route of 1792–3 on that updated map (figure 4.4 and plate 4).[81] Thompson's pre-1795 maps and surveys appear to have been incorporated into Turnor's 1794 map, which was the basis of Aaron Arrowsmith's map of 1795.[82]

For the few years after 1795, the London Committee appears to have been ill informed about Fidler and Thompson's activities. The committee wrote on 31 May 1797 that "it is with great astonishment that neither Mr. D. Thompson nor Mr. P. Fidler has sent us their Journals for two Years past nor their remarks to their different Charts."[83] The committee must have been unaware that Fidler had in fact done little surveying or mapping since 1793, but had been occupied during those years with commanding trading posts. In fact, the committee probably did have the official post journals completed by Fidler (but was unable to keep track of all of the journals that arrived each year). Thompson had done more surveying than Fidler had, but he may have submitted all that he could. Perhaps he had submitted the map of his 1794–5 surveys without any observations or remarks.[84] It would have been very difficult for him to have delivered the journals of his expedition to Lake Athabasca, carried out in the summer of 1796, to either Churchill or York Factory before the ships of 1796 returned to London. So, as of May 1797,

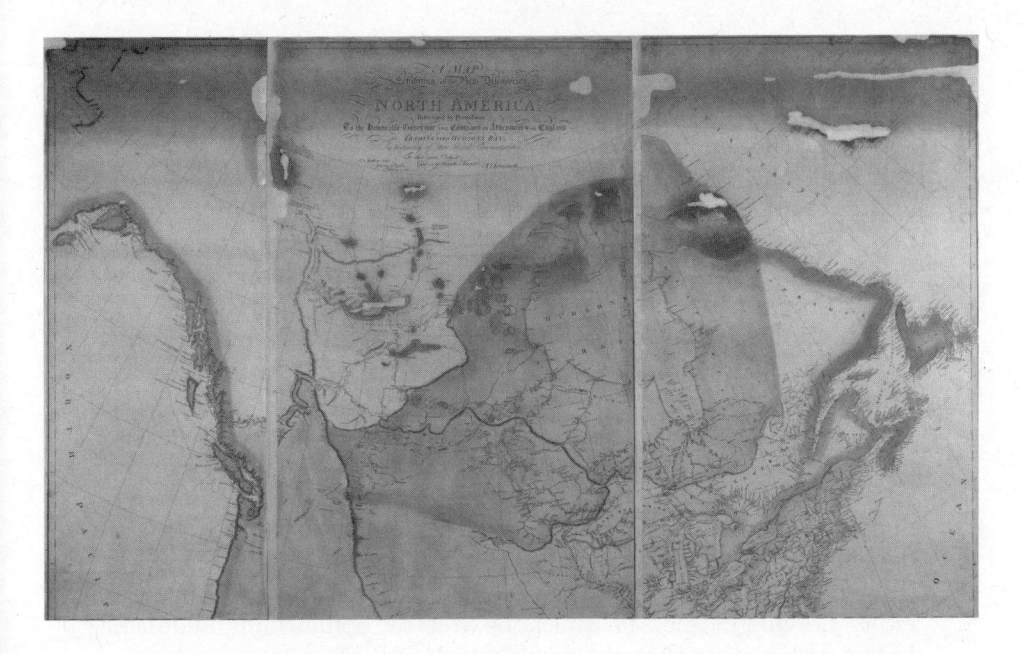

4.4 "A Map Exhibiting all the New Discoveries in the Interior Parts of North America, Inscribed by Permission to the Honorable Governor and Company of Adventurers of England Trading into Hudson's Bay in Testimony of their Liberal Communications to their Most Obedient and Very Humble Servant A. Arrowsmith." by Aaron Arrowsmith, 1795, with additions to 1802 (122 × 146.5 cm; Sections 1–6). The original 1795 edition of Aaron Arrowsmith's map incorporated the results of the surveys of Samuel Hearne, Philip Turnor, and Peter Fidler. This updated 1802 edition added much new information received from the HBC. Arrowsmith's frequently updated maps were among the most respected maps of North America at the time, thanks in part to the fact that the HBC's Governor and Committee continually supplied Arrowsmith with new information. This helps explains why Arrowsmith dedicated successive editions of the map to the Governor and Committee of the HBC. Note that the locations of the mouths of the Mackenzie and Coppermine Rivers were placed much farther north than their true locations. For more detail of this map, see colour plate 4. Source: HBCA G. 3/672.

the HBC may have had little justification for complaint against Fidler or Thompson. It soon did have more legitimate cause to be annoyed with Thompson. Unbeknownst to the Governor and Committee, he had deserted the HBC for the NWC on 21 May 1797, ten days before they wrote their letter complaining of his failure to submit his reports.[85] Furthermore, Thompson appears never to have submitted the journals of his 1796 expedition to Lake Athabasca to the company's directors in London. The fact that the HBC actually had Fidler redo those surveys in 1807 supports this conclusion.

With Thompson's desertion, Peter Fidler became the company's only remaining surveyor. In 1799–1800 he surveyed the route between the Churchill River and Lac La Biche, and between Greenwich House and the confluence of the Athabasca and Lesser Slave Rivers. He also climbed a tall tree from which he thought he saw the banks of the Lesser Slave Lake, about whose existence he had been told. The results of these surveys appeared on Arrowsmith's 1802 map (figure 4.4).[86] By the time he completed that survey, it was understood that no commercially practical Northwest Passage existed, and that no two navigable rivers, flowing in opposite directions from the continental divide and linked with a short portage, existed north of the South Saskatchewan River. James Cook's explorations had shown how far west the outer coast of North America was, and George Vancouver's explorations had found no deep inlet or large navigable river along the northwest coast of North America (his expedition failed to note the Columbia, Fraser, and Skeena Rivers.) Results of explorations from the east were no more encouraging. Turnor and Fidler had shown that Lake Athabasca was much farther east than Peter Pond had placed it. Alexander Mackenzie's explorations also showed that no rivers offered practical communication between Lake Athabasca and the Pacific. Everyone also knew by 1799 that neither the Athabasca nor the North Saskatchewan River would fill the bill. The little-understood South Saskatchewan River was the northernmost candidate left. That, more than anything, explains why Fidler was sent to establish Chesterfield House in 1800.[87]

The South Saskatchewan River was another disappointment. It was too shallow and it passed through northern Great Plains territory that was judged too dangerous (because of the populous and well-armed plains peoples) to permit its use as a major transportation route. Fidler returned to Chesterfield House for a second season, but he abandoned it in 1802.[88] In 1801 and 1802, he collected maps drawn by several aboriginal people that depicted the western plains from the North Saskatchewan River to New Mexico. These suggested that no navigable river routes existed across the continent. One of those maps was conveyed to Aaron Arrowsmith, who incorporated it into his 1802 map of North America.[89]

Between 1802 and 1806 Fidler spearheaded an attempt by the HBC to expand its operations into the Athabasca country – an attempt thwarted by the aggression of the NWC. During that time, Fidler did little surveying, although this was the initiative that depended so heavily upon the Turnor and Fidler surveys of 1790–2. Then, from June to August 1807 he surveyed the route between Cumberland House and Lake Athabasca via Frog Portage.[90] Turnor and Fidler's confirmation that Lake Athabasca was much farther east than Peter Pond had placed it meant that it would surely be difficult to find an easy route between Lake Athabasca and the Pacific Ocean, but that it might be feasible to develop a transportation route between Churchill and Lake Athabasca. Fidler was ordered to explore that possibility (essentially repeating Thompson's survey of 1795 and 1796). He travelled via Frog Portage to a lake "which I have denominated Lake Wollaston in Honour of Mr Wollaston a member of the Committee of the Honourable Hudson's Bay Company," and arrived at Lake Athabasca on 29 June 1807.[91] Unfortunately for Fidler and the HBC, his surveys showed that the most practical transportation route between Hudson Bay and Lake Athabasca was the route via the Methye Portage (see figure 4.1) that had been pioneered by the Canadians in the 1770s.

During the winter of 1807–8 Fidler was stationed at Swan River. He evidently used his time to draft some maps, for in 1808 Fidler sent to the Governor and Committee, "a 12 Sheet Map of the Communication between ... Frog Portage and the Athapescow Lake ... also the Track from the Lower End of Deers River, down thro' the Missinnipe or Churchill River to that Factory," and a small-scale version of the same map on one page.[92] The twelve-sheet map has disappeared. Ruggles was probably right when he wrote that "It seems logical to assume that the multi-sheet version was turned over to Aaron Arrowsmith, for the details are inserted on his maps of British North America, at least those after 1811. Although Arrowsmith may have not returned the multi-sheet map to the company, the smaller version, at least, remained at Hudson's Bay House."[93]

After wintering at Swan River in 1807–8, Fidler surveyed the rivers from the Swan River, over to and down the Qu'Appelle River to the Assiniboine and Red Rivers to Lake Manitoba, and then up the east shore of Lake Winnipeg to the entrance of the Nelson River in 1808.[94] Fidler stayed with the company for another fourteen years after these surveys, and he continued to try to determine the locations of the posts at which he worked, but he never again conducted any major surveying expeditions. However, Fidler made one more major contribution to British mapping. During his furlough in England in 1811–12, he drafted a series of regional maps which he submitted to the London Committee in March 1812.[95] Unfortunately, these regional maps – perhaps the closest Fidler came to David Thompson's magisterial map of the northwest – no longer exist. Ruggles has suggested that "the company likely turned them over to Arrowsmith for his use, as was its custom at the time."[96]

Fidler had no special instructions from the company to gather or submit scientific specimens or observations aside from his surveys. However, his scientific interests were broad, ranging from astronomy and meteorology to natural history, and he wrote many more documents that would have been of considerable interest to intellectuals in his time, some of which he did submit to his employers. However, since there is no evidence that any of these documents were conveyed to learned people, and since the focus of the present study is on knowledge that *was* shared, the content of those documents are of little relevance here. They are worthy of note, however, because they hint at the significant amount of scientific research carried out by HBC traders that was never shared with contemporary savants.[97]

Aside from Thompson and Fidler, several other HBC traders supplied maps and survey material that eventually contributed to British maps. Joseph Howse became in 1810 the first HBC officer to cross the Rocky Mountains. He travelled along the west side of the Rocky Mountains as far south as the vicinity of Flathead Lake, and even into the upper Missouri River region (in present-day Montana). Unfortunately, he did not take any surveying instruments (he probably did not have the requisite training) but he provided several maps that may account for the portrayal of the river systems to the west of the Rocky Mountains on the 1814 and 1818–19 editions of Arrowsmith's map.[98] Howse distinguished himself years later as a linguist of some note. In the 1830s, some years after his retirement from the HBC in 1815, he secured the patronage of the Royal Geographical Society and the Church Missionary Society to publish a remarkable grammar of the Cree language.[99] He also subsequently published vocabularies of several indigenous languages in the 1850s.[100] Another HBC employee, James Clouston, provided maps of the region east of Hudson Bay and north of the Eastmain River that were incorporated into Arrowsmith's maps.[101] In 1814, the company began requiring annual district reports, which were to include maps of each district. Few of the company's employees were actually up to the task, but the company did evidently see the initiative as a way to fill in gaps in Arrowsmith's maps, and at least one resulting map – Donald Sutherland's 1819 map of the Berens River district – influenced the content of subsequent editions of Arrrowmsith's map.[102]

Scholars have yet to acknowledge the significance of the HBC's contribution to the mapping of much of the North American interior. Between 1845 and 1850, when writing his memoirs, David Thompson defended his 1797 decision to desert the HBC and join the NWC by writing:

> how very different the liberal and public spirit of this North West Company of Merchants of Canada; from the mean and selfish policy of the Hudson's Bay Company styled Honorable; and whom, at little expense, might have had the northern part of this Continent surveyed to the Pacific Ocean, and greatly extended their Trading Posts; whatever they have done, the British Government has obliged them to do.[103]

On one hand, Thompson pointed to obvious facts. The HBC did not maximize the use of its surveyors when he worked for the company. There is no doubt that between 1778 and 1821 the NWC supported many more spectacular explorations than did the HBC. Also, at the time that Thompson deserted the HBC, the HBC's generosity with its geographic and other knowledge had only recently become evident. When Thompson made his move, he probably did not know that Hearne's journals and Arrowsmith's map had been published. But when Thompson actually wrote his words, it was abundantly obvious that the HBC had been much more liberal and public with the results of its surveys than the NWC had ever been.[104]

The fact is that the HBC and NWC each supported the kinds of exploration and travel that suited its own circumstances. The profitability of the NWC (and other Canadian companies) depended on it being able to operate without competition in fur-rich regions. Some of the profits from operating in those regions were necessary to subsidize the company's operations in other areas where it faced competition. So, the NWC needed to support risky explorations such as those undertaken for it by Peter Pond, Alexander Mackenzie, David Thompson, and Simon Fraser in order to outpace its competition. Given the great strategic value of this knowledge, and the cost at which it was obtained, it is no wonder that the partners of the NWC shared little of the knowledge gained and maps drawn by its men, especially after its decision to submit maps in 1785 failed to reap the intended results.

The HBC was a very different organization from the NWC. Governed by a London Committee that never set foot in North America, the company depended upon accurate information set down on paper to manage its operations. But thanks especially to its lower operating costs, it was far easier for the HBC than for its Montreal-based competition to operate profitably in a competitive environment. Thus, it was unnecessary for the HBC to undertake the risk and expense of exploring unknown lands. Instead, the HBC's efforts were focused on what one historian of science has described as "scientific travelers": those who "measured exactly what explorers had reported inaccurately."[105] The HBC hired its first scientifically trained surveyor in 1778, almost two decades before the NWC hired its first (and the NWC acquired David Thompson in 1797 only when Thompson deserted the HBC). Most of the work of Philip Turnor, Peter Fidler, and David Thompson (during the time Thompson was with the HBC) was aimed at surveying what had been explored, but was only imprecisely understood.

Although the HBC's explorations were considerably less spectacular than the NWC's, its efforts between 1778 and 1821 were considerable nonetheless. As historian of science Raymond Stearns argued,

> the cooperation of the Hudson's Bay Company materially assisted Fellows of the Royal Society and their scientific colleagues throughout the world to improve their knowledge of geography, to correct maps of the Hudson's Bay area and the Arctic regions, to enlarge their knowledge of oceanography, astronomy, meteorology, natural history, and their information regarding the native peoples of the New World – all of which added priceless increments to learning as a whole, to the problems of classification and nomenclature of the natural production of the Arctic regions of North America, and to the formulation of natural laws which regulate them.[106]

The careers of men such as Turnor, Fidler, and Thompson (while he was in the HBC employ) reflect the HBC Committee's belief that exploration and mapping were always to be subservient to the company's commercial interests. Although they were surveyors, in the eyes of the company they were always traders first. But after his career in the fur trade ended – notwithstanding his statements to the contrary – Thompson must have been acutely aware that the results of Turnor and

Fidler's surveys had been conveyed to the pre-eminent geographers and cartographers in London.

Contrary to Thompson's assertion, the HBC exhibited a more "liberal and public spirit" than the NWC. His (and the NWC's) explorations and surveys were more extensive than Fidler's (and the HBC's), but for most of his career Thompson explored, surveyed, and mapped for the NWC. Not only were the NWC's Montreal headquarters less well located than the HBC's to facilitate cooperation with British cartographers, but the NWC was necessarily far more secretive with its surveys and maps than was the HBC, and as an unchartered company, the NWC did not face the same level of public and political scrutiny and expectation that the HBC did.

There is no evidence that the NWC willingly shared its geographical knowledge with people outside the company after 1785. It is not surprising that the company showed no interest in sharing Thompson's surveys and maps with outsiders. This was especially because Thompson's explorations, surveys, and maps represented one of the partnership's significant strategic advantages over the HBC, who explored considerably less.

Individual Nor'Westers did share some of their knowledge with British cartographers. For example, when Alexander Mackenzie published his journals in 1801, they included a map drawn by Aaron Arrowsmith of Mackenzie's 1789 and 1793 explorations. Some of Thompson's surveys appear to have been included on that map, which, of course, were thereafter incorporated into the 1802 edition of Aaron Arrowsmith's map of North America. In that way, the results of at least some of Thompson's explorations and surveys were disseminated, although it is not clear whether Thompson cooperated with Mackenzie or not. And Mackenzie, hardly the most loyal Nor'Wester (he officially defected to the XY Company, a Canadian competitor to the NWC, in 1802 after cooperating with them before that), almost certainly shared his knowledge with Arrowsmith on his own initiative. During their expedition, Lewis and Clark possessed a 1798 map by David Thompson, and perhaps other geographical knowledge about the Missouri bend area, but it is unclear how they got it.[107] The results of some of Thompson's surveys also appear to have gotten to British maps through Peter Fidler, perhaps without Thompson's collaboration. For example, Fidler collected a map from the Canadian Jacques-Raphael (Jaco) Finlay, who had blazed a trail across the Howse Pass and built a canoe in the Pacific drainage for David Thompson in 1806.[108] Fidler also collected the results of some of Thompson's surveys. For example, he once noted that "Mr Thompson fixes the source of the Bad river in 50° 57' & 115° W."[109] Fidler also drew a map in 1812 (now lost) entitled "The Rocky Mountains from Acton House to Howse's House and Great Fall, based on James Whitway and David Thompson," that Arrowsmith must have used (together with information supplied by Joseph Howse) to fill in gaps in the 1814 edition of his map of North America that also incorporated the results of Lewis and Clark's explorations.[110]

Generally speaking, it seems that little of Thompson's work left the possession of the NWC before 1817, and it is difficult to know how British cartographers got hold of the little that did. After Thompson had retired to Canada, he completed his "Great Map," commissioned by the NWC, for which he is so famous. This map was entitled "Map of the North-West Territory of the Province of Canada from actual Survey during the years 1792–1812." There were at least three versions of this map. The first appears to have been delivered in June 1814. Updated and corrected versions followed. One copy, likely the first, was hung in the Great Hall of the NWC's building at Fort William.

As Barbara Belyea argued, the fact that Thompson wanted to sell his own maps suggests "that he wanted to compete, not collaborate, with the London cartographer [Arrowsmith]."[111] Between 1814 and 1817, Thompson decided that he would try to publish a copy of his great map. To that end, he tried to sell subscriptions to the map. That is when disaster struck for him. The HBC took at least one of Thompson's maps when it seized the NWC's Fort William in 1816–17, and conveyed that map to Arrowsmith in London.[112] In 1817 Thompson sent a version of his map to England, in the hopes of publishing it. But it was too late. Arrowsmith had already updated his map using information from the map that the HBC had confiscated. So, portions of Thompson's map did contribute significantly to Arrowsmith's 1817 map, but for much of the western interior, Thompson's information duplicated information that Arrowsmith had previously acquired from the HBC, and Thompson did not get credit for the new information incorporated into that map.

The NWC and its explorers deserve credit for being the first to explore and survey vast areas of the western North America. Thompson also deserves credit for his impressive map of the western interior of the continent. However, his maps made relatively little impression on British knowledge until 1817, and even in 1817, he was given little credit for the information he did provide. It was not until the twentieth century that historians and the public really began to appreciate David Thompson. Thompson is justly renowned for his remarkable map of the northwest – certainly a more impressive map than any map that Fidler (or any other HBC man) could have drawn of the same area. Nonetheless, when Fidler died in 1822, his impact, and the impact of Turnor and the HBC, on the British cartography of British North America was considerably greater than the impact of Thompson and the NWC.

Thompson's surveys take on an additional and ironic significance because on at least one occasion, when Thompson's remarkable surveys for the NWC were recognized, the HBC sometimes got credit for them. In 1859, when the HBC was again under some public scrutiny, the president of the Royal Geographical Society Sir Roderick Murchison published an address about the expedition of John Palliser that included a laudatory summary of the work of David Thompson. Speaking about knowledge of some of the passes in the Rocky Mountains, Murchison noted that the British Member of Parliament and director of the HBC, Edward Ellice, had informed him "that the geographical position of these passes was laid down many years ago upon a MS. Map, at the instance of the Hudson Bay Company, by Mr. David Thompson."[113] Murchison carried on implying that the entire career of "this great but little-known explorer" was spent in the service of the HBC.[114] It is impossible to know whether Ellice knowingly misled Murchison about Thompson's true relationship with the HBC, but few in 1859 knew enough about Thompson to set the record straight.

Just as had been the case in the 1750s, the British intellectuals who had benefitted from their association with the HBC were generous with their tributes. Thomas Pennant, like his "very intimate" friend, George Edwards, was among the naturalists whose reputation depended heavily on his efforts to describe the fauna of northern North America.[115] Pennant first became prominent for his *British Zoology*. After publishing its third volume in 1766, he became, by his own account, "desirous of forming a zoology of some distant country."[116] Although he long had his heart set on writing a zoology of North America, the American Revolution deflected his attention instead to the Arctic. The result was the multi-volume *Arctic Zoology*, which earned him, perhaps ironically, an election to the American Philosophical Society at Philadelphia.[117] Even half a century after it was published, John Richardson would describe *Arctic Zoology* as the "fullest account of the birds of Arctic America which has hitherto been published."[118] Thus, a substantial portion of Pennant's long-lasting renown arose from help he got from the HBC.

Pennant duly inserted a note explaining that "at the time this sheet was printing, I had the good fortune to meet with *Mr. Hutchins*, surgeon, a gentleman many years resident in *Hudson's Bay*; who, with the utmost liberality, communicated to me his MS. observations, in a large folio volume; in every page of which his extensive knowledge appears. The benefits which this work will, from the present page, receive, is here once for all gratefully acknowledged."[119] At the beginning of the second edition, Pennant paid tribute in detail:

> To the Late Mr. THOMAS HUTCHINS, a gentleman greatly distinguished for his philosophical enquires, I was unspeakably obliged for his judicious remarks made during sixteen years residence in *Hudson's Bay*, of which he most liberally indulged me with the perusal.
>
> To Mr. SAMUEL HEARNE, the great explorer by land of the *Icy Sea*, I cannot but send my most particular thanks, for his liberal communication of many zoological remarks, made by him on the bold and fatiguing adventure he undertook from *Hudson's Bay* to the *ne plus ultra* of the north on that side.
>
> Mr. ANDREW GRAHAM, long a resident in *Hudson's Bay* obliged me with numbers of observations on the country, and the use of multitudes of specimens of animals transmitted by him to the late Museum of the Royal Society, at the instance of that liberal patron of science, my respected friend the Honorable DAINES BARRINGTON.[120]

Pennant's social intelligence was obviously very much appreciated. In his own journals, published in 1795, Samuel Hearne wrote of "my respected friend Mr. Pennant, who with a candour that does him honour, has so generously acknowledged his obligations to all to whom he thought he was indebted for information when he was writing his Arctic Zoology."[121]

Cartographers and others interested in exploration were as generous with their tributes as was Pennant with his. The exertions of men like Captain William Christopher, Moses Norton, and Samuel Hearne had been enough to win the company support from some influential people. In his introduction to the very popular official journals of James Cook's expeditions, published in 1784, John Douglas wrote that

> if obstructions were thrown in the way of Captain Middleton, and the Commanders of the Dobbs and California, the Governor and Committee

of the Hudson's Bay Company, since that time, we must acknowledge, have made amends for the narrow prejudices of their predecessors; ... every thing has been done by them, that could be required by the Public, toward perfecting the search for a North West passage.[122]

Elsewhere Douglas wrote that "Mr. Wegg besides sharing in the thanks so justly due to the committee of the Hudson's Bay Company, for their unreserved communications, was particularly obliging to the Editor, by giving him repeated opportunities of conversing with Governor Hearne, and Captain Christopher."[123] Subsequently, Aaron Arrowsmith, who had been given privileged access to HBC information, dedicated successive editions of his "Map Exhibiting all the New Discoveries in the Interior Parts of North America" to the HBC, "In testimony of their liberal Communications."[124] In 1789, Dalrymple wrote that "if ever a charge could have been made with justice against that Company for mysterious concealment, nothing of this nature can be imputed to the Present Managers."[125] And when Hearne's journals were published in 1795 the introduction defended the HBC by asserting that "that air of mystery, and affectation of secrecy, perhaps, which formerly attended some of the Company's proceedings in the Bay, might give rise to those conjectures, and the unfounded assertions and unjust aspersions of Dobbs, Ellis, Robson, Dragge, and the American Traveller, the only Authors that have written on Hudson's Bay."[126] Hearne had become one of the company's defenders. Others were convinced. When the *Gentleman's Magazine* reviewed Hearne's journals in 1796, it stated that "our readers are no strangers to the merits of Mr Hearne in the line of discovery, or to the exertions of his employers, the Hudson's bay company, in promoting discoveries, or to the misrepresentations of them by travellers and navigators."[127] It was another friend of the company, John Reinhold Forster, who helped spread the reputation of the HBC internationally when he published a German translation of Hearne's journals in 1797 and a Dutch translation in 1798. In 1790, Edward Umfreville published *The Present State of Hudson's Bay*, a book reminiscent of Arthur Dobbs' books, accusing the HBC of neglecting its obligation to explore Rupert's Land. And later in the same decade, partners of the NWC used similar grounds to appeal to the British government to revoke the HBC's charter.[128] But the HBC had positioned itself well enough that the company easily weathered the storm. Thanks to their contributions to science between 1774 and 1821, particularly to

British cartography, the HBC possessed public tributes that went far towards addressing the very accusations that its critics had made in previous decades: its secrecy and its lack of commitment to exploration. The leadership of the HBC understood that knowledge shared could be as useful for the company as knowledge hoarded.

PART II

The Hudson's Bay Company and Science, 1821–1870

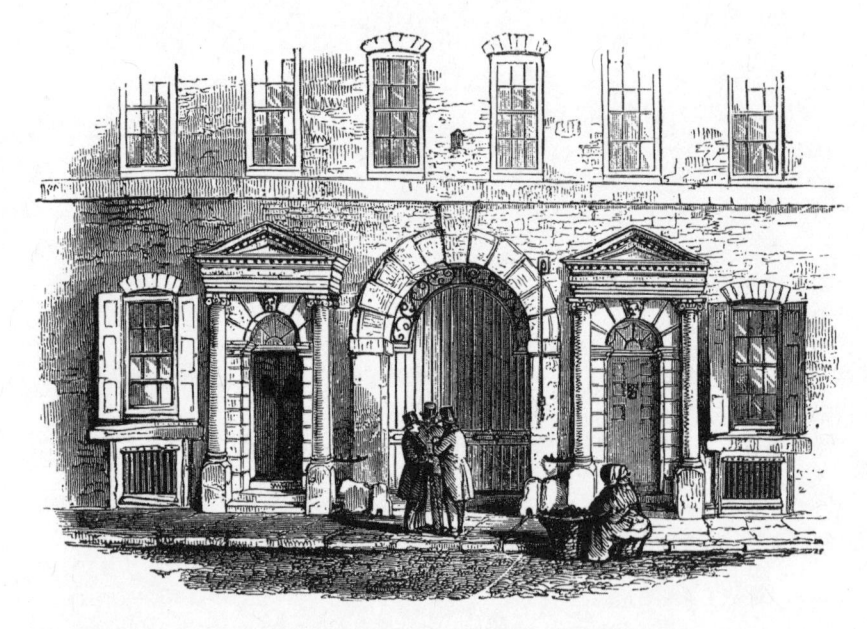

Hudson's Bay Company's House, Fenchurch Street, 1843. This wood engraving of Hudson's Bay House on Fenchurch Street was made by John Jackson (1801–48) for publication in the sixth volume of Charles Knight's guide to London, published in 1844. Source: Look and Learn

5 "Benevolent Intentions": The Hudson's Bay Company, the Royal Navy, and the Search for the Northwest Passage, 1818-1855

If the Transit of Venus of 1769 marked the first great turning point in the history of science in the HBC, 1821 marks a second. After 1821, the HBC's circumstances were dramatically different than they had been between the 1760s and 1821. During the years before 1821, commercial competition between the HBC and NWC became an increasingly important and costly preoccupation of the directors of the HBC. After 1821, the HBC faced significant commercial competitors only along the fringes of its vast territories. The lack of competition meant that after a few years of adjustment, the company enjoyed unprecedented prosperity between 1825 and 1863.[1] But the company was not invulnerable. Between 1821 and 1856 the growing strength of liberalism (as well as the emergence of a humanitarian movement in the 1820s) in Great Britain created an economic and political climate increasingly hostile to monopolies. Under these circumstances, the company's directors understood the importance of cultivating the sympathies of officials, politicians, and other influential people.[2] Supporting, and being seen and recognized to support ostensibly disinterested scientific activity was one way in which the company created a positive corporate image in Great Britain.

The HBC's participation in the search for the Northwest Passage between 1818 and 1838 illustrates important aspects of the company's strategy towards science. The paramount importance of being perceived to be a patron of science and exploration presented the HBC directors with a dilemma. Historian of science Trevor Levere has hinted at this quandary by noting that "science was often a spur to arctic exploration, but not always an aid to it; and more than once, the demands of scientific research and those of geographical discovery came into conflict."[3]

From the perspective of the directors of the HBC, the efforts of the British Navy in overland arctic exploration garnered the Admiralty unjustified accolades from the British government and public. The praise may have been attributable to the emphasis placed – before they left England and when their results were published – on the accomplishment of the scientific aims of those expeditions, and to the evident difficulty of travelling through the harsh environment of the North American subarctic and arctic. Fortunately for the HBC, the company itself earned praise for its cooperation with these expeditions. Still, the mediocre precedent set by the Royal Navy in the 1820s and 1830s presented the HBC with a golden opportunity to outperform the Admiralty in the realm of exploration, particularly if it subordinated the scientific objectives of its expeditions to their exploratory goals. But science could not be ignored. The conduct of the HBC Governor and Committee suggests that their strategy was to relegate the scientific dimensions of their expeditions to a second priority in order to ensure the success of the exploratory aims, and yet to realize maximum benefit from these journeys by emphasizing publically the scientific results of their expeditions. In this regard, the company was remarkably successful.

Evidence of how the merger of the HBC and NWC improved the capacity of the HBC to contribute to science and exploration can be found in the HBC's involvement in the search for the Northwest Passage and its support for science, which formed an important auxiliary to that search. The year 1818 marks a significant turning point in the history of arctic exploration, because in that year the Northwest Passage, the north magnetic pole, and the arctic more generally, became the renewed focus and interest of naturalists, explorers, government, and the public.[4] Despite its proximity to Europe, Europeans knew little about the North American arctic in 1818. They knew that there was an archipelago north of North America's mainland, but few of its islands had been mapped. Furthermore, along the entire mainland between Hudson Bay and Icy Cape near the Bering Strait Europeans had visited only the mouths of the Mackenzie and Coppermine Rivers, and their surveys of the mouths of those rivers were highly suspect.[5]

Because the search for the Northwest Passage attracted so much official and public interest, it offered ideal ways for the HBC to bolster its reputation among politicians and the public, and for arctic explorers individually to attain great public renown. Few by 1818 believed that the discovery of the Northwest Passage would have much commercial significance. But advocates of costly and risky expeditions emphasized other reasons for exploration, including national prestige and scientific discovery. This very well-known but ostensibly disinterested quest was an ideal one for the HBC to be seen to support, especially when exploration could contribute directly to the expansion of its fur trade operations.

The HBC did not, at any time before 1870, have trading posts on the arctic shores, and until 1821 its monopoly rights did not extend to the arctic watershed. Still, its subarctic posts and transportation system were bound to be crucial for overland expeditions headed for the arctic coast of the North American mainland. Between 1819 and 1836 the HBC lent logistical support and personnel to several expeditions led by the Royal Navy, expeditions whose contributions to science were significant despite their disappointingly small contributions to the search for the Northwest Passage, especially considering their cost in human lives. Indeed, convinced by the ineffectiveness of those expeditions, the Governor and Committee directly sponsored its own expedition between 1836 and 1839. In terms of the amount of coastline surveyed, the HBC expedition was considerably more successful than Royal Navy expeditions, although that success came by using methods – travelling lightly and fast – that undermined its ability to contribute to other branches of science.

The British had first sought the Northwest Passage via the Hudson and Davis Straits in the days of Martin Frobisher, John Davis, and Henry Hudson in the late sixteenth and early seventeenth centuries. Interest via the Hudson Strait was renewed in the eighteenth century with efforts by James Knight and Christopher Middleton. Tantalizing misinformation circulating after Middleton's expedition kept up interest, and in 1745 the British Parliament offered a £20,000 reward to the discoverer of the Northwest Passage.[6] Despite doubts about the accuracy of Samuel Hearne's report that the mouth of the Coppermine River was at 71° 54′ North, the results of his trip to the Coppermine River in 1771–2 seemed to prove that any Northwest Passage would be impractical as a shipping route. At any rate, the United States War of Independence (1776–83), followed shortly thereafter by a generation of warfare (between Great Britain's declaration of war with France in 1793 and the end of the War of 1812 in 1815), kept those in the Royal Navy sufficiently occupied so that few in the Admiralty took interest in the Northwest Passage.

But when peace returned to Europe in 1815, the British Navy confronted a large surplus of seasoned mariners. The number of seamen paid by the British government plummeted from about 113,000 in 1812 to 24,000 in 1816.[7] In order to keep as many mariners as possible employed, John Barrow, the Secretary of the British Admiralty and Great Britain's emerging champion of imperial science, campaigned for a new commitment to the search for the Northwest Passage.[8] He cast doubt again on the accuracy of Hearne's surveys. Barrow, for example, pointed out that if the mouth of the Coppermine River was really as far north as Hearne had claimed it was, the Sun could not have been as high in the sky as Hearne implied it was.[9] Hearne's evident error also reminded people of the perplexing problem of magnetism and compass variation, which became especially troublesome in the region near the north magnetic pole.

So, in 1818 those issues had become important enough once again to attract the attention of naturalists, bureaucrats, politicians, and the public. Two events of that year illustrate the breadth of British interest in the arctic. The first is the publication of Mary Shelley's *Frankenstein*. The second is that the British Parliament renewed its reward for the discovery of the Northwest Passage.[10] "Animated more by the vision of British greatness than by practical considerations of commercial and strategic gain," as scholar Franklyn Griffiths has argued, the passage had become "as much a cultural artifact as an Arctic navigation route," and "as much a metaphor for human perseverance and ingenuity as a physical reality."[11] Griffiths might well have added a reference to the importance of science in the reimagining of the Northwest Passage. As part of his campaign to gain support for the search for the Northwest Passage, in 1818 John Barrow published a history of the efforts to find the elusive strait. In his conclusion to that book he emphasized the scientific value of the search. Pointing not only to geographical questions but also to questions related to meteorology and natural history, Barrow argued that the "enterprise itself may be truly characterized as one of the most liberal and disinterested that was ever undertaken, and every way worthy of a great, and prosperous, and enlightened nation; having for its primary object that of the advancement of science, for its own sake, without any selfish or interested views."[12]

Notwithstanding Barrow's assertion, the search had – beyond yet-hoped-for commercial value – great significance for states and governments, institutions, and individuals. Indeed, only a few pages before his argument that the search was "disinterested," Barrow had noted "that the discovery of a north-west passage to India and China has always been considered as an object peculiarly British," but that "Russia ... nay, a private individual of Russia [Count Nikolai Petrovich Rumyantsev, Chancellor of Russia], has recently fitted out a ship at his own cost, for the discovery of a communication between the two oceans by a passage round North America."[13]

Otto von Kotzebue, chosen by Rumyantsev to conduct the exploration, had taken with him as official naturalist Adelbert Von Chamisso, as well as a surgeon (Johann Friedrich von Eschscholtz) with a particular interest in entomology.[14] Kotzebue returned to Europe in August 1818 with new ethnological information and specimens of several plants entirely new to science. Volumes 2 and 3 of his *Voyage of Discovery* included scientific analysis by Chamisso. In that context Barrow invited his British readers to imagine the shame if "England had quietly looked on, and suffered another nation to accomplish almost the only interesting discovery that remains to be made in geography, and one to which her old navigators were the first to open the way."[15] As Barrow's case for arctic explorations suggests, although everyone understood by 1818 that the search for the Northwest Passage was wrapped up in matters as mundane as keeping mariners employed during a period of economic and social strain in Great Britain and as lofty as the honour and prestige of the world's pre-eminent maritime nation, it was yet a search from which might flow other unanticipated practical benefits.

What role did Barrow imagine for the HBC in this enterprise? Of the HBC, Barrow repeated the then familiar history: "There can be little doubt that the Hudson's Bay Company were for a long time exceedingly jealous of their monopoly; ... but of late years the governors of this Company have liberally communicated whatever information may have been sent to them respecting the geography and hydrography of Hudson's sea and lands adjoining, as Mr. Arrowsmith can testify."[16] In a quest as prominent as the search for the Northwest Passage, both opportunity and obligation demanded that the HBC respond willingly.

The first nineteenth-century British expedition to search for the Northwest Passage was a Royal Navy expedition led by John Ross, accompanied by First Mate William Edward Parry, who would subsequently lead several of his own expeditions. Ross's expedition explored north into Baffin Bay, a great distance from HBC operations, and thus the HBC could do little to assist it. However, the published journals and reports of Ross's expedition did buttress Barrow's argument that arctic explorations had significant scientific objectives. Ross's instructions made the search for the Northwest Passage the central aim, but they stipulated "at the same time, that it may likewise be the means of improving the geography and hydrography of the Arctic Regions, of which so little is hitherto known, and contribute to the advancement of science and natural knowledge."[17] So, "at the recommendation of the President and Council of the Royal Society," geophysicist and Royal Artillery officer, Edward Sabine, who was "represented to us as a gentleman well skilled in astronomy, natural history, and various branches of knowledge," was to accompany the expedition. According to Ross's journal, Sabine was "to assist you [Ross] in making such observations as may tend to the improvement of geography and navigation, and the advancement of science in general."[18]

In the end, the scientific contributions of Ross's expedition may have been its most substantial. In terms of exploration its few new discoveries did not make up for Ross's mistaken conclusion that Lancaster Sound was a bay, but members of the expedition collected plant and animal specimens (including several species new to science) and gathered observations on aurora borealis, magnetism, navigation, geology, and mineralogy, much of which was published in appendices to Ross's privately published *Voyage of Discovery* (1819).[19] As the first British expedition of the nineteenth century to search for the Northwest Passage, Ross's expedition confirmed that the British would follow Kotzebue's lead and ensure that science and exploration would be partners in this search.[20]

The HBC's involvement with the first overland expedition of John Franklin between 1819 and 1822 showed how difficult it was for the HBC (or NWC) to contribute to science and exploration in the arctic before their merger. The fact that British maritime expeditions in search of the Northwest Passage turned north into Davis Strait and Baffin Bay meant that the HBC was not in a position to cooperate with those expeditions. In contrast, overland expeditions were utterly dependent on the cooperation of fur traders. Franklin's first expedition was sent by the Royal Navy primarily to survey the northern coastline of the North American mainland eastward from the mouth of the Coppermine River, but its scientific aims were clear and explicit. Franklin's own instructions included the taking of various scientific observations, and assisting the surgeon Dr. John Richardson, who "to his professional duties, was to add that of naturalist."[21] Richardson was instructed "to collect and preserve specimens of minerals, plants, and animals."[22] Midshipmen and artists George Back and Robert Hood were also instructed to sketch natural history.[23]

In his published journal, Franklin later explained that "I was to be guided by the advice and information which I should receive from the wintering servants of the Hudson's Bay Company, who would be instructed by their employers to cooperate cordially in the prosecution of the Expedition."[24] Those words were published in 1825, when Franklin no longer needed to cultivate the good graces of the NWC, but more accurately, the Franklin Expedition was to depend not only the HBC but also on the NWC. Franklin's expedition departed London in 1819 and travelled on HBC ships from London to York Factory. From there it continued overland with the HBC brigades via Fort Chipewyan to Fort Providence on Great Slave Lake, inauspiciously passing through one of the epicenters of HBC-NWC hostility. From Fort Providence the expedition was to move in the spring of 1821 beyond the realm in which either company operated and was to employ aboriginal guides to find its way overland to the mouth of the Coppermine River. Only then, two years after leaving London could it begin its main task: mapping the arctic coast eastward of the Coppermine River as far as possible during the short arctic summer of 1821 before returning to London. For the crucial exploratory portion of its trip, the expedition would be reliant more on the help of the NWC than the HBC, because in that northern region the NWC was far more established and experienced than was the HBC (see figure 5.1).

During most of the first Franklin expedition, the struggle between the HBC and NWC was at its peak. So although the HBC Governor and Committee and the NWC partners all agreed to cooperate with the expedition, the animosity between the trading companies interfered with their ability to fulfil their promises.[25] In fact, when the NWC captured Colin Robertson, in charge of the HBC's operations in the Athabasca country, in the spring of 1820, George Simpson had to travel to the Athabasca country to replace him. On 29 September 1820, Simpson nevertheless wrote Franklin from Fort Wedderburn, that "on the part of the Hudsons Bay Coy. I beg leave to assure you, that I shall be happy to render the Expedition every assistance in my power connected with this Department."[26] Simpson's assurances were not idly made; they were rooted in orders from the company's directors. In January 1820, the HBC Governor in Chief William Williams issued a circular pertaining to Franklin. He wrote to all officers that "I must particularly enjoin you all individually and collectively to give this Gentleman every possible assistance he may require, in supplies, Provisions, men dogs Sledges

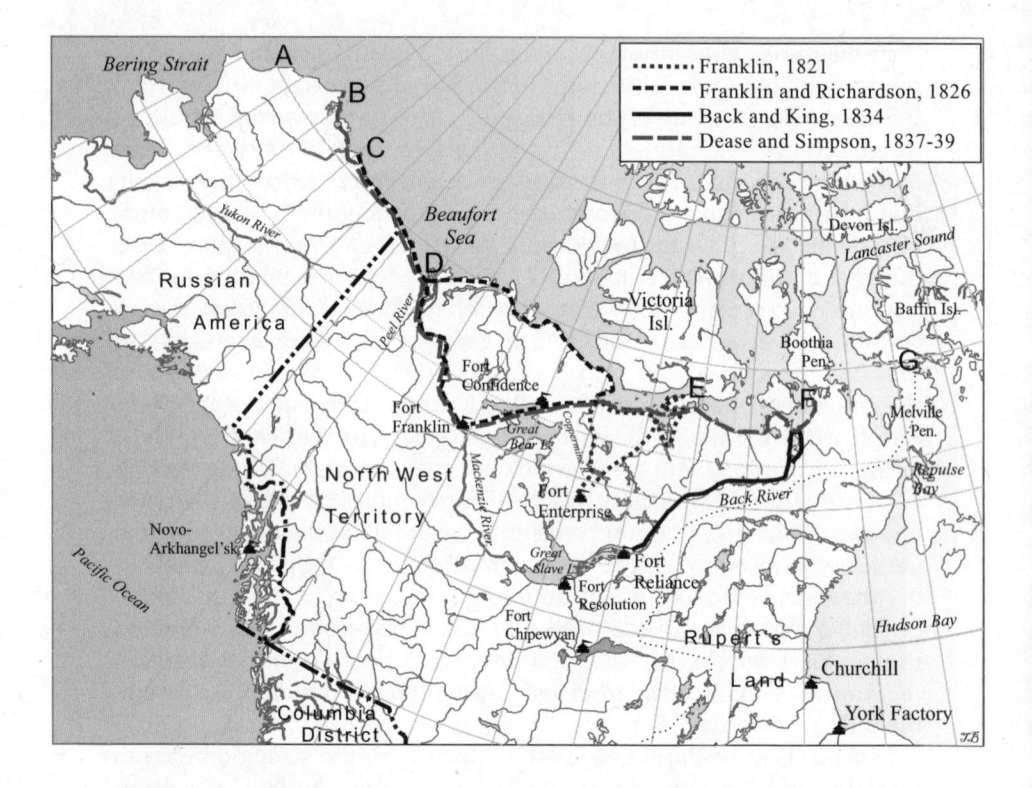

5.1 Arctic Explorations Sponsored or Assisted by the HBC, 1821–1838.
Neither Hearne's trip to the mouth of the Coppermine River in 1771, nor Alexander Mackenzie's to the mouth of the Mackenzie River in 1789 produced reliable estimates of the locations of the Arctic coast. Thus, as late as 1821 the entire coast from Icy Cape (A), surveyed by James Cook in 1778, to Fury and Hecla Strait (G) was virtually unknown to Europeans. In 1821, Franklin's first expedition surveyed the coast from the mouth of the Coppermine River to Point Turnagain on the Kent Peninsula (E). In 1826 his second expedition surveyed from the mouth of the Mackenzie River (D) to Return Reef (C), while the Beechey Expedition surveyed east from Bering Strait as far as Point Barrow (B). At the same time, Richardson surveyed eastward from the mouth of the Mackenzie to the mouth of the Coppermine. Back and King explored the Back River but surveyed only a small section of the Arctic Coast in 1834. Dease and Simpson surveyed the stretch between Return Reef and Point Barrow in 1837, and the remaining uncharted coast as far as Spence Bay (F) in 1839. Source: T. Binnema

Hunters, in fact every thing that may be deemed necessary to render as facile as possible an undertaking so arduous and interesting."[27]

Simpson's promises and Williams' orders notwithstanding, the Franklin expedition was heading into a very difficult situation. Williams had already written Franklin that "from our deficiency of men and dogs I am sorry I cannot wholly supply you for your intended journey, but in part, it will therefore be necessary that you should apply to the NW house for 2 sledges, 2 trains of dogs and two Canadians to accompany them."[28] Williams also wrote that the HBC traders were "not *to make observations respecting the trade before them* [members of the Franklin expedition] *in conversation or otherwise.*"[29] And once they got to the Athabasca country, the members of the expedition appear to have found it impossible to avoid becoming embroiled in the HBC-NWC rivalry. Whether he was correct or not, by 8 February 1821 Simpson jealously concluded that the members of the expedition, instead of being neutrals in the HBC-NWC conflict, "evince a strong party feeling and consider themselves no where at home except in a N.W. Fort."[30]

As the Franklin party prepared for its explorations during the winter of 1820-1, it became increasingly obvious that the HBC was not going to meet all of its requests. In January 1821, Simpson referred to Franklin making "a demand of a supply of Goods which I am unable to comply with."[31] Simpson defended himself to George Back in January 1821 by telling him that "my inability I trust will not be construed into an unwillingness to accommodate the Expedition, as it is not only the positive instructions of the Committee that every assistance and facility should be rendered, but to myself individually nothing could be more gratifying than having it in my Power to supply it's present exigencies."[32] To his colleagues, Simpson was more frank. On 26 January 1821 he wrote Robert McVicar at Great Slave Lake that he believed that officers in the company were jeopardizing the company's interests with their "liberal supply," and stating that George Back "seems to think that every thing must give way to his demands, ... it must however be perfectly understood that altho' the Company are anxious to meet the views of those Gentlemen, their necessities are a very secondary consideration to our own difficulties."[33]

It was during this time of tension and mistrust that George Simpson penned his famously derisive assessment of John Franklin and his expedition. On 8 February 1821, he wrote that:

> Mr. Back paid me a visit preparatory to his departure; from his remarks I infer there is little probability of the objects of the expedition being

> accomplished, not so much on account of any serious difficulties to be apprehended, but from a want of unanimity amongst themselves; indeed it appears to me that the mission was projected and entered into without mature consideration and the necessary previous arrangements totally neglected; moreover Lieut. Franklin, the Officer who commands the party has not the physical powers required for the labor of moderate Voyaging in this country; he must have three meals p diem, Tea is indispensible, and with the utmost exertion he cannot walk above *Eight* miles in one day, so that it does not follow if those Gentlemen are unsuccessful that the difficulties are insurmountable.[34]

Simpson's assessment of Franklin was not entirely fair, but the fact that the expedition did become a disaster has given Simpson's prediction an air of prescience. Aside from the fact that the HBC and NWC did not support the expedition as fully as the expedition's members had hoped, the failure of the expedition was caused by inadequate planning, inexperience on the part of Franklin and his men with overland travel and subsistence in the arctic, Franklin's poor physical fitness and indecisiveness, and bad luck. The party did map a considerable stretch of the arctic coast east of the Coppermine River, but at the cost of the lives of eleven of the twenty men, and with rumours of murder, execution, and cannibalism.[35]

The British public ignored the rumours. Franklin arrived home a hero – the most compelling example of the arctic variant of the Romantic scientific traveller as hero.[36] This fact, combined with Franklin's lavish praise for the HBC, meant that the HBC's assistance was prominently repaid in a published volume that quickly sold out: "From Joseph Berens, Esq., the Governor of the Hudson's Bay Company, and the gentlemen of the Committee, I received all kinds of assistance and information, communicated in the most friendly manner previous to my leaving England, ... and I most cheerfully avail myself of this opportunity of expressing my gratitude to these Gentlemen for their personal kindness to myself and the other officers, as well as for the benefits rendered by them to the Expedition."[37]

As Franklin's *Narrative* was being prepared for publication, Franklin already hoped to lead another overland expedition to explore more of the arctic coast of North America, so his magnanimous tribute was certainly judicious, if not wholly truthful. Furthermore, by mentioning the HBC Committee specifically, and mentioning the governor by name, Franklin was flattering the very men whose support he was seeking.

Richardson was as tactful as Franklin. An enemy of the HBC might have seized upon the fact that Franklin's expedition proved definitively that Hearne's placement of the mouth of the Coppermine River was far off (they found the mouth of the Coppermine River at 67° 48') to embarrass the HBC, but John Richardson in 1836 defended Hearne. "When we consider the hardships which Hearne had to endure," Richardson argued, "the difficult circumstances in which he was frequently placed, the utter insufficiency of his old and cumbrous Elton's quadrant as an instrument for ascertaining the latitude, particularly in the winter, with a low meridian sun, and a refraction of the atmosphere greatly beyond what it was supposed to be by the best observers of the period, and the want of any means of estimating the longitude, except by dead reckoning ... we shall not be inclined to view with severity the errors committed."[38]

Despite the fact that geological, zoological, and botanical specimens had to be abandoned during the expedition's most desperate days, the first Franklin expedition contributed to various branches of scientific knowledge. The published *Narrative* included "Geognostical Observations," and remarks on the aurora borealis by John Richardson.[39] Thus, although science was not the primary motivation behind Franklin's expedition, it did contribute to science, and in ways that garnered the HBC some high-visibility credit. But to those in the know, the expedition revealed how difficult it was for arctic explorers to depend on the HBC and NWC for assistance before the merger of the companies.

By the time Franklin returned to England from his first expedition, the HBC and NWC had already merged, placing the HBC in a much better position to render assistance to Franklin's second expedition. In arguing for the need for another expedition, Franklin in 1823 predicted that such an exploration would be valuable in "advancing geographical knowledge and the fur trade, as well as preventing the encroachments of Russia."[40] In his published narrative of that second expedition he claimed that having shown the Admiralty that "the objects to be attained were important at once to the naval character, scientific reputation, and commercial interests of Great Britain," he received its support.[41] When he did so, the scientific goals of the expedition were explicitly emphasized. Once again, John Richardson was to accompany the expedition, and Franklin's instructions stated that "the principal object of Dr. Richardson accompanying you, is that of completing as far as can be done, our knowledge of the natural history of North America." This time another naturalist, Thomas Drummond, was to accompany the expedition for part of its travels as assistant naturalist before turning towards the Rocky Mountains to undertake botanical explorations there.[42] Franklin also wisely sought the help of HBC officers to accomplish the scientific aims of the expedition. According to John

Richardson, "previous to our setting out on the Second Expedition, Sir John Franklin addressed letters to many of the resident Chief Factors and Traders of the Hudson's Bay Company, requesting their cooperation with our endeavours to procure specimens of Natural History, and their ready acquiescence with his desire was productive of much advantage to us."[43]

It would have been far more impolitic for the HBC to have refused to cooperate with this expedition than it would have been before the merger.[44] Indeed, to cooperate suited the needs of the HBC in some very direct ways. The expedition had the potential to pre-empt Russian expansion in northwestern North America and supply information that the company could use to expand its trading operations.[45]

Franklin's second overland expedition was planned in such a way that it might connect with two British maritime expeditions. William Edward Parry was to attempt to traverse the Northwest Passage in ships from the east via Hudson Strait, and Frederick William Beechey from the west via Bering Strait. Franklin's expedition was to travel overland to the mouth of the Mackenzie River, from where Franklin was to lead some men westward along the coast, and John Richardson was to lead the rest of the men eastward as far as the Coppermine River. The hope was that at least part of Franklin's overland expedition might meet up with one of the maritime expeditions. To that end, Franklin's party and Beechey's party were instructed to try to rendezvous along the arctic coast between 15 and 20 August 1826.[46]

Franklin's second expedition was more successful than his first, not only because Franklin had learned important lessons from his first expedition but also because the HBC was in a much better position to assist the second expedition.[47] The difference was obvious. In July of 1826 Franklin thanked the Governor and officers by writing that "I am happy to say these [provisions and other essential articles] have been most abundantly provided by the Gentlemen in charge of the different departments. ... We are now in possession of much more Pemmican than we were at any time of my last voyage."[48] Franklin's expedition travelled to the northwest via New York and Upper Canada, establishing a base on Great Bear Lake called Fort Franklin, from which they travelled down the Mackenzie River to the Arctic Ocean. From there Richardson successfully mapped the arctic coast from the Mackenzie River to the Coppermine River by 8 August, and was back at Fort Franklin by 28 August. (Parry's maritime expedition got no further than Fury and Hecla Strait.)[49]

Thomas Elson, mate on Beechey's expedition, took a barge as far east as Point Barrow in present-day Alaska on 23 August 1826, but did not connect with Franklin. Franklin's party fell about 280 kilometres, or five days' travel, short of his rendezvous with Elson. When he realized that he would be unable to rendezvous with Beechey's expedition, his party turned back at "Return Reef" on 18 August, and arrived at Fort Franklin on 21 September.[50]

Meanwhile, Thomas Drummond stayed behind as planned after the party reached Cumberland House in June 1825, reuniting with it at the same place in July 1827. He botanized around Cumberland House until 20 August 1825, when he accompanied the HBC brigade up the North Saskatchewan River, collecting and observing all the way. At Edmonton House, he joined the Columbia brigade as it travelled from Edmonton to Fort Assiniboine on the Athabasca River, and from there to the Rocky Mountains, travelling as far as Boat Encampment on the Columbia River. He collected about 1500 botanical specimens, 150 birds, 50 mammals, and many insects.[51]

The scientific world evidently sought news of Franklin's second expedition. Even as the expedition was in the field, the British naturalist William Jackson Hooker, chair of botany at the University of Glasgow from 1820 to 1841, published an article in the January 1827 issue of *Edinburgh Journal of Science* with updates on the scientific work of Drummond (to 26 April 1826) and Richardson (to 26 November 1826).[52] When the expedition was complete, the published narrative included several appendices by John Richardson on topography, geology, meteorology, solar radiation, magnetism, the aurora borealis, and other topics.

Once again, the HBC and its personnel were prominently thanked in the 1828 published narrative. In fact, Franklin did not limit his long published tribute to the company and its London Governor and Committee, but mentioned several HBC officers by name:

> Mr. Pelly, the Governor of the Hudson's Bay Company, and Mr. Garry, the Deputy-Governor, as well as every Member of its Committee claim my most sincere thanks for their unremitting endeavours to promote the welfare of the Expedition through its whole progress; and I feel truly obliged to Mr. Simpson, the Governor of the Fur Countries; to Mr. McTavish, Mr. Haldane, Mr. McDonald, Mr. Leith, Mr. Stuart, and Messrs. James and George Keith, Chief Factors, who, acting in the spirit of their instructions, were very assiduous in collecting provisions and stores for the use of my party, and in forwarding all our supplies.[53]

Franklin also described the policies and conduct of the HBC as beneficial to the indigenous peoples, arguing that the end of competition in 1821 and the establishment of an effective monopoly for the HBC had improved the company's ability to protect the interests of indigenous peoples.[54] Given that Franklin had travelled through the country shortly before and shortly after the merger, his opinion, highly acceptable to the HBC, was bound to carry some weight with politicians and the public at home.

John Franklin concluded the narrative of this second edition echoing John Barrow's assertion of the detached and scientific nature of the search for the Northwest Passage: "Arctic discovery has been fostered principally by Great Britain; and it is a subject of just pride that it has been prosecuted by her from motives as disinterested as they are enlightened; not from any prospect of immediate benefit to herself, but from a steady view to the acquirement of useful knowledge, and the extension of the bounds of science."[55] Notwithstanding Franklin's assertions, as he had previously predicted, the expedition did produce some commercial benefits for the HBC. Acting on the discoveries of the expedition, in 1827, Peter Warren Dease was sent to re-establish a presence at Fort Good Hope (occupied by the NWC between 1805 and 1815).[56] This was the beginning of the HBC's expansion into what would become the very lucrative lower Mackenzie River region.

John Ross, whose first voyage in 1818 had inaugurated Britain's nineteenth-century search for the Northwest Passage, returned to the arctic in 1829. His naval expedition became trapped in ice for four years before escaping in 1833.[57] The HBC was not directly involved in that expedition, but while the fate of the Ross expedition was still unknown, the London Committee believed that a proposed overland expedition to search for Ross was so "humane and philanthropic" that they agreed to support it.[58] This expedition was led by the naval officer and veteran of the two previous Franklin expeditions, George Back.

Even when the expedition was a rescue mission, Back's instructions were to map coasts unknown to Europeans, and to make "other scientific observations as your leisure will admit."[59] To that end, Dr. Richard King was sent along as naturalist. News of the safe return of Ross's expedition to England reached Back before he left Great Slave Lake. Once its original main goal became irrelevant, the expedition was able to direct all of its energies to exploration and science.

The directors of the HBC may have supported Back's mission, but privately several HBC men disparaged Back in ways reminiscent of

George Simpson's earlier assessment of Franklin. In 1834, when Back was still in the field, William Mactavish wrote of the expedition that "you'll learn what a fine story they'll make of this bungle, they will you may be sure take none of the blame themselves ... They will return next summer and like all the other Expeditions will do little and speak a great deal."[60] Once it had returned, some, at least, considered it a failure. The expedition did map the river now named after Back from near its source to the Arctic Ocean, but it mapped only a small part of the coast itself. As he himself was in the midst of his own exploration with Peter W. Dease, the ungenerous Thomas Simpson wrote that "Back's expedition turned out much as I anticipated. My judgment was however founded more on moral than on physical considerations. Back I believe to be not only a vain, but a *bad* man; and his failure is retributive justice. I think with you [James Hargrave] that *his* adventures are closed; having a wide field, which cannot be fully explored by their *present* expedition."[61]

Back and his second in command, Richard King, were a study in contrasts during and after their expedition. Compared with King, Back may have been an ineffective leader, but he clearly had the social intelligence to cultivate the respect of those in the HBC and the government. An indication of the level of respect he continued to hold is provided by the fact that in 1851 he served on the Arctic Council established to advise the Admiralty on the search for Franklin's lost expedition.[62] When Back published the journals of his 1833–5 expedition, his tribute to the HBC characterized the company not only as "liberal" but also as "benevolent" and "zealous," thus adding words that would resonate with the Romantics and the Christian humanitarians (whose influence was growing in England).[63] In the published journals of the expedition, Back wrote of the Governor and Committee of the HBC that "I should be indeed ungrateful, if I were not to add that their benevolent intentions were zealously fulfilled, and their judicious arrangements carried into complete effect by Mr. [George] Simpson, the resident Governor, and the various officers in the service of the Company."[64] More than two decades later, he spoke very highly of the HBC during his testimony before the Parliamentary Committee struck to investigate the company in 1857.[65]

Back gave all the credit for the natural history contributions of the expedition to Richard King, his second in command.[66] The published narrative included appendices on the scientific contributions of the expedition written by such luminaries as John Richardson (on zoology),

John George Children, the founding president of the Entomological Society of London (on spiders and insects), W. J. Hooker (on botany), William Henry Fitton, former president of the Geological Society of London (on geology), Richard King (on botany, entomology, and meteorology), and George Back (on meteorology, aurora, and magnetism).

Either unaware or oblivious to how impolitic it was to do so, King published a competing narrative of the expedition in the same year as Back. King clearly believed that his scientific collections had not been properly treated and that Back's poor leadership was to blame for the expedition's failure to map more than a small part of the arctic coast.[67] In his preface King noted that he had decided "while yet tented on the shores of the Polar Sea," to propose to lead an expedition of his own. It was not to be. King was critical not only of Back but also of the HBC, which he portrayed scathingly as exploitative of aboriginal people.[68] Thomas Simpson, a cousin of Governor Simpson, characterized the book as "the most venomous thing I have read for a long time."[69] Coming when it did, just as the HBC was applying for a renewal of its licence (its application for renewal was submitted on 10 February 1837), and as the attention of humanitarians in London was focused on the circumstances of aboriginal peoples in the British Empire, the attack was particularly unwelcome. Although King continued to seek support to launch another expedition at least until 1842, "having made himself very obnoxious" to the HBC's directors, Alexander Simpson noted in 1845 that "his return to their country in command of an expedition would have been exceedingly distasteful to them."[70] Fortunately for the company, King had annoyed enough people in the Admiralty that the HBC Governor and Committee were not ever likely in the awkward position of having to refuse a government request to permit his return to their territories.

Although the HBC and the Admiralty prevented King's return to the HBC territories, King became, after his return to England in 1835, one of the HBC's main critics. King became an important contributor to the wave of humanitarian concern for aboriginal peoples that swept through London in the late 1830s, and the founder of Britain's first organization dedicated to the scientific study of humanity. Before he participated in the Back expedition, King had been a medical student of Thomas Hodgkin at Guy's Hospital in London. A Quaker humanitarian, Hodgkin became the most influential advocate for aboriginal peoples in Great Britain in the late 1830s. More than anyone else, he was responsible for convincing the British Parliament to establish its Select

Committee on Aboriginal Peoples (1836), and he was the most important founder of the Aborigines Protection Society (APS [1837]). Hodgkin apparently arranged for King to appear before the Select Committee in July 1836, an opportunity that King used to condemn the HBC's treatment of aboriginal people.[71] His testimony was especially damaging because King was the only non-HBC person to testify about the HBC territories before the committee.[72] Although not a Quaker, King then assisted his mentor in establishing the APS.[73]

The influential historian of anthropology George W. Stocking described the APS as "the oldest lineal ancestor of modern British anthropological institutions."[74] That assessment is supported by the fact that in view of its goal "to assist in protecting and promoting the advancement of defenceless or uncivilized Tribes," the founders decided that "the first object of the Society will be, to collect Authentic Information concerning the Character, Habits, and Wants of the Uncivilized Tribes, and especially those in or near the British Colonies."[75] Thus, from its establishment, the APS was both a humanitarian organization dedicated to the protection of aboriginal peoples in the British Empire and an organization dedicated to applied "ethnology."[76]

Although many members of the APS saw the scholarly and humanitarian goals of the APS as entirely reconcilable, it appears that from its early days some were more committed to the humanitarian mission and others more interested in the scholarly purposes, and that some of the former believed that the two goals were incompatible. King was obviously among those who sensed a tension, for in 1842, while serving as secretary of the APS, he issued a prospectus for the foundation the Ethnological Society of London (of which King became secretary upon its establishment in 1843 and president in 1844).[77] The Ethnological Society and the APS were thus both products of Christian humanitarians who emphasized the unity of humanity (monogenesis), although their membership and their orientations diverged fairly quickly.[78]

King very quickly began publishing scholarly articles on the Inuit.[79] These articles reveal King's admiration for the Inuit, but also come across as informed and sophisticated scholarship. In that sense they had the potential to establish King as an independent and disinterested observer, but if King intended them to do so, they appear to have been ineffective in overcoming King's poor reputation in official circles.[80] Still, King's ethnological publications and his testimony before the Select Committee on Aboriginal Peoples in 1837 and the Select Committee on the Hudson's Bay Company in 1857 suggest that he long remained committed to both the scholarly and the humanitarian impulses that had animated the APS between 1837 and 1842.[81]

The public acclaim generated by the modest accomplishments of Franklin's and Back's overland expeditions sponsored by the Royal Navy spoke eloquently of the tremendous public-relations value of contributing to arctic exploration at a time when the British public was very interested in the explorations. And so in 1836 the London Committee of the HBC informed the Colonial Office of its intentions to mount its own expedition – the first HBC expedition to the arctic region since Samuel Hearne's – and instructed George Simpson to make the arrangements.[82] The committee's letter to Simpson specifically mentioned the "much interest in the public mind," before stating their belief that "the habits of the Gentlemen employed in the [HBC] service, their knowledge of the Indian character and their being more or less accustomed to the privations and difficulties usually met with travelling thro' the Wilds," made them "well qualified for conducting such an expedition."[83] Here, even before the first expedition was sent out, the HBC's directors already hinted at what would become the hallmark of HBC explorations: the adoption of indigenous methods of travel and subsistence.

The Company's sponsorship of this expedition was directly related to its broader goals. According to Alexander Simpson (cousin of Governor Simpson, brother of Thomas Simpson, and himself employed by the HBC from 1827 to 1843) the company was motivated partly by a desire to prevent Richard King from ever getting a chance to come back to their territories. But, he added, "another more weighty reason was the necessity which they felt of doing something to place themselves in a position to make an application to the British Government for a gratuitous renewal of their grant of the exclusive trade of the region which was to form the field of the operations of the expedition."[84] Their licence, granted in 1821, was due for renewal in 1842, but the Governor and Committee applied earlier, in February 1837. As historian Ken Coates argued, the company confronted a rising tide of free-trade sentiment in Britain: "Chartered enterprise, once the engine of the Empire, no longer found favour with the British public, and in the eyes of many politicians and government officials, the Hudson's Bay Company was becoming increasingly anachronistic. ... the time was right to claim a greater share of the glory, and as much of the credit as possible. The Board of Trade and the Colonial Office, it was hoped, could not help but take notice of the Company's unselfish commitment to scientific adventure."[85]

At the same time, the HBC was seeking redress for actions on the part of the Russian American Company (RAC), the RAC having illegally prevented the HBC from establishing a post near the mouth of the Stikine River in 1834. As E. E. Rich has explained, by mounting the Dease-Simpson expedition the company "was staking a claim for support from the British government."[86]

George Simpson chose Peter W. Dease and Thomas Simpson to lead the expedition. The two men were a study in contrasts. George Simpson probably considered Dease's most important qualifications to be his sound judgment, his easy rapport with aboriginal peoples and his own subordinates, and his ability and willingness to undertake arduous travel.[87] Dease, a Canadian, possibly of mixed British Mohawk parentage, had been employed in the Athabasca and Mackenzie River districts since he was thirteen years old. He served with the XY Company (a Montreal-based competitor of the NWC) until its merger with the NWC in 1804, and then with the NWC until the 1821 merger, when he was appointed a Chief Trader with the HBC. As early as May 1820, while still with the NWC, Dease had provided ethnological and geographical knowledge to the first Franklin expedition.[88] At Franklin's request, the HBC appointed him to assist the second Franklin expedition for three years, a service for which he was promoted to Chief Factor in 1828.[89] He had been stationed at Fort Good Hope from 1827 to 1830, before being put in charge of New Caledonia, which he managed with considerable success.[90] Dease had impressive credentials.

Dease was accompanied by Thomas Simpson, a cousin of Governor George Simpson. Having been educated at King's College in Aberdeen, he was, according to E. E. Rich, "admirably equipped for the scientific side of the expedition."[91] He was also an extraordinarily capable traveller. He joined the company in 1828, and was soon willing and able to undertake arduous journeys. For example, between 10 February and 10 March 1830 he travelled about 1,100 kilometres by dogsled from York Factory to Red River. And in 1837, in order to join Dease for this expedition, he walked an average of over 30 kilometres per day – in winter – from Red River to Fort Chipewyan. But Simpson was arrogant and mentally unstable. Although twenty years Dease's junior, he exhibited little respect for Dease. He also harboured a deep contempt for Native people.[92]

Dease and Simpson were instructed to survey in only two years more than all of the three Royal Navy overland expeditions combined had been asked to explore. It was possible for them to accomplish so much more than the Royal Navy did because Royal Navy expeditions had to depart from England (requiring three years of travel for each summer of arctic exploration), while Dease and Simpson left from a base in the Mackenzie River District. But the instructions were ambitious nonetheless. George Simpson's instructions required them to complete the survey of the northern coast of North America, first by travelling in 1837 from the mouth of Mackenzie River to Point Barrow, and then, during a following summer, by travelling eastward from the mouth of the Coppermine River. In an interesting juxtaposition that implied a conscious connection between exploration, naming, and scientific activity on one hand, and legal possession on the other hand. Simpson noted that:

> The necessary astronomical & surveying instruments are provided to enable you to take observations and to make surveys, in which you will be as accurate as possible, and you will be pleased to prepare a full and particular journal or narrative of the voyage, likewise a chart of the Coast, and to take formal possession of the country on behalf of Great Britain in your own names, acting for the Honble Hudson's Bay Company, at every part of the coast you may touch, giving names to the different headlands, mountains, rivers and other remarkable objects you may discover. It is also desirable that you make a collection of minerals, plants, or any specimens of natural history you may fall in with, that appear to be new, curious, or interesting.[93]

The scientific achievements of the Dease-Simpson expedition were small, but the expedition's surveys certainly bore up the London Committee's assumption that the company's men were particularly well qualified to succeed in the main goals of the expedition.

During the summer of 1837, Dease and Simpson went down the Mackenzie River intending to survey to the west using small boats. Their journals suggest that they were eager to invite comparison with Franklin's and Back's expeditions, and thereby to claim a place among Great Britain's traveller-heroes, although as those who, like Hearne, succeeded by employing indigenous practices. Travelling lightly and quickly, befriending Inuit along the way and borrowing their technologies – an umiak for example – Dease noted with obvious pride that "the [Esquimaux] ladies declared that our party were true Esquimaux and not 'Kabloonan' [Whites]." After retracing Franklin's route of 1826, they continued on to reach their goal, Point Barrow, in early August 1837.[94] Thus, they completed the last gap (about 280 kilometres long)

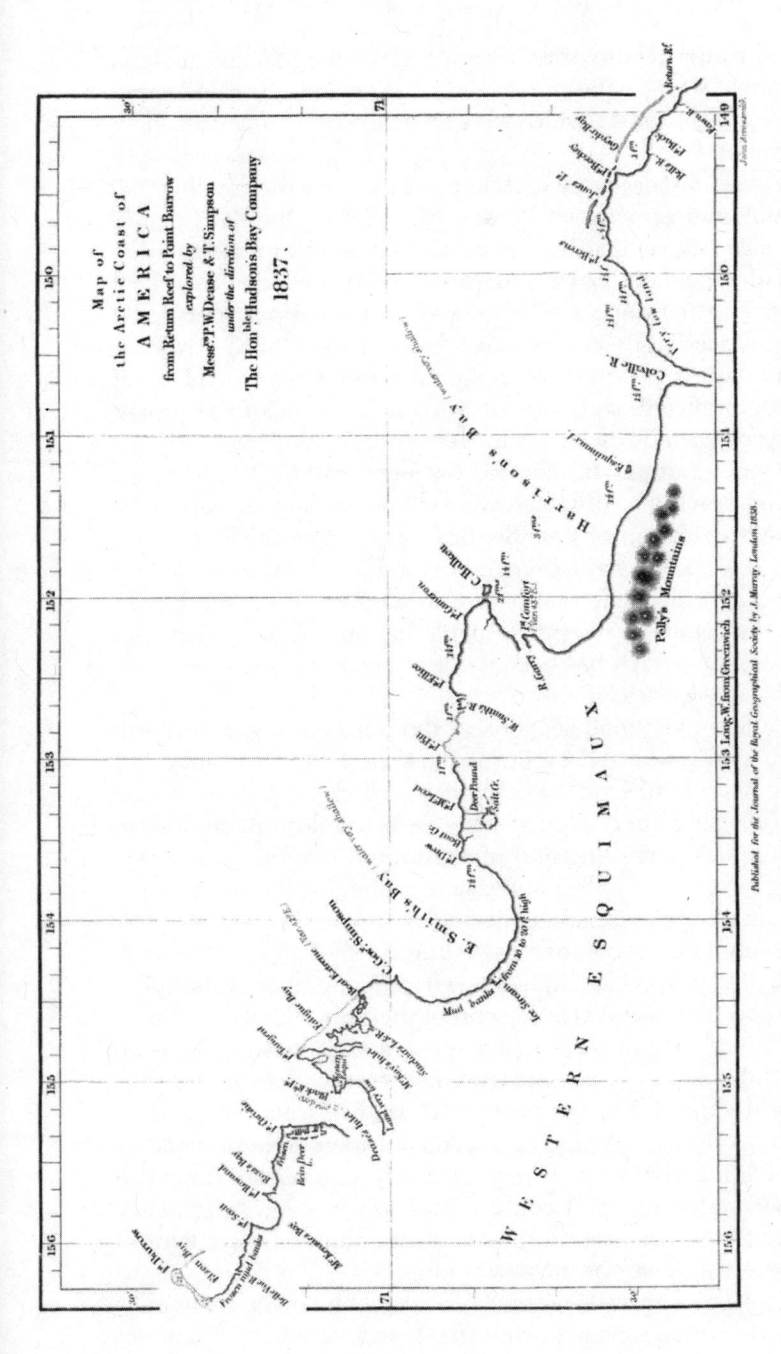

5.2 Map of the Arctic Coast of America, 1837. This map of the Arctic coast between Return Reef and Point Barrow, drawn by John Arrowsmith and published in the *Journal of the Royal Geographical Society of London*, reveals that Dease and Simpson took seriously their instructions to give names to geographical features. Many of the places were named after governors and officers in the HBC. The map, and the article in which it appeared, prominently mentioned the role of the HBC in these explorations. Source: *Journal of the Royal Geographical Society of London* 8 (1838).

in the surveys of the Northwest Passage west of the Mackenzie delta between the most westerly point reached by Franklin's overland expedition in 1826 and the most easterly point reached by Thomas Elson, in the same summer.

The next two summers were directed at exploring the roughly 500 unsurveyed kilometres between Point Turnagain, on the Kent Peninsula, and Spence Bay, on Boothia Peninsula.[95] Poor weather prevented them from achieving their goal in 1838, but they succeeded in 1839. Lest the significance of their success be missed, in their report published in the *Journal of the Royal Geographical Society* in 1840 (HBC governor J.H. Pelly and deputy governor Nicholas Garry had been members of the Royal Geographical Society since its foundation in 1830), they noted that they had completed "by far the longest voyage performed in boats on the Polar Sea," and playfully noted finding a cache left by George Back five years previous, "of which we took possession as memorials of our having breakfasted on the identical spot where the tent of our gallant, though less successful, precursor stood."[96] They concluded by boasting hyperbolically that "we rejoice in having ... secured to our country and the company the indisputable honour of discovering the north-west passage, which has been a object of search to all maritime nations for three centuries."[97]

Despite the fact that small sections of the northern coast of North America remained uncharted by Europeans, the HBC had supported a remarkably successful land expedition in search of the Northwest Passage. Dease's biographer argued that "Dease's logistical abilities in organizing supplies, recruiting and maintaining discipline among his men, keeping peace among the natives, and managing the swift movement with a simplicity of equipment while living off the land in so far as possible assured the success of these arduous expeditions."[98] Indeed, the party had always travelled lightly, even to the point of abandoning their tents to speed their travel at the end of their final season of travel.

The expedition well illustrates the tension that existed between exploration and science – the emphasis on the success of the exploratory mission required that the party relegate the scientific goals to a distant second priority. As a consequence of their style of exploration, the expedition contributed only modestly to science. The party had no trained naturalist, and could afford to carry few specimens. In the report of their first season they admitted that "in the botanical kingdom scarcely a flower or moss was obtained. ... In zoology, reindeer, arctic foxes, one or two[lemmings], seals, white owls, snow buntings, grouse (Lagopus salicite et rupestres), and various well-known species of water-fowl were the only objects met with."[99] In the reports of the subsequent seasons, contributions to science were even fewer. Still, Thomas Simpson's journals, published posthumously in 1843, did include an appendix by W. J. Hooker listing the plants collected by the expedition. The plants collected were "beautifully preserved," but the report could not hide the fact that most of the specimens were collected during the first two summers, and mostly in the vicinity of their base, Fort Confidence, with only a few specimens gathered along the arctic coast in 1838 when the expedition was trapped in ice. None of the plants were new to science, and only one species had not been previously collected by Richardson.[100]

The results of Dease and Simpson's explorations certainly were useful to the HBC and to the men involved in planning and executing it. In terms of positive public exposure for the company, reports of the expedition were "published in most of the leading papers of the civilized world ... and attracted much attention," probably in large part because the two men so ably conformed to the still-influential Romantic image of the solitary scientific traveller.[101] Thomas Simpson's published journals, which denigrated Dease and aboriginal people (especially mixed-blood people), nevertheless lauded the HBC's "zealous and effective co-operation" with northern explorations and "geographical science," as well as its treatment of Native peoples.[102] On 30 May 1838, during the last year of the expedition, the British government granted the HBC the renewal of its licence.[103] At an individual level, in 1840 Queen Victoria announced that she would grant both Dease and Simpson a pension of £100 per year, "for their exertions towards completing the discovery of the North West Passage," and George Simpson a knighthood and John Pelly (London governor of the HBC from 1822 to 1852) a baronetcy for their support of arctic exploration. The Royal Geographical Society awarded Dease and Simpson its gold medal, although Simpson was dead before he could learn of his awards and distinctions.[104] Overall, the investment had returned handsome dividends for the company. "The Company might justly be proud," argued its historian, E. E. Rich; they "had shown the liveliest interest throughout, they had voted supplies, and rewards to the tune of £1,000, they had proclaimed that their interest was purely scientific, and only when the westward portion of the task was complete did they begin to hope that the discoveries might bring in increases of furs or speed up the trade of Mackenzie River."[105]

Following the successful expedition, Dease went to England on furlough and retired to Lower Canada, never returning to the HBC territories. But Thomas Simpson declined a leave of absence, pushing instead for the London Committee to support another expedition under his leadership, this time to explore the remaining unknown coastline from the mouth of the Back River (Great Fish River) in the Gulf of Boothia to the northern tip of Melville Peninsula departing from Fort Reliance, and travelling via the Back River. The London Committee approved the request, but Simpson died, under mysterious circumstances, before learning of the decision.[106] His death postponed the company's efforts to complete the mapping of the remaining coast of North America until 1845.

The Dease-Simpson expeditions also served the fur trading interests of the HBC by facilitating the expansion of HBC operations northwestward, even beyond 141° W – territory that was, by the Anglo-Russian Convention of 1825, officially part of Russian America. Dease and Simpson's explorations suggested that the company's fur trade operations might be expanded westward from the Mackenzie River. John Bell, Peter Dease's son-in-law, was sent to explore this possibility in 1839, and together with a young mixed-blood man who was destined to become one of the company's most energetic critics, Alexander Kennedy Isbister, he travelled up the Peel River that summer, and established the Peel River House (Fort McPherson) in 1840. That led to further highly profitable expansion in the region. La Pierre's House was established in 1846, and Fort Yukon (in Russian America) in 1847.[107]

The search for the Northwest Passage convinced the HBC Governor and Committee to invest the company's own money in expeditions like that of Dease and Simpson (which cost the company £1000).[108] It also stimulated further scientific endeavours not directly related to exploration. For example, John Richardson's connections with the HBC began with John Franklin's first overland expedition, but he used the cooperation of the HBC to expand his research in the 1830s. During that time, his interests expanded to include permafrost and meteorology. Richardson wrote the Governor and Committee in 1834 to ask them to assist with his efforts to study the permafrost of northern North America.[109] In his published report, he wrote that "the Governor, now Sir John H. Pelly, Bart., ... and the other members of the Committee of the Company, with the zeal for the advancement of science for which they have long been distinguished, early in 1835 transmitted copies of my letter to the several Chief Factors in charge of districts in the fur countries, with instructions for them to comply with the directions therein expressed."[110] The request was met with more or less diligence by several traders scattered throughout the HBC territories, and Richardson was able to publish an article on the state of permafrost in 1841 and

1851.[111] Richardson also published two papers on meteorology, one of them co-authored by HBC officer Murdoch McPherson.[112] In November 1840, Peter Dease corresponded with John Richardson regarding cultivation of cereal crops at HBC posts. In another co-authored paper, John Richardson and Peter Dease again reported on cereal crops in 1841, and Richardson expanded the discussion with commentary on livestock and potatoes in his book, *Arctic Searching Expedition*, in 1851.[113]

In June 1843, the HBC's North American Governor, George Simpson, reported to the London Governor and Committee that if Americans did not already outnumber HBC men in the Willamette Valley, they soon would.[114] During the American election campaign of 1844, many Democrats argued that the United States should annex all of the Columbia District (Oregon Country). After the election of the Democratic presidential nominee, James K. Polk, and especially after Polk's inaugural address on 4 March 1845, Simpson grew convinced that war between Great Britain and the United States was inevitable.[115] Polk's State of the Union Address of 2 December 1845 maintained the state of anxiety. Assuming that the HBC would be the biggest loser should war erupt, George Simpson was in London from November 1844 to April 1845, often dealing with top British diplomats.[116] During the same months, he planned an expedition that could bolster the HBC's profile among politicians and the public.

The 1837–9 Dease-Simpson explorations had contributed significantly to the delineation of a Northwest Passage by surveying the northern coast of mainland North America. In 1844 Simpson turned to the Orcadian John Rae to complete this work. Rae was chosen to survey the last uncharted section of North America's coastline, from the west end of Boothia Peninsula to the Fury and Hecla Strait, which separates Melville Peninsula from Baffin Island (see figure 5.3).

Rae was the right man for the job. His family was well connected to the HBC. His father was an agent for the HBC on the Orkney Islands, where the HBC hired many of its labourers, and two of his older brothers worked for the HBC.[117] John Rae studied medicine in Edinburgh before he joined the company as a surgeon in 1833, but his athleticism, his remarkable physical and mental toughness, combined with his superb leadership skills were his most outstanding attributes. Dr. Rae more than any other HBC explorer embodied the hyper-masculine scientific traveller – a well-educated British man who seemed utterly indifferent to discomfort and danger, while wholly committed to discovery. Rae claimed that "by the time I was fifteen, I had become so seasoned

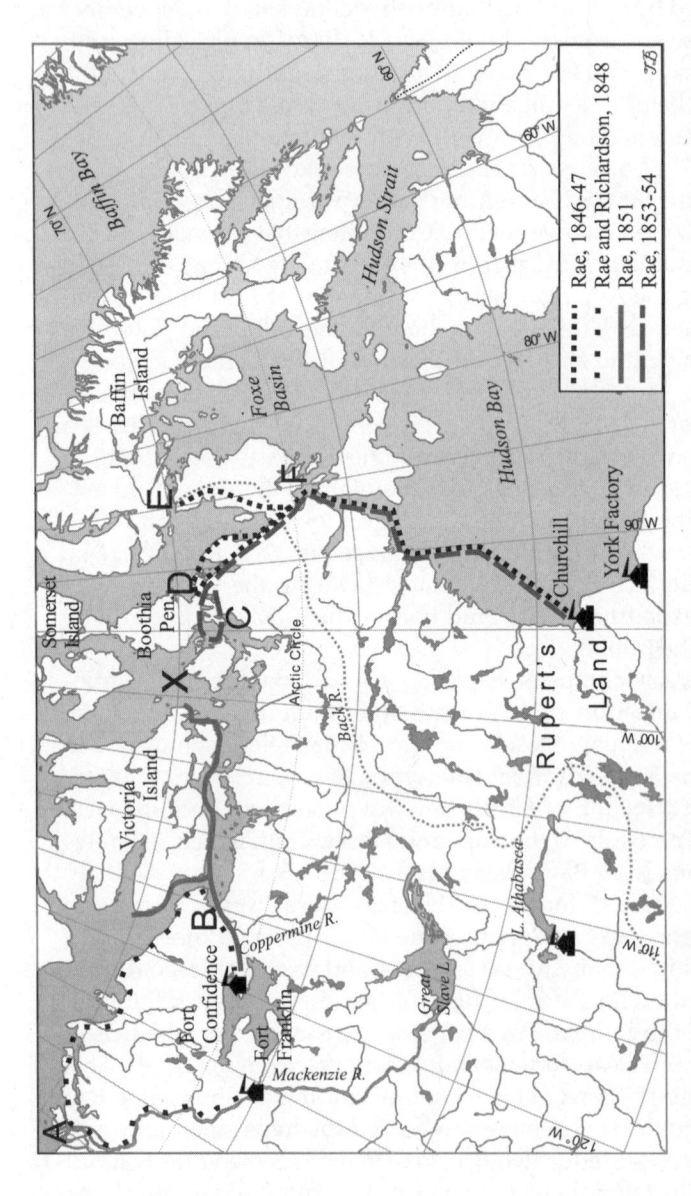

5.3 **Arctic Explorations Sponsored or Assisted by the HBC, 1845–1854.** HBC surgeon, trader, and explorer John Rae led several land-based explorations between 1845 and 1854. Operating from a base at Repulse Bay (F), in 1846 and 1847 he surveyed the Arctic coast between the base of Boothia Peninsula (D) and Fury and Hecla Strait (E). His subsequent expeditions were increasingly influenced by the search for John Franklin's third expedition, which had departed England in 1845. With John Richardson, Rae searched along the Arctic coast between Mackenzie River (A) and Coppermine River (B) in 1848, and he reconnoitred the southern coast of Victoria Island in 1851. Then in 1854, after wintering again at Repulse Bay, he surveyed as far east as Rae Strait (C), where he secured from local Inuit conclusive evidence of the fate of the 129 members of Franklin's expedition. The approximate location of Franklin's trapped ships is indicated by an X. Source: T. Binnema.

as to care little about cold or wet, had acquired a fair knowledge of boating, was a moderately good climber among rocks, and not a bad walker for my age, sometimes carrying a pretty heavy load of game or fish ... on my back."[118] A perusal of the journals of Rae's arctic explorations suggests that his self-assessment was not immodest. His ability to "stand an immense amount of fatigue," became legendary in his own time.[119] Rae served for a decade at Moose Factory, by which time he had become, according to one of his colleagues, "the best and ablest snow-shoe walker not only in the Hudson Bay Territory but also of the age."[120] Rae ably demonstrated his ability to travel when he walked from Upper Fort Garry to Sault Ste. Marie during the winter of 1844-5, before carrying on to Toronto to be given the requisite training in surveying by John Henry Lefroy, who had just returned from his own expedition through the HBC territories (see chapter 6). Thus was the HBC repaid for the assistance it had previously given to Henry Lefroy.

In 1840, when George Simpson wrote his letter to inform Thomas Simpson that the London Committee had accepted his proposal to complete his survey of the arctic coast, he made it clear that, if the preparations were not already underway for an expedition out of Fort Reliance, that the expedition should leave instead from Hudson's Bay, travelling northward via York Factory and Churchill.[121] Simpson died before he could undertake the expedition, but Rae was similarly instructed in 1844 to depart from Churchill rather than Great Slave Lake.[122]

An indication of the importance placed on this expedition can be seen in the correspondence sent to James Hargrave, Chief Factor at York Factory. In December 1844, while in London, Governor Simpson wrote to Hargrave that the "expedition will be placed under the charge of Dr. Rae, and it is intended that he shall start from Churchill at the opening of the navigation (June or July) 1845, if time admit of the necessary preparations being completed. ... You will understand that, the Expedition is to be entered upon if possible next season, and that Dr. Rae has a *carte blanche* to draw on our resources in order to carry it into operation."[123]

On 31 March 1845, only three weeks after James K. Polk's inauguration, Henry Hulse Berens, from the HBC's London office, wrote to James Hargrave stating that "we are all in great anxiety here respecting the speech of Mr. President Polk who seems desirous of going the whole hog for the Oregon Territory. Our Government are fully alive to the subject and I am in great hopes will not allow themselves to be bullied out of their right by a people whose sole ground of claim is the desire of aggression and our wish for peace."[124]

No one drew the connection between scientific and corporate interests more explicitly than Donald Ross in April 1845. Donald Ross, one of George Simpson's most trusted men since the 1820s,[125] wrote to James Hargrave regarding the preparations for the Rae expedition that

> I really would wish to see this long pending question [the Northwest Passage] finally set at rest under the auspices of the Company – my own opinion is that we shall soon require the assistance of the "strong arm" of Government, on both sides of the Rocky mountains, and it can only be by the liberal use of the means and power we possess, towards the advancement of religion, knowledge, civilization, science, and Discoveries, throughout our wide spread possessions, that we can hope to receive any sympathy or support from the government or the nation at large, in our true character of "monopolists" we need expect but little favour, and I confess that I would not be at all surprised to see our "venerable Charter" consigned to the "tomb of all the Capulets" ere many years go over our head.[126]

Ross was right; the company soon needed the "'strong arm' of Government, on both sides of the Rocky mountains." Sectional divisions in the United States had made expansion controversial, but in 1844, by tying the annexation of Oregon with the annexation of Texas, Polk was able to unite expansionists in the American south and north. The American appetite for new lands seemed suddenly very large. In 1844 Major Davezac addressed the 1844 New Jersey Democratic convention saying "Make way, I say, for the young American Buffalo – he has not got land enough; ... I tell you we will give him Oregon for his summer shade, and the region of Texas for his winter pasture. ... He shall not stop his career until he slakes his thirst on the frozen ocean."[127] More famously, in July 1845 the editor of the *Democratic Review* John O'Sullivan rued the influence of European nations (including England "our old rival and enemy"), which were interfering, he believed, in "the fulfilment of our manifest destiny to overspread the continent allotted by Providence for the free development of our yearly multiplying millions."[128]

Upon receiving the text of President Polk's address to Congress of 2 December 1845, George Simpson asked the British government to send military forces to the Columbia District and to Red River, to protect them in case of war between the United States and Great Britain.[129] Simpson argued that should war break out, the Americans would attempt to seize not only the Oregon Country but also the Red River Colony.[130] In 1819, the American army had established Fort Snelling, at the confluence of the Mississippi and Minnesota Rivers, and following a treaty with the Sioux in 1837, Americans began settling in that district. Soldiers from Fort Snelling could easily conquer an unprotected Red River Colony. If Red River were taken, it would be only a matter of time before the United States took much of the rest of the HBC territories. The government responded to Simpson's request by deploying soldiers of the Sixth Regiment of Foot to Red River in June 1846 and keeping them there until 1848 – the first time since 1697 that the British government assisted the HBC militarily.[131]

Still, the Oregon Treaty, signed on 15 June 1846, only two weeks before the collapse of Robert Peel's Conservative government, was a bitter disappointment to the HBC. The British ceded much more of the Oregon Country than the HBC hoped, leading James Douglas to remark that "the closing measure of Sir Robert Peels administration was the cession of the Columbia to a foreign Power, a monstrous injustice to the parties who have explored and settled the country."[132] But the HBC did not have the luxury to anger its own government. Until 1870, the company continued to need the support of the government to help it defend its private claims to land (as stipulated by the treaty) in the territories ceded to the United States.[133]

Rae's explorations, like the earlier expeditions of Dease and Simpson, reflect the tension between science and exploration in the Hudson's Bay Company. Simpson's letter of instruction to Rae emphasized the gravity of the exploratory aims of the expedition: "the eyes of all who take an interest in the subject are fixed on the Hudson's Bay Company; from us the world expects the final settlement of the question that has occupied the attention of our country for two hundred years; and your safe and triumphant return, which may God in His mercy grant, will, I trust, speedily compensate the Hudson's Bay Company for its repeated sacrifices and its protracted anxieties."[134]

The first priority was for an HBC expedition to achieve its exploratory aims without any embarrassments like those that attended previous Royal Navy expeditions. Thus, scientific activities were not to jeopardize the main goal. On the other hand, Simpson understood that at least some scientific results were important to ensuring the most advantageous response to the company's efforts. Simpson's instructions to Rae betray a strong emphasis on geographical, navigational, and meteorological goals, but Rae was told to

> do your utmost, consistently with the success of your main object to attend to botany and geology; to zoology in all its departments; to temperature

both of the air and of the water; to the soundness as well with respect to bottom as with respect to depth; to the magnetic dip and the variation of the compass; to the aurora borealis and the refraction of light. You will also to the best of your opportunities, observe the ethnographical peculiarities of the Esquimaux of the country. ... collecting, at the same time any new, curious, or interesting specimens.[135]

Given the way that George Simpson intimidated many within the company, Rae's response reveals both his self-confidence and his sense of humour: "You appear to think that I have got a head stuffed with all sorts of knowledge, if I may judge by my letter of instructions. The head is big enough certainly outside, but whether there is a large quantity of bone in it or not I have not yet tested."[136] The correspondence suggests a tacit agreement that Rae was going on this expedition primarily as an explorer, not as a collector; that he would gather scientific evidence without jeopardizing the main goal of the expedition.

On 5 July 1846, the day after he received final instructions from Simpson, Rae left Churchill in two boats to "complete the geography of the northern shore of America."[137] In stark contrast to Royal Navy expeditions, but in line with the precedent set by Hearne, Dease, and Simpson, Rae's party left with enough food to last only four months even though they intended to be gone for between fifteen and twenty-seven months.[138] Rae's expedition would have to live and travel much as aboriginal people did. He arrived at Repulse Bay, which had been discovered in 1742 by Christopher Middleton, on 25 July and immediately crossed the isthmus to Committee Bay, hoping thereafter to reach the farthest point surveyed by Dease and Simpson, but was prevented from doing so by ice. Instead, he made a reconnaissance along the western coast of the Melville Peninsula before returning to Repulse Bay. Then, unlike all previous expeditions, which had returned to a sub-arctic base at the end of a season of surveying, Rae stayed in the arctic; he and his men wintered uncomfortably in a shelter they built of rock and wood near an Inuit community at Repulse Bay.[139] In 1851, in a published description of Rae's expedition, the experienced Arctic traveller John Richardson singled out that accomplishment. With an obvious sense of awe, he told his readers of Rae's "boldness and confidence in his own resources that has never been surpassed."[140]

Come spring, Rae once again attempted to survey westward from Committee Bay, but convinced that Boothia was a peninsula, he had an Inuit man in that region draw him "a chart which he made of the bay," which, "agreed very closely with one drawn by the natives of Repulse

Bay."[141] Rather than continue a survey he now regarded as pointless, he turned back on 19 April to complete a survey of the Melville Peninsula. After walking nearly to Fury and Hecla Strait (confirming that there was no strait south of it), he returned to Repulse Bay, and was back at Churchill on 31 August.

Inevitably, Rae's highly successful explorations earned the HBC more renown than his scientific work. The journals reveal his impressive skills with aboriginal people, his capacity for overland travel, and his tolerance of – and even indifference to – cold and privation, combined with his ability to maintain an esprit de corps among his men despite the hardships. The scientific results were meagre. Still, his published journals made the most of them. They included appendices by W. J. Hooker (on plants), James Tennant (on rocks), and J.E. Gray (on zoology), charts of magnetic phenomena, and weather observations. Rae also included natural history observations in his journal, and collected a number of specimens that were submitted to the British Museum – although at a time when natural historians were growing more interested in the range and variation of species, his failure to note the exact location where each specimen was acquired badly undermined their usefulness.[142]

While the HBC assumed the leadership of efforts to map the northern coastline of North America after 1836, the Royal Navy continued to probe the Arctic Archipelago. These efforts included John Franklin's most famous expedition, the one in which he and all the other members perished – "so many gallant victims to science," as John Richardson, described them in 1851.[143] Since that expedition, last seen in May 1845, carried with it a three-year supply of food, the Admiralty did not begin searching for the expedition until 1848, but thereafter the search for Franklin did as much to spur arctic exploration as did the search for the Northwest Passage itself.[144] Knowing that a party irretrievably trapped in ice might be forced to make for the North American mainland, some of the rescue attempts were made overland. Given the financial backing and reward offered by the British government, and the renown that would redound to anyone who could find Franklin, the HBC was motivated to join the search.[145] And Dr. John Rae's experience and expertise made him the HBC man best positioned to lead the company's efforts.

After his first expedition (1846–7), Rae went on furlough to London, where he met Dr. John Richardson. Richardson, worried about his friend John Franklin, invited Rae to join him on a rescue expedition. This search was guaranteed to become a well-publicized one that the HBC could not afford to ignore, despite the fact that the company was not its official sponsor. When the expedition left Montreal, George

Simpson ensured that it was accompanied by expert Canadian voyageurs to guide the canoes.[146] The inclusion of Rae in a Royal Navy expedition also represented an acknowledgment of the skills and abilities of this HBC employee. When Richardson explained his choice of Rae in his published account, he wrote that Rae was "thoroughly versed in all the methods of developing and turning advantage the natural products of the country, a skilful hunter, expert in expedients for tempering the severity of the climate, and accurate observer with the sextant and other instruments."[147]

The Admiralty's instructions to John Richardson made no mention of any scientific objectives – perhaps it would seem insensitive to order men to botanize on a rescue mission – but Richardson intended to indulge his scientific interests.[148] To that end, Richardson noted that he carried with him "an ample supply of paper for botanical purposes."[149] Contrasting with Rae's 1846–7 expedition, where the explorations were a success, in this case the expedition failed in its main goal to find Franklin's expedition. There is little doubt that, had it succeeded, the title of the published journals of the expedition would have emphasized the main goal of the mission, but instead, *Arctic Searching Expedition* focused on its not insignificant scientific achievements. The published account of the expedition includes many detailed scientific observations within the narrative, ethnographic chapters on the Inuit, Kutchin (Gwich'in), Chipewyan, and Cree, and a lengthy appendix with sections devoted to geography, climate and phenology, plant distribution north of the 49th parallel, insects (by Adam White of the British Museum), and Inuit and Dene vocabularies.

On quite a different note, *Arctic Searching Expedition* included another valuable endorsement of the HBC's monopoly from a respected man of science. Richardson wrote that "without entering into the question of the chartered rights of the Hudson's Bay Company, or the propriety of maintaining a monopoly of the fur trade it is my firm conviction, founded on the wide-spread disorder I witnessed in times of competition, that the admission of rival companies or independent traders into these northern districts would accelerate the downfall of the native races."[150] The London Governor and Committee must have welcomed this endorsement, published as it was during a decade of rapid economic liberalization in the British Empire, and when the company was once again under public scrutiny.

In 1851, Rae set off once again to look for Franklin's party, this time on orders from George Simpson. He found two pieces of wood that he assumed to be part of Franklin's ships, but was unable to learn more.

While doing so, however, he "added about 500 miles of coast to our geographical knowledge ... along the shores of Wollaston & Victoria Lands."[151] In 1852, the Royal Geographical Society awarded him its Founder's Gold Medal in recognition of his explorations of 1846–7 and 1851.

Finally, at Rae's own request, the HBC sent Rae on yet another expedition to map the last remaining unsurveyed portion of the coast of North America, a section from Bellot Strait (between Boothia Peninsula and Somerset Island) that Rae himself had reached in 1846, to the western end of Boothia Peninsula, which Dease and Simpson had surveyed in 1839.[152] He spent the winter at Repulse Bay (in a snow shelter modelled on those of the Inuit, instead of the cold and clammy structure he built in 1846) before completing his surveys in 1854. In 1854, a party of Inuit told Rae about the fate of Franklin's expedition, and Rae was able to secure a few artefacts undeniably connected with it. Rae's reports that the Inuit believed that the crew of the Franklin expedition resorted to cannibalism were met with horror and, in some circles, disbelief. The revelation may have been particularly bitter for those in the Admiralty, for its irony emphasized the stark difference between modes of arctic travel preferred by the HBC and the Royal Navy. Historians and contemporaries have noted that the Royal Navy disapproved of Rae's highly successful use of Inuit methods of subsistence and travel, despite their obvious practicality – "anybody can succeed *if* he is willing to go native," Vilhjalmur Stefansson later claimed Rae's critics to have said.[153] But now, the information supplied by the Inuit suggested that the Franklin crew – exploring as well-prepared civilized British men were expected to – had exhibited the kind of behaviour that the British expected to find only in the most savage of peoples, while the HBC expeditions, emulating Inuit practices, had not lost a man. If, as one historian has argued, "Rae was no enthusiast for the Admiralty nor for the naval explorers," the reasons are not hard to see.[154]

Despite the controversy, Rae and the HBC were rewarded for their efforts. As E.E. Rich has argued, Rae's "explorations and successful search for the remains of Franklin's expedition brought the greatest credit to himself and his Company."[155] A contemporary author reported that "a gratuity of 400*l*. was awarded Mr. Rae, by the Hudson's Bay Company, for the important services he had thus rendered to the cause of science."[156] Rae was also eventually awarded £10,000 from the Admiralty for discovering the fate of the Franklin expedition, although Rae's confirmation of the death of Franklin and all of his men also marked the end of

5.4 *John Rae,* **by Stephen Pearce**. Typical of many HBC officers, Dr. John Rae did not wish to convey an impression that his years in Indian country had diminished his civility. Obviously proud of his ability to withstand cold, hunger, and isolation, John Rae nonetheless donned his sumptuous academic robes and posed as an aristocrat when he sat for this portrait. Stephen Pearce, the artist, portrayed Rae here in the 1850s as an energetic man whose distant gaze is so resolutely fixed on his mission that he remains unaffected by the presence of the viewer. Also see colour plate 5. © National Portrait Gallery, London, England, (NPG 1213).

official British and HBC expeditions to the arctic until after 1870. In 1880 Rae was elected a fellow of the Royal Society of London, and was invited to attend meetings of the Royal Geographical Society, the British Association for the Advancement of Science, and the Canadian Institute.[157]

John Rae subsequently emerged as an important defender of the HBC. For example, Rae sharply criticized the British naval officers who cast aspersions on the HBC's treatment of aboriginal peoples. He responded by writing that "these self-sufficient donkeys come into this country, see the Indians sometimes miserably clad and half-starved, the causes of which they never think of enquiring into, but place it all to the credit of the Company."[158] He also appeared in 1857, on behalf of the HBC, before the British Select Committee to inquire into the company.[159] According to William Mactavish, so dogged was Rae's defence of the company that "scurrilous Canadian newspapers," referred to him as "Sir George's Henchman."[160] His defence of the company extended to articles he published in learned journals. In an article in the *Transactions of the Ethnological Society of London* in 1865, Rae argued that the Natives along the coast of Hudson's Bay were, "said to be increasing in number on this part of the coast, partly, I suppose, because at Churchill, as at all the Hudson's Bay Company's posts where it can be done, an extra stock of provisions is specially kept, to provide against starvation in seasons of scarcity, to which all places in Northern America are exposed, where animal food forms the sole means of subsistence of the natives. These provisions are always given away or sent to a distance gratuitously when required."[161] Rae offered a good example of how a company man who became famous could use his stature to defend his employer's reputation in the scholarly as well as in the political, realm.

Rae also vociferously defended the Inuit, with whom his own reputation was so closely connected.[162] In the same 1865 article in which he defended the HBC, he argued that "there are few races of men about whom a greater diversity of opinion has been expressed than the Esquimaux."[163] But Rae had argued that "the more I saw of the Esquimaux the higher was the opinion I formed of them, and that they were much more susceptible of civilising influences than any of the Indians tribes I had met."[164] Among the Inuit, he claimed, "the wife is treated as an equal, and indeed generally rules the establishment, which are said to be *signs of civilisation.*"[165] In another article published in the *Journal of the Society of Arts* in 1882, he added that "I may be laughed at for saying that the Eskimos are scientific, but I think I can prove the fact." Thereafter, he supported his argument by describing the ingenuity of Inuit snow goggles, igloos, sledge runners, and kayaks, and the Inuit treatment of frostbite, which contrasted sharply with British treatment at the time.[166] Mactavish may have resented the portrayal of Rae as Governor Simpson's pawn, but Rae was certainly a tenacious defender of the HBC.

If Rae was Simpson's henchman, another accomplished HBC man became one of the most important critics of the HBC in the 1840s and 1850s. Alexander Kennedy Isbister was born at Cumberland House in 1822, the son of HBC assistant trader Thomas Isbister from the Orkney Islands and his mixed-blood wife Mary Kennedy.[167] Thomas Isbister, who had collected birds for John Richardson in the 1830s, ensured that his son received a good education, first at the Orkney Islands (Thomas's homeland [1829–34]), and then at the Red River Academy (1834–6).[168] Alexander Isbister then worked for the HBC between 1837 and 1841, serving at Norway House, Fort Simpson, and Fort McPherson (Peel River House). The young man showed considerable initiative. In 1839, building upon knowledge acquired by Dease and Simpson, the HBC instructed John Bell and Isbister to establish a post on the Peel River (initially called Fort McPherson). Isbister wrote that when, thanks to John Bell and Thomas Simpson, he "unexpectedly found [himself] in possession of the means of making a survey," he undertook to survey the Peel River.[169]

Isbister apparently grew disillusioned by 1841 when he concluded that, despite his education and energy, he stood little chance as a mixed-blood person of ever being promoted to the HBC's officer ranks.[170] He left the HBC territories for the last time in 1842 when he went to Scotland to study at King's College (Aberdeen) and the University of Edinburgh. According to Barry Cooper, Isbister's biographer, Isbister did not emerge as opponent of the HBC before 1846.[171] Even after he emerged as a "respectable critic" of the company, as Cooper dubbed him, he always argued that his opposition to the company was not rooted in any personal grudge. Still, the language he used in 1845 when he published his surveys of the Peel River in the *Journal of the Royal Geographical Society of London* seems in hindsight to foreshadow the much more critical posture he would later take. He introduced his article by claiming that "under happier auspices, and had the sanction of the Hudson's Bay Company been extended to the undertaking, I should have been enabled to devote more time to those botanical and geological researches which confer so much value upon the narratives of my predecessors in Arctic discovery."[172] The article included a map of the Peel River and some fairly detailed observations on the Native communities, botany, and zoology of the Peel River region. This was the first of his many scientific publications. Cooper perceptively assessed the significance of these scientific publications: "the 'detached' pursuit of science served to establish Isbister's credentials as an authority on the land and peoples ruled by the Hudson's Bay Company."[173] Building upon his new credentials, Isbister applied to the Colonial Office in February 1846 (during the Oregon Crisis) to travel to the Columbia District to provide "a scientific description of the vast region under the sway of the Hudson's Bay Coy and of its interesting inhabitants." To support his application, he offered to provide recommendations from "the highest names in science in Scotland." The Colonial Office declined the offer.[174]

Between 1846 and the 1860s, often working through the Aborigines Protection Society, Isbister acted as a formidable opponent of the HBC, a staunch defender of the rights (as he saw them) of the natives of the HBC territories and the Red River colony, a firm believer in liberal British imperialism, and a respected educator in England. His glancing blow in his Peel River article was soon supplemented by more direct and repeated attacks on the company in the form of lobbying, delegations, letters, petitions, and pamphlets, the first of which was *A Few Words on the Hudson's Bay Company* (1846). In *A Few Words* Isbister attacked the HBC, "the only survivor of the numerous exclusive bodies which at one time depressed almost every branch of British commerce," in a way reminiscent of the previous attacks of Dobbs, Robson, and Umfreville (Isbister even quoted Robson and Umfreville), but drew additionally on the ideas of Adam Smith.[175] Isbister hoped that his efforts, which showed "a marked lack of discrimination between facts and falsehoods," according to John Galbraith, would undermine the company's efforts to renew its licence.[176] But his efforts availed little. Isbister's attacks did inspire William E. Gladstone and the Earl of Lincoln to raise probing questions in the House of Commons in 1848 and 1849 that led the House of Commons to direct the Colonial Office to inquire into the legality of HBC's charter.[177] Informed by the attorney general and solicitor general in January 1850, that "we are of opinion, that the rights ... claimed by the Company do properly belong to them," and finding no one willing to pursue the question in the courts, officials in the Colonial Office found no way to investigate the matter further.[178] In fact, thanks to the January 1850 opinion – of which HBC officials reminded subsequent colonial secretaries – the HBC's legal position was stronger after Isbister's attacks than before. Although few elected politicians or newspapers came to the defence of the company, officials in the Colonial Office seem consistently after 1850 to have dismissed Isbister's (and others') attacks.[179] Herman Merivale, the permanent undersecretary in the Colonial Office, described one attack on the company published by the Aborigines Protection Society in 1856 as "a violent and unscrupulous attack on the Hudson's Bay Company and appeal to Canadian interests and passions of the most reckless kind."[180]

Isbister seemed eager to complement his overtly political campaign against the HBC with scientific publications that certainly must have added to his reputation as a disinterested scholar.[181] In 1848 he published three short ethnological articles on Dene-speaking peoples of the Mackenzie River valley.[182] They are startlingly disparaging of those peoples, particularly when compared with Joseph Howse's *Grammar of the Cree Language* published four years earlier (see chapter 4). For example, Isbister wrote that Chipewyan "is exceedingly meagre and imperfect – not only barren, as most Indian languages are, of abstract and general terms, but singularly deficient in the means of expressing the commonest objects of nature; thus, there is but one word for *a kettle, a stove, a spade, a spoon,* and *a tin dish* – *because these articles are all manufactured from iron*."[183] But his depictions of these communities are particularly noteworthy for how obviously they are intended to reflect upon the HBC. For example, he argued that "Hare Indians" had "the most intimate connection with the Hudson's Bay Company of all the Chippewyan tribes, and they show the effects of that connection. Their condition is the most wretched and deplorable that can be imagined. Cannibalism, almost justified by the extreme necessity of the case, exists to a frightful extent. … It is from this wretched tribe that the Hudson's Bay Company draw nearly all the profits of their trade in this quarter."[184]

Meanwhile the "Dog-Ribs" who were "almost entirely independent of the whites," were "well-clothed in the skins of the rein-deer, and have all the elements of comfort and Indian prosperity within their reach. They are a healthy, vigorous, but not very active race, of a mild and peaceful disposition."[185]

In the mid-1850s, Isbister also published on the geology of the Mackenzie River Valley. The evidence for that work came largely from the various arctic exploring expeditions, although it also drew on fossils collected by Dr. Robert Kennedy at Moose Factory, and rocks collected by George Barnston. Aside from its scientific interest, that work was important for highlighting the presence of previously little known significant coal and mineral deposits in the HBC territories – something upon which opponents of the company could and did seize.[186]

Notwithstanding the attacks of Alexander Isbister and the Aborigines Protection Society, the HBC collected a number of very valuable tributes in connection with its support of northern science between 1818 and 1855. Just as assistance from the HBC had been prominently acknowledged in the groundbreaking taxonomic work of George Edwards in the 1740s and Thomas Pennant in the 1780s, the company's contributions were fully recognized in the pre-eminent taxonomic collection of the 1820s and 1830s.[187] The tributes to the company and its directors and employees were getting longer and more detailed. No publication paid more eloquent or prominent tribute to the HBC's sponsorship of science during the first half of the nineteenth century than the six-volume report on the zoology and botany of northern North America, spearheaded by John Richardson. The natural history specimens collected during the second Franklin expedition, "being too numerous for a detailed account of them to be comprised within the ordinary limits of an Appendix to the narrative of the proceedings of the journey," Richardson resolved to report on them in a separate work.[188] Aided by a grant of £1000 from the British government, John Richardson's illustrated four-volume *Fauna Boreali-Americana* (1829–37) (one volume each on "quadrupeds," birds, fish, and insects) and William Jackson Hooker's two-volume *Flora Boreali-Americana* (1833–40) became seminal publications in their fields for over thirty years.

Although these volumes relied most heavily upon specimens collected by John Richardson and Thomas Drummond themselves during Franklin's second expedition, Richardson was eager to make known the many ways in which the directors and employees of the company had helped make his publication possible. He paid tribute to many people, from the HBC Governor and Committee to the traders in the field. For example, in the third volume, on fish, Richardson expressed his "obligation in an especial manner to Captain Pelly, Governor of the Hudson's Bay Company, and [Nicholas] Garry, Esq., Deputy Governor, for the liberality with which they have always promoted my endeavours to illustrate the zoology of the fur-countries."[189] In the first volume, on "Quadrupeds," Richardson detailed some of the ways in which the London office had helped him. He thanked the Governor and Committee "for granting me free access to their museum, and to the manuscript accounts of the Fur Countries [obviously going as far back as those of Hutchins], in their possession, and for the strong recommendations they transmitted to the resident Chief Factors and Chief Traders, to forward the views of the Expedition, with respect to Natural History."[190] He further explained that "much information was ... derived from frequent visits to the museum of the Hudson's Bay Company, and from repeated examination of the specimens imported by that Company from their posts on James's Bay, on the Columbia, and in New Caledonia, and presented by them to the Zoological Society and British Museum."[191]

Richardson also acknowledged the help of traders in the field. In the second volume, on Birds, Richardson explained that specimens "have continued to be transmitted annually to London up to the present time, as presents either to the Governor and Committee or to the personal friends of the parties. The former, besides forming a museum of the Hudson's Bay productions, which is liberally open to the public, have presented numerous specimens to the Museum and Zoological Society."[192] He also mentioned a moose sent from Churchill to "His late Majesty."[193] Some of the specimens described in the volumes came as the direct result of collections made by HBC officers and submitted to the members of Franklin's second expedition. For example, "a collection of birds and quadrupeds, of much interest, made at Fort Nelson on the River of the Mountains [Liard River], a branch of the Mackenzie, was forwarded to us by Mr. Macpherson [Murdoch McPherson], together with some valuable specimens obtained in the same quarter by Mr. [Edward] Smith, Chief Factor of that district. Mr. [Thomas] Isbister also had the kindness to prepare for us copious collection of birds at Cumberland House."[194] In various other places Richardson acknowledged HBC collectors by name including John Haldane, James Leith, John Pruden, Robert McVicar, Murdoch McPherson, Nicholas Garry, the deputy Governor, John Scouler, Meredith Gairdner, and Peter Warren Dease (particularly with fish).[195] In the first sentence of the introduction of the second volume, co-authored with William Swainson and published in 1831, John Richardson announced that "science is indebted to the exertions of the Hudson's Bay Company for almost all that is known of the ornithology of the American fur countries."[196] Richardson's accolades even extended to earlier traders of the HBC. He praised Samuel Hearne's account of the beaver, and quoted it at length.[197]

The HBC's contributions to the search for the Northwest Passage earned it highly visible and positive publicity in Great Britain. Not only was the company recognized for its assistance to the Royal Navy, it also earned recognition for its own highly successful expeditions. Furthermore, the company received the tribute of many men of science, none more prominent than that which it garnered in the pages of *Fauna-Boreali*. These tributes reinforced earlier testimonies to the company's contributions to geographical knowledge and added attestations to the company's benevolence and zeal for natural history. However, the company's contributions to science in the arctic in this period, as we shall see, were supplemented by its assistance to scientists farther south.

6 "The Liberal Spirit": David Douglas, Edinburgh, and the Douglas Legacy, 1823-1870

The search for the Northwest Passage was one of the most high-profile ways in which the HBC supported science after 1818, but it was not the only way. Scientific organizations with which the company might cooperate proliferated in Great Britain in the 1820s. One scholar has described the period between 1820 and 1870 in Britain as the "heyday of natural history," during which "natural theology made the study of natural history not only respectable, but almost a pious duty."[1] Many individual scientists and scientific organizations in England and Scotland after 1823 turned to the northwest coast of North America as a source of a wide range of plants (particularly evergreen trees) for their ornamental, scientific, and commercial value. The networks that linked the HBC and its directors, elite metropolitan scientists (such as William Jackson Hooker), scientific travellers (such as David Douglas), and HBC officers in the peripheries (such as George Barnston) illustrates how many motives, from prestige and profit to religious devotion, encouraged the development and maintenance of scientific brotherhoods, in which shared fascinations with nature tied men in what appears to have been highly emotionally satisfying friendships.[2] The history of the HBC's cooperation with British naturalists also exemplifies how important social intelligence could be for sustaining those networks. David Douglas's contributions to British knowledge of the natural history of the Columbia District were significant in and of themselves, but it is also thanks to his ability to infect others with a passion for science that his remarkably large influence on scientific activity in the HBC territories endured well beyond his own tragic death in 1834.

The HBC particularly supported natural history in the Columbia District (Oregon Country) in the 1820s and '30s. This region was important to the company because it was only in 1821 that the HBC's licence was extended to this region, and because the British hold on the territory was tenuous. During wide-ranging negotiations relating to the boundary between the British and American possessions in North America following the War of 1812, British and American diplomats were unable to decide on a border west of the Rocky Mountains, so they agreed in 1818 on a joint occupation of the Columbia District (Oregon Country) for ten years. In 1821, with the merger of the NWC and HBC, the British government essentially handed its authority in the region to the HBC. Thus, the HBC inherited an uncertain tenure in a highly profitable fur-trading region about which its directors knew little. Although the most profitable fur-bearing region was in the northern part of the district (known as New Caledonia), the more southerly regions had far greater agricultural potential, and the Columbia River became the lifeline of the entire district.[3]

In 1824, with only four years left in the joint occupancy agreement, and negotiations between the British and American governments soon to be resumed, the London Committee wanted to learn as much as possible about the Columbia District. In 1823 it had ordered that Peter Dease be sent to explore the Finlay River and other rivers west of the Rocky Mountains and north of the Fraser, but Dease had been unable to complete the expedition. The company then ordered "that Mr. [Samuel] Black be directed to prosecute the original object of the expedition towards the Frozen Ocean, in the discovery of whatever may tend to promote Science and encourage Mercantile speculation and that he be provided with such assistance and facility as he may require."[4] More importantly, it ordered George Simpson to visit the Columbia District.[5] Reporting later on that visit, Simpson wrote that "the Columbia presents a wide field for botanical research as there is a great variety of Plants to be found every where; I regret exceedingly that my ignorance of that interesting branch of Science prevents my attempting any description of them." He immediately added, implying the need for a naturalist to be sent to the district, that "any one of experience in the study of natural history generally would add much to his stock of knowledge therein by a visit to this part of the World." He reported that in the meantime "specimens of every kind within our reach will this season be sent Home as I have given directions to that effect at the different Establishments."[6]

Traders at the various trading posts evidently responded with alacrity to Simpson's orders, for a letter addressed to the Chief Factor at Fort Vancouver, Dr. John McLoughlin, in September 1826 noted that "curiosities" that had been submitted "were very acceptable and have made a considerable addition to a small museum now forming here, and we have to desire that any interesting specimens of natural history which may be collected should be sent home."[7] The letter implies that the HBC's museum at its headquarters in London was established shortly after the merger of the HBC and NWC, and may have been created to house the collections arriving from the Columbia District and the arctic. The museum, which was "liberally open to the public," was another means by which the company could support science and public interest.[8] In 1829 and into the 1830s, John Richardson mentioned having been given frequent access to the HBC museum.[9]

If the HBC had reason to learn more about the Columbia District, so did British naturalists and horticulturalists. British and European interest in exotic plants for their pharmaceutical, commercial, botanical, and ornamental interest long predated this period, but interest surged in Britain after the War of 1812.[10] In the ensuing decades estate owners and prestigious horticultural societies, including the Horticultural Society of London (now the Royal Botanical Society), prized rare and exotic flowers, shrubs, and trees from the Americas, Asia, and Africa.[11] In 1821, the Horticultural Society of London (founded in 1804) secured a thirty-three acre parcel at Chiswick for the purposes of an experimental garden geared particularly to ornamental plants.[12] The botanical discoveries made by Archibald Menzies (as part of George Vancouver's voyage) and by Lewis and Clark hinted at the ornamental value of plants along the northwest coast of North America, a region with a moist temperate climate much like that of Great Britain.[13] Thanks to the HBC, the Horticultural Society of London's Chiswick Gardens was quickly stocked with hundreds of plants from the Columbia District previously unknown to European and American gardens.

In 1823, on the recommendation of W. J. Hooker, chair of botany at the University of Glasgow, the Horticultural Society of London hired David Douglas, who was then working under Hooker's supervision at the Glasgow Botanic Garden.[14] On behalf of the society, Douglas made a trip to the eastern United States in 1823. He later explained how he came then to be chosen to travel to the Columbia District. According to Douglas, "the Horticultural Society of London, desirous of disseminating among the gardens of Britain the vegetable treasures of those widely-extended and highly diversified countries, resolved on sending a person experienced in the modes of collecting and preserving botanical subjects, and of transmitting seeds to England."[15] In the spring of 1824, Joseph Sabine, the secretary of the society, asked the HBC to transport Douglas to the Columbia District.[16] On 24 June 1824, the HBC Governor and Committee offered to the Horticultural Society of London free passage to the Columbia District for their collector, and in July wrote a letter instructing John McLoughlin, who was in charge of the district, to "afford every assistance in promoting the objects of his Mission," an order with which McLoughlin readily complied.[17]

When Douglas boarded the HBC ship *William and Anne* in July 1824, he joined an "old friend & zealous botanical associate," John Scouler.[18] But while Douglas came aboard as a guest of the HBC, Scouler came as a new employee – a surgeon-scientist to serve aboard the *William and Anne*. Glasgow-born Scouler had evidently not applied for the job, but had been offered the position upon recommendations by W. J. Hooker and John Richardson.[19] Hooker, when later recalling these events, remarked that "the Hudson's Bay Company, with a liberality that reflects the highest credit upon them, made application and provision for a surgeon to one of their ships, who, to his medical knowledge, should have added the acquirement of natural history, particularly of botany."[20] For his part, when he published an account of the trip he and David Douglas made from London to Fort Vancouver, Scouler explained that "the Hudson's Bay Company, with an honourable zeal to advance the knowledge of those extensive regions which are within the sphere of their commercial exertions, were anxious to have a surgeon ... who, in addition to his professional acquirements, was qualified to make collections in the various branches of natural history."[21] After accepting the position, Scouler met with Archibald Menzies (who had accompanied George Vancouver's voyage) and John Richardson in London, before departing for the Pacific Northwest.[22]

The *William and Anne*'s voyage around South America afforded the two young men opportunities to collect together during stops in Madeira, Brazil, Juan Fernandez, and the Galápagos Islands. After the long voyage Scouler and Douglas collected their first specimens in the Columbia District on 9 April 1825.[23] While at Fort Vancouver, the young men had recourse to a library of sixty volumes that included, along with at least one book by their mentor, W. J. Hooker, the four volumes of George Vancouver's *Voyage of Discovery To The North Pacific Ocean, And Round The World In The Years 1791–95* (1798), and other books on natural history, exploration, chemistry, medicine and pharmacy, astronomy, mathematics, and physiology.[24]

Scouler then botanized in the environs of Fort George and Fort Vancouver until the end of May when he accompanied the *William and Anne* on a trading expedition to Haida Gwaii (the Queen Charlotte Islands),

Vancouver Island, and the Strait of Juan de Fuca. Although he was on a five-year contract, Scouler then soon left the company. He left Fort Vancouver permanently on 20 September, although not before raising the ire of the local Chinook people by raiding one of their graveyards for skulls.[25] Hinting at the importance of their companionship, Douglas wrote Hooker, "I felt very lonely for some weeks after he had sailed."[26] Scouler, meanwhile, acknowledged that "to the encouragement of the company, and the cheerful assistance I obtained from their servants, I am entirely indebted for the numerous excursions and extensive collection I was enabled to make."[27]

Although Scouler left the HBC after only a few months, his time in the Pacific Northwest may have been an important turning point in his life. One of his contemporaries noted that "on his return to Glasgow he settled down to the practice of medicine; but in a short time a more congenial sphere of labour, although a less lucrative one, was opened up to him by his being appointed Professor of Natural History in the Anderson University."[28] He remained professor (and curator of the museum) at Anderson University, Glasgow, before moving on in 1833 to a professorship in mineralogy at the Royal Dublin Institution.[29] In 1848 he published a paper on the aboriginal peoples of the northwest coast of North America.[30] His experience in the Columbia District obviously was an important early step in his scientific career, and one that continued to influence him twenty years later.

Chief Factor Dr. John McLoughlin was crucial to the reception of all scientists in the Columbia District. McLoughlin, now known affectionately as the Father of Oregon, was trained as a doctor in Lower Canada before joining the NWC. After the merger of the NWC and HBC, he rose quickly to Chief Factor, a position he held from 1826 to 1846. During those years he was stationed primarily at Fort Vancouver.

Relatively little evidence survives relating to McLoughlin's direct contributions to science. He apparently submitted seeds directly to W. J. Hooker.[31] And in 1835 and 1838, the prominent French scientist and politician François Arago published an analysis of meteorological journals originally kept by McLoughlin at Fort Vancouver in 1832 and 1833 and from April 1836 to March 1837.[32] He also assisted his visitors. John Scouler similarly wrote of McLoughlin that "from him I experienced the utmost politeness & to his kindness was indebted for some curious specimens of the rocks of the Rocky Mountains."[33]

But McLoughlin is most famous (and controversial) for the hospitality he showed to many who came to Fort Vancouver – including scientists – regardless of nationality. Although Douglas was a visitor, not an employee, he was welcomed warmly at Fort Vancouver. David Douglas acknowledged that McLoughlin "received me with demonstrations of the most kindly feelings, and showed me every civility which it was in his power to bestow."[34] In fact, in 1826, after deciding to extend his stay by a year, Douglas wrote George Simpson to say, "with infinite gratification do I mention the friendly attentions and assistance I have experienced from every person connected with the Company in particular John McLoughlin Esqr."[35] Almost certainly at Douglas's prompting, the Horticultural Society of London rewarded McLoughlin with its silver medal on 11 May 1826 "for his assistance rendered Mr. David Douglas, whilst making his collections in the countries belonging to the Hudson's Bay Company in the Western part of North America."[36] Douglas was also able to benefit from the work of traders such as John Work, who had begun collecting specimens, perhaps in response to Simpson's orders of 1824, before Douglas arrived.[37]

Douglas's collecting between 1825 and 1827 took him from the mouth of the Columbia River to the dry interior region, and from the Willamette River to Puget Sound, before he tagged along with HBC brigades that took him across the HBC territories to York Factory (figure 6.1). During that trip across the continent, Douglas experienced the fraternity that existed within the relatively small community of British scientific travellers. He made sure to meet with his old acquaintances who were returning to England after the second Franklin expedition in 1827, Thomas Drummond at Carlton House and John Richardson at Cumberland House, and took a month-long side trip to Red River, collecting all the way. Drummond showed Douglas his collection and generously pointed out where Douglas might collect promising seeds near Carlton House.[38] Douglas, Drummond, and other members of the Franklin expedition (although John Franklin himself returned to England via Montreal) then travelled back to England together on a HBC ship sailing from York Factory.[39] While crossing the Rocky Mountains, Douglas also took advantage of an opportunity to pay tribute to fellow scientists by naming Mounts Hooker and Brown, the first after his mentor and friend, and the second after Robert Brown, "the illustrious Botanist, a man no less distinguished by the amiable qualities of his mind than by his scientific attainments."[40]

Douglas garnered considerable attention in Britain as a result of this trip. One of his friends noted that "his company was now courted[;] ... he seemed for a time as if he had attained the summit of his ambition."[41] Soon after his return, he was elected to membership in the Linnaean

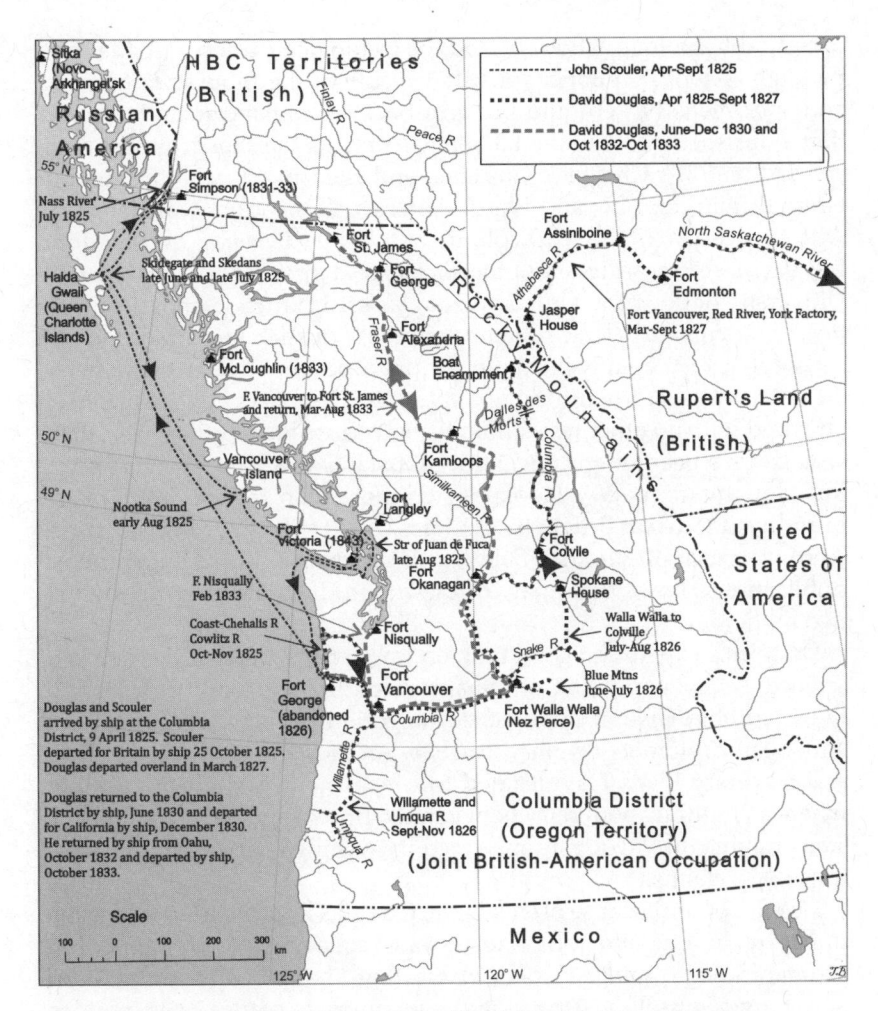

6.1 The Columbia District, 1821–1838. The Columbia District (Oregon Territory) was the focus of a great deal of horticultural interest beginning in 1824. The map shows the travels of John Scouler and David Douglas. Scouler served as a surgeon and naturalist aboard the *William and Anne* during its trading expedition to Vancouver Island, Haida Gwaii (Queen Charlotte Islands), and the Nass River during the summer of 1825. David Douglas spent several years in the Columbia District. He travelled along the Columbia River between its mouth and Fort Colville several times (typically botanizing in the lower country in winters and higher country in summers). The dates of his most remarkable travels away from the Columbia River are indicated. Source: T. Binnema.

Society, the Zoological Society, and the Geological Society.[42] Aside from the eight scientific publications that resulted, the thousands of seeds and specimens he sent and brought back to England included many that flourished as ornamental plants, first in the Horticultural Society of London's Chiswick Gardens, and then in gardens throughout Great Britain, Europe, and North America. The society was delighted with the results. In the 1830 volume of *Edward's Botanical Register*, John Lindley, assistant secretary of the society and professor of botany at the University of London, argued of Douglas's introduction of the "purple-flowering currant" (*Ribes sanguineum*) that "of such importance do we consider it to the embellishment of our Gardens, that if the expense incurred by the Horticultural Society in Mr. Douglas's voyage had been attended with no other result than the introduction of this species, there would have been no ground for dissatisfaction."[43] But Douglas introduced many more. W. J. Hooker listed more than 150 species of plants introduced to Great Britain by Douglas as a result of his 1825–27 trip.[44] And a perusal of *Edward's Botanical Register* and other contemporary publications shows that many of these introductions were highly valued at the time.

Douglas's expedition soon took on importance beyond the world of science. As the expiration of the agreement for the joint occupancy of the Columbia District approached, negotiations between British and American diplomats resumed in 1825. As they dragged on, the Colonial Secretary R. W. Hay decided to consult with Douglas to ask him where a "natural" boundary between the British and American possessions might run.[45] Douglas answered by writing a letter that defended the HBC's claims.[46]

Despite of (or because of) his fame, Douglas appears to have become unhappy in England. His friends and colleagues soon found him arrogant and irritable. Hooker wrote that "happy as he unquestionably found himself in surveying the wonders of nature in its grandest scale, ... it was quite otherwise with him during his stay in his native land."[47] "His temper," Hooker admitted, "became more sensitive than ever, and himself restless and dissatisfied; so that his best friends could not but wish, as he himself did, that he were again occupied in the honourable task of exploring North-west America."[48]

Douglas and his friends soon got their wish. Douglas returned to North America in 1830, again with an HBC offer of free transportation to the Columbia District, and again with payment from the Horticultural Society of London, but, tellingly, this time with an offer from the Colonial Office to pay part of the expenses of the trip in exchange for geographical information, maps, and other information.[49] The connection between science and geopolitics was clear. Fortunately, Douglas's acquaintances in the Columbia District found him as cheerful and affable as ever.[50] He and his friends learned first hand that for some, research is a mental and emotional tonic. Without doubt, they were neither the first nor the last to learn this.

Douglas spent the summer of 1830 in the Columbia District before taking a HBC ship to Alta California in December to botanize in Mexican territory.[51] He returned to the Columbia District and collected at Fort Vancouver during the winter of 1832–3. In the spring he embarked up river with the HBC brigade hoping that he might travel through New Caledonia to Russian America. When he realized that travel from Fort St. James to Russian America via Fort Simpson (on the northwest coast) was probably impossible, he returned instead to Fort Vancouver, but not before losing four hundred specimens in an accident on the Fraser River. When the dispirited Douglas arrived at Fort Vancouver, he was rejuvenated by meeting the newly hired HBC surgeon-naturalists Meredith Gairdner and William Fraser Tolmie. Douglas wrote to W. J. Hooker that "science has few friends among those who visit the coast of North-West America, solely with a view to gain. Still with such a person as Mr. McLoughlin on the Columbia [Gairdner and Tolmie] may do a great deal of service to Natural History."[52] Douglas left the Columbia District in October 1833 to collect in Hawaii, but he died there on 12 July 1834, mysteriously trampled to death by a bull at the bottom of a cattle trap.[53] He was only thirty-five years old.

According to his biographers, although he died young, "Douglas's harvest of plants and seeds established a record for species introduced by an individual into Britain, the leading country in botanical research. The gardens of the Horticultural Society were overwhelmed, and recourse was had to private nurseries."[54] Hooker noted in 1836 that "not only in this country [Great Britain], but throughout Europe, and in the United States of America, there is scarcely a spot of ground deserving the name of a *Garden*, which does not owe many of its most powerful attractions to the living roots and seeds which have been sent by him to the Horticultural Society of London."[55] Ironically, a society focused on introducing ornamental plants rather than economically valuable plants was also responsible for identifying and introducing conifer species that have become some of the most valuable trees in Britain's (and Europe's) forest industry, including the Douglas fir (*Pseudotsuga menziesii*), noble fir (*Abies procera*), grand fir (*Abies grandis*), Sitka spruce (*Picea sitchensis*), and lodgepole pine (*Pinus contorta*).[56]

Had Douglas lived long enough to publish his journals during the 1830s, the HBC would almost certainly have enjoyed tributes in another prominent and widely read publication. But he did not. A full edition of Douglas's journals was not published until the twentieth century. Still, the HBC did benefit from Douglas's thanks during the 1830s. In 1836 W. J. Hooker published a long memorial to Douglas, a student who had become a dear friend, incorporating extracts from his letters and journals, and including a section in which Douglas effusively thanked the company and its London Governors:

> I must beg leave to return my grateful thanks to John Henry Pelly, Esq., Governor, and Nicholas Garry, Esq., Deputy Governor of the Honourable the Hudson's Bay Company, for the kind assistance I, on all occasions, experienced at their hands, and for much valuable information received both before and after my arrival in England. To the enlightened zeal with which these gentlemen forward every enterprise for the advancement of Science, ... I am happy to have this occasion of bearing my grateful, though feeble testimony. I also beg leave to thank the different residents, partners, and agents of this Company, both individually and collectively.[57]

These words of tribute cannot have had the same impact as they might have had they been published in a volume of Douglas's journals, which undoubtedly would have sold well, but the London Governor and Committee could use them nonetheless to support the company's application for the renewal of its licence. The words "enlightened zeal" reflect a combination of intellectual curiosity and religious devotion that appears to have animated many men of science in Great Britain at the time. David Douglas also took the opportunity to name the Garry oak (*Quercus garryana*), and the silk tassel bush (*Garrya elliptica*) after the HBC's deputy governor, who, Douglas had written, had been "exceedingly kind to me."[58] The Governor and Committee could also console themselves with the fact that Douglas's large botanical collection was described in Hooker's *Flora Boreali*, and his zoological specimens were described by Richardson in *Fauna Boreali-Americana* and (somewhat less prestigiously) by James Wilson in *Illustrations of Zoology* (1831).

Douglas encouraged science in one other important way. Jack Nisbet has correctly argued that "not least among Douglas's gifts was an ability to communicate his enthusiasm to the astonishing variety of people that he met on his journeys."[59] Nisbet did not elaborate on this argument, but Douglas's infectious enthusiasm left a legacy in the HBC that helps explain the scientific activities of men such as Andrew Murray, Robert Wallace, John Jeffrey, and George and James Barnston.

Indigenous people helped both Douglas and Scouler in their collecting. That Native people acted as guides, paddlers, and freighters hardly needs mentioning. They were important throughout the HBC territories for those purposes. But Natives also helped with collections. Scouler and Douglas offered only glimpses into aboriginal collecting, but in 1825 John Scouler noted that "my herborising yesterday had attracted the notice of the Indians & one of the most intelligent among them brought me a *Monoecius* plant I had not before detected. To encourage this disposition I gave him a few presents & I doubt but he may bring many more interesting plants."[60] Douglas hired Native boys near Fort Walla Walla to capture lizards. "A slight reward," he wrote, "would put them in ecstasies, and they would again scamper off for renewed captures."[61] Douglas also paid a native man in tobacco for a specimen of North American tobacco (*Nicotiana*), and information on how to cultivate it.[62]

When malaria decimated Native populations in the lower Columbia River region in the early 1830s, John McLoughlin and the HBC seemed eager to hire new surgeon-naturalists to replace the departed Scouler. After turning again to John Richardson and W. J. Hooker for recommendations, in 1832 the HBC's London Committee hired William Fraser Tolmie and Meredith Gairdner as surgeon-naturalists, each on five-year contracts.[63] Both men arrived at Fort Vancouver in the spring of 1833. During two stints with the HBC (1833–41 and 1843–71) Tolmie spent time at Fort Nisqually, Fort McLoughlin, Fort Vancouver, and Victoria. It is impossible to know how many people and institutions received collections from Tolmie, a very serious and curious man, but he submitted at least several natural history collections and collections of aboriginal artefacts to Great Britain (including to John Scouler and to a museum in Inverness, Scotland) and the United States (the Smithsonian Institution).[64] For example, in an 1841 article on the aboriginal peoples of the northwest coast, from Russian America to the Columbia District, John Scouler acknowledged that his information came overwhelmingly from Tolmie.[65] More than forty years later, after British Columbia had joined Canada, Tolmie collaborated with George M. Dawson of the Geological Survey of Canada to write *Comparative Vocabularies of the Indian Tribes of British Columbia with a Map Illustrating*

Distribution (1884). That publication exemplifies the transfer of knowledge from fur traders to the government of Canada after British Columbia joined the Canadian federation in 1871. In his introduction to the volume, Dawson explained that Tolmie's "constant intercourse with the Indian tribes, while an officer of the Hudson's Bay Company and subsequently, give special value to the results of his investigations in linguistic matters."[66] Unlike Scouler, Tolmie made North America his permanent home, eventually entering British Columbia politics.

Meredith Gairdner was hired the same year as Tolmie, but his career was much shorter than Tolmie's. Gairdner had respectable scientific credentials before he joined the company, having published a book on mineral springs in 1832. He was a Scottish doctor and a student of Robert Jameson, professor of natural history at the University of Edinburgh and editor of the *Edinburgh New Philosophical Journal*.[67] While Tolmie was hired as a ship's surgeon, Gairdner was posted as a surgeon at Fort Vancouver. He, like Tolmie, arrived at Fort Vancouver at the end of April 1833, but stayed in the Columbia District only until late September 1835.[68] Gairdner spent most of his time at Fort Vancouver, but following what he claimed was a sudden onset of illness (pulmonary consumption) at the end of March 1835, he went up to Fort Walla Walla in May to try to improve his health.[69] Still unwell, he left for Hawaii in November, hoping to recover there, but not before repeating Scouler's indiscretion and raiding a Chinook graveyard for a skull – this time the skull of Chief Concomly. He eventually sent the skull, with a letter, to John Richardson, who deposited it at the museum at the Royal Naval Hospital Haslar in Gosport.[70]

Two excerpts of letters, probably letters Gairdner sent to Jameson, were published in the *Edinburgh New Philosophical Journal*, including one that describes the bitterly cold winter of 1833–4, during which the Columbia River was frozen over for three weeks, and another that describes two eruptions of Mount St. Helens in the early 1830s.[71] Gairdner also sent seeds home in 1833.[72] John Richardson acknowledged him as an "able naturalist" and as an employee of the HBC when he named the steelhead trout (*Salmo gairdnerii*) after him.[73]

George Simpson apparently grew sceptical of Gairdner's intentions when Gairdner was attempting to recuperate in Hawaii. In June 1836, Simpson wrote McLoughlin confidentially that "should Dr. Gairdner determine on quitting the Service it may be well to discourage his coming across the mountains, but that he should return to England by the Ship, as I understand he is collecting materials for the press, and it is not desirable that our trade should be brought under public notice."[74] McLoughlin never had to act on Simpson's caution because Gairdner's health continued to decline, and he died in Hawaii in March 1837, but the letter hints at the caution that continued to affect the directors of the company.

In various ways, much of the scientific work done in the Hudson's Bay Company territories after 1834 built upon the work of David Douglas. In direct and indirect ways, he was the inspiration for the scientific endeavours of Robert Wallace and Peter Banks, John Jeffrey, Andrew Murray, and George and James Barnston. Much of this work, like that of Douglas, was focused on the northwestern coast of North America, largely because plants introduced from that region flourished in British outdoor gardens. Scientists hoped that many new species might yet be found that Douglas had not collected. Gardeners were motivated by the fact that many plants collected by Douglas remained rare and expensive as nursery stock in Great Britain. However, although Douglas's precedent inspired subsequent expeditions, it may also have raised unrealistic expectations for them.

In 1838, Sir Joseph Paxton, the head gardener at Chatsworth Gardens (one of the foremost gardens in England at the time, on the estate of William Cavendish, 6th Duke of Devonshire), sent two naturalists, Robert Wallace and Peter Banks, on a three-year trip to the Pacific Northwest to follow up on Douglas's work.[75] Paxton sought to enlarge the already-impressive conifer garden on the estate, but was frustrated by the scarcity and expense of exotic confers in England.[76] His solution was to send Wallace and Banks to the Pacific Northwest. Contrary to advice from friends and from the HBC directors, themselves who recommended that the two men travel aboard a ship directly to the Columbia District as Douglas had, Paxton had the two men travel to New York on a HBC ship, then overland across the continent with a special HBC brigade that left from Montreal in the spring of 1838 (see figure 6.2).[77] The HBC gave the botanists a letter introducing them to its officers. It said that "the Governor and Committee feel much interested in the object of their mission, which is patronised by several persons of high distinction in this country. I have therefore to request that every assistance and facility that may be required by Mr. Wallace and Mr. Banks ... be afforded them."[78] This assistance was to include gratis "maintenance while at the establishments, for which no charge is to be made."[79]

The Wallace and Banks expedition ended tragically. On 22 October 1838, triggered by Wallace's own panic, the brigade's boat overturned

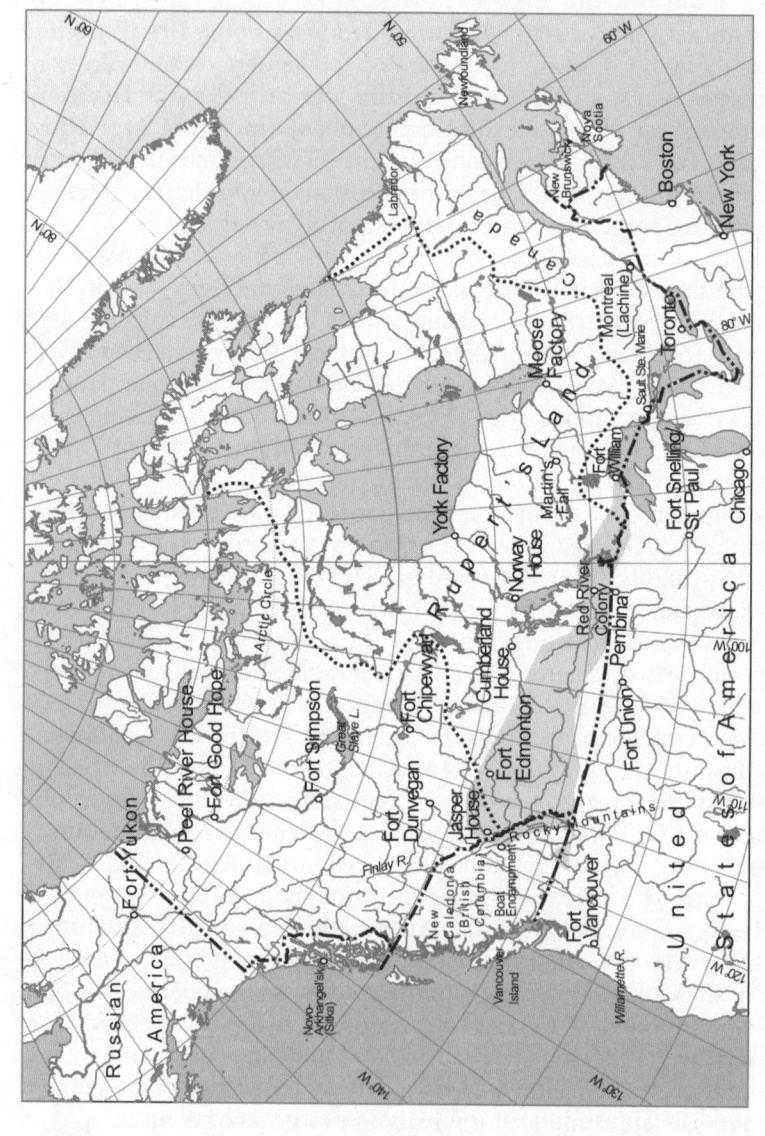

6.2 Hudson's Bay Company Territories, 1838–1870. Several scientists were active in the HBC territories between 1838 and 1870. The locations of places mentioned in chapter 6 are indicated on this map. The swath of grey between Fort William and the Rocky Mountains shows the region which the Palliser and Hind expeditions surveyed. Source: T. Binnema.

on the infamous *Dalles des Morts* (Death Rapids) on the upper Columbia River, causing the death of twelve people, including both naturalists and Wallace's wife. Still, the company received tribute when the notice of their deaths said that "we may embrace this occasion to declare publicly the handsome manner in which the gentlemen officially connected with the Hudson's Bay Company afforded every assistance in their power towards the accomplishment of our enterprise. At all times ready to sanction and further any measure tending to the advancement of science, these gentlemen have manifested a disposition to advance the ends of this expedition which is beyond all praise."[80]

After the death of Wallace and Banks, several years passed before British naturalists were sent again to the Columbia District. When they were, Sir W. J. Hooker, having been knighted in 1836, was behind it. His connections with the HBC dated back to the early 1820s, shortly after he had taken up the chair of botany at the University of Glasgow. Hooker expanded the botanical garden at Glasgow into a well-respected institution, complete with an American department.[81] But by the early 1830s he aspired to the directorship of the Royal Botanical Gardens at Kew, despite the fact that the position at the ill-managed gardens would require a pay cut.[82] After finally landing the position in 1841, he spearheaded a program that enlarged the gardens from eleven acres at his appointment to three hundred acres at his death, in the process transforming them into the pre-eminent botanical garden in the world.[83]

Hooker's ambitious program could not succeed without a global network of support, of which the HBC was a reliable part. Hooker was sufficiently well connected with the HBC that its governors instructed its employees to supply his wants. In September 1843, for example, Archibald Barclay, the HBC's secretary in London, wrote John McLoughlin that "Sir W. Hooker is extremely desirous to obtain some of the Nuts of the magnificent chestnut of California, the leaves of which I recollect are of a golden color on the underside, and if you have the means of procuring any, it would be as well to forward a few for him."[84] This reference to the giant chinkapin (then called *Castanea chrysophylla*, but now *Castanopsis chrysophylla*), also known as the golden chestnut, hints at what appears to have become an obsession for Hooker. On 9 October 1826 David Douglas first saw the "golden chestnut," and was able to preserve a specimen of its leaves but was unable to get any viable seeds.[85] The beauty of the leaves appears to have spurred W. J. Hooker's two-decade effort to secure a living specimen of the plant. According to one scholar, by 1843 "for twenty years or so he had been bombarding every likely traveler with requests for it [the golden chestnut], and all had failed him."[86] Although Hooker hoped the company's officers might search for the chestnut, he also hoped a collector of his own might do so. In 1843, he published a short article in the *London Journal of Botany* about the tree, "which in the beauty of its evergreen foliage far exceeds any hitherto known to us."[87] He explained that he was calling attention to the plant at that time because "it has been my good fortune, under the liberal patronage of the Governor and Directors of the Hudson's Bay Company, to unite with the Earl of Derby in sending out an able collector, (Mr. Burke) to North-western America and California, through whom we have every prospect of seeing this splendid tree introduced to our own pleasure-grounds and plantations."[88]

Hooker must have learned about Joseph Burke through Edward Smith-Stanley, the thirteenth earl of Derby. The eccentric Derby had turned away from politics in 1839 to concentrate fully on his immense menagerie, aviary, museum, and gardens at Knowsley Hall near Liverpool. He had sent Burke on a successful expedition to South Africa from 1839 to 1842, before retaining him as gardener at Knowsley Hall. As a former head of the Linnean Society of London, and as president of the Zoological Society of London from 1831 to 1851, Derby must have been known to Hooker. By early 1843, Hooker had taken the lead in a plan to send Burke to the HBC territories to collect plants, especially oaks and conifers (but also specifically the golden chestnut) for the Kew Gardens, and animals (including live ones) for the menagerie, aviary, and museum at Knowsley Hall.[89] Burke was introduced to the HBC officers by a letter from the Governor and Committee that said that

> in conformity with the request of Sir William Hooker a passage will be granted to a gentleman engaged for the purpose of making botanical researches, and to him also it is our desire that attention should be shewn and the necessary facilities afforded. Sir Wm. Hooker will be charged in account with the services of such people as may be employed in accompanying him in his researches, and with any supplies he may receive; no charge however in any such cases is to be made for hospitalities afforded at the establishments.[90]

If careers and reputations could be made on collecting expeditions, they could also be destroyed or damaged on them. Burke's expedition apparently went badly partly because of errors made by Hooker and

Burke. Burke, however, was also the victim of bad luck. In any case, the expedition appears to have damaged Burke's career as a collector.

Hooker had benefitted from his connections with the HBC since the 1820s, but, although he knew much about the plants of northern North America, he knew remarkably little about the company or its territories. Sending large live animals from the HBC territories to England would have required tremendous effort and expense that Burke could not have accomplished without considerable help. Furthermore, as had been the case with Wallace and Banks, HBC personnel urged Hooker to send Burke to the Columbia District on the company's ship due to depart in September 1843, but Hooker elected instead to send him on the June ship bound for York Factory, probably unaware of how long and arduous the overland trip to the Columbia District was. Burke's trip did not end tragically, as Wallace and Banks' trip did, but it did end unhappily, especially for Burke.

Burke arrived at York Factory aboard the *Prince Rupert* in August 1843. There, Letitia Hargrave, wife of the Chief Trader, was evidently not particularly impressed by Burke. She judged that Burke "is evidently not a literary character. … I dont think he will publish his travels."[91] Burke immediately accompanied the HBC brigade to the interior, wintering at Edmonton House in 1843–4. Bad luck plagued him. He departed for Jasper House in late winter, evidently hoping to cross the Rocky Mountains immediately. If so, it is difficult to understand how he failed to learn that HBC brigades would not cross the mountains until summer, when the passes of the Rocky Mountains cleared of snow. So he spent the spring and early summer of 1844 at Jasper House, stymied by exceptionally wet and cold weather.[92] Burke wrote despairingly that "I have been nearly twelve months from England & have not done any thing."[93] When Burke finally arrived at Fort Vancouver, John McLoughlin wrote that "the wishes of the Governor and Committee, in regard to Mr. Burke, will be met with every attention, and his expenses will be charged, according to your directions, to the account of Sir. W. Hooker."[94] But it was not going to be easy for traders in the Columbia District to help. In light of the surge of American immigrants to the region, the HBC had decided not to send any more trapping parties south of the Columbia River.[95] Burke collected few specimens. When he received a scolding letter from Hooker while he was at Fort Walla Walla in October 1846, Burke immediately resigned and returned to London the same way he had come.[96] Only weeks earlier, on 20 August 1846, Burke had gathered several seeds of the coveted golden chestnut.[97]

After Burke returned to London by October 1847, the dispute between him and Hooker over his conduct and his pay had to be settled by an arbitrator. Although Hooker was normally a warm, kindly man, he was ungenerous in this case. He inaccurately wrote to Derby that in "less than half the time" that Burke had spent in the same territories, David Douglas, Thomas Drummond, Thomas Nuttall, William Tolmie, and Meredith Gairdner had "sent home lots of rarities," but that Burke had "sent home *nothing* of the least value, not a Pine, not an Oak, nor any forest Tree, nor even a single Shrub … for these he was expressly sent. All I can get from him is an account of the difficulties he encountered and that he did what he could. That, again, I said, 'you know is *nothing.*'"[98] The arbitrator sided with Hooker. Burke, he decided, deserved only half his wages on the grounds that he "most grossly misspent" his time in North America.[99] But for reasons that are not clear, he actually got considerably more.[100] Historians have been sympathetic to Burke. One defended Burke by saying that he sent home many specimens, including seeds of the "golden chestnut" – one of which germinated for Derby at Knowsley Hall.[101] Burke *was* more successful than Hooker first argued, although there is no evidence that Burke appreciated how important it was to get along with the HBC officers who could have made his expedition so much more successful than it was. Hooker was accustomed to working with collectors; it seems unlikely that Hooker deserves all of the blame. At any rate, Burke's trip to the HBC territories undermined his career as a collector, although he did find in his visit to North America a new homeland. He soon returned to Oregon Territory, spent time in California during the gold rush, settled permanently in the United States, and continued to collect for scientists, although his bad luck seems to have continued to dog him into the 1850s.[102]

Hooker forged ties with another botanist even as he was involved with Joseph Burke. Karl (Charles) Andreas Geyer was one of the many peripatetic, "practical botanists" who attempted to establish a living and a name by collecting seeds of uncommon ornamental plants for buyers in the eastern United States and Europe. Originally from Dresden, Germany, Geyer arrived in North America in 1834.[103] Some people saw collecting seeds as an enjoyable way to earn a little extra income while pursing their hobby or adventure, but for Geyer, life as a roving botanical collector in North America was neither an avocation nor an end in itself. He obviously hoped that he would be able to parlay his experience as a practical botanist, and the connections with established men of science that could come with that experience, into a more remunerative and stable career as an estate gardener or a botanist in Great Britain or the continent. He must have seen it as a promising break when the French surveyor Joseph Nicolas Nicollet, who had been hired by the United States government to chart the Mississippi River region, hired Geyer as botanist to accompany his 1838–9 expedition.[104] Unluckily, many of Geyer's specimens collected during those two years were lost in an accident. Geyer then found employment with an 1841 survey of the Des Moines River led by John C. Frémont (a fellow veteran of Nicollet's surveys).

By 1842, apparently without stable employment as a naturalist, Geyer ended up in St. Louis where he became connected with a fellow German, George Engelmann. Engelmann was a physician and respected naturalist who coordinated a network of metropolitan scientists (especially W. J. Hooker and Hooker's American friend Asa Gray, from Harvard University), collectors, and naturalists in the United States, many of them fellow German speakers like Ferdinand Lindheimer, Augustus Fendler, and Friedrich G.L. Lüders from Germany, and Karl Friedrich Mersch of Luxemburg.[105] Geyer spent 1842 collecting seeds in Illinois and Missouri that were sold by Engelmann to buyers in the eastern United States and Great Britain, with the assistance of Hooker and Gray.[106] Judging Geyer to be "an excellent collector," Engelmann signed a contract with him in 1843, in which Geyer promised to go to the Rocky Mountains and Oregon to collect for Hooker and Gray.[107]

The American Fur Company denied Geyer passage up the Missouri River on its steamboats in the spring of 1843. The reason for their refusal, Geyer surmised, "is plain to those who know their illegal dealing with intoxicating liquors, to which they of course, want as few witnesses as possible."[108] So, Geyer, Lüders, Mersch, and the Scottish collector Alexander Gordon left St. Louis instead with Scottish adventurer Sir William Drummond Stewart, who had previously travelled to Fort Vancouver with Americans Nathaniel Wyeth, Thomas Nuttall, and John K. Townsend in 1834 (see chapter 7).[109] Stewart was familiar with the American West, having travelled to the rendezvous of the Rocky Mountain Fur Company each year from 1832 to 1838. Perhaps inspired by a mix of Romantic interest and Scottish patriotism, Stewart collected Native artefacts, natural history specimens, and live animals and plants throughout the west to adorn his Scottish estate, Murthly Castle.[110] After spending 1838 to 1842 at Murthly Castle, Stewart returned to North America in 1842, and was on his way to the fur trade rendezvous at Wind River when Geyer and his botanical colleagues joined him.

Geyer accompanied Stewart as far as Wind River, at which point he and Lüders joined another party travelling westward to the Oregon Country. Before the two men separated, Stewart gave Geyer a letter of introduction to John McLoughlin.[111] Stewart wrote to McLoughlin that

> Mr. Geyer a German of considerable Eminence as a Botanist will Visit [Fort] Vancouver I Believe next Autumn and I am sure will Recommend himself to you by his knowledge of Gardening and Specimens of plants he can give you of which I have spoken to him and of your taste for Gardening I think of hireing him to superintend my Garden in Scotland and if he has any Difficulty in settling for a passage home to England he may have what assistance you may consider Reasonable on my account.[112]

Stewart's letter ended up serving as Geyer's ticket, at Stewart's expense, for a trip back to Europe on an HBC ship, although as a character reference it might have been less successful. Geyer wrote that "as soon as I became acquainted with some of the Hudson's Bay Company factors, I learned that Sir Wm. Stuart did not stand in the highest estimation here, on account of crude violations of hospitality at [Fort] Vancouver."[113] Still, Geyer must have behaved better than Stewart had. When he left the Columbia District in November 1844, Geyer held letters of recommendation to W. J. Hooker from John McLoughlin and Archibald McDonald.[114] But then Geyer erred. Having travelled to London by ship, rather than returning to St. Louis as planned, Geyer submitted all of his specimens directly to Hooker, rather than to Engelmann as his contract stipulated, failing to understand the awkward position that this placed Hooker in relation to Engelmann.[115] Hooker and Geyer may have resolved the issue, but Geyer did not land a job through Hooker. Having had his trip home paid for by Stewart on speculation that Geyer might be Stewart's gardener at Murthly Castle, Geyer also went to Scotland, only to find that Stewart was not there. By the early fall of 1845 he appears to have exhausted all leads in Great Britain and returned to Germany where, after failing to find work as a botanist or gardener there as well, he established his own nursery near Berlin, founded a respectable horticultural journal, and gave private lessons in botany and English.[116] Geyer's North American travels may not have landed him the career of his dreams, but he had done better than some of the itinerant scientific collectors with whom he had travelled.

The HBC and its employees benefitted from Geyer's tributes. A long report of Geyer's trip was published in the *London Journal of Botany* by its editor, W. J. Hooker. Referring to the HBC, Geyer wrote "the liberality of that body of gentlemen is too well known, especially in the scientific world, to require any encomium from me, yet I may be allowed to make special mention of the kindness and assistance I received from the Chief Factors, Macdonald [Archibald McDonald], at Fort Colville, [Archibald] McKinlay, at Fort Walla-Walla, and especially from Chief Factor [James Douglas] and Governor McLoughlin, at Fort Vancouver."[117]

Interest in Oregon did not diminish after the trips of Burke and Geyer. In 1849, some Scottish estate holders, horticulturalists, and scientists established the Oregon Botanical Association (later renamed the British Columbia Botanical Association), based in Edinburgh. The association was chaired by John Hutton Balfour, who had succeeded Hooker as professor of botany at the University of Glasgow in 1841, but had become by 1849 professor of botany at the University of Edinburgh and the head of the Royal Botanical Gardens in Edinburgh.[118] The association sought to build upon the botanical work of David Douglas, although the economic potential of the effort was not missed. Commenting on it in March 1850, the *Constitutional Perthshire Agricultural and General Advertiser* asserted that "when we consider that there are only four or five forest trees indigenous to this country, and that the others have been introduced from different parts of the world, we have every reason to hope that the trees of a district, bearing great resemblance in point of climate, to our own, will prove in the highest degree beneficial."[119] The association decided to seek permission to send John Jeffrey, a gardener at the Royal Botanic Gardens, to the HBC territories. Like Archibald Menzies and David Douglas, Jeffrey was from Perthshire.[120] But, as one of his biographers has suggested, "perhaps the association expected too much of him; he was not another David Douglas."[121]

The Oregon Botanical Association appears to have become as disappointed with Jeffrey as Hooker had been with Burke. Like Burke seven years before him, Jeffrey travelled on an HBC ship to York Factory, arriving in August 1850. Also like Burke, Jeffrey immediately accompanied the HBC brigade inland, but he spent the fall and early winter at Cumberland House before joining the winter express to Jasper House, where he arrived in March. Then, in 1851 he did what Burke was unable to arrange. He travelled from Jasper to Fort Colville in the spring. He then botanized until 1853 on Vancouver Island, in the Similkameen Valley, and in Willamette Valley before moving on to San Francisco in 1853.[122] Jeffrey submitted hundreds of specimens, but the association was disappointed that he enclosed no diary. They dismissed Jeffrey in November 1853.[123] He probably never learned of his firing, because he mysteriously disappeared in 1854, never to be heard from again.[124] No diary of his travels was ever found, so the HBC received little publicity for supporting his work.[125] He did, however, submit seeds of many trees, the most important being the western hemlock.[126]

The work of the Oregon Botanical Association inspired further efforts by the Scottish scientist Andrew Murray. Murray had abandoned a career in law to become a scientist. His interest in science apparently began when he became secretary of the Oregon Botanical Association during the late 1840s, at the time it sent Jeffrey to the HBC territories.[127] In 1852 he published his first paper in entomology. In 1857 his path from law to science continued with a term appointment in natural history at New College in Edinburgh. Then he served in the Botanical Society of Edinburgh in 1858–9, before relocating to become the assistant secretary of the Royal Horticultural Society in London in 1861.[128] According to his biographer, "many of his publications were a testament to 'network research': he relied on family, friends, and colleagues to provide specimens and information from the Americas, Africa, and Asia."[129] Murray himself explained that following Jeffrey's disappearance,

> in studying the route followed by Jeffrey, I had the enormous extent of their [the HBC's] territory forced strongly upon my attention – thousands of thousands of miles still inhabited only by the "wild"; and all this territory dotted over by the trading or hunting stations of the Company. I found also, in the occasional correspondence I had with the officers stationed at some of these remote posts, that they were obliging and intelligent. I imagined that many of them (from their hunting propensities, which may have led them to the life they followed) must have an instinctive taste for natural history; and when I put all this together, I felt that here was a great opportunity for enlarging our knowledge of the natural history of a considerable portion of the globe, which was lying fallow only because no one advanced his hand to seize it.[130]

Instead of travelling to the HBC territories himself, or sending another emissary, Murray contacted the Governor and Committee of the HBC for permission to circulate a request for assistance in gathering natural history specimens. Murray explained that Edward Ellice Jr. and the committee "not only sanctioned the distribution of my circulars, but charged themselves with it, and undertook to forward to me any collections that might be made."[131] Five hundred copies of the circular distributed in 1856 or 1857 soon reaped six cases of natural history objects.[132]

Just as Engelmann's network was populated by Germans, Murray's was dominated by Scotsmen. John McKenzie at Moose Factory, for example, sent correspondence and specimens from which Murray noted that he "received much satisfaction."[133] Mackenzie also attempted to ship a live beaver to Murray, but it died before reaching Scotland.[134] James Hargrave sent caribou specimens.[135] Ulster Scot Bernard Rogan Ross, Chief Trader in charge of the Mackenzie River District, submitted over a hundred specimens to Murray and corresponded with him as late as 1862, as a detailed letter about Canada geese on 1 June 1862 attests.[136]

Murray reported on the natural history specimens in two long articles in the *Edinburgh New Philosophical Journal* in 1858 and 1859, dealing primarily with mammals, but also with birds and fish.[137] When he did, he paid tribute to the HBC in a particularly effusive way. He opened his "Contributions to the Natural History of the Hudson's Bay Company's Territories" (1858) by quoting the HBC charter, and arguing that the charter imposed "no condition that the Company should do anything for science, or future expeditions, or discoveries. Whatever was the motive which led to the charter being granted, the grant itself was unfettered by any restriction or condition relating to such matters."[138] The company had nonetheless acted as if they had "an express obligation on them to do everything in their power to foster researches in the dominions so conferred on them."[139] Elaborating, he explained that

> the extent to which the assistance of the Company has thus been given to science cannot be estimated; but it is not too much to say, that no public or private expedition was ever conducted through their territories which did not draw largely upon the liberality and assistance of the Company. Their own numerous explorations, their extensive geographical surveys, and the able and ready help which they have given to the search after Franklin and his crew, are instances which it is scarcely necessary to recall to the mind of the reader. I have, however, had special opportunity of seeing the liberal mode in which they extend their assistance to scientific objects, on the occasion of a botanical expedition being sent out a few years ago by an association formed in this city, to procure seeds of new and valuable hardy trees and plants from Oregon and the neighbouring districts. I acted as secretary to that association, and conducted the negotiations

with the Hudson's Bay Company for securing their assistance to the collector (Mr Jeffrey). The liberal spirit in which I then found that the Company looked at things impressed me no less than the extent of the power they possessed.[140]

It is difficult to imagine that Murray wrote those words unaware that the HBC and its charter was then at the centre of great interest. With its licence due to expire the very next year, and with politicians once more deliberating on the future of the HBC's licence (in the context of a liberalizing economy), it must have seemed timely to the HBC's directors to have a scientist expound so repetitively on the HBC as a liberal corporation.

Later, from 1863 to 1866, almost certainly with Andrew Murray's involvement, Robert Brown of Caithness – not the famous Scottish naturalist of the same name, but a botany student at the University of Edinburgh – undertook another expedition to British Columbia on behalf of the British Columbia Botanical Association of Edinburgh. His expedition reaped fewer specimens than John Jeffrey's, and fewer than Murray had received directly from the HBC officers.[141]

Depending on HBC officers to collect for him turned out to be an economical and effective way for Murray to acquire specimens. Instead of spending money on travel expenses, Murray offered in 1860 to reimburse the HBC for "anything curious" the officers might purchase from Indians on his behalf.[142] This arrangement must have suited all parties: Murray must have gotten his specimens far more cheaply than he could have by sending another collector, and the HBC and its traders could use the funds strategically to buy the curiosities from their favourite Native trading partners.

Scottish officers in the HBC also contributed to the collections of the Industrial Museum of Scotland (now the Royal Scottish Museum) in Edinburgh between 1858 and 1862. George Wilson, who was appointed director of the museum in 1855, faced the challenge of amassing exhibits for the one-year-old museum. It was natural that a man like Wilson – president of the Royal Scottish Society of Arts, and also appointed in 1855 as professor of technology at the University of Edinburgh – sought to assemble a collection of technology from around the world.[143] To begin his collection, he sought the help of members of the Scottish diaspora. Through his older brother Daniel, who had moved to Canada to take up a professorship at the University of Toronto in 1853, George Wilson contacted George Simpson. Simpson was eager to help. He wrote to George Barnston (and probably other HBC officers), asking him to assist in collecting "specimens of Indian manufactures" for the museum and reminding him that "as it is the wish of the company to be useful in advancing works of national or scientific Interest, I have to request the favor of Your Collecting such articles as You may think likely to be useful or appropriate to be forwarded to London, with a Letter of advice as to the destination for which they are intended."[144] Several Scottish officers, including Robert Campbell, Roderick Mac-Farlane, Joseph James Hargrave, James Anderson, and George Barnston responded by sending ethnological specimens, including entire sets of clothing, acquired from the Gwich'in (Kutchin or Loucheux), Blackfoot, and Inuit peoples.[145] But the largest number by far were sent by Bernard Rogan Ross, who submitted material gathered by MacFarlane, Julian Stewart Onion (Camsell), the Rev. W. W. Kirkby, and Nicol Tayor.[146] Soon after George Wilson died in 1859, the connection between the HBC and the Industrial Museum appears to have been severed, perhaps because there was no will on either side of the Atlantic to sustain it. A letter from Robert Campbell to Wilson may explain why Hudson's Bay Company officers seemed so unenthusiastic about ethnological research. He wrote that "even in this Northern District, the Indians appreciate the convenience of the articles of civilized usage so much, that hardly a trace now remains of their former dress, domestic utensils or weapons of war, or the chase, all have already fallen into disuse among them."[147] Europeans influenced by Romantic notions of aboriginal peoples would have been disappointed by his words.

Andrew Murray's interest in the natural history of the HBC territories appears to have been inspired originally by a desire to build upon the work of David Douglas, and George Wilson may have been influenced by Murray, but the work of George Barnston was even more directly a legacy of Douglas. Barnston was born in Scotland and evidently trained as a surveyor and army engineer before joining the NWC in 1820.[148] Documents show that Barnston was an emotionally unstable man, prone to bouts of deep depression, particularly before the mid-1830s. It is unclear whether Barnston's gloominess was the product of or cause of his slow career advancement in the 1820s. His own letters suggest that his unhappiness was not caused merely by being passed over for promotion. One letter, written in 1828 to his friend James Hargrave, suggests that his deepening bouts of depression were caused by boredom: "I have told you before now that my mind found happiness only when in a state of excitement from novelty of circumstance, or change of scene, but even these have lost their affect except when they required from me a constant stretch of attention or exertions of thought. When these leave me, gloom and misanthropy again usurp their reign."[149] George Simpson's characterization of him reinforces the impression that Barnston battled depression, particularly in his early years with the HBC, and that his depression may have interfered with his career development. In his 1832 "Character Book," Simpson described Barnston as "so touchy & sensitive that it is difficult to keep on good terms or to do business with him. ... and so much afflicted with melancholy or despondency, that it is feared his nerves or mind is afflicted."[150] In fact, Simpson wrote that Barnston was "sometimes of gloomy disponding turn of mind and we have frequently been apprehensive that he would commit suicide."[151]

Books offered Barnston some relief. In 1834 he wrote Hargrave that "our tastes differ in many particulars, but in the love of reading I think they are similar, nor have the occupations in which we are engaged, yet rendered us insensible to the pure and rational pleasure of cultivating the mind."[152] But science, more than anything, appears to have changed Barnston's gloomy disposition.[153]

Barnston first met David Douglas in the fall of 1826 at Fort Vancouver, and the two soon formed a close friendship that lasted as long as Douglas lived.[154] When Douglas returned to Fort Vancouver in 1830, Barnston wrote that "we were glad to see again amongst us an old friend, with his noble countenance, and agreeable hearty manner unchanged."[155] A year later, he wrote that in Douglas he had found "a man after my own Heart."[156] Barnston actually quit the HBC for a time in 1831, only to return in 1832, in large part because it allowed him to retain "my connections with those friends who are so deservedly dear to me."[157] David Douglas was among these friends. In 1834, he noted receiving "two well filled sheets from Douglas," in which Douglas reported on his scientific endeavours in California. Barnston added that Douglas "possesses a Warm Heart, and Quick feelings, and I really believe has done more for the promoting of Botanical Science, than any one of late time with the exception of Sir Joseph Banks & the Celebrated Humboldt."[158]

Barnston's metamorphosis from a vicarious or occasional student of science to a lay devotee may actually have been prompted by news that Douglas had died. Douglas's death obviously affected the sensitive Barnston deeply. On 5 February 1836, upon hearing of Douglas's death, he wrote with raw emotion:

Unhappy Douglas! thy Shade is now before me, now shrieks for assistance – I yet hear the dismal and heart rending moans of my ill fated friend.

Was there no hand to help? Could no Arm Save? Alas! the Almighty! his decrees are just and good, and the weary wanderer has been taken home. He had obtained a noble Conquest over warm and Powerful passions, his mind so often fixed on the wonderful works of his Maker, had melted into love and good will towards the whole creation, his spirit had caught the eternal flame which dieth not, and he was hurried from a troubled seam to rest in happiness & peace.[159]

Almost exactly a year later, Barnston wrote Hargrave a letter that suggests that he had recently discovered a new hobby, and regretted not taking it up sooner. He wrote that

> A Tolerable Collection of the Insects of the Country which I have made, I may yet attempt to class and describe. At the end of our store I have nearly 100 different species of caterpillars and Maggots in their winter quarters, and I only await the return of warm weather to be kept in constant employment watching them through their various stages of metamorphosis. This is not so dry a study as Geology by half, and I believe will henceforth as a favourite occupation keep its hold upon my mind. How much I regret not having paid attention to it when on the Columbia. The Company's affairs would just have gone on as well, without my having been so deeply interested in them – perhaps better for too much anxiety often defeats its own object, and the only return I got was to be told that I had been of little or no service to the Concern. Verily every man shall have his reward, and I have had mine.[160]

And Barnston appears to have been rewarded in more ways than one. The changes in Barnston's state of mind, and in his career prospects, were apparent to his friends. After he developed his interest in entomology, Barnston appears to have become much happier, and his relationship with George Simpson improved.[161] In 1839 James Douglas wrote Hargrave to say that "I am glad to hear of a break in the clouds, that for some years have obscured our friend Barnstons prospects, through which the sun begins to peer upon him."[162] In 1840 Barnston was promoted to Chief Trader, and seven years later to Chief Factor.

Years later in 1860, Barnston recalled that it was "when at Martin's Fall [where he arrived in mid-1834, and remained for six years] I first took a liking to Entomology, as it appears to be a good Locality for flies. I bred lots of *Lepidoptera* [moths and butterflies] and a few *Cimbicidae* and *Tenthredinae* [sawflies]."[163] Clearly then, Barnston's interest in entomology, and probably his active practice of science, began between 1834 and 1836, perhaps stirred directly by David Douglas's death.

By 1840, Barnston opened correspondence with John Richardson, who arranged for the publication of some of Barnston's papers.[164] Unfortunately for Barnston, "my Eyes were hurt at M. Falls by a too careless use of the Microscope, which caused me to give up studying Insects."[165] While in England on furlough in 1843–4, Barnston met Richardson, who gave him a tour of the British Museum. Barnston wrote George Simpson that after handing "over without reservation all my collections of insects," he learned that "one half of my specimens nearly are new."[166] He later wrote, "I found three kind friends in that Institution [the British Museum] Messrs. [John Edward] Gray [keeper of zoology at the museum], [Edward] Doubleday [a specialist in *Lepidoptera*] and Adam White [assistant in the zoology department at the British Museum, and another specialist in entomology], and the same feelings that prompted me then still draw me towards every labourer in Science and every Lover of Nature."[167]

The significance of Barnston's work was recognized in his own lifetime and in recent times. In 1851 Adam White of the British Museum implied that Barnston's contributions to the museum compared favourably with a collection of insects at the British Museum that had been accumulating for over thirty years, "while Mr. Barnston's was formed in three months, on one spot and under almost unheard-of disadvantages, counterbalanced, however, by an enthusiasm not easily deterred by difficulties."[168] White added that "as Mr. Gray's Catalogues of the collections of the British Museum ... are published, it will be seen how valuable are Mr. Barnston's and Sir John Richardson's collections to our acquaintances with the articulated animals of British North America, especially in its more northerly parts."[169] In a late twentieth-century scientific article that cited Barnston's work, two scientists noted that when Barnston abandoned his study of insects, almost a century passed before the next person took up a scholarly study of stoneflies in what became the Canadian province of Ontario.[170]

George Barnston transmitted his love of science to his mixed-blood son, James, born at Norway House on 3 July 1831. Barnston made sure that his son was well schooled, first at Red River and Montreal, then, in 1847 Edinburgh. In 1847, although underage, James passed the final exam in medicine "with the highest honours." He practiced in Scotland, Paris, and Vienna before opening a practice in Montreal in 1853. Upon arrival in Montreal, he grew convinced that Canada was "rich in Plants yet to be made known."[171] He soon became the curator and librarian of the Natural History Society of Montreal, becoming, according to its journal, "one of its most valued members, and foremost and most active friends. He read many interesting papers, and delivered many delightful and instructive lectures, before its members."[172] In 1857 he became the first chair of botany at McGill University, but died shortly thereafter on 20 May 1858 at only twenty-seven years of age.[173] His obituary lamented that "among those of his own age, whom he has left behind, we fear the Society will find few upon whom his mantle will fall."[174] The fact that George Barnston published his "Abridged Sketch of the Life of David Douglas" in 1860 suggests that the early death of his own son James prompted George Barnston to write his memorial to Douglas.

George Barnston's associations with science were founded on personal relationships. Just as his initial pursuit of science may have been inspired by Douglas's death, his subsequent associations with scientific institutions were connected with his son. During his son's time in Great Britain, George Barnston conveyed many butterflies, moths, and skippers to the British Museum through James.[175] Barnston also contributed to the Royal Industrial Museum of Scotland.[176] Soon after James moved to Montreal, George appears to have become affiliated with the Natural History Society of Montreal, McGill University, and the Canadian Geological Museum.[177] After James's death, George Barnston spent a furlough in Montreal, during which his involvement in the Natural History Society of Montreal intensified.[178] Barnston also submitted specimens to the United States in the 1850s. At the request of the foremost American entomologist of the time, John Lawrence LeConte, he submitted a collection of Coleoptera (beetles) to Philadelphia from locations near the north end of Lake Winnipeg, along the Saskatchewan River system, and the Mackenzie River.[179]

Barnston retired from the HBC in 1863, although he remained active in science. He served as the president of the Natural History Society of Montreal in 1872–3, and became a fellow of the Royal Society of Canada in 1882. Between 1857 and 1876, he published several articles in the society's journal drawing upon his experiences gained during his HBC career.[180] His descriptions of natural history convey something of his humane sentimentality, lacking the detached tone of most scientific writing. For example, when describing the Canada goose, he wrote that

> When the long and dreary winter has fully expended itself, and the Willow Grouse have taken their departure for the plains [tundra] of the North,

there is frequently a period of rank starvation to many, who are on their way from their wintering ground to the Trading Posts. The first call, therefore, of the large Canada or Grey Goose is heard with a rapture known only to those who have endured great privations. The tents are filled with hope, to which joy soon succeeds, when the happy father or hopeful son and brother throw down their grateful load.[181]

Scientific activity proliferated in the HBC after 1821 because of the development of networks that tied the HBC Governor and Committee, a range of newly emergent scientific organizations, several prominent men of science, scientific travellers, and HBC personnel in mutually beneficial relationships. Foremost among the promoters of science was David Douglas. Douglas's importance in the history of science rests not merely in his own tireless work during his two trips to the HBC territories in the 1820s and 1830s. He could not have accomplished what he did without the help of others. He was as successful as he was because of the way he could inspire those around him. The size and value of his collections, and his ability to instil in others a passion for science, directly or indirectly, inspired and influenced the work of many others, even decades after his death. These certainly included Wallace, Banks, Burke, Jeffrey, Murray, and the Barnstons, though none so poignantly as George Barnston, who grieved so deeply for Douglas, only to see his own son die at an even younger age than did Douglas. Some of his successors, Burke and Jeffrey most notably, were less successful than him, evidently and primarily because they lacked the ability to enlist the enthusiastic assistance of others. These examples reveal the importance of socially aware scientists who were able to cultivate the zeal of those around them.

7 "Disinterested Kindness": The Hudson's Bay Company and North American-Based Science, 1821-1870

Whether they were searching for the Northwest Passage along the bleak arctic coast or for new species of trees in the temperate rainforest of northwestern North America, British scientists found a steady patron in the HBC after 1821. The HBC also lent assistance to scientists based in the United States and Canada. But the company's assistance to North American scientists had results and repercussions far more dangerous to the company than its support for British science had. Directors of the company had always known that their support for science could backfire. The trouble unleashed by Christopher Middleton's contributions to science in the eighteenth century (discussed in chapter 2) and Richard King's contributions (chapter 5) proved that seemingly innocuous scientific activity could have unforeseeable negative consequences for the company. The HBC's support to scientists based in North America in the 1830s and 1840s offered other examples. Cooperation between the HBC and scientists contributed to expansionist movements in the United States and Canada.

At no time between 1821 and 1859 did the HBC actually help American scientists travel through the HBC territories, but several American scientists who travelled to the Columbia District (Oregon) on their own initiatives did receive generous welcomes from John McLoughlin, the HBC's officer in charge of the district. McLoughlin's hospitality to American scientists in the Columbia District may have been only a minor contributor to the emergence of an annexationist movement in the United States, but that movement led to the loss of a large part of the district to the United States in 1846.

Because Canadians were fellow British subjects, the HBC was more inclined to support scientists based in Canada than those based in the

United States. Still, the HBC's links with the small scientific community in Canada were tenuous until after the Royal Artillery established the Toronto Geomagnetical and Meteorological Observatory in 1841, the flagship observatory in the British contribution to the so-called "Magnetic Crusade."[1] This massive international geomagnetic survey is little known today because its scientific accomplishments were modest, especially given the high expectations of its advocates and its enormous cost. However, it is historically significant for many reasons, not the least of which is that it was to date by far the largest scientific project ever attempted. It was the first scientific venture that involved an international network of widely scattered permanent (or at least semi-permanent) observatories, extensive scientific travel, expenses and coordination so large and complex that the project required state sponsorship, and management and direction by professional scientists. Among its many important legacies is the fact that it strengthened the cooperative relations between the HBC and departments of the British state, particularly the British Artillery, and also indirectly influenced the evolution of other networks of which the HBC was a part. The geomagnetic survey also illustrates, as clearly as any aspect of the history of science in the HBC, that, although historians have often emphasized how scientists and companies acted as agents of empire, empires and companies were at least as likely to act as agents of science.[2]

Almost immediately after John Henry Lefroy, a young artillery officer, arrived in Toronto to take charge of the Toronto observatory in 1842, planning began for an expedition to the HBC territories. The expedition, which took place between May 1843 and November 1844, permitted Lefroy to conduct geomagnetic research in some of the remotest regions of the company's domain. It also helped incorporate Toronto into the HBC's knowledge network. Upon his return, Lefroy helped arrange for the young Toronto-based artist Paul Kane also to make an extended visit to the HBC territories.

The HBC's sponsorship of Canadian-based travellers was almost certainly crucial to the rise of an expansionist movement in Canada in 1856. The HBC's directors had no reason to regret the emergence of an expansionist movement in Canada because they did not cling unrealistically to the belief that they might be able to retain their monopoly in Rupert's Land much longer. In fact, they welcomed the prospect of being able to sell the company's rights to Rupert's Land to a willing buyer within the British Empire. But Canadian expansionists adopted virulently anti-HBC rhetoric in their campaigns to acquire the HBC territories without having to compensate the company. In its support of American and Canadian science then, the company unleashed unwelcome forces beyond its control.

Unable to agree on a boundary between the Continental Divide and the Pacific Ocean, Great Britain and the United States agreed in 1818 to a ten year "joint occupation" of the region west of the great divide between 42° N and 54'40"° N. (The joint occupation actually lasted until 1846.) Neither country's government actually administered the area. The American government had no official presence there at all, and Great Britain handed jurisdiction over the area to the HBC in 1821. It was not because of ineffective fur trading that the HBC eventually lost the southern portion of the territory in 1846. Various strategies (creating a "fur desert" in the Snake River region, establishing permanent posts along the northwest coast, and underselling all American competitors in the region even if it had operate at a loss) allowed the HBC to dominate the fur trade of the region throughout the period of joint occupancy.[3] However, when missionaries and permanent agricultural settlers began arriving in the district in the mid-1830s, they were overwhelmingly American, and by 1844 American agricultural settlers so dominated the Willamette Valley that HBC officials conceded that that part of the Columbia District would inevitably be ceded to the United States.[4]

Relationships between American scientific travellers and HBC men in the Columbia District became significant during the 1830s, when American efforts to compete commercially with the HBC seemed doomed, but before American settlement seemed very threatening. The reception that Americans received in that region was attributable to Chief Factor John McLoughlin, the very man who had assisted David Douglas and John Scouler in the 1820s and 1830s. In most of the HBC territories, travel by people outside the company remained difficult unless assisted by the company, until at least 1870. However the HBC's London Committee saw Americans as a threat.[5] George Simpson's anti-Americanism appears to have softened considerably as the years passed, but was vehement in his earlier years.[6] As late as 1845 he implied that he would assist visitors to the HBC territories only if they were British subjects.[7] The joint occupancy agreement made it impossible for the HBC to prevent Americans from arriving and settling in the Columbia District, but any missionaries, settlers, or scientists who arrived in the 1830s benefitted significantly when assisted by the HBC traders. Under those conditions their reception by local HBC officers mattered much more than the policy of distant governors. That is why Chief Factor John McLoughlin's demeanour mattered so much.

John McLoughlin was remarkably hospitable to Nathaniel J. Wyeth, an American who later became a prominent promoter of American settlement in Oregon. Wyeth first arrived at Fort Vancouver in October 1832 after travelling overland from Boston via St. Louis and the Rocky Mountain fur trade rendezvous. He had hoped to establish a fur trading and fishing business in competition with the HBC, but those plans were ruined by the desertion of most of his men along the way. Even if his men had not abandoned him, the HBC had the resources to undersell Wyeth until his enterprise failed. Thus, McLoughlin must have seen Wyeth as more to be pitied than feared. At any rate, McLoughlin welcomed him to Fort Vancouver, where Wyeth also met and conversed with David Douglas.[8] Wyeth left Fort Vancouver in the company of the HBC brigade in February 1833, carrying botanical specimens and expressing gratitude to John McLoughlin.[9] When he returned to Boston later that year, he showed the plants he collected to the British American botanist Thomas Nuttall at Harvard University. Nuttall not only published a report on the collection but also resigned his position at Harvard to travel with Wyeth to Oregon in 1834.[10]

On 16 September 1834 Wyeth returned to Fort Vancouver in another attempt to establish a fur trading business in competition with the HBC. This time he was accompanied by Nuttall, the American ornithologist and collector John Kirk Townsend, and the Scottish aristocrat Sir William Drummond Stewart of Murthly Castle.[11] According to one plausible account, Nuttall, hoping to guarantee a cordial welcome at Fort Vancouver, "wrote to the Governor of the Hudsons Bay Company for protection and hospitality ... but received a very unsatisfactory reply, which Nuttall said was not much more than he expected, as the subordinates, in such cases, he had always found to sympathize with his objects more than the officials."[12] Stewart, however, did carry a letter of introduction written by Edward Ellice, deputy governor of the HBC, to George Simpson and the officers in the HBC.[13] If the men were at all apprehensive of getting a friendly reception at Fort Vancouver, their fears were soon dispelled. Townsend wrote that upon arrival at the fort, he and Nuttall met John McLoughlin, "a large, dignified and very noble looking man, with a fine expressive countenance, and remarkably bland and pleasing manners."[14] He continued, "we were greeted and received with a frank and unassuming politeness which was most peculiarly grateful to our feelings. He requested us to consider his house our home, provided a separate room for our use, a servant to wait upon us, and furnished us with every convenience which we could possibly wish for. I shall never cease to feel grateful to him for his disinterested kindness to the poor houseless and travel-worn strangers."[15] Wyeth hoped to put his business, the Columbia River Fishing and Trading Company – a potential competitor with the HBC – on firm footing, but the commercial aims of his expedition were far less successful than the scientific aims. Wyeth proved unable to compete with the HBC along the lower Columbia River.

All accounts show that John McLoughlin did not treat Wyeth, Nuttall, and Townsend as interlopers, with hostility, but with generosity. Nuttall and Townsend collected specimens in the vicinity of Fort Vancouver with the help of McLoughlin. In fact, McLoughlin employed Townsend as post surgeon for almost a year, beginning when Meredith Gairdner left for Hawaii in October 1835, and while William Tolmie was absent.[16] When Tolmie returned, Townsend and Tolmie became friends, and Townsend even later attempted to name a species after Tolmie.[17] Given the congenial circumstances, it is not surprising that Townsend did not leave Fort Vancouver until November 1836. When he did, he wrote affectionately that "I took leave of Doctor McLoughlin with feelings akin to those with which I should bid adieu to an affectionate parent. ... Words are inadequate to express my deep sense of the obligations which I feel under to this truly generous and excellent man, and I fear I can only repay them by the sincerity with which I shall always cherish the recollection of his kindness, and the ardent prayers I shall breathe for his prosperity and happiness."[18]

McLoughlin's assistance to American scientists reflected the growing level of cooperation between metropolitan scientists in Great Britain and the United States at the time. American and British scientists worked increasingly easily together by the 1830s, "giving substance," according to Bruce Sinclair, "to the conception of science as an intellectual pursuit that transcended national boundaries."[19] For example, the prominent American botanist Asa Gray formed lifelong friendships with his British counterparts W. J. Hooker and his son Joseph Dalton Hooker. And the British geophysicist, astronomer, and naturalist Edward Sabine got along well with the American physicist Alexander Dallas Bache.[20] Prominent American scientist Joseph Henry, while in Great Britain to attend the annual meeting of the British Association for the Advancement of Science in 1837, said "truth and science should know no country."[21]

The HBC's dominance in the southern Columbia District waned significantly in the late 1830s and 1840s, but McLoughlin's hospitality did not diminish. In the early 1840s, a nucleus of American settlers in the Willamette Valley, south of Fort Vancouver, made it easier for Americans to travel to and stay in the Oregon Country without the assistance of the HBC. Nevertheless, during that period John McLoughlin continued to show hospitality to Americans arriving in the district. According to his biographer, McLoughlin "lacked the ruthless streak that was part of Simpson's make-up, and saw no reason why a trading rival should necessarily be regarded as a personal enemy."[22] If this was true of rival fur traders, it was certainly true of scientists. The same biographer noted that

> McLoughlin's treatment of the American immigrants who were flowing into the Willamette valley was much criticized. He was a humanitarian at heart; he received the immigrants kindly and provided the needy with seeds, implements, and supplies, often on credit. By the spring of 1844 several hundred settlers had received advances totalling £6,600, a sum that alarmed the governor and the London committee. Both they and Simpson failed to realize that, apart from other considerations, McLoughlin's policy was realistic, for settlers were not likely to starve quietly with the well-filled warehouses of Fort Vancouver near by.[23]

Realistic or not, it seems that John McLoughlin was not inclined to be a loyal partisan either for his employer or his country. The staunchly conservative George Simpson once noted that McLoughlin "would be a Radical in any Country – under any Government and under any circumstances."[24] After his adopted home in the Columbia District became part of the United States, McLoughlin applied for American citizenship as soon as he was eligible.[25]

Not all HBC men welcomed American collectors and scientists. Indeed, some tired of them. In 1837, Peter Skene Ogden wrote John McLeod that "last summer ... by the Snake country we had ... five more Gent. as follows 2 in quest of flowers 2 killing all the birds in the Columbia & 1 in quest of rocks and stones all these bucks came with letters from the President of the U. States and you know it would not be good policy not [sic] to treat them politely they are a perfect nuisance."[26] But McLoughlin remained hospitable. In 1841, when members of Lieutenant Charles Wilkes's United States Exploring Expedition (1838–42) arrived in the Columbia District, its members penned descriptions of their reception at Fort Vancouver that were reminiscent of those penned by Townsend in the 1830s. Wilkes wrote that McLoughlin "gave us that kind reception we had been led to expect from his well-known hospitality," and William D. Brackenridge, the expedition's horticulturalist, wrote that McLoughlin, "in the most friendly manner showed me around his gardens."[27] Later, in November 1843, Captain John C. Frémont wrote that upon his arrival at Fort Vancouver after an overland trek, McLoughlin "received me with the courtesy and hospitality for which he has been eminently distinguished and which makes a forcible and delightful impression on a traveller from the long wilderness from which we had issued."[28]

The historian E. E. Rich has noted that "so frequent did the calls upon McLoughlin's hospitality become that he was taken to task by the Governor and Committee, and charged with keeping an open house at the expense of the Company."[29] Although, the company's directors do not appear to have resented McLoughlin's generosity towards scientists specifically, they may nonetheless have been caught in the crossfire. The company's directors evidently grew annoyed with the conduct of some arrivals, for in 1844 Karl Geyer feared that he was going to be denied assistance returning to Europe: "no one hereafter is to be helped, runs an order of the general council in London." The cause of the policy, explained Geyer, was not the behaviour of scientists, but "the ingratitude experienced by the Hudson's Bay Company in return for much help to one missionary, and to missionaries in general."[30]

By 1844, when Geyer reported on the HBC's order that no one be helped, the Oregon Crisis had already erupted. Because Nathaniel Wyeth had been among the most prominent promoters of Oregon in the eastern United States, it did not take a particularly perceptive person to understand that the assistance that McLoughlin had afforded Wyeth, Nuttall, and Townsend contributed at least in some measure to Americans' "Oregon Fever," and to the subsequent movement to annex the region to the United States.

Evidence of the continued sympathy of the HBC for scientists is reflected by the fact that George Simpson himself displayed some generosity towards American scientists even during the Oregon Crisis. In April 1845 John James Audubon wrote to Simpson explaining that he was "now engaged in the publication of a work on the quadrupeds of North America."[31] Requesting specifically musk oxen, foxes, "barren ground bears" and any other animals unique to the north, Audubon assured Simpson that should any additional specimens be submitted that were not previously known to science, "I will if you allow me to publish them take pleasure in naming them according to your wishes."[32] The letter received no response, but probably because Simpson was very busy in London occupied with issues related to the Oregon Crisis at the time, and not because of hostility towards Americans or Audubon, for he responded enthusiastically to another request made in November on behalf of Audubon by his agent in Montreal.[33] On that occasion Simpson wrote enthusiastically that "it affords me much pleasure to be able to assist you in the completion of your valuable work, the importance of which to the Scientific world I highly appreciate."[34] Simpson explained that he had written the trader at Churchill (William Sinclair) instructing him to attempt to obtain a musk ox, but explained why it would be difficult and time consuming actually to obtain and send a specimen of the musk ox, and other specimens requested by Audubon, preserved according to Audubon's specifications.[35] Audubon's fame, and the fact that Audubon requested specimens rather than permission to visit the HBC territories probably explain Simpson's willingness to cooperate with him.

Audubon's advancing senility, deteriorating health, and death (in 1851) made it impossible for him to complete the planned work on quadrupeds on his own, but John Bachman and Audubon's own sons, particularly John Woodhouse Audubon, were instrumental in publishing the three-volume *Quadrupeds of North America* (1846–54), two volumes of which were published after Audubon's death.[36] The second and third volumes paid generous tribute to the HBC. In the second volume, after contrasting the "narrow-minded policy pursued towards us" by the "directors of the National Institute at Washington," with the generosity of various European and British scientific institutions, they added that "when the Hudsons Bay Company received an intimation that we would be glad to obtain any specimens they could furnish us from their trading posts in the arctic regions, they immediately gave orders to their agents and we secured from them rare animals and skins, procured at considerable labour and expense, and sent to us without cost, knowing and believing that in benefitting the cause of natural science they would receive a sufficient reward."[37] In the third volume, they added that "it has been an unmixed gratification to have with us the sympathies and assistance of gentlemen like Sir George and many others, and of so powerful a corporation as the Hudson's Bay Fur Company."[38]

Audubon and Bachman also thanked the HBC for their efforts, even if they were unsuccessful, to acquire specific species. For example, in their entry for the musk ox, they explained that they had to resort to a poor specimen held by the British Museum.[39] However, they then noted that

> Sir George Simpson, of the Hudson's Bay Fur Company, most kindly promised some years ago that he would if possible procure us a skin of the Musk-Ox, which he thought could be got within two years – taking one season to send the order for it to his men and another to get it and send the skin to England. We have not yet received this promised skin, and therefore feel sure that the hunters failed to obtain or to preserve one, for during the time that has elapsed we have received from the Hudson's Bay Company, through the kind offices of Sir George, an Arctic fox, preserved in the flesh in rum, and a beautiful skin of the silver-gray fox, which were written for by Sir George at our request in 1845, at the same time that gentleman wrote for the skin of the Musk-Ox.[40]

And in the entry for the black fox, they wrote that "it gives us pleasure to render our thanks to the Hon. Hudson's Bay Company for a superb female Black or Silver-gray Fox which was procured for us, and sent to the Zoological Gardens in London alive, where J. W. Audubon was then making figures of some of the quadrupeds brought from the Arctic regions of our continent for this work."[41] The HBC even offered the live animal to John Woodhouse Audubon, but he was unable to accept the gift, a fact that invited the plaintive query, "When shall we have a Zoological Garden in the United States?"[42]

Even as the HBC cooperated with American scientists, the Oregon Treaty of 1846 turned much of the Columbia District over to the United States, although those in the HBC resented the British government at least as much as the American government for the terms of the treaty. For the HBC, the loss of the southern portion of the Columbia District in 1846 was a significant blow, but its hold on other portions of its territories also seemed to be loosening at the time. For example, the Red River Colony and portions of southern Rupert's Land seemed to be falling into the United States' orbit. Separated by thousands of roadless kilometres from Canada – the nearest settlements in British America – the Red River Colony became linked to the United States. In 1844, the Canadian-born American trader Norman Kittson established a trading post at Pembina, strategically located just inside US territory directly south of the Red River Colony. Although some furs made their way south from Red River to the United States before then, the flow grew significantly after Kittson established his post. Soon, hundreds of carts of furs headed south to St. Paul, Minnesota, each year, rather than to London on HBC ships. The settlement frontier in Minnesota also crept northwestward towards Red River in the late 1830s, and transportation and communication routes between St. Paul and Red River gradually improved during the 1840s and 1850s. No wonder that, when American expansionist thought and rhetoric reached a peak between 1844 and 1848, many Americans considered it probable that the United States could soon possess the entire North American continent.

In some respects, trends after 1848 seemed only to further to confirm beliefs in the United States' manifest destiny. The population of Minnesota, barely 6,000 in 1850, swelled to 172,000 in 1860, by which time Minnesota had acquired statehood.[43] Links between Minnesota and Red River improved dramatically over that time. A United States post office was opened at Pembina in 1857, and steamboat transportation on both the Mississippi and Red Rivers was inaugurated by the end of 1859.[44] It was hard to avoid the conclusion that St. Paul had become the natural metropolis of Rupert's Land and the great northwest. Members of the small American community in Red River, expansionists in Minnesota, and other proponents of American northward expansion hoped that the United States might yet be able to annex Red River and the great northwest.[45]

Threats to the HBC's hold on what remained of the HBC territories west of the Rocky Mountains also grew in the 1850s. A gold rush along the lower Fraser River in 1858 brought thousands of Americans, Canadians, and other non-native people to the Fraser Canyon, near the southern border of the British and American territories, raising the spectre of American filibusters serving as the advance forces of American expansionists on the Pacific slope. The influx prompted the British government in August 1858 to assume direct rule of what remained of the Columbia District (renamed British Columbia in 1858).[46]

Even on the northwestern plains the effectiveness of the HBC monopoly was gradually being eroded. Before the 1830s, HBC posts along the Saskatchewan River system were acquiring through aboriginal trading partners furs gathered in United States territory. Then the flow of furs reversed. During the 1820s and 1830s, the American Fur Company (AFC) established posts along the Missouri River that eventually traded furs from southern Rupert's Land. The navigability of the Missouri River (steamboats travelled as far as Fort Union as early as 1832), allowed the AFC to offer better prices to indigenous people than the HBC could. One could even sense that the British colony of Canada might become part of the United States when a short-lived movement advocating the annexation of the colony of Canada by the United States flared up within Canada in 1849.[47]

Fortunately for the HBC and for those Canadians who did not wish annexation by the United States, the support for territorial expansion in the United States diminished significantly with the controversies unleashed by the Mexican–American War (1846–8) and the intensifying internal divisions of the 1850s. Not until after the American Civil War (1861–5) was the United States again in a position to acquire significant new territories. Curiously, even as enthusiasm for expansion collapsed in the United States, an aggressive expansionist movement emerged in the colony of Canada. The HBC's support for Canadian-based travellers in the 1840s appears to have been instrumental in the development of that movement.

The HBC had little to do with Canadian science before 1821.[48] The HBC turned to Canada as a source of employees after 1810 when it began to compete more aggressively with the NWC, but because Montreal was the headquarters of the NWC, the HBC had little incentive to cooperate with Canadian-based travellers. Furthermore, as a small colony, Canada had few people with scientific credentials, and no scientific institutions of any significance before 1821.

The potential for scientific cooperation between Canadians and the HBC improved shortly after the merger of the HBC and NWC. With the reorganization that followed the merger of the two companies, Montreal lost its position as a major fur-trading centre, although the city did become the headquarters of the HBC's Southern Department. Confronted by the impossibility of crossing the Atlantic Ocean from any location in Rupert's Land except on the occasion of the arrival and departure of the annual supply ships, Governor Simpson established his headquarters in Montreal in 1826.[49] Thereafter, Simpson soon joined the ranks of the city's English-speaking commercial elite.[50] Politically conservative and aristocratic, Simpson fit into that community well.

In 1827 members of Montreal's anglophone, Protestant, professional, and commercial community spearheaded the formation of the Natural History Society of Montreal, and the society soon asked Simpson to help it gather information and specimens of aboriginal people, geography, and natural history from the HBC's territories.[51] The company did transmit its traders' written answers to the society's queries, although specimens gathered for the society appear to have remained in London, rather than being forwarded to Montreal.[52] Apart from those tentative beginnings, the HBC's cooperation with Montreal scientists was slight before the 1850s. The lack of assistance may be attributable primarily to the fact that the Natural History Society of Montreal failed to flourish until the mid-1850s. In 1855 the Nova Scotian scientist John William Dawson arrived in Montreal to take up the principalship of McGill College (now McGill University), and in 1857 he also became president of the Natural History Society of Montreal. Thanks to Dawson's energetic leadership both languishing institutions soon began to flourish.[53] The society inaugurated its first journal, *The Canadian Naturalist and Geologist*, in February 1856, and moved into an impressive new building in 1859.[54] By the early 1860s, its membership was growing dramatically.[55] This rejuvenation in the late 1850s and early 1860s helps explain the connections forged between some of the HBC traders and the society (discussed in chapter 6) at the time.

The HBC's established reputation as a patron of British science encouraged new requests and opportunities for cooperation that also added Toronto to the HBC's scientific network in the 1840s. The company's reputation certainly influenced the scientists who came to seek the help of the company with a major international survey of terrestrial magnetism in the 1840s. The so-called "Magnetic Crusade" of the late 1830s and 1840s was, in several respects, an important milestone in the history of science. At the broadest level, it inaugurated and served as a model for "big science": large-scale expensive scientific enterprise requiring multi-year state sponsorship and management by professional scientists and government departments.[56] It was the first scientific project based on simultaneous and continuous observations at widely scattered locations.[57] But the magnetic survey had many other important local and specific effects. Of particular relevance to this study is the fact that the project strengthened the links that had developed during the search for the Northwest Passage between the HBC and departments of the British state. The project also did much to establish Toronto as a scientific centre of some weight. It was also crucial in adding Toronto as one of the nodes in the scientific network of which the HBC was a part. The connections that also developed between Toronto scientists and American scientists such as Joseph Henry almost certainly also played a part in the connections that thereafter developed between American scientists and the HBC.

Until the British geomagnetic survey began, the British state only sporadically supported science. Because an understanding of much of the significance of geomagnetic data was dependent upon systematic analyses of data gathered in many places, research in terrestrial magnetism before the 1830s was conducted primarily in countries such as France, where state support permitted that kind of research. Research in Great Britain lagged well behind until 1834. Most British geomagnetic research of the 1820s and early 1830s occurred as part of the expeditions to search for the Northwest Passage (the proximity of the north magnetic pole inspiring such studies). Those expeditions particularly piqued the interest of James Clark Ross and Edward Sabine in terrestrial magnetism.[58] For various reasons, the perceived importance of geomagnetism also grew dramatically inside and outside of scientific circles during the 1820s and 1830s.[59]

In the early 1830s, a group of influential British scientists began to urge the British government to support an ambitious international study of terrestrial magnetism that would include a worldwide network of geomagnetic observatories. This lobby was based in the British Association for the Advancement of Science (BAAS), a new scientific organization founded in 1831.[60] To add weight to their campaign, scientists adopted a strategy that was used throughout the life of the project: scientists were recruited to write letters internationally. For example, prompted by Edward Sabine, Alexander von Humboldt wrote supportive letters in 1836 to the Royal Society and the British government.[61] British scientists likewise petitioned the American government to support observatories in the United States.[62] Their unsuccessful efforts led in 1848 to the formation of the American Association for the Advancement of Science (modelled on the BAAS), which not only lobbied for government support for observatories at American universities but also petitioned the British government to continue supporting its observatory in Canada.[63] Joseph Henry, the foremost American physicist of the time, also lobbied the British government on behalf of the Canadian observatory.[64]

In 1839, the British government, on the recommendation of the Royal Society, did commit itself to a network of colonial observatories and to expeditions to study terrestrial magnetism, both under the direction of the eminent geophysicist Edward Sabine. To that end, a global network of observatories was established. Geomagnetic research may not easily capture the public imagination, but scientists and politicians boasted of its importance. In the conclusion to his chapter on the history of magnetism, William Whewell of Trinity College, Cambridge, argued that "the manner in which the business of magnetic observation has been taken up by the governments of our time makes this by far the greatest scientific undertaking which the world has ever seen. The result will be that we shall obtain in a few years a knowledge of the magnetic constitution of the earth which otherwise it might have required centuries to accumulate."[65]

The intensity of geomagnetism in the HBC territories made them particularly suitable for geomagnetic research, but the lack of any cities in the territories made them impractical for an observatory. Even before 1830, Sabine had identified the region around Great Slave Lake as

> the field for observations of the very highest importance on the subject of the magnetism of the globe; and as it is traversed annually under the direction of the Hudson's Bay company, we may confidently hope, from the ready disposition which that company has shewn in so many instances to promote scientific researches, that much time will not elapse, before that really important journey will be performed by some person, properly qualified by previous practise, to observe with the precision necessary on so particular occasion.[66]

As the largest and most accessible city in Canada, Montreal was the most obvious location for Britain's North American observatory, but because of the magnetic nature of the rocks near Montreal and along the entire St. Lawrence River, Toronto was chosen over Montreal in 1841 as the location of the observatory.[67] Toronto, Upper Canada, was then a small city of under 20,000 people, but the establishment of the observatory helped make Toronto something of a Canadian scientific centre. Toronto's observatory was accompanied in North America by others (unfunded by government) in Philadelphia, Washington, and Cambridge in the United States, and in Novo-Arkhangel'sk (Sitka), Russian America.[68]

The prominence of the geomagnetic research within scientific and government circles, and the ease with which the HBC could assist the research made it natural for the directors of the HBC to want to be connected with it. Edward Sabine later explained how the HBC got involved in this project:

> the good offices which the Hudson's Bay Company had contributed to the success of the geographical expeditions, undertaken by the British Government, for the purpose of tracing the American rivers and coasts of the polar sea, and their liberality in originating expeditions of the same nature, and in executing them at their own cost and by their own officers, justified

the hope that assistance might be given by the Hudson's Bay Company, which should render an undertaking feasible, which undoubtedly would not have been so without their aid.[69]

Fortunately for Sabine, the HBC's London governor J. H. Pelly and its North American governor George Simpson responded enthusiastically. Thanks to the HBC's assistance, John Henry Lefroy of the Royal Artillery was able to conduct far more geomagnetic research in northern North America than Edward Sabine had anticipated. Accompanied by William Henry, also of the Royal Artillery, Lefroy left Lachine in May 1843 with the HBC brigade and travelled westward along the normal HBC brigade routes through Fort William, Norway House, and Cumberland House to Fort Chipewyan during the spring and summer of 1843, and wintered at Fort Chipewyan in 1843-4. During the winter he made side trips as far north as Fort Good Hope, just south of the Arctic Circle. In 1844 he travelled up the Peace River as far as Fort Dunvegan before returning to Toronto in November 1844.[70] He returned with observations from thousands of locations in the Hudson's Bay Company territories.[71]

In his published report of this expedition, Lefroy noted that "by the exertions of Mr. [Colin] Campbell, – to whose kindness, as well as to that of Mr. Lewis [John Lee Lewes], the Chief Factor resident at Fort Simpson, and to Sir George Simpson, the Governor of the Hudson's Bay Company, I have to acknowledge the greatest obligations."[72] The kindness to which Lefroy referred may well have been elicited by his own splendid conduct while in the territories. According to the HBC officer, John McLean, "few gentlemen ever visited this country who acquired so general esteem as Mr. Lefroy; his gentlemanly bearing and affable manners endeared him to us all."[73] Earlier, he wrote that Lefroy's "zeal for scientific discovery neither cold, nor hunger, nor fatigue, seems to depress."[74] And with obvious admiration, McLean wrote that "this gentleman seems equal to all the hardships and privations of a voyageur's life, having performed the journey from Athabasca hither [Fort Simpson], a distance of at least six hundred miles, on snow-shoes, without appearing to have suffered any inconvenience from it; thus proving himself the ablest *mangeur de lard* we have had in the country for a number of years: there are many of our old winterers who would be glad to excuse themselves if required to undertake such a journey."[75]

Making such a positive impression on HBC officers, Lefroy was well placed to seize the opportunity to request further assistance from them. Lefroy, for example, canvassed HBC traders to assist him in developing

7.1 Paul Kane's Portrait of John Henry Lefroy (1845–1846). John Henry Lefroy poses in front of a dog sled that carries his surveying equipment. Lefroy and the man behind him are both depicted wearing capotes, assomption sashes, leggings, and moccasins characteristic of French Canadian voyageurs and Metis in the northwest. The artist, Paul Kane, was a friend (and later became a brother-in-law) of Lefroy. Thanks in part to Lefroy's intervention, George Simpson also ensured the HBC's support for Kane's visit to the HBC territories between 1845 and 1848, although Lefroy posed for this painting in a Toronto studio, with Kane adding the background from his imagination. In 2002 this painting set a new record for a Canadian painting by selling at auction for $5.1 million. Credit: Paul Kane, Canadian 1810–71. *Scene in the Northwest – Portrait of John Henry Lefroy*, c. 1845–6. Oil on canvas. 55.5 × 76 cm. Art Gallery of Ontario. The Thomson Collection © Art Gallery of Ontario.

an estimate of the aboriginal population of the Hudson's Bay Company territories, the results of which he published in Toronto in the 1850s.[76] He also convinced several HBC clerks and officers, including W. L. Hardisty, Bernard Rogan Ross, James Stewart Clouston, and Colin Campbell, to keep and submit auroral journals in 1850 and 1851. When he added to them journals he got from Joseph Henry at the Smithsonian,

Lefroy had the valuable data he needed for an article on the aurora borealis published in 1852.[77]

Scientific research and a culture of learning appear to have reinforced one another. By the time of Lefroy's expedition to the HBC's territories, the Mackenzie District had been the focus of geographical and scientific interest for two decades. That fact (combined with the long cold dark nights of those latitudes) may explain why officers in the district sought to improve the intellectual climate there. At any rate, in March 1852, James Anderson, Chief Trader in charge of the Mackenzie River District, wrote a circular to the officers in charge of the posts of the district, explaining that "the men and officers" had agreed to establish a subscription library in the district "quite distinct from the officers proprietary library." He told them that the "books will be periodically exchanged, and after they have gone through the District, will be deposited at Headquarters, lists of the Books there deposited will be forwarded to all the Posts so that any person can get the book or books he may require thence, exclusive of the annual supply."[78] He continued by writing, "I persuade myself that many officers will not hesitate to subscribe: – great care has been taken in selecting useful and entertaining works, and I have sent a list for a large supply to Gov [Andrew] Colville, in the full expectation, that every one in the district will contribute his mite towards such a laudable undertaking."[79] This initiative was intended to appeal to all of the men at the posts, including the illiterate labourers. "I beg," Anderson urged the officers, "you will endeavour to persuade all the men – whether they can read or not – to subscribe, as they will be much benefitted by such an institution."[80] Elaborating further, Anderson clarified that "many of the works sent for are illustrated, so that even those who cannot read will derive some information and entertainment from them."[81] Adhering to what had evidently always been their policy, the London Governor and Committee did not charge for shipping and delivery of books from London.[82] During the 1850s, officers and men in the Mackenzie District enjoyed the kind of access to books that few denizens of the interior of North America could match.

Lefroy's very positive experience with George Simpson and the HBC in 1843 and 1844 put him in a good position to help the budding Toronto artist Paul Kane make connections with George Simpson and the HBC. Evidently inspired by George Catlin and Catlin's belief that the western Indians were doomed to extinction, Paul Kane naively left Toronto in June 1845 with, according to his biographer, "no money, no resources, no promises of assistance," and very little knowledge of the HBC territories.[83] "My principal object in travelling among the Indian tribes of the Far West," he later explained, "was to obtain accurate sketches of their chiefs, medicine men, &c., and representations of their most characteristic manners and customs."[84] He got no farther in 1845 than Sault Ste. Marie, Upper Canada, where the HBC's Chief Trader John Ballenden advised him to turn back and apply to George Simpson for permission to visit the HBC territories. To that end, Ballenden wrote a letter of recommendation to Simpson on Kane's behalf. Ballenden was obviously impressed with Kane's determination. He wrote Simpson that Kane had "coasted the southern shore of Lake Huron to Manatowening in a bark Canoe with only one man."[85] Kane did return to Toronto, and further armed with an introductory letter written by Lefroy, he travelled to Montreal to meet Simpson.[86] Simpson not only gave Kane permission to travel to the HBC territories but also issued a circular that instructed HBC officers that "you will be pleased to afford Mr. Kane passages from post to post in the Company's craft – free of charge."[87]

In the spring of 1846, Kane left with the HBC brigades, arriving, in December of 1846, at Fort Vancouver. Using Fort Vancouver as his base for the next six months, Kane made trips to the Willamette Valley and Vancouver Island before slowly making his way from Fort Vancouver to Edmonton House in the second half of 1847. While he was in the field, probably in 1847, Simpson wrote Kane to request a dozen "sketches of buffalo hunts, Indian camps, Councils, feasts, Conjuring matches, dances, warlike exhibitions, or any other scenes of savage life that you many consider likely to be attractive or interesting."[88] After wintering at Edmonton House, Kane returned to Toronto on 13 October 1848 with hundreds of field sketches (see figure 7.2).[89] Even as he made his sketches, Kane clearly did hope, as he later claimed, that his work would be accepted not merely as art, but as exact and true depictions of reality. To that end, he even went so far as to collect letters from HBC traders testifying to the accuracy of his sketches. James Douglas's letter described the landscapes as "faithful," and the portraits and depictions of clothing as "perfect." William McBean praised the "exact and strong similitude to the original" of Kane's art. And John Lee Lewes commented on the sketches of Indians as "most true and striking, their manners and customs are depicted with a correctness, that none but a Master hand could accomplish."[90]

Historical literature already acknowledges that the period between 1846 and 1850 marked an important transition in Canada's relationship with Rupert's Land, but does not posit the probable importance

"Disinterested Kindness" 217

7.2 Paul Kane, Field Sketch: *Kee-a-kee-ka-sa-coo-way: Head Chief of the Crees*, **1848**. Paul Kane executed this sketch (watercolour over pencil) while in the field at the HBC's Fort Pitt in the summer of 1848. It was item number 84 (of 240) in the Toronto City Hall exhibition of his work in November 1848. The fact that he gathered fur traders' testimonies to that effect, it was obviously important for Kane that the public accept paintings such as this one as "truthful" representations of reality. Also see colour plate 6. Source: LAC C-114386, 1981-055 X PIC.

of Lefroy and Kane in stimulating Canadian interest in the territories.[91] When Lefroy travelled to Rupert's Land, he was not yet well connected with Upper Canadian society. But his place in Toronto society changed dramatically soon after he returned to Canada in late 1844. In 1845 he began courting Emily Robinson, the daughter of John Beverley Robinson, the chief justice of Upper Canada. When he and Emily married on 16 April 1846, it was in a dual ceremony in which Emily's sister Louisa married Lefroy's friend George William Allan, the son of William Allan. William Allan was a rags-to-riches Scottish immigrant who had arrived in Toronto in the 1790s.[92] By the 1840s, he was the "unquestioned doyen of Upper Canadian business" and one of the wealthiest men in Canada.[93] Augusta, another of Emily's sisters, was married to the son of John Strachan, the Anglican bishop of Toronto and leader of the "Family Compact" (the informal network of conservative elite that had dominated Upper Canada between 1812 and 1837). "These marriages," as Suzanne Zeller noted, "brought Lefroy into the network of the Family Compact and consolidated his stature within Toronto society."[94]

Kane too, was soon prominent in Toronto. Kane's art was exhibited to widespread acclaim at Toronto's City Hall, beginning on 9 November 1848 – less than a month after he returned from the HBC territories.[95] Given that reviewers commented on the "truthfulness" and accuracy of the sketches, Kane himself probably emphasized his argument that he assumed no artistic licence, likely even by exhibiting the letters that he had received from HBC traders attesting to his art's careful faithfulness to reality.[96] Soon, none other than George W. Allan, Lefroy's brother-in-law, became his friend and patron. Allan first met Kane around the time of Kane's exhibition.[97] In 1853, shortly after returning from an extended time in Europe, Allan agreed to buy a cycle of one hundred paintings from Kane for $20,000 (figure 7.3).[98] The exhibition must have played a significant role in increasing awareness of the British North American West among Canadian elite, and in that way must have contributed to the brief show of interest in Canada in the HBC territories at the time.[99] Evidence of the respect Kane had achieved among politicians rests in the fact that in 1851 he acquired a £500 commission for twelve paintings for the province's House of Assembly.[100] Historian Doug Owram has argued that the period between 1846 and 1850 was important in the history of Canada's westward expansion, for although those years did not produce any annexationist rhetoric, "the first words in what would later become an expansionist torrent came in this period."[101]

The interest in the HBC territories displayed by Torontonians during the late 1840s following the visits of Lefroy and Kane to the West

7.3 Paul Kane, *Kee-A-Kee-Ka-Sa-Coo-Way, Man Who Gives the War Whoop* **(Oil on Canvas), 1848–56.** This studio painting was among the paintings Kane sold to his wealthy friend and patron, George William Allan, and was reproduced in his *Wanderings of an Artist* in 1859. It was eventually donated to the Royal Ontario Museum by Sir Edmund Osler (1845–1924), a Toronto businessman and philanthropist who made much of his wealth through investments in western Canada. Compare this image with the field sketch (fig. 7.2) upon which it is based. Also see colour plate 7. Published with permission of the Royal Ontario Museum © ROM. (ROM 912.1.42)

waned in the early 1850s, but suddenly surged in the form of an annexationist movement in 1856. This movement, including its scientific dimensions, has attracted considerable scholarly attention.[102] Historian Doug Owram has noted that "from the Canadian perspective, the most noticeable characteristic of this successful assault on the position of the Hudson's Bay Company was the extremely small number of individuals who acted as its spearhead."[103] Owram attributed the rise of the expansionist movement to a number of forces operating in Canada at the time, but neither he nor any other scholar has acknowledged the probable role of Toronto's first influential scientific organization, the Canadian Institute, in nurturing this movement.[104] The Canadian Institute was the organization that linked those Toronto-based travellers who had been to the HBC territories, and other Canadian scientists and politicians, with those men who came to lead the expansionist movement.

In 1855 the scientific community in Toronto had only very recently acquired a level of organization and coherence that allowed it to shape the beliefs and attitudes of Canadian politicians and the Canadian public. According to historian of science Richard A. Jarrell, as late as 1851 "for anyone with an interest in science, Toronto seemed a wasteland."[105] The scientific establishment in Canada was indeed tiny before the 1850s, and was dominated by the Natural History Society of Montreal. Canada's scientific profile grew significantly in 1841 with the establishment of the Geomagnetical and Meteorological Observatory in Toronto and the establishment of the Geological Survey of Canada by the government of the newly created United Province of Canada (uniting the former provinces of Upper Canada and Lower Canada).[106] A public scientific organization – the Canadian Institute – joined these others in 1851. The Canadian Institute was established in 1849 as a professional organization, but was transformed in November 1851 by a royal charter that defined it as "a Society for the encouragement and general advancement of the Physical Sciences, the Arts and Manufactures."[107] In August of 1852, the institute published the first issue of its monthly periodical, the *Canadian Journal*.[108] But the institute did not really flourish until the mid-1850s.

One of the most important people – perhaps the single most important person – in ensuring the success of the Canadian Institute in the 1850s was not a scientist at all, but Paul Kane's patron, George W. Allan. Allan later claimed that he had been connected with the Canadian Institute from the receipt of its charter.[109] He was elected its secretary in

1853, its second vice-president in 1854, and its president in 1855 (when he was also Toronto's mayor).[110] By that time, Allan was familiar with tragedy and grief. By 1854, both of his parents, nine of his ten siblings, and his wife Louisa had died. As the only surviving son of one of Toronto's wealthiest citizens, George inherited his father's entire substantial estate, which included one of the long, narrow hundred-acre "park lots" north of present-day Queen Street in downtown Toronto.[111] His commitment to support science and the arts in Toronto appears to have grown significantly after his father's death in July 1853. In June 1854, less than a year after his father's death, George W. Allan offered to donate a lot on George Street to serve as the location for a permanent home for the Canadian Institute.[112] In 1855 the Canadian Institute merged with the Toronto Athenaeum, bringing to the institute the Athenaeum's library, museum, and membership list, and in January 1856 the institute inaugurated a more ambitious new series of the *Canadian Journal*.[113] Fortunately for the institute, Toronto also became the capital of the colony of Canada in 1856. Soon, politicians and government officials helped swell the ranks of the institute's members.

It seems that by December of 1855 the Canadian Institute had acquired enough prestige that not only Canada's scientists but also many of Canada's political, economic, and social elite (particularly those in Toronto) wanted to be members. In 1855–6, when George W. Allan served as its president, its members included all of Canada's scientific elite. They also included men connected with the HBC territories, from Paul Kane and J. H. Lefroy (although Lefroy had relocated to England) to George Simpson. Many politicians and officials (both liberal and conservative) including W. W. Baldwin, Robert Baldwin, John Beverley Robinson, Egerton Ryerson, Francis Hincks, and George-Étienne Cartier were members. Canada's most influential railway propagandist Thomas Keefer was on the rolls. And many men from the extremely small group that would soon launch the expansionist movement, including George Brown, Henry Youle Hind, Alexander Morris, William McMaster, William P. Howland, Sandford Fleming, and Philip M. VanKoughnet, were members.[114] In short, the meetings of the Canadian Institute became, in the year before the annexationist movement burst onto the scene, *the* place in which scientific travellers, politicians of all stripes, promoters, and future expansionists interacted. Not only that, in the year before expansionists launched their public campaign to annex the HBC territories, its members listened to, saw, and read much about the HBC territories.

In the 1850s, Kane's scientific profile increased significantly, thanks almost certainly to the encouragement of George W. Allan. He was elected a member of the Canadian Institute in January 1855. A paper by Kane on the Chinook people of the Northwest Coast was read at the institute's meeting of 14 March 1855, and published in the *Canadian Journal* that July and in the conservative Toronto newspaper, *The Daily Colonist*, in August.[115] Much of the meeting of 31 March 1855 was also devoted to Kane's travels. At that meeting George W. Allan read another paper by Paul Kane on the Chinook people, and "various articles of dress worn by the Chinook Indians, specimens of their bows and arrows, spears, cooking utensils, and a skull taken from one of their graves, were exhibited. Several admirable oil paintings, executed by Mr. Kane, illustrated many important features of the lives and characters of the Chinook Indians."[116] In 1857, "in accordance with an invitation of the Council of the Canadian Institute," Kane published a longer, more detailed, and polished paper on the Chinook.[117] On 13 November 1855, Kane's paper on the "half-breeds" of Red River was read. That paper was published in January of 1856.[118] Finally, on 5 April 1856 Kane's paper on the Walla Walla people was presented.[119] The dates of these presentations seem so significant because they are followed almost immediately by the sudden emergence of an organized and public movement for the Canadian annexation of the HBC territories, and because, according to George Simpson, Kane's work published before 1858 "attracted considerable attention."[120]

Kane did not publish in the *Canadian Journal* after 1858, but his international reputation was established in 1859 when his *Wanderings of an Artist* was published in London, England.[121] Kane's standing among scientists was further enhanced by the noted scientist Daniel Wilson. Born in Edinburgh, and schooled there and in London, Wilson had taken up a position at the University of Toronto in 1853 and had developed a friendship with Paul Kane shortly thereafter.[122] In 1857, Wilson presented a paper about Kane's paintings and artefacts at the annual meeting of the American Association for the Advancement of Science.[123] Then, in his *Prehistoric Man* (1862) Wilson listed Kane among the "scientific friends" to whom he was obliged, cited Kane often, and included colour reproductions of some of Kane's sketches.[124]

The transformation of Paul Kane from travelling artist to respected scientific reporter cannot have been a simple or straightforward one. Kane was not very well educated. His field notes show that he had neither the writing ability nor the detailed knowledge necessary to pen his articles or *Wanderings of an Artist* without the help of a ghostwriter. Ian

MacLaren has argued convincingly that "Kane was not in control of the process that was metamorphosing him from traveler into author, and it may be that he had little say in the matter."[125] Whoever helped him write his work "knew all the names of the fur trade that Kane did not have in his field notes, including the names of small lakes, of portages, and so on."[126] That fact led MacLaren to the reasonable suggestion that George Simpson might have secured a ghostwriter for Kane.[127] But the evidence suggests that Simpson actually played only a minor role in Kane's career after Kane returned to Toronto.

Kane's relationship with Simpson was evidently respectful but probably never close. Kane did defend the HBC and thank some of its personnel, but not in a way that suggests that Kane felt particularly indebted to Simpson. For example, Kane published generous tributes to the company without mentioning Simpson:

> Without entering into the general question of the policy of giving a monopoly of the Fur trade to one company, I cannot but record as the firm conviction which I formed from a comparison between the Indians in the Hudson's Bay Company territories and those in the United States, that opening up the trade with the Indians to all who wish indiscriminately to engage in it must lead to their annihilation. For while it is the interest of such a body as the Hudson's Bay Company to improve the Indians and encourage them to industry, according to their own native habits, in hunting and the chase, even with a view to their own profits; it is as obviously the interest of small companies and private adventurers to draw as much wealth as they possibly can from the country in the shortest possible time, although in doing so the very source which the wealth springs should be destroyed. The unfortunate craving for intoxicating drinks, which characterises all the tribes of Indians, and the terrible effects thereby produced upon them, render such a deadly instrument in the hands of designing men. It is well known that, although the laws of the United States strictly prohibit the sale of liquor to the Indians, it is impossible to enforce them, and whilst many traders are making rapid fortunes in their territories, the Indians are fast declining in character, wealth, and numbers, whilst those in contact with the Hudson Bay Company maintain their numbers, retain their native characteristics unimpaired, and in some degree share in the advantages which civilization places within their reach.[128]

Kane's endorsement of the company, one of the most laudatory it ever got, came at a crucial time in HBC history. The looming expiry of its licence and the scrutiny the company was undergoing and would continue to undergo meant that such tributes were rarely more valuable than they were in 1856.

As valuable as Kane's tribute was for the company, the lack of any mention of Simpson was noteworthy, especially in light of the many published tributes that named Simpson explicitly. If Simpson had assisted Kane significantly, Kane would have been foolhardy to have neglected thanking him as enthusiastically as he defended the HBC. But Kane's silence was not an oversight. Kane later wrote, "I entertained great respect for Sir George Simpson, but I felt that he was in no sense my patron."[129]

An incident described by Kane in *Wanderings of an Artist*, in which Kane was left behind in Makinaw by an HBC steamer with Simpson aboard, might have generated hard feelings between Kane and Simpson even at the beginning of Kane's trip to the HBC territories.[130] Then, within a week of returning to Toronto in October 1848, Kane wrote to Simpson about the possibility of publishing a book about his travels. Simpson's response, though encouraging, included no offer of direct assistance, and Kane acted on none of Simpson's suggestions.[131] If Simpson's response disappointed Kane, Simpson may also soon have been annoyed with Kane.[132] In January 1849, after learning of the success of Kane's exhibit in Toronto, Simpson generously offered to help organize a similar show in Montreal. But Kane demurred, pleading that "my circumstances at present do not permit it."[133] It is difficult to imagine that a man of Simpson's temperament would have received Kane's injudicious response as anything but a snub. Also in January 1849, Kane delivered ten of Simpson's requested paintings to Simpson, evidently with an invoice.[134] In 1861, Kane recalled that he had offended Simpson at that time because "I had not furnished him copies of my Indian painting[s] *gratis* in recognition of his official protection."[135]

If Kane and Simpson's relationship was very warm one would expect to find evidence of the fact in Kane's publications. But in only one of his scientific articles (in passages that were repeated in *Wanderings of an Artist*) did Kane defend the HBC or thank any of its employees. In the very 1856 article that he defended the HBC, Kane also thanked Alexander Christie, the HBC's governor at Red River, "whose many acts of kindness and attention I must ever remember with feeling of grateful respect."[136] Although Simpson is mentioned several times in *Wanderings of an Artist*, there is no similar tribute to Simpson. In 1861, Kane explained why. He wrote, "such favour as I had received from Mr. Christie and others was in no way due to his [Simpson's] example."[137]

Plate 1 *Hudson's Bay House – Fenchurch Street (London)* by Thomas Colman **Dibdin, 1854.**
Source: HBCA, PAM 1987/363-S-25/T78

Plate 2 *Sir George Simpson* by Stephen Pearce, 1857.
Source: HBCA, PAM 1987/363-S-25/T78

Plate 3 *Ring-Tail'd Hawk* **by George Edwards, 1743–51.**
Source: HBCA, PAM 1987/363-S-25/T78

Plate 4 (overleaf spread)   Detail of "A Map Exhibiting all the New Discoveries in the Interior Parts of North America, Inscribed by Permission to the Honorable Governor and Company of Adventurers of England Trading into Hudson's Bay in Testimony of their Liberal Communications to their Most Obedient and Very Humble Servant A. Arrowsmith," by Aaron Arrowsmith, 1795, with additions to 1802.
Source: HBCA G.3/672 (portion 2)

(map)

Methy Carrying Place 11 Miles
Methy Lake
Washacummow or Clearwater L.
Buffalo L.
Bear Hill
La le Crofs
Lake Cross Lake or Min rus tick a Tock aha twan
Primo L.
Knee L.
Grass R.
Black Bear Islands Lake
Beaver Riv.
R. of log da Plonge
Pike R.
Rapids
Messin a ne pe or Churchill Riv.
Canadian Sett.
into Churchill Riv.
House
Fox L.
Burntwood
Pelican L.
File L.
Rood L.
Beaver L.
Goose L.
S
T
I
N
E
A
U
Pine Island L.
Bluff or a Pine
Middle Track
Sturgeon R.
R. Sepanach
R. Basquiau
Canadian R.
Cumberland Ho.
Saskashawm R.
Falls
Turtle R.
Pike Lake
Shell Cr.
Setting R.
Islands
Hudson House
Battle R.
Rafberry L.
Red Berry Cr.
South Branch House
Ceder L.
Red Deer L.
Red Deer R.
Etow we mah-me L.
Shoal R.
Swan L.
Swan Riv. Ho.
Swan R.
Thunder Hill
Somerset Ho.
Ft. Dauph
Red River
Carlton Ho.
Marlboro Ho.
White Mud R.
Gods L.
CREE INDIANS
Eaglehill Cr.
Can. Ho.
ASS
The most Northern Bend of the Missesourie Riv.
Villages of the Indians &
Knife R.
S N A K

Plate 5 *John Rae* **by Stephen Pearce, 1853.**
© National Portrait Gallery, London, England (NPG 1213).

Plate 6 Field Sketch: *Kee-a-kee-ka-sa-coo-way: Head Chief of the Crees* **by Paul Kane, 1848.**
Source: LAC C-114386, 1981-055 X PIC.

Plate 7 Paul Kane, *Kee-a-kee-ka-sa-coo-way: Man Who Gives the War Whoop* **(oil on canvas), 1848–56.**
Published with permission of the Royal Ontario Museum © ROM (ROM 912.1.42).

George W. Allan, not George Simpson, was responsible for Kane's transformation to a man of scientific standing. The 1848 genesis of Allan and Kane's friendship has already been noted. For Kane, the canvases that Simpson had requested in 1847 must have paled in significance beside the one hundred paintings for which Allan subsequently paid $20,000. The fact that at least Kane's first presentation to the Canadian Institute (and perhaps all of them) was read by George Allan (before Simpson was elected a member of the institute) suggests that Allan began assisting Kane early to project a learned image of himself. Similarly, the fact that the last of Kane's presentations took place in April 1856, only a month before Allan left Toronto for an extended trip to England (he left Toronto in May 1856 and did not return until February 1859) suggests that Allan was instrumental in promoting Kane within the Canadian Institute.[138]

Allan continued to help Kane while Allan was in England. He, with the help of his brother-in-law J. H. Lefroy, probably helped Kane more than anyone else when it came time to try to publish *Wanderings of an Artist*. In 1858 Kane, Lefroy, and Allan were all in London (Allan's brother-in-law J. H. Lefroy was based in Woolwich at the time) trying to negotiate the publication of Kane's manuscript with the very publisher that had put out Lefroy's 1855 book.[139]

Before he left Canada Kane asked Simpson for assistance. Simpson did not offer any direct aid to Kane, but did write a letter of introduction on Kane's behalf to the board of the HBC, "so that you may be in a position to refer to them when necessary (as a well known and influential Corporation) in your negociations with publishers & others."[140] To the HBC's London Secretary, William G. Smith, Simpson wrote that

> Mr. Kane has always gratefully acknowledged the favors shewn him by the company and the kindness of their officers. He has been almost their only advocate in Toronto since the Anti-Hudsons Bay cry arose: and both verbally & in writing has defended them from the indiscriminate attacks made upon them. He was looked on as an impartial witness and his writings attracted considerable attention. I have no doubt his Journal will represent his views in the most favorable light for the Company, while from the interest that would attach to the illustrations, it would be likely to be generally read.[141]

Had Simpson's letter to the HBC's board been instrumental to the publication of Kane's manuscript, Kane would surely have mentioned it, but when *Wanderings of an Artist* was published Kane did not dedicate it to Simpson, but to George William Allan, "as a token of gratitude for the kind and generous interest he has always taken in the author's labours, as well as a sincere expression of admiration of the liberality with which, as a native Canadian, he is ever ready to foster Canadian talent and enterprise."[142] The dedication's emphasis on Canadianness, and the fact that Kane claimed to have been born in Canada when in fact he was born in Ireland, suggests that Kane was at least as eager to advance the interests of Canada as he was to promote the HBC.[143] Kane later wrote that Simpson was offended "that I had omitted to dedicate my recent book, describing my wanderings, to himself," although Kane's evident pro-Canadian sentiments expressed in the dedication would have done nothing to endear him to Simpson.[144] There appears to have been no reconciliation before Simpson's death in 1860. In March 1861, while trying to secure HBC assistance to travel to Labrador, Kane wrote: "I made formal application on two occasions to the late Sir George Simpson, but as I had the misfortune to incur the Governor's displeasure, I understand that he had given orders that I was not to be countenanced by any officers in the service."[145] The HBC never did assist Kane again.

The activities of Kane, Allan, and the Canadian Institute almost certainly influenced the sudden emergence of the Canadian expansionist movement in 1856. Until 1856, few people considered the possibility that Canada might annex Rupert's Land. To be fair, Allan Macdonell had promoted Canadian westward expansion as early as 1851, but without generating any significant public or political interest.[146] Until then, opponents of the HBC tended to advocate for the revocation of the company's charter (without clearly explaining how the jurisdictional void might be filled), or the creation of a separate Crown Colony. Between 1821 and 1856 even most Canadians thought of the HBC territories as distant, subarctic fur-trading territory that had been completely lost to Montreal traders with the merger of the HBC and NWC in 1821. Few then imagined it as a territory suitable for agricultural settlement.[147] Even the reports of Lefroy and Kane had not changed that impression.

But the situation changed suddenly and significantly during 1856. On 26 April 1856, three weeks after Kane's presentation on the Walla Walla, the *Montreal Gazette* published an article that asserted that the HBC's illegitimate charter hindered Canada right to expand westward.[148] Then on 29 July Toronto newspapers began discussing the same issue. The Toronto *Globe* became the most active voice. On 19 August, the *Globe* published the first in a series of long letters written by "Huron" (Allan Macdonell) that stridently attacked the charter of the HBC and asserted Canada's superior right to the territories.[149] Soon, that small but influential group of Canadians primarily in Toronto (rather than Montreal, which had far deeper historical connections with the West) took up the argument that Canada should acquire the HBC territories.[150] Theirs was an ambitious project in which they commonly portrayed the HBC as the enemy: the tyrannical chartered monopoly construed by Alfred Roche and others as "a foreign body" that for centuries had jealously guarded its illegal exclusive trading privileges by concealing the true potential of its territories.[151] It was a variation on an argument that critics of the company had taken since the first half of the eighteenth century. For a time, the Canadian government adhered to an argument that the colony of Canada, as successor to New France, had inherited France's allegedly stronger claims to the West (as taken over and expanded by the North West Company) even as far west as the Pacific Coast.[152] In September 1856, the president of the Executive Council (and member of the Canadian Institute), Philip VanKoughnet, declared that "no charter – no power could give to a few men exclusive control over half a continent. That vast extent of territory stretching from Lake Superior and the Hudson's Bay belonged to Canada – or must belong to it."[153] Essentially then, the HBC's assistance to Canadian-based scientists in the 1840s had backfired in ways that no one in the company's offices could have predicted. Although Lefroy and Kane did not present any evidence that could be used against the HBC, and although Kane evidently defended the HBC consistently, their HBC-supported travel ironically drew Canadian attention to the HBC territories, attention that eventually developed into an anti-HBC Canadian expansionist movement.

If Canadians continued to believe that Canada was separated from British Columbia by territories that were suitable only for the fur trade, the Canadian expansionists' transcontinental dream would be but a chimera.[154] They needed to convince Canadians and the British government that the territory was valuable for far more than its furs. Between 1856 and 1869, they were spectacularly successful. In those years, they transformed the dominant portrayal of Rupert's Land in Canadian and British writing from that of a bleak subarctic and arctic fur-trading region to that of a resource-rich temperate region of great agricultural potential, and as they did so, "Canada increasingly arrogated to itself the role of trustee of the North West in the name of the British Empire and civilization."[155]

To support their arguments about Rupert's Land, Canadian expansionists turned to any available evidence, including scientific evidence. Unfortunately for them, the well-grounded relevant scientific literature was scanty and ill-suited for their purposes. The work of Lefroy and Kane had stimulated Canadian interest in the HBC territories, but it provided Canadian annexationists with little ammunition. Most of the other scientific literature on the HBC territories was little better for their purposes. Most described the more northerly subarctic and arctic regions dominated by boreal forests and tundra, not the grasslands and parkland (the prairie-forest ecotone). And there were no studies of the climate, soils, or agricultural potential of the grassland or parkland regions.

Serendipitously for Canadian expansionists, useful, albeit somewhat speculative, scientific theories were becoming public just before the Canadian expansionist movement emerged, and when American expansionism was in hiatus. Canadian expansionists could and did turn to Alexander Kennedy Isbister's 1855 article presenting evidence of coal and mineral deposits in the HBC territories, as they could also turn to his earlier attacks on the HBC.[156] More importantly, scientists were transforming perceptions of the climate of western North America, pointing to evidence that summer isotherms trended towards the northwest throughout much of North America. In September 1855, the Toronto *Globe* referred to Americans who had called attention to the fact that "isothermal or climatic lines bend far away to the north as we go west towards the Rocky Mountains. If we mistake not, it is nearly as warm at the north bend of the Missouri as it is at Chicago."[157] Americans and Canadians discussed the implications of this theory in some depth. By August 1856, the Toronto *Globe* pointed to the evidence that towards the west, "the isothermal line which passes through Quebec ... is deflected northward." The article continued by noting that "a territory exists therefore on the north shores of Lakes Huron and Superior, varying in breadth from 70 to 160 miles, having a more favourable temperature than Quebec, and consequently sufficiently moderate to admit the successful cultivation of cereal grain."[158]

If the portrayal of the HBC territories was becoming more optimistic, the portrayal of the American West was not. Scientists at the Smithsonian Institution saw the American high plains as a bleak desert. Joseph Henry, head of the Smithsonian Institution, included an isothermal map of the United States in a long article published in the Agricultural Report of the United States Patent Office for 1856. In the introduction to his report, he announced that "we think it will be found a wiser policy to develop more fully the agricultural resources of the States and Territories bordering on the Mississippi, than to attempt the further invasion of the sterile waste that lies beyond."[159] Lorin Blodget, another Smithsonian Institution physicist, sounded a similar note in his *Climatology of the United States and the Temperate Latitudes of the North American Continent* in 1857. After noting the "absence of attention heretofore given" to "the plains east of the Rocky Mountains," Blodget noted the existence of the Red River Colony, the grain grown along the Saskatchewan River system, and the fact that the elevation of the plains decreased towards the northern lakes before concluding that "in every condition forming the basis of national wealth, the continental mass lying westward and northwestward from Lake Superior is far more valuable than the interior in lower latitudes, of which Salt lake and Upper New Mexico are the prominent known districts."[160] It is important to note that scientists in the 1850s had access to no hard climate data from the HBC territories with which they could argue that summer isotherms trended northwest throughout the interior of North America.[161] But, added to the widespread belief that clearing forests and cultivating soils actually improved the climate for agriculture, the new portrayal of the northwest made it possible to imagine prosperous agricultural communities in large parts of Rupert's Land. Blodget's report was quickly followed by James Cooper's survey of the forests of North America which similarly showed the northern limit of forests and the distribution of tree species trending northwestward across the continent (see figure 7.4).[162] Suddenly the HBC territories were highly attractive to expansionists – American or Canadian. Had the internal political circumstances in the United States not made it nearly impossible for the United States to acquire the HBC territory either by purchase or conquest, American expansionists might have been able to use the work of Blodget and Cooper to effect an American acquisition of those territories. Only Canadian expansionists were in a really good position to seize on this new work in the late 1850s and early 1860s.

Thus, in 1856 the HBC faced a significant challenge. Critics in Canada and Great Britain were arguing that the company had stood in the way of progress by misrepresenting the suitability of Rupert's Land for agricultural settlement. These were critics the company could not ignore, especially since the criticism came as the next expiration of the company's exclusive licence approached (it was due to expire in 1859).[163] To better understand its options, the British House of Commons appointed

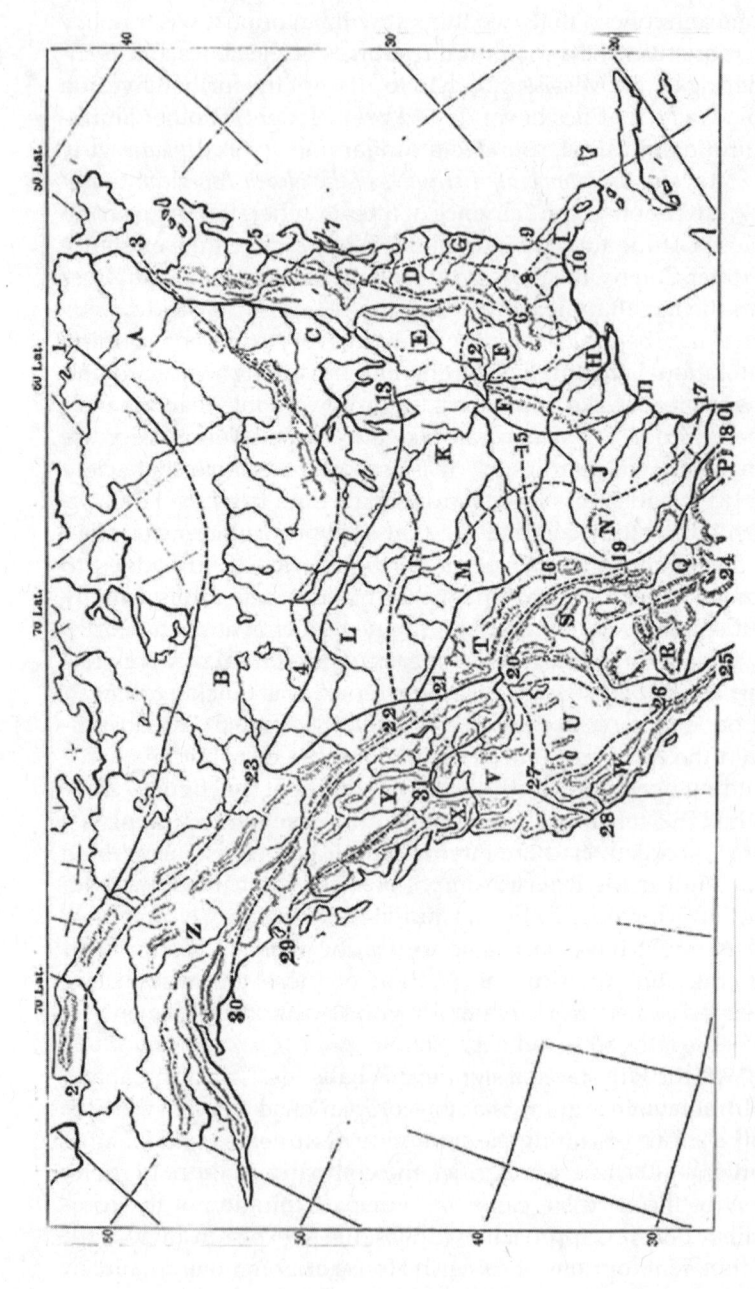

7.4 Distribution of Forests. In 1858, the American scientist James Cooper published a map that showed the forest regions of North America trending towards the northwest throughout the territory west of the Great Lakes. Cooper thus joined several other scientists whose work suggested that the climate of northwestern North America was better for agriculture, and that the resources of the HBC territories (from forests to minerals) were more valuable than had previously been assumed. This scientific work inspired expansionists in both the United States and Canada. Source: James Cooper, "On the Distribution of the Forests and Trees of North America, with Notes on its Physical Geography," *SIAR*, 1858.

a select committee of nineteen members which met from February to June 1857 "to consider the State of those British Possessions in North America which are under the administration of the Hudson's Bay Company, or over which they possess a License to Trade."[164] A review of the committee's report suggests that the committee placed a great deal of weight upon the testimony of the witnesses who were regarded as independent of the company (neither its employees nor its opponents).

Among the twenty-four witnesses to appear before the committee were two Canadians (both of whom projected a much more moderate and conciliatory tone before the committee than the strident language used in Canada by Roche and VanKoughnet), two company men (including George Simpson, who testified very ineffectively), a number of opponents of the company (including Alexander Isbister), and a number of men who, in the words of historian A. S. Morton, "might be expected to give impartial testimony."[165] Those "impartial" witnesses included Sir John Richardson, J. H. Lefroy, George Back, and Richard King.[166] The conclusions reached by the committee suggest that most committee members accepted the testimony of Richardson, Lefroy, and Back: that the company had not misrepresented the agricultural potential of its territories, that the company's treatment of aboriginal peoples was benevolent, and that the HBC's monopoly was crucial to the well-being of the Native population.[167] There is little doubt that the company benefitted greatly from the testimony of scientists whom it had supported over the years.

The committee's report, issued on 31 July 1857, suited the HBC's interests very well.[168] It recommended that the company's licence be terminated on Vancouver Island, and shortly in other regions west of the Rocky Mountains, where settlement pressures were already building.[169] Significantly for Canadian expansionists it also argued that "it is essential to meet the just and reasonable wishes of Canada to be enabled to annex to her territory such portion of the land in her neighbourhood as may be available to her for the purposes of settlement."[170] The committee declined to suggest how such an annexation might be effected or what compensation the HBC might deserve. On the other hand, it recommended that "whatever may be the validity or otherwise of the rights claimed by the Hudson's Bay Company, under the Charter, it is desirable that they should continue to enjoy the privilege of exclusive trade, which they now possess, except so far as those privileges are limited by the foregoing recommendations."[171] That meant that the company should retain its rights in those regions "in which, for the present at least, there can be no prospect of permanent settlement, to any extent, by the European race."[172]

The committee's recommendations were largely adopted. In 1859, when the HBC's licence expired, the British government confirmed the permanence of its direct control over the newly created colony of British Columbia and the older colony of Vancouver Island. The company's licence to the Athabasca and Mackenzie Districts was renewed in 1859, but by that time everyone, including the directors of the HBC, had accepted the fact that the days of the HBC's control over those districts and over Rupert's Land itself were numbered.[173] If nature seemed to be determined to make Rupert's Land part of the United States, the British government increasingly saw Canada as the most promising heir of the territory.

Until 1857, scientific expeditions to the HBC territories had been geared towards understanding the territories *as they were*. However, in 1857 two expeditions were proposed not merely to answer questions but to confirm hopes of what the territories *might become*. The first expedition was originally proposed by John Palliser, sponsored as an expanded scientific expedition by the Royal Geographical Society, and reluctantly funded by the British government.[174] He was instructed to explore "that portion of British North America which lies between the northern branch of the River Saskatchewan and the frontier of the United States, and between the Red River and the Rocky Mountains."[175] Among his orders were instructions to note the territory's "principal elevations, the nature of its soil, its capability for agriculture, the quantity and quality of its timber, and any indications of coal or other minerals."[176]

Palliser, an Irish country gentleman and sport hunter, had spent almost a year sport hunting in the United States' West in 1846 and 1847. Compensating for Palliser's own lack of scientific credentials, his 1857–60 expedition was staffed by respected men of science, including Eugène Bourgeau, Thomas Wright Blakiston, James Hector, and John William Sullivan.[177] From 1857 to 1860 they explored southern Rupert's Land from Lake Superior to the passes of the Rocky Mountains. Palliser concluded that much of the prairie land of southern Rupert's Land was an extension of the "Great American Desert" (this region has become known in Canada as "Palliser's Triangle"), but that a "fertile belt" along the northern fringes of this desert was suitable for agriculture and raising stock. Compared with subsequently more optimistic evaluations of the agricultural potential of the northern plains and parkland, Palliser's assessment seems cautious, even pessimistic, but his was the first

British study of Rupert's Land to echo American literature that argued that portions of Rupert's Land were suitable for agricultural settlement. The expedition's report was also important for changing perceptions within the British government of the potential future of Rupert's Land.

It would have been highly impolitic for the HBC not to assist the Palliser expedition. But given that the company's control was certain to end soon, its directors had every reason to assist an expedition whose findings might help increase the value of Rupert's Land as real estate. And so the company and its officers did cooperate. When the journals of the expedition were published, Palliser wrote, "I must avail myself of this opportunity to express my thanks to the officers of the Hudson Bay Company for the assistance they have always afforded in furthering the objects of the Expedition."[178]

The Palliser expedition's research was not dispassionate or disinterested, but Palliser did not particularly want, and the British government did not *need*, evidence of the agricultural potential of Rupert's Land the way Canadian expansionists did. Evidently not willing to wait until the British Parliamentary Committee reported, Canadian expansionists convinced the Canadian Legislative Assembly to hold its own 1857 investigation. Then they convinced the government of Canada to mount its own scientific expedition to Rupert's Land. That expedition was officially led during its first season by the retired, mixed-blood, European-educated HBC trader George Gladman, but in its second season by Henry Youle Hind, with whom the explorations have been most closely associated. Hind was a chemistry and geology professor at the University of Toronto, a member of the Canadian Institute, and the editor of the *Canadian Journal* from 1852 to 1856. He was also sympathetic to the expansionists.[179]

Perhaps because they had convinced themselves that the HBC's directors refused to accept the inevitable end of their privileges, many Canadian expansionists imagined that the HBC resisted the very idea that Canada would acquire Rupert's Land.[180] In fact, the directors of the HBC were quite willing to sell their rights to Rupert's Land to Canada – or anyone else – at an advantageous price. Internally, the Governor and Committee members and various officers had discussed the inevitability of the end of their exclusive privileges as early as 1845.[181] In 1848 Donald Ross wrote to George Simpson that

> We can no longer hide from ourselves the fact, that free trade notions and the course of events are making such rapid progress, that the day is

certainly not far distant, when ours, the last important British monopoly, will necessarily be swept away like all others, by the force of public opinion, or by the still more undesirable but inevitable course of violence and misrule within the country itself – it would therefore in my humble belief be far better to make a merit of necessity than to await the coming storm, for come it will.[182]

According to both E. E. Rich and John Galbraith, by 1856 Ross's 1848 opinion had become the accepted wisdom of the company's directors. George Simpson and the London Committee had concluded that the HBC's charter had become worthless, but believed that the British government was not aware of the fact. On 2 August 1856 George Simpson pondered selling the charter "to make a merit of necessity."[183] So, by 1856 the company's concern was with a practical question: how to sell (rather than simply lose) its monopoly. The British government was not an interested buyer. So although Canada's interest in the HBC territories came relatively late, that interest in and of itself was welcome. No wonder that the HBC was on record in 1857 as agreeing that land of agricultural potential in Rupert's Land should go to Canada.[184] What the directors of the HBC resisted and resented were the expansionists' adherence to the old argument that the HBC's charter was illegal (thus implying that Canada should not have to pay for the territory), and the expansionists' portrayal of the HBC as a benighted abuser of its exclusive privileges.

Expansionists expected the Hind expedition to provide ostensibly dispassionate scientific evidence to support their arguments. The expedition was instructed to ascertain "the kind and quality of the soil and its fitness for agriculture," and "the kinds of timber and their commercial value."[185] Given the Canadian adoption of the old argument that the HBC had no legal rights to Rupert's Land – an argument defended by a memorandum prepared by Canada's Crown Lands Department (a hotbed of expansionist support) – the Canadian government did not consult with the HBC before dispatching its hastily organized expedition.[186] As one historian has noted, however, "it was discovered almost immediately upon arrival at Fort William that essential travel assistance could only be secured with the co-operation of the officers of the Hudson's Bay Company – a co-operation which was readily extended."[187] Before the second season of surveys, Hind convinced the Canadian Provincial Secretary to ask George Simpson for letters of introduction to the HBC's officers. Simpson responded by writing that the HBC would be happy to "forward the objects of the exploring expedition with the same cordiality with which they are ever anxious to co-operate with the Government of this Province."[188] At the risk of annoying expansionists, Hind even visited Simpson in Montreal before setting off for his 1858 survey.[189]

When Hind published his report, Canadian expansionists got what they wanted. The report announced that the expedition had not only found a "broad strip of fertile country," amenable to permanent agricultural settlement, but that

> It is a physical reality of the highest importance to the interests of British North America that this continuous belt can be settled and cultivated from a few miles west of the Lake of the Woods to the Passes of the Rocky Mountains, and any line of communication, whether by waggon road or railroad, passing through it, will eventually enjoy the great advantage of being fed by an agricultural population from one extremity to the other.[190]

Lest any of his readers miss the significance of this finding, Hind drew upon the work published by the Smithsonian Institution and claimed that "any railroad constructed within the limits of the United States must pass, for a distance of twelve hundred miles west of the Mississippi, through uncultivable land, or, in other words, a comparative desert."[191] Canadian expansionists now had the words of a scientist who declared that the geographical and environmental obstacles to the creation of a transcontinental nation in British America were actually less daunting than those faced by the United States.

Hind's expedition was tendentious. According to the historian W. L. Morton, when Hind published his conclusions, "the explorer and scientific observer had all too clearly become the publicist and the promoter."[192] And according to Suzanne Zeller, the expedition and its report were "driven more by enthusiasm than by concrete evidence."[193] But if Hind's report was poor science, it was great propaganda. The façade of scientific rigor certainly contributed to its value for Canadian expansionists.

In the end, Hind's report included tributes to the HBC, to George Simpson, and the officers of the company. Hind wrote, probably more generously than many of his fellow expansionists would have preferred, that he took

> much pleasure in tendering my warmest thanks to Sir George Simpson, not only for the letters of introduction with which he favoured me to the

officers of the Hon. Hudson's Bay Company's service in Rupert's Land, but also for his personal efforts when at Fort Garry, to facilitate the progress of the expedition by every means in his power. The assistance rendered by Sir George Simpson was of the greatest use to me, and the courteous manner in which it was granted increases my indebtedness to him.[194]

If Hind was a second-rate scientist, he was a wise strategist. If Canada was ever going to acquire the HBC territories, it was unlikely that it could avoid negotiating with the HBC. Hind's tributes would not have hurt the relationship between Canadians and the HBC.

The responses of aboriginal people to government-sponsored exploring expeditions suggest that some indigenous people understood the connection between the Palliser and Hind expeditions and the future of their homelands. When John Palliser met an Ojibwa leader along Rainy River, the chief wanted Palliser "to declare to us truthfully what the great Queen of your country intends to do to us when she will take the country from the Fur Company's people."[195] The same man also "objected to Mons. Bourgeau collecting plants, and requested that Dr. Hector should not take away any mineral specimens as long as we were in his territories."[196] The response to Palliser was similar to that of a Native leader – perhaps the same man – to Hind. This man, obviously familiar with developments in the United States, objected to the activities of the Canadian government expedition. He said to Hind that "we see how the Indians are treated far away. The white man comes, looks at their flowers, their trees, and their rivers; others soon follow him: the lands of the Indians pass from their hands, and they have a home nowhere."[197]

Some observant American expansionists were much affected by the way Canadian expansionists were re-imagining northern North America. In 1857, the prominent American Republican and proponent of United States' northward expansion, William H. Seward, wrote that "hitherto, in common with most of my countrymen, ... I have thought Canada, or to speak more accurately, British America, a mere strip lying north of the United States, ... right soon, to be taken on by the Federal Union."[198] But, continued Seward, after visiting Labrador and Canada in the summer of 1857

> I have dropped the opinion as a national conceit. I see in British North America, ... a region grand enough for the seat of a great empire. ... [T]he policy which the United States ... pursues is the infatuated one of rejecting

and spurning vigorous, perennial, and ever-growing Canada, while seeking to establish feeble States out of decaying Spanish Provinces on the coast and in the Islands of the Gulf of Mexico. I shall not live to see it, but the man is already born who will see the United States mourn over this stupendous folly.[199]

In 1867, as Secretary of State, William Seward orchestrated the United States' purchase of Russian America. If his words of 1857 are any indication, Seward may have seen that acquisition as a defensive move against an emerging northern empire. At any rate, the fact that Seward's words were reprinted by Hind in 1863 suggests that Canadian expansionists hoped that they were prophetic.

Rarely have the consequences of the HBC's support for science been as obvious and dangerous to the HBC as when the company supported American and Canadian science between 1830 and 1859. However the company's directors could not have anticipated all of the political implications of the activities carried out in the HBC territories by Nathaniel Wyeth, Thomas Nuttall, J. H. Lefroy, and Paul Kane. Neither Lefroy nor Kane became enemies of the company. Still, the work of American and Canadian scientists appears to have encouraged expansionists that threatened the company's interests. Wyeth later became a prominent advocate of American settlement in the Columbia District. In the case of Canada, the work of the travellers that the company supported spawned an expansionist movement that seized on evidence, including scientific evidence, to justify the Canadian acquisition of the territory. For the company, the expansionist movement by the late 1850s was a double-edged sword. On one hand, thanks to Canadian expansionists, the HBC finally had in 1856 a potential buyer for its rights. On the other hand, those expansionists were trying to acquire the territory without compensating the company.

8 "Knowing the Liberal Disposition": The Hudson's Bay Company and the Smithsonian Institution, 1855–1868

By 1855, the Hudson's Bay Company's contributions to British science had been growing for several decades, but the years between 1855 and 1868 were probably the most productive in the history of the HBC. The most remarkable change of the period however, was that the Smithsonian Institution, rather than scientists in the British world, became the primary beneficiary of the HBC's patronage of science. This development could not have been predicted in 1855. HBC men in the Columbia District began assisting visiting American scientists such as John K. Townsend and Thomas Nuttall in the 1830s, and George Simpson, the company's North American governor, helped John James Audubon in the 1840s, but the HBC's contacts with Americans remained sporadic until the mid 1850s. But by the early 1860s, the Smithsonian Institution (figure 8.1) had managed to channel in its direction the lion's share of the HBC's scientific contributions. It is impossible to confirm that the contributions to scientists in Great Britain actually declined as those to the Smithsonian grew, but it is almost certain that the flow of specimens to the United States dwarfed the flow to Great Britain during the 1860s.

Ties between the HBC and Smithsonian started desultorily when individual HBC traders submitted specimens to the Smithsonian in the mid-1850s, but strengthened after the Smithsonian's directors formally requested the assistance of Governor Simpson in 1857. The cooperation then burgeoned after the Smithsonian sent one of its scientists, Robert Kennicott, to the HBC territories in 1859. Quite apart from Kennicott's own collecting efforts between 1859 and 1862, HBC men from the Mackenzie River District submitted on average about 1700 specimens per year to the Smithsonian during the first six years of the 1860s.[1] In all, the Smithsonian Institution acquired over 12,000 natural

8.1 Smithsonian Institution Building from the Northeast with Flowers along Mall (1860s). Despite its rural appearance in this photograph, the Smithsonian Institution's building was set in central Washington, DC, walking distance from Congress and the White House. During the 1860s, many thousands of specimens arrived here from the Hudson's Bay Company's territories. Source: Smithsonian Institution Photo MAH-9748A

history specimens, many meteorological journals, abundant ethnological material, and several volumes of field notes from the HBC between 1858 and 1868.[2] The relationship between the HBC and the Smithsonian is an appropriate one with which to conclude this study, for in this well-documented relationship we can identify most clearly many of the ingredients that made scientific networks flourish throughout the history of science in the HBC, and in the history of hinterland science generally. Several people within the Smithsonian, including its secretary Joseph Henry, assistant secretary Spencer Fullerton Baird, and scientist Robert Kennicott, are most responsible for the success. Their task was made much easier by the existence of a flourishing scientific and intellectual network and corporate culture within the HBC and in the Mackenzie River District in particular. But the ability and willingness of the Smithsonian's scientists to cultivate and reward contributors from the HBC encouraged many company employees – from governors to officers in the field – to devote tremendous energies (much of which might otherwise have been devoted to British scientists, or simply remained untapped) towards meeting the desires and needs of the scientists at the Smithsonian.

Given the obvious expansionist motives that drove the Canadian scientific expedition to Rupert's Land in 1857 and 1858, and the history of American expansionism in the 1840s, one might be tempted to assume that American scientific activity in the HBC territories between 1857 and 1870 was driven by American notions of manifest destiny. But that was almost certainly not the case. In a book-length study of the HBC's cooperation with the Smithsonian Institution, historian of science Debra Lindsay found that Smithsonian scientists "were not concerned with territorial expansion, transportation routes, resource exploitation, or the suitability of the north for settlement."[3] But this does not mean that this science was disinterested or apolitical. It is certainly remarkable that a London-based corporation like the HBC cooperated with the Smithsonian Institution, located in the heart of Washington, DC, during years of deep tensions between the British and American governments. But the fact that the HBC and Smithsonian transcended these strains does not mean that their cooperation was disinterested or apolitical. The same range of corporate and individual interests (and many of the same individuals) that drove scientific cooperation before 1855 explain the collaboration between the HBC and the Smithsonian between 1855 and 1867. The HBC and the Smithsonian (and personnel within them) had interests quite separate from those of the nations in which they operated, and, although the cooperation may not have been expected to produce direct commercial or political advantage, it was driven by many motives beyond purely scientific ones.

The history of HBC-Smithsonian cooperation is noteworthy because it peaked at a time of tense and volatile Anglo-American relations. During the American Civil War (1861–5), the British government exhibited clear sympathies for the Confederacy – even to the point of offering assistance irreconcilable with neutrality. For their part, the Union provocatively violated the neutrality rights of Great Britain by arresting Confederate diplomats on their way to Great Britain aboard a British mail ship, the *Trent*, in the fall of 1861. The *Trent* Affair erupted in November 1861 and climaxed in January 1862 with hawks in both countries calling for war.[4] Had war been declared, as it appeared it would be in early 1862, the HBC's operations would have been immediately, and perhaps permanently, disrupted. British troops sent at the HBC's request to protect Red River from American aggression in 1846 had been withdrawn in 1848, so the Union army would have had little trouble seizing the tiny and isolated Red River Colony. But even if the Americans had not invaded Red River, an Anglo-American war would have cut off the HBC's newly adopted supply lines. In 1859, a year after Minnesota achieved statehood and a year before the Civil War began, the HBC began shipping the main outfit for its western interior posts through the United States rather than via Hudson Bay and York Factory. But faced with the possibility of war, the HBC sent duplicate shipments to North America in 1862, one by rail through St. Paul, Minnesota, and a second one via its old route to Hudson Bay.[5]

A declaration of war would also have disrupted scientific cooperation. In January 1862, George Barnston predicted that "if War should break out, I suppose there will be an interruption to all Communications, for the practice and Science of War are essentially obstructive."[6] As it turned out, war did briefly interrupt communications, but it was an unforeseen war. In August 1862, Dakota bands in Minnesota took up arms against the United States. American forces quickly crushed the uprising, but some of the Dakota combatants fled to the HBC territories, raising the possibility that the United States might use the presence of the rebels as a pretext for an invasion.[7] Notwithstanding war, tension, and disruption, the HBC continued to arrange transportation of specimens from its territories to the Smithsonian Institution throughout the Civil War (although the Dakota War did interrupt the shipments of 1862 from the border post at Pembina, Minnesota, to Washington, DC).

Even if scientists and collectors might have wished their work to be apolitical, it could not be so. One might argue that the very fact that scientists cooperated with one another when war loomed was an inherently political statement – a statement that, as it was put by James Smithson, the Englishman whose money funded the establishment of the Smithsonian Institution, "the man of science has no country; the world is his country – all men are his countrymen."[8] The directors of the Smithsonian projected a similar attitude. Joseph Henry tried to steer the Smithsonian away from research related to volatile political and racial debates.[9] Spencer Baird appears to been so silent on issues related to politics and religion that even his biographers found it impossible to identify in his many letters his position on the Civil War, slavery, or the theories of Charles Darwin.[10] In their correspondence with one another, the HBC officers and Baird rarely brought up political or ideological issues. There were a few interesting exceptions. In 1860, while discussing whether caribou might be domesticated, the HBC trader Bernard Rogan Ross gratuitously implied that he was a polygenist when he wrote to Baird that "I have myself little doubt of the perfect facility with which the Raindeer [caribou] could be domesticated, but the process would require many years to finish – as in a similar manner it requires Several generations to civilize Savage man, though I much doubt if Mr. [Andrew] Murray would agree that there is any distinctive difference requiring a *separate Adam* between even the Negro and the White."[11] Smithsonian scientists were also made aware of some of the politics internal to the HBC territories. In 1858, Donald Gunn, a former HBC employee residing at Red River, volunteered to Baird that "we are determined to do all in our power to blot out the last remnant of the Despotic rule of the House of Stuart, and to have this country emancipated from the Iron bondage of the Fenchurch Street Nabobs, to Unite it to Canada and have it opened up to civilization with all it's ameliorating blessings."[12] But such commentary was rare. Traders otherwise occasionally wrote things that suggested that the scientific community was properly an international one. For example, George Barnston wrote to Spencer Baird, in 1860 that "I am disposed to believe that, even perhaps in the next Generation, there will be a *General Convention* of men of Science, a deputation from each civilized Country, to determine not only the best principle on which to classify, but even to attempt the allotment to each species its proper place."[13] But for the most part, they corresponded without mentioning the political dimensions of science.

It is worth recalling that there was nothing unusual about international cooperation among scientists in this period. Previous chapters have shown that the history of science in the HBC reflected the growing cooperation among British and American scientists that became obvious by the early 1830s. The Smithsonian and its directors were part of the international networks. The annual reports of the Smithsonian Institution show that it routinely exchanged specimens internationally in the 1850s, including exchanges with the Montreal Natural History Society and the University of Toronto.[14]

It is not that scientists were not patriots, only that international cooperation between the Smithsonian Institution and the HBC seems to have transcended nationalist sentiment. In 1841, George Simpson expressed his fear of the United States and its "dangerous principles of Republicanism," but those fears were not enough to cause him to spurn all American scientists.[15] Joseph James Hargrave described Robert Kennicott as "a staunch patriot, and any expression of a slighting nature leveled at these 'United States,' ... called forth his indignant remonstrances."[16] Yet Hargrave disparaged the American community at Red River, rather than Kennicott: "He gladly availed himself of any opportunity to show civility to his compatriots resident in the [Red River] colony, some of whom in allusion to his scientific proclivities, so incomprehensible to them, bestowed on him the elegant professional *alias* of 'Bugs'"[17] George Barnston also alluded to national loyalties, probably with tongue in cheek, when he wrote to Baird, "I gave my whole collection to the British Museum, in which perhaps You may consider me to have been too patriotic, but I do not regret that."[18] In fact, Barnston had donated his collection to the British Museum in 1843 or 1844, before the Smithsonian had even been created. Barnston was never a major contributor to the Smithsonian, but that was probably not because of any anti-American feeling. Recall that he had submitted specimens to John L. LeConte in Philadelphia the 1850s. His small contributions to the Smithsonian are probably better explained by age and retirement. In 1860, Barnston explained that he was getting "so old, and withal so blind," that he was not able to promise any more contributions to science.[19] Furthermore, in 1862, Barnston left the fur trading territories for the last time.

The non-scientific significance of the Smithsonian-HBC cooperation went beyond vague internationalism. James Smithson's decision to leave his money to the government of a country he had never visited may be rooted in resentment of British customs that curtailed opportunities for illegitimate children of the nobility, like himself.[20] The wording of the bequest – it said nothing more than that the money should be used "to found at Washington, under the name of the Smithsonian Institution, an Establishment for the increase and diffusion of knowledge among men," – helped ensure that politicians would decide what was actually established.[21] The United States Congress accepted the windfall gift of over $500,000 in 1836, but politicians wrangled for a decade over the proper use of the money.[22] When the Smithsonian was finally created, the Board of Regents of the institution included many members of the House of Representatives and Senate, and its *ex officio* members included the president and vice president of the United

States, the secretaries of state and of war, the attorney general, the chief justice, and other federal officials, but only one professional scientist.[23] Its headquarters were established on the Mall, a convenient walking distance from both Capitol Hill and the White House. Knowledge and power were not to be separated. Sir Francis Bacon would have understood.

Given the context of and domestic popularity of American expansion before, during, and after the Smithsonian's cooperation with the HBC, it would be impossible to disentangle the scientific and non-scientific importance of the institution's efforts to amass the largest collections and the greatest expertise relating to the entire North American continent. Henry and Baird *were* building a scientific empire. William Healey Dall of the United States Geological Survey wrote of Baird that his "ambitions and endeavors were leading toward the establishment of a national museum in fact, if not in name. Multitudinous expeditions were set on foot for Pacific railway routes, military surveys, the coast survey, the routes for an Isthmian canal, the exploration of the Hudson Bay territory, Lower California, and Alaska."[24] Debra Lindsay echoed this argument by noting that "Baird was a politician and a lobbyist, but his politics were Smithsonian politics and his special interest was the advancement of Smithsonian science."[25] More recently, a historian of American science wrote that Baird "surely meant for the Smithsonian's holdings to become definitive for the natural history of the entire continent. And in realizing his imperial dream, Baird had a military strategy. The army and to some extent the navy were to be enlisted."[26] If there be doubts about that argument, in 1860 Joseph Henry himself pointed to the Smithsonian's efforts in the HBC territories as examples "of the efforts of the Institution to enlarge its boundaries," efforts more successful in their geographical reach than the "manifest destiny" of the American nation more broadly.[27] Clearly, Baird's efforts in the HBC territories were connected with important questions about what kind of scientific institution the young Smithsonian Institution would become. The connections between the HBC and the Smithsonian may have been relatively unconnected with Britain–United States relations at the time, but they certainly were closely related to the interests of the two bodies, and to the prestige and careers of individuals who participated in the cooperation. The moment an institution is created, it begins to defend its own interests. Neither the Smithsonian nor the HBC were simply arms of government. Both bodies, and the men of science and business that directed them, had interests of their own. But, given the dependence of the Smithsonian on the government of the United States for its funding, and the reliance of the HBC on its government for its licence, the men who ran them also had to remain conscious of their obligations to the state.

The first secretary of the Smithsonian was one of the most renowned physical scientists of the day, Joseph Henry. However, it was not until the Smithsonian appointed Spencer Fullerton Baird as its assistant secretary in 1850 (accepting at the same time his large natural history collections), that the Smithsonian began a program of original research.[28] Before Baird arrived, there was no reason to believe that the Smithsonian was likely to emphasize research in the natural sciences or ethnology. But, as two historians of the Smithsonian have explained, "the appointment of Spencer Baird and the acceptance of his collections were to change completely the concept of the Institution as envisioned by Henry."[29] By directing his efforts towards the collection of natural history specimens that could capture significant public attention when displayed, and that could contribute significantly to knowledge in natural history for relatively low cost, Baird sought to enhance the reputation of the institution, thereby attracting more funding from (but at the cost of greater dependence on) Congress. Baird's work earned him a reputation as the foremost native-born natural historian in the United States.[30] His position also gave him considerable influence, even outside the scientific community. Two historians have noted that "enjoying the backing of the many congressional supporters of the museum and the recognition of American and European scientists, he held a position that was both politically and scientifically important."[31]

The outbreak of the Civil War was particularly inopportune for the Smithsonian's program in natural history. The need to house collections made by the United States Exploring Expedition (1838–42) had been an important factor in the establishment of the Smithsonian in 1846, and during the 1850s the institution's natural history collection swelled again thanks to further submissions made by other US military exploring expeditions.[32] To display these collections, in 1858 the United States National Museum was established, apart from the Smithsonian but under the same management and under Baird's curatorship.[33] Then contributions from the military stopped when the Civil War interrupted all of the government's exploratory and surveying efforts. Contributions from the seceded states also evaporated. Moreover straitened wartime finances meant that the Smithsonian could not afford to support its own expeditions, and policies forbade the institution from paying for specimens.[34] The Civil War, then, magnified the importance of contributions from unpaid private donors, not only from within the Hudson's Bay Company territories, but also from Mexico, South America, the Caribbean, and elsewhere. That meant that the institution needed to maintain an effective strategy for getting specimens from volunteers.[35]

The HBC had its own interests to protect. Even before the HBC's exclusive trading licence was renewed in 1859, the HBC's directors, like most knowledgeable people, including some of the aboriginal people of the territories, could well foresee the end of their monopoly. To the HBC directors themselves, Rupert's Land was increasingly valuable as real estate rather than as a source of fur. That fact became abundantly clear in June 1863 when the International Financial Society took control of the company.[36] The change in ownership was a major turning point in the history of the company, for the new shareholders of the company were not interested primarily in the fur trade but in the company's title to Rupert's Land itself. It now suited the interests of the HBC for the natural endowments of their territories to be as widely known as possible. Given the fact that the British government conceded more of the Columbia District to the United States in 1846 than the HBC's directors had thought necessary, and that it was plausible that Rupert's Land could fall into American hands entirely, it did not require a particularly vivid imagination to see why it might also be useful in the future for the company to have influential friends in Washington, DC. Meanwhile, given the acceptance that science was properly an international enterprise, the HBC's cooperation with the Smithsonian could also contribute to the company's prestige at home.

No one was more responsible for the development of the Smithsonian's natural history program than Spencer Baird (figure 8.2). Immediately upon assuming the position of assistant secretary in 1850, Baird set to work to emphasize the importance of natural history, noting in his first report the inadequacy of scientific knowledge of natural history collections in North America at the time.[37] He was successful very quickly. Already in 1855, Henry asserted that the Smithsonian had the richest collection of United States natural history.[38] But its collections relating to the northern portion of the continent were small. The Smithsonian's interests in the HBC territories were influenced by the fact that scientists were expanding beyond their interests in natural history (identification, cataloguing and classification of species), to more analytical questions related to the distribution, variation, and morphology of plants and animals.[39] Interest was also fueled by the fact that

8.2 Spencer Fullerton Baird, 1867. This photograph of Spencer Baird, taken in 1867, suggests firmness and determination, but also hints at his warmth and affability. Source: Smithsonian Institution Archives MAH-46853.

many birds that wintered in the United States nested in the HBC territories, that the northern limits of the ranges of many species resident in the United States were unknown, and that northwestern North America would be crucial to any attempt to compare the flora and fauna of North America and Asia.[40]

Cooperation between individuals in the HBC territories and the Smithsonian Institution began several years before formal cooperation between the directors of the two organizations. Baird's interest in the HBC territories can be traced back as far as early 1850, although Baird clearly had not yet developed an appropriate strategy for acquiring specimens from those territories. On 19 February 1850, shortly after he was hired, he wrote to a colleague in the Smithsonian that he had "a plan for getting specimens of arctic animals from the posts of the Hudson's Bay Company; as also for enlisting the different missionary associations in our behalf. A correspondence with Methodist and Baptist Union has already been started, with this end."[41] This letter shows that Baird had decided to try to get natural history specimens from the HBC territories several years before he contacted the directors of the company itself, but that his first instinct was to contact missionary organizations (apparently unaware which missionary organizations were active in the HBC's territories at the time), rather than company officials. These early efforts by Baird probably explain why natural history specimens began trickling, but not flooding, into the Smithsonian from the HBC territories by the mid-1850s. For example, the Smithsonian Institution's annual report for 1855 acknowledged contributions from William F. Tolmie.[42] Donald Gunn, a resident of the Red River Colony, had not been an HBC fur trader since 1822, but his first contacts with the Smithsonian clearly also occurred some time before 1855.[43]

Donald Gunn's efforts show how much more difficult it was for people unaffiliated with the company to contribute to science than it was for those inside the company. Gunn had worked for the HBC from 1813 to 1822, but lost his job after the merger of the North West Company and the HBC. Thereafter, like many other laid-off traders, he settled in the Red River Colony. Although Gunn was a willing collector, he did not have the resources to assist as much as an HBC officer could. It was expensive for him to travel, acquire goods (including everything from the alcohol necessary to preserve specimens to the lint with which to stuff them), and ship or deliver specimens to Pembina, Minnesota, from where the United States Post Office shipped them gratis to the

Smithsonian Institution.[44] As a colonist he either had to pay for travel and shipping out of his own funds, or at the Smithsonian's expense. Thus, through Gunn the Smithsonian could acquire samples only from the Red River region, and only at considerable expense to Gunn or the Smithsonian.[45] Still, Gunn provided natural history specimens, answered queries, and sent fossils at least to the end of the 1860s.[46] He also published two articles in the Smithsonian's annual report in 1867–8.[47]

Official correspondence between the Smithsonian Institution and the directors of the HBC appears to have begun in November 1857, when Joseph Henry, the secretary of the Smithsonian, signed a letter (probably written by Baird) to George Simpson that said that "we are now engaged in fitting up our large hall for the reception and exhibition of a complete collection of the animals of North America as far as they can be procured." He then explained that "knowing the liberal disposition which the Hudson's Bay Company has ever shown towards the interests of science we have thought it probable that you might indicate some way of obtaining through your postes a part or all of our more important & considerate [needs]."[48] He specifically indicated that the Smithsonian sought samples of musk oxen, barren ground caribou, mountain goats, and "barren ground brown bears."[49] Henry shrewdly reminded Simpson that the Smithsonian was founded by "the liberality of a subject of England."[50] Simpson's enthusiastic response must have been heartening: "The Hudson's Bay Company are ever ready to render assistance towards the pursuit of scientific research in their territory. It will afford them peculiar satisfaction to further the objects of the Smithsonian Institution in which the English cannot but take a warm interest. I shall accordingly send a copy of your letter to the Officers in charge of the remote district of MKenzie River, bordering on the Arctic Sea, from whence only the specimens you desire can be procured."[51]

With this initiative, the Smithsonian was ushered into the vibrant intellectual and scientific culture that had already developed among officers in the Mackenzie River District. The subscription library established by James Anderson in 1852 continued to augment the sizable private book collections held by officers in the district. In 1861, Bernard R. Ross wrote to John Richardson that he had "a library of above 700 volumes, besides the use of a public one of about the same size."[52] The library grew during the 1860s, thanks in part to donations made by the Smithsonian. Roderick MacFarlane later remembered it as "a fine library of some two or three thousand volumes, which enabled them [HBC men] to somewhat enliven the tedium and monotony pertaining to the life itself and the long northern nights of the winters in that remote region of the Great North West."[53]

In November 1858, Bernard Rogan Ross, the Chief Trader in charge of the Mackenzie River District wrote to Henry acknowledging receipt of a copy of Henry's November 1857 letter which Simpson had forwarded to him during the previous summer.[54] Ross, alluding to his previous scientific contributions, assured Henry that "it gives me at all times much gratification to advance the cause of Science, ... and in the present case the pleasure is doubly enhanced by affording any assistance to an Institution of such worldwide renown as the Smithsonian."[55] This letter was almost certainly the first that Ross sent to the Smithsonian.[56]

After receiving another request for specimens of specific animals in April 1859, Ross redoubled his efforts and sent several cases of samples to the secretary of the Smithsonian from Methye Portage (Portage La Loche), during the summer of 1859.[57] He followed that shipment up with a letter on 30 November 1859, assuring the secretary that he would "not merely restrict myself to these particular objects of research, the whole field of either science or curiosity will be considered in all contributions which I may hereafter forward to your collection."[58] Ross explained that his duties prevented him from keeping a meteorological register, but that he would delegate the task to "Mr. Andrew Flett, a very careful and intelligent person."[59] Soon the trickle of specimens from the HBC to the Smithsonian became a flood.

Joseph Henry contacted the HBC directors again in early 1859, this time not only to ask for more cooperation but also to ask for permission to have one of its scientists, Robert Kennicott, actually travel to the HBC territories. (Henry was wise enough to have Lord Francis Napier, the British ambassador to the United States, also write a letter to Simpson, requesting that Simpson cooperate with the Smithsonian.)[60] Governor Simpson once again responded positively. He assured Henry that "the Hudsons Bay Company are always happy to be instrumental in promoting the interests of science; and on their behalf I beg to state it will afford them much pleasure to extend their hospitalities to Mr. Kennicott at such of their establishments as he may visit, and to be useful in facilitating the objects of his expedition."[61]

In March 1860, Simpson sent a letter actually drafted at the Smithsonian to the HBC officers, urging the officers to respond positively to the Smithsonian's requests.[62] He also included a circular from the

8.3 Robert Kennicott Shortly before Going to the HBC Territories. Accounts suggest that Robert Kennicott was a frail man. This photograph, taken before he departed for the HBC territories, shows him unremarkable in his dress or hair. Source: Smithsonian Institution Archives SIA 2010-1593.

Smithsonian.[63] Joseph Henry's "Circular to Officers of the Hudson's Bay Company," began by explaining that "The Smithsonian Institution has been engaged for several years in the prosecution of researches relative to the climatology and natural history of North America. ... a serious obstacle has been experienced in the lack of sufficient data from the region north of the boundary line of the United States, especially from its more northern regions."[64] The circular was accompanied by instructions for the proper collection of specimens and the keeping of meteorological registers.[65]

The man chosen for the Smithsonian's expedition, Robert Kennicott, was the ideal choice to strengthen the connections between the HBC and the Smithsonian (see figure 8.3). Raised on the grounds of a private garden developed by his own father in Northfield, Illinois, Kennicott was encouraged by his father to become a naturalist.[66] To that end, his father sent him in 1852 and 1853 to study under Jared P. Kirtland in Cleveland, Ohio, and Philo Romayne Hoy in Racine, Wisconsin.[67] Kennicott was soon submitting specimens to the Smithsonian, and began corresponding with Baird by November 1853.[68] He twice undertook studies in medicine between 1855 and 1857, but apparently "his health, which was always best in the field, declined under the confinement of the lecture-room so as to compel a discontinuance of study."[69] By that time, Kennicott had turned firmly towards a career in natural history. From August to November 1857, the already-respected young naturalist, in his capacity as the curator of the Museum of Natural History at Northwestern University, visited the Red River Colony where he collected specimens for the university and the Smithsonian.[70] There he met Donald Gunn, who was already contributing to the Smithsonian, and HBC Chief Factor, William Mactavish, who later became instrumental in expediting the transportation of specimens from the HBC territories to the Smithsonian.[71] According to a biography published shortly after his death, and written after consultation with Spencer Baird, it was on this trip that Kennicott "learned of the practicability of penetrating northward, and of the probable co-operation of the Hudson's Bay Company in any scheme of exploration which might be undertaken."[72] Shortly after that trip, Kennicott travelled to Washington, DC, where he worked at the Smithsonian during the winters of 1857–8 and 1858–9.[73] The timing suggests that it was probably thanks to Kennicott's advice that Joseph Henry contacted Governor Simpson directly in November 1857 and again in early 1859.

The timing of the Smithsonian's request to Simpson might also be related to the fact that the American Association for the Advancement of Science held its annual meeting in Montreal from 12 to 20 August 1857. This was the first time that the association had held its meeting outside the United States. Many prominent American scientists, including Joseph Henry and Alexander Dallas Bache, attended and presented papers. So did John Rae, formerly of the HBC. Given the way that the meeting is known to have inspired greater cooperation between scientists in Canada and the United States, it is possible that the meeting might have influenced Joseph Henry's decision to approach Simpson.[74]

Kennicott departed for the HBC territories soon after the Smithsonian received permission to send him (see figure 8.4). He left Washington in early April 1859, travelling via Cleveland, to visit Jared Kirtland, and Ann Arbor, to assist with the arrangement of a new collection at the University of Michigan (which was supporting his expedition financially), before spending time with his family in Illinois. Kennicott and Simpson had discussed the probability that Kennicott would depart Chicago for St. Paul by rail on 20 April and travel from there to Red River, but Kennicott instead boarded the steamer *Foundation City* on 28 April bound for Collingwood, Upper Canada, where he met George Barnston – "a man of no small scientific acquirements," Kennicott noted respectfully.[75] After a quick trip to Toronto to visit John Rae and Simon James Dawson, veteran of the Red River Exploring Expedition (Hind Expedition), he returned to Collingwood in time to board the HBC's steamer *Rescue*, bound for Fort William, on 4 May.[76] At Fort William, Kennicott had the opportunity to meet Chief Trader John McIntyre, who had accompanied George Simpson on his trip around the world in 1841–2.[77] Kennicott quickly learned that HBC officers "are generally of good families, and well educated; and that they are hospitable and companionable, I can testify of all whom I met."[78]

If the HBC officers were amiable men, so was Kennicott himself. Kennicott's disarming personality earned quick dividends, for he succeeded in convincing Barnston, who had never heard of the Smithsonian before meeting Kennicott, to donate some of his natural history specimens to the Smithsonian, "chief among these ... a skin of the reindeer in superb condition, and now mounted in the museum; also a nearly complete skeleton of the same animal."[79] Then, in June, as he continued his journey, Kennicott met George Simpson at Norway House, where Kennicott not only represented the interests of the Smithsonian but also encouraged Simpson to establish a museum in Rupert's Land (a suggestion Simpson considered impractical), and volunteered to help Simpson develop his private museum in Montreal.[80] According to Kennicott, Simpson "expressed much interest in my success and ...

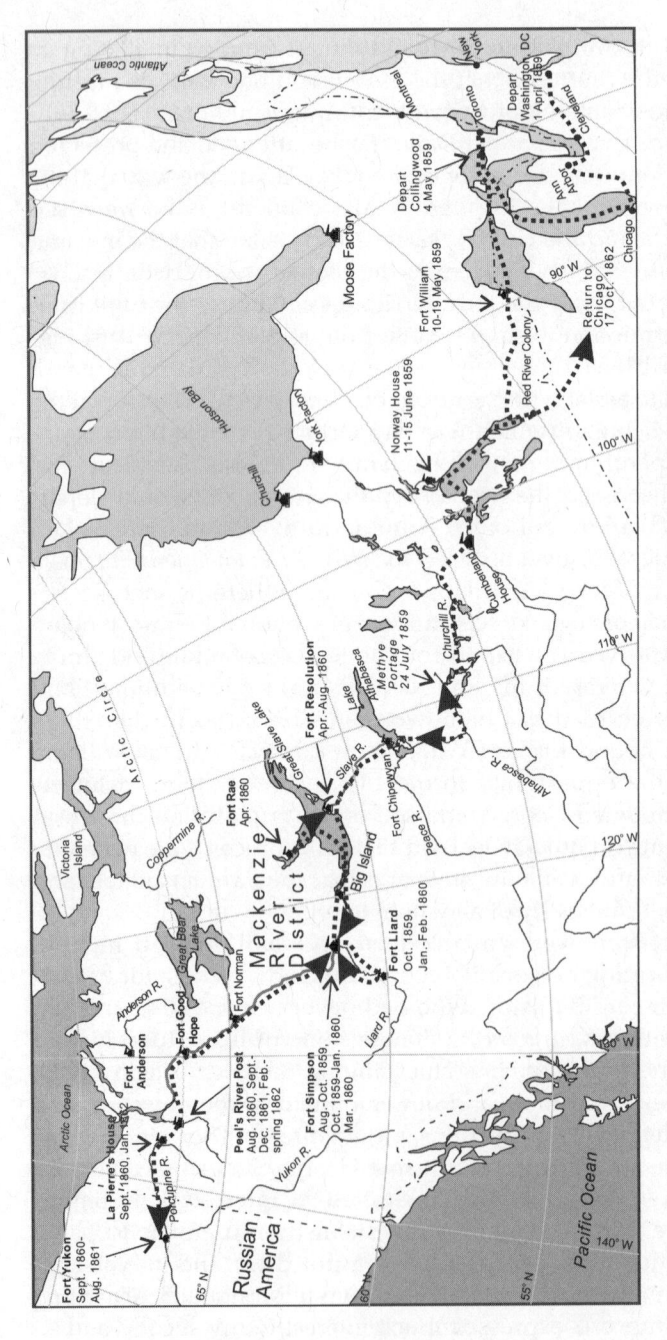

8.4 **Robert Kennicott's Travels in the Mackenzie River District.** In roughly three years spent in the Mackenzie River District, Robert Kennicott spent most of his time at Fort Simpson, headquarters of the HBC's Mackenzie River District (with Bernard Rogan Ross in charge) and Fort Yukon (James Lockhart in charge), but he also spent time at Fort Rae, (Lawrence Clarke), Fort Resolution, (W. L. Hardisty), Peel's River Post (Fort McPherson [with Charles P. Gaudet]), Fort Liard, and La Pierre's House. He met Roderick MacFarlane several times while MacFarlane was at Fort Good Hope and hoped to spend time with him at Fort Anderson, but news of his father's illness cut short his visit to the HBC territories. Constantin Drexler, also from the Smithsonian Institution, was active primarily in the vicinity of Moose Factory. Source: T. Binnema.

told me that while at any of the companies posts, my living would cost nothing but that for provisions for the route, for my passage and for any special service on which I employed men I would pay myself."[81]

From Norway House Kennicott accompanied the northbound HBC brigade. While travelling up the Saskatchewan River on his way to Fort Simpson, Kennicott met Eugène Bourgeau, the French botanist who had taken early leave of the Palliser Expedition, then still in the field.[82] Finally, in late July at Methye Portage, Kennicott met the Chief Trader in charge of the Mackenzie River District, Bernard Rogan Ross. "Barny" and "Kenny" must have hit it off immediately.[83] Within days of their first meeting, Kennicott penned a letter to George Simpson, requesting permission to extend his stay by two years.[84] Ross had obviously shared the same intentions, as he stated in a letter to Henry in November 1859 that "every facility will be given to Mr. R. Kennicott to collect and forward specimens of natural history; free passage will be allowed him from post to post throughout the district, and all his plans the various officers under my command will, I am sure, gladly render assistance."[85]

Just as the Smithsonian contacted the HBC "knowing the liberal disposition" of the company, the HBC's cooperation with the Smithsonian soon led to requests from other American scientists. In May 1860, as Kennicott was en route to the Mackenzie River District, Alexander Dallas Bache of the American Coast Survey contacted the HBC, mentioning that, as a member of the Board of Regents of the Smithsonian, he had heard of the HBC's cooperation with that institution.[86] He proposed to send an expedition to the coast of Labrador to observe the solar eclipse of 18 July 1860.[87] One day later, Charles Henry Davis, commander in the United States Navy, and superintendent of the Nautical Almanac, wrote to the HBC to ask to send "a small party of scientific gentlemen" under the authority of the American secretary of the Navy, into HBC territories "in July next for the purpose of observing the total eclipse of the Sun; and I take the liberty respectfully to request your assent to this proposal, and also information as to the most convenient mode of reaching Cumberland House or Fort York; both of which stations are near the central line of the Shadow."[88] Simpson replied that "I have to state that, the Company are always desirous of rendering their assistance to any such undertaking as that you propose, having for its object the promotion of the interests of science."[89] The HBC did assist with an American party led by William Ferrel that travelled to Rupert's Land but it failed to make any observations. Travel delays prevented them from reaching Cumberland House in time for the eclipse, and clouds prevented them from viewing the eclipse. The experience led Samuel Hubbard Scudder, a naturalist with the expedition to lament of the "three thousand miles of constant travel occupying five weeks, to reach by heroic endeavor the outer edge of the belt of totality; to sit in a marsh, and view the eclipse through the clouds!"[90] Curiously, although the expedition to Rupert's Land produced little of scientific literature, the 1860 eclipse evidently did provide Jules Verne with the inspiration for his adventure novel, *The Fur Country*.[91]

When Kennicott arrived in the Mackenzie River District, he hoped to realize his dream to travel from there to Russian America. However, he was dissuaded. Ross resisted Kennicott's desire to travel down the Yukon River, on the grounds that "it may not be for the interest of the H.B. Fur Co to have the Yukon explored and known."[92] (Ross knew full well that the HBC's highly profitable post at Fort Yukon was situated well within the boundaries of Russian America, as established by treaty in 1825.)[93] By the end of 1860, the generous opportunities afforded him at the HBC posts, the assistance of HBC officers, and the challenges of making it to Russian America led him to abandon his aim of going there.[94] Instead, Kennicott travelled extensively in the Mackenzie River District (spending most of his time at Fort Simpson, Fort Resolution, Fort Liard, Fort Rae, Peel River House, La Pierre's House, and Fort Yukon), benefiting immensely from the HBC's assistance (figure 8.4). He stayed in the HBC territories until the spring of 1862, finally leaving Fort Simpson for Chicago on 1 June 1862 after learning that his father was ill.[95] After visiting with his family during the fall, he spent the winter at the Smithsonian, returning to Illinois in time to be at his father's deathbed on 4 June. He became in 1864 curator of the museum of the Chicago Academy of Sciences, with part of his job being to accept and process some of the specimens from his northern expedition, which were given by the Smithsonian to his institution.[96]

Kennicott was clearly much changed by his experience in the HBC territories. Photographs taken of Kennicott before he left depict a man who fit in well in urban American society, but in portraits taken in 1862, after he returned, his long hair and clothing clearly set him apart not only from most of his urban contemporaries, but also from most HBC officers (figure 8.5). As the various illustrations show, when they sat for portraits, HBC traders normally dressed formally according to urban fashions of the day, with nothing more than the occasional fur pelt as a prop.

"Knowing the Liberal Disposition" 257

Rob't Kennicott.

8.5 "Robert Kennicott, Explorer, in Field Outfit." In this picture taken after his return from the HBC Territories, Robert Kennicott obviously wished to be portrayed as active, healthy, and alert, yet contemplative. He is dressed here

(*Continued*)

Kennicott's subsequent experiences showed him how exceptional the HBC's assistance had been. His dream of travelling to Russian America materialized soon after he returned to the states. After the failure of several transatlantic telegraphs, the Western Union Telegraph (WUT) decided to build an overland telegraph from North America to Europe via the Bering Strait. In 1864, the WUT hired Kennicott to serve as the head of the scientific corps of the Russian-American division of its project.[97] His experience in the HBC territories led him to suggest that the WUT build its telegraph through Red River, Fort Edmonton, and Fort Simpson, with HBC cooperation and investment.[98] If the WUT had adopted that proposal, the HBC's expenses incurred in supporting Kennicott and the Smithsonian would have been repaid very handsomely. Instead, the WUT's rejection of his advice in favour of a line that would pass through British Columbia and Russian America foreshadowed his later experience with the company. Kennicott seemed constantly frustrated with his superiors' lack of support for scientific research. Soon after he reached Russian America in the fall of 1865, his dream of visiting there became a nightmare. Shortly after Kennicott's death, his memorialists at the Chicago Academy of Sciences claimed that

> accustomed to battle with nature and not with men, Kennicott, who had scarcely ever had an enemy in his life, found himself suddenly thrown

---

8.5 (*Continued*)
very much as Canadian and Metis voyageurs in the northwestern territories dressed, and evidently as he normally dressed while in the HBC territories. After encountering Kennicott with Bernard Ross at Methye Portage, the Roman Catholic missionary Émile Petitot wrote that Ross "was followed by a very *refined*, exuberant young American naturalist by the name of Kennicott, who spoke horribly through his nose. He also wore a Great North outfit on which he had embroidered in white ribbon a lizard, a butterfly, a turtle, and a snake, the insignias of his profession. We might say that he looked like a clown from the new Circus. The lovers of science are, we say, all original and a bit cracked, but the naturalists that we meet in the wilds of America seem to march to the beat of their own drums." Émile Petitot, *En Route Pour La Mer Glaciale* (Paris: Letouzey et Ané, 1887), 273–4. Although some HBC officers may have dressed this way in the field, they did not normally pose for photographs dressed this way. Debra Lindsay discussed Kennicott's Romantic interest in savagery in *Science in the Subarctic*, 59. Source: Smithsonian Institution SIA2011-0145.

among uncongenial spirits, and forced by business to have intercourse with them; and the natural disgust and aversion excited in his mind by these circumstances, aggravated by the provoking delays which occurred in San Francisco, told heavily upon his delicate nervous organization.[99]

One historian has noted that the WUT valued Kennicott's knowledge for its own purposes, but "it became apparent that there would be little time, and even less desire on the part of the company, to collect flora and fauna."[100] Another pointed out that Kennicott's activities for the WUT were far less successful than his expedition with the HBC, because, for the WUT, "the Smithsonian science program remained a remote second priority," such that "Kennicott became overwhelmed by the problems of running a huge exploration party in unknown country where logistical support like that provided by the HBC in the Mackenzie was absent."[101] Kennicott died near Nulato in Russian America of an apparent suicide on 13 May 1866 at only thirty-one years old.[102] Like Douglas, he left behind grieving friends in the HBC territories. Donald Gunn wrote that "science has, in him, lost one of its ablest and most indefatigable students, and all who had the felicity of being intimately acquainted with him, lost an amiable and warm hearted friend."[103]

Kennicott has been compared with John James Audubon and Alexander von Humboldt, but a more apt comparison seems to be with David Douglas.[104] Both men seemed destined for prominence, but both died before they reached their potential. Both were remarkable for their abilities to get along with fur traders and to motivate them to become naturalists and collectors. Both spent their happiest days and most successful years gathering natural history specimens in the HBC territories. According to one account of Kennicott, "one of his bitterest opponents and rivals confessed, long afterward, that one glimpse of Kennicott in the field gave him a totally new and different opinion of the man. 'If I had *known* him sooner,' said he 'we should have been always friends!'"[105]

As had been the case with Douglas, science appears to have been key to Kennicott's happiness. Apparently in poor health as a child, Robert Kennicott did not attend school regularly, but spent a lot of time outdoors to maintain his health.[106] His father wrote Simpson that because his son was "always rather feeble in physical organization, we had great fears on that account."[107] But was his frailty primarily physical or mental? Debra Lindsay has noted that "any physical ailments that Kennicott might have had disappeared miraculously while he was living at the Hudson's Bay Company's northern posts, and he often referred to his newfound physicality in letters home."[108] So striking was this evidence of physical vigour that Lindsay argued that Kennicott's "frailties were emotional rather than physical."[109] He was indeed subject to depression for much of his life. But he appears to have been happy in the HBC territories. His letters from the HBC territories frequently mention lethargy, but his emotional state declined significantly after he left. He plunged into prolonged depression after his father's death in 1863.[110] Letters written during his time with the Western Union Telegraph reveal a very frustrated and gloomy man. A life in the field, collecting flora and fauna, may always have been the key to Kennicott's physical and mental health.

The productivity of the Smithsonian's partnership with the HBC was not only related to the effectiveness of those within the Smithsonian. It was also dependent upon the cooperation of the directors of the company. The importance of the role of governor is illustrated by the governorships of George Simpson, Alexander Grant Dallas, and William Mactavish. Simpson's support was crucial at the outset. Debra Lindsay has correctly argued that "the hospitality, fraternity, and cooperation Kennicott met with at posts scattered throughout the Mackenzie River District were ... in large part, due to Simpson's interest and support."[111] But Simpson died in September 1860. Mactavish, who succeeded him as interim governor until 1862, was similarly supportive. He arranged shipments from the Mackenzie River District to Washington, "free of charges."[112] Alexander Grant Dallas, governor from 1862 to 1864, was evidently less generous. He appears to have wanted specimens for a museum he hoped to establish in Scotland.[113] When Roderick Ross MacFarlane wrote Governor Dallas that he had sent specimens of grizzly bears to the Smithsonian, Dallas responded, "The specimens you allude to as having been sent to the Smithsonian Institution will no doubt be acceptable; but I should prefer that you send them to my address to the Hudson's Bay House London."[114] When Mactavish became governor in 1864, his generous support was once again instrumental to the relationship between the HBC and the Smithsonian.[115] The London Governor and Committee appear to have been unstintingly supportive of the cooperation.

The success of the Smithsonian with the HBC was certainly influenced by the efforts of Baird and Kennicott, but was also attributable to the fact that, from the directors to the officers, there was already a culture of science within the HBC. The fact that Joseph Henry mentioned believing that it was "probable" that the company would assist the Smithsonian "knowing the liberal disposition which the Hudson's Bay Company has ever shown towards the interests of science," speaks to the established reputation of the HBC by that time.[116] And George Simpson's introductory letter urging his officers to assist the Smithsonian began by noting that "you are well aware of the desire of the Company to promote the interests of science by all the legitimate means in its power."[117] They were indeed well aware. Many of the same individuals who Simpson addressed and who subsequently assisted the Smithsonian had previously contributed to science and continued to contribute in other ways even as they assisted the Smithsonian. In 1859, Kennicott described William Mactavish as "really much interested in science, working daily at his microscope."[118] In November 1860 Mactavish apologized to Baird that he would not contribute as much to the Smithsonian as he could because "I am under a promise of long standing to send any collections of plants to an Edinburgh Botanist who is now in Canada."[119] Bernard Rogan Ross, one of the most important HBC contributors to the Smithsonian, also contributed natural history and ethnographic material to the Industrial Museum of Scotland, to Andrew Murray, and to John Edward Gray, keeper of zoology at the British Museum,[120] and sent "fossils and minerals" to Baird to be conveyed to "Sir W. Logan."[121] On another occasion, Ross thanked Baird for forwarding some duplicate specimens to the Montreal Natural History Society.[122] George Barnston wrote in November 1861 that Ross had "just received his Diploma of Membership from the Montl Nat Hist Socy."[123] Roderick MacFarlane, the most prolific contributor to the Smithsonian in this period, asked that some of the surplus specimens that he sent to the Smithsonian be forwarded to the Montreal Natural History Society, Oxford University, and the Edinburgh Museum of Science and Art (as the Industrial Museum of Scotland was renamed in 1866).[124] In 1886 MacFarlane became a member of the British Royal Geographical Society and the Royal Colonial Institute (London), in 1890 he became a member of the American Ornithological Society, and in 1910 he became a Fellow of the National Geographic Society (Washington).[125] The HBC-Smithsonian relationship was not merely a bilateral one. The traders in the Mackenzie River District were part of larger scientific networks.

Scientific networks were fragile. In many – perhaps most – cases, collection efforts did not long survive the departure or death of the naturalist or scientist at their centre. Even when the principal members of a network remained in place, the relationships needed nurturing. Kennicott's expedition was so remarkable and so successful because of the collecting that continued, and even accelerated, after his departure from the HBC territories. In 1863, about 3000 lbs. of boxes and packages arrived in Washington, the results of two years' collection – the 1862 shipments having been held over because of the Dakota uprising in Minnesota.[126] In 1864, Roderick Ross MacFarlane single-handedly submitted thirty large cases of specimens.[127] He continued to submit specimens at least until 1870.[128] In total, MacFarlane submitted at least 5,000 specimens to the Smithsonian.[129] The Smithsonian's program in the HBC territories flourished because of the social acumen of the men within the Smithsonian Institution.

Although Joseph Henry, a physical scientist, was unenthusiastic at best about developing a natural history program at the Smithsonian when he was appointed to its head in 1846, and although he never made natural history an important priority, the evidence relating to the Smithsonian's cooperation with the HBC supports the argument made by two historians that Joseph Henry supported the efforts of his assistant secretary in that field.[130] His signature at the bottom of many of the most important letters to the HBC shows that Henry was aware of Baird's efforts and willing to put his weight behind them. Indeed, it must have occurred instantly to Henry that any contact the Smithsonian institution would make with the HBC would have great potential to contribute to his great research interest: meteorology. In a letter written on 24 October 1854 – thus before Baird had made any overture to the HBC – Henry responded to a query by Major R. Lachlan of the Canadian Institute to say that "the system of winds which prevail in this Continent can never be properly understood until a series of simultaneous observations are made at intervals from the Gulf of Mexico to near the arctic circle; and no greater favour could be conferred on the science of Meteorology than the establishment of a series of observations in the British possessions in North America."[131] To that end, Henry agreed with Lachlan "that aid be asked by the Canadian Institute from Parliament, and the Hudson's Bay Company to procure the necessary instruments; that intelligent persons who have a taste for science, residing in different parts of the country be invited to co-operate; that observations be made at all military and trading posts; and that the Returns be reduced and published under the direction of the Canadian Institute, as fully as the means which may be obtained would warrant."[132] Given that the Canadian Institute had not initiated a network like that which Henry envisioned, there is every reason to believe that Henry welcomed the opportunity to make connections with the HBC. Still, although Henry's support for the natural history program was required, Spencer Baird's efforts made it a success.

Baird has been aptly described as a "collector of collectors."[133] One scholar described his methods: "by cajolery and presents, scientific reports, preserving alcohol, even novels and poetry, Baird sustained the zeal of the company men and ensured the flow of shipments."[134] In his youth Baird himself had been a lay collector for scientists (including Audubon) and trader in natural history specimens, so he knew firsthand what motivated collectors.[135] The determined but mild image captured of Spencer Baird in about 1860 (figure 8.2) reinforces the impression conveyed by the documents that he knew how to combine encouragement, flattery, and dogged persistence to develop and maintain a network of contributors to the Smithsonian. American zoologist Henry Fairfield Osborn later wrote that Baird "drove men before him with a quiet force."[136] Letters were the goad with which Baird "drove" his collectors; he claimed to have written 3050 letters in 1861![137] In 1861, Kennicott wrote to Baird regarding James Lockhart, remarking that "if your letters to the other officers did one half the good the one to Lockhart did, you will have effected more for science by them than I shall in a years work. Lockhart was pretty well primed for zoological operations, but your letter 'touched him off.' And as quick as the spring boat was off for the outfit of the post he began and has been working not less eagerly than myself ever since."[138] Similarly, in 1889, Roderick MacFarlane recalled that "while the friendly and rather extensive correspondence carried on for years with many of the foregoing [officers of the HBC] by the late eminent and much lamented Professor Spencer F. Baird, of the Smithsonian, evinced his own deep love for science, it did much to intensify their interest in, and desire to meet more fully perhaps than was otherwise possible, the views and objects of that obliging and well conducted Institution."[139]

Because MacFarlane had assisted several scientists and scientific organizations by the time he wrote that passage, his comment reinforces the impression that Baird and the Smithsonian were exceptionally adept at cultivating their lay contributors.

But Kennicott also deserves credit for the Smithsonian's success. His ability to inspire lay collectors not only while alive but even after his death is reminiscent of that of David Douglas. More than anything it was Kennicott's personality that made him so successful. Kennicott's infectious enthusiasm for natural history meant that the Smithsonian received vastly more specimens because of his efforts than it ever could have if Kennicott had collected alone. Several of Kennicott's contemporaries described why he was so successful. After meeting Kennicott only once, George Simpson remarked on "his agreeable manners and amiable disposition."[140] Simpson's secretary, Edward Hopkins noted, "It is not easy part to play, going as a stranger into a territory inhabited by men bound to a foreign government and with exclusive views on many points. But Kennicott knew how to meet the circumstances."[141] And one of Kennicott's colleagues noted that, for northern traders, the arrival of Kennicott, "young, joyous, full of news of the outside world, ready to engage in any of their expeditions or activities and to take hardships without grumbling was an event in their lives. When he taught them how to make birdskins and collect Natural History objects and showed them how by means of their collections, their names would become known in the civilized world and even in books, they seized on the project with enthusiasm."[142]

After describing Kennicott as "able" and amiable," Roderick MacFarlane wrote that Kennicott "managed to infuse into one and all with whom he had any intercourse, more or less of his own ardent, zealous and indefatigable spirit as a collector."[143] And in 1861, Lawrence (Laurence) Clarke attributed an intensification of his collecting efforts to "a further acquaintance with Mr Kennicott, who's zeal in the pursuit of science cannot be too much applauded, admiration for his many estimable qualities, regard for his amiable character, and a consequent wish to aid him furthering the objects of his journey to the far North."[144] Quite a number of HBC men, including Clarke, appear to have contributed specimens only because of Kennicott's presence and direct urging, for some stopped collecting once Kennicott left in 1862.

After Kennicott left the territories, Clarke wrote to him that "you are much missed, and all were regretting your departure from amongst us last Autumn."[145] Clarke's words express how well liked Kennicott was when he was in the HBC territories, but they also hint at the connections he maintained after he left. Kennicott set a strong foundation for the continued contributions. Joseph James Hargrave later wrote of Kennicott that he had "himself been highly successful in his collecting efforts," but "his great zeal in the work had also communicated itself to almost all the officers in the northern districts who, with unwearied diligence, assisted him in his huntings, and have, since his departure, maintained a close connection with the institution which employed him."[146] The fact that Clarke, Ross, MacFarlane, and Lockhart corresponded with Kennicott after he left the Hudson's Bay territories attests to the personal friendships he was able to forge when he was in the country.[147] Those personal links help explain why Roderick MacFarlane's contributions peaked only in 1865, well after Kennicott was gone. In April 1864, about two years after he left the Mackenzie River District, Kennicott wrote to MacFarlane,

> upon my word Macfarlane you and Lockhart quite make me ashamed of the little work I did in the [M.] R. District. I would rather have had the honor of contributing what you and Lockhart have to the history of Arctic zoology than to be a chief factor in the Hudson's Bay Company, or a member of Parliament. The latter would be jolly during life but in the former case my name would be immortal among naturalists. Indeed your names are already on record in many a public museum in Europe as well as America.[148]

The significance of Kennicott's affability is underscored by a comparison with his colleague Constantin Drexler. In January 1860, Henry wrote Simpson to say that "we may try to have some one visit James Bay this spring for the purpose of collecting eggs of birds for our great work on North American oology, if it is possible to reach Moose Factory in season, and your consent is given to it."[149] During the final preparations for the expedition, Baird wrote, "Mr. Drexler, though not very polished in his manners, is yet well behaved, intelligent, and obedient to instructions, and is highly accomplished in all that relates to his business."[150] But Drexler did not flourish in the fur trade territories. Not long after arriving at Moose Factory in late May 1860, Drexler wrote to Baird that "this is the wors place for egging i have seen yet, everything is said to breed further north, and if Mr. [Chief Factor John] McKenzie dooes not send me further i could had better staid at home and colected at the Smithsonian. ... i will not stay at this infernal post if otherwise can be helpt, as it is shure wher ther ar no bird, ther can be no Eggs."[151] John McKenzie did, in fact, arrange for Drexler to travel farther north to Fort George. By September, when Drexler left, he and John McKenzie had collected at least 600 specimens.[152] Eight boxes of them were sent immediately by ship to London from where Joseph Henry asked that they be shipped to the Smithsonian, with the assurance that the Smithsonian would pay tribute to the company: "The investigation in Meteorology and Natural History now in course of prosecution by the Institution, with the support and cooperation of officers of the Company, promises to furnish in a very few years results of the greatest moment to science;

and it will always be a pleasure independent of duty, to bear testimony to this countenance on all possible occasions."[153] In reply, Thomas Fraser, secretary to the Governor and Committee in London, wrote, "As the Collections in question are for scientific purpose, I am directed by the Governor and Committee to say that they will not charge the Smithsonian Institution with the freight from Hudson's Bay," billing it only for "the outlay made by the Company for Customs & Port Charges in London, amounting to £1.10.2."[154] He added that "the Governor and Committee feel obliged by your acknowledgement of the facilities afforded to that Gentleman by the Company and its Agents and they direct me to assure the Managers of the Smithsonian Institution that it will at all time afford them pleasure to give any aid and encouragement in their power to those who visit their country for purposes of scientific research."[155]

The Smithsonian, true to its promise, did pay tribute to the HBC. The annual report for 1860 noted that "the cooperation of the gentlemen at the posts visited, enabled him [Drexler], with the small means at his command, to accomplish results of great interest and magnitude. The collections made by Mr. Drexler were also taken from Moose Factory to London, free of expense, in the ship belonging to the Hudson's Bay Company, and then transmitted to this country."[156] Clearly then, Drexler's expedition was not unsuccessful, but he lacked the infectious enthusiasm that might have inspired others to build upon the work he had started. Notwithstanding the enthusiasm of the directors, the results of Drexler's expedition pale beside those of Kennicott.

Once Kennicott was gone, the cooperation of HBC officers in the Mackenzie River District was essential to the maintenance of the relationship between the HBC and the Smithsonian. A discussion of a few of the most prolific contributors reveals much about lay collectors in the period.

One of the earliest and most ambitious contributors to the Smithsonian was Bernard Rogan Ross (figure 8.6). Educated at Royle College, Londonderry, Ireland, Ross had joined the HBC in 1843 upon the recommendation of Sir George Simpson, a distant relation.[157] Ross's contributions to science predate and extend beyond his support of the Smithsonian in 1858 and his acquaintance with Kennicott in 1859. In 1850 he was trained to survey by John Rae and William John Samuel Pullen in anticipation that he might be able to determine the precise locations of HBC posts in the Mackenzie River District.[158] He also kept a meteorological and auroral journal for Colonel Lefroy in 1850–1.[159]

8.6 Bernard Rogan Ross. Bernard Rogan Ross hoped that his contributions to science would earn him "a name." He succeeded insofar as Ross's Goose is named after him. He may have spent most of his time at isolated HBC posts in the northwest, but when he posed for this photograph in about 1865, he dressed in formal tie, collar, vest, and jacket, and wore the "friendly mutton chops" then fashionable. This is in stark contrast to the way Émile Petitot encountered him in the HBC territories. Petitot described him "dressed as follows: Scottish merino shirt with red and green squares bordered with

(Continued)

By 1858 he had risen to Chief Trader, stationed at Fort Simpson, at the confluence of the Mackenzie and Liard Rivers, but in charge of the entire Mackenzie River District. From Fort Simpson, he made contributions to many scientists. Apart from his contributions to the Smithsonian, he also contributed to John Richardson and Andrew Murray in Edinburgh, George Wilson at the Industrial Museum of Scotland (Edinburgh), John E. Gray at the British Museum, William Logan at the Geological Survey of Canada, and the Academy of Natural Sciences in Philadelphia.[160] Ross also wrote several articles in British and Canadian scientific journals between 1859 and 1862.[161] Although occasionally quite speculative, these articles betray considerable knowledge and familiarity with scientific ideas and literature, particularly in natural history, but also include the growing preoccupation with race in the field of ethnology. Before and during his cooperation with the Smithsonian, Ross also formed associations with other organizations. He was a member of the Natural History Society of Montreal by 1859. In 1863, he became a founding Fellow of the Anthropological Society of London, and of the Royal Geographical Society in 1864. He was also connected with learned societies in the United States, including the New York Historical Society and the Academy of Natural Sciences of Philadelphia (in 1861).[162]

As the man in charge of the Mackenzie River District, Ross was important for motivating the officers in the field to collect specimens and to relay them to him for shipment to the Smithsonian and elsewhere. He was eventually credited with submitting over 2,200 specimens to the Smithsonian.[163] These included natural history specimens as well as native clothing. For example, he collected a caribou-hide

---

8.6 (*Continued*)

yellow ribbon tucked into a pair of white pants with purplish red stripes. Around his waist a multi-coloured assomption sash. Just below his knees were garters embroidered with glass beads and adorned with tufts of red silk that hung like scalps. No jacket. On his head a Florentine toque of green velvet ringed with silk embroidery. No shoes, no boots but immaculate white Chipewyan moccasins embroidered with silk of many colours. This fantastic costume, very elegant, is characteristic of the Great North. It was worn well by a little Irish officer with black eyes and great sideburns." Émile Petitot, *En Route Pour La Mer Glaciale* (Paris: Letouzey et Ané, 1887), 273. Source: HBCA 1987/363-E-700-R/118.

"Loucheux" (Gwich'in) tunic for Kennicott.[164] The 1860 annual report of the Smithsonian also indicates that Ross donated twelve years' worth of meteorological records from Fort Simpson to that institution.[165]

Bernard Ross's collecting also offers a rare hint at the role of women in the collections. In 1860, shortly after he married Christina Ross, the daughter of Chief Factor Donald Ross, Bernard Ross wrote to Baird, "I do not find that marriage interferes in the least with my scientific occupations. My better half is becoming quite interested in the thing and proposes making a collection of butterflies."[166] "Mrs. W.L. Hardisty" was also acknowledged as a donor to the Smithsonian.[167]

Another enthusiastic supporter of science among HBC servants in the wave of interest between 1859 and 1866 was clerk Roderick Ross MacFarlane (figure 8.7).[168] MacFarlane, a Scot from the Outer Hebrides and nephew of Chief Factor Donald Ross, joined the company in 1852.[169] Kennicott trained MacFarlane in the proper preparation of specimens in the late winter of 1860 while at Fort Simpson, but MacFarlane was a minor contributor until he was sent to establish Fort Anderson in 1861.[170] Fort Anderson was the northernmost HBC post at the time, established east of the Mackenzie River near the Arctic Ocean. From 1861 to 1866, MacFarlane collected thousands of specimens there.[171] On 28 July 1862, he wrote to Spencer Baird of the Smithsonian that as soon as he had arrived at Fort Anderson in May of that year, he "at once set such Indians and Esquimaux as were about the place, to collect; and have since neglected no opportunity of directing their attention to the matter."[172] His submissions stopped briefly in 1866 after Kennicott's death and the closing of Fort Anderson, but resumed thereafter. Between the late 1860s and the late 1880s MacFarlane contributed little to science, but he became active again in the late 1880s. He continued to contribute sporadically until his retirement in 1894.[173] Although Baird encouraged him to publish his knowledge, he initially demurred: "I have no ambition to figure as an author, or to be thought other than I am – a mere collector."[174] But MacFarlane was not a "mere" collector; he was the most prodigious collector in the company, contributing 5,700 specimens to the Smithsonian in the 1860s.[175] And eventually he did also become an author.[176]

The annual report of the Smithsonian for 1868 acknowledged "sixteen boxes, eleven packages, and one keg of specimens of natural history and ethnology" donated by MacFarlane in the previous year, and donations of insects from Lockhart.[177] The report notes that

8.7 Chief Factor Roderick McFarlane (age 35, 1870). By the time Roderick MacFarlane posed for this photograph in 1870, he had been named as the most prolific contributor of scientific and ethnographic specimens to the Smithsonian Institution. Source: HBCA 1987/363-E-700-Mac/13.

the last invoice from Mr. Macfarlane is fully equal to those with which he has favored the Institution in previous years, and entitles him to the credit of being the largest contributor to the Smithsonian collections, and of having done more than any other person in making known the productions and character of the regions he has explored. The record of specimens bearing his name already amounts to over ten thousand entries, including some of the choicest contributions to natural history and ethnology.[178]

A number of other officers in the Mackenzie River District who had met Kennicott subsequently stood out as collectors. Lawrence Clarke, who joined the company in 1851, became an avid collector, primarily around Fort Rae, on Great Slave Lake. Kennicott noted that Clarke said that he did not "care a fig for the credit of working for science," yet when suspicious that Bernard Ross was trying to get credit for specimens he had collected, complained that Ross "is too fond of getting others to work and he getting the credit."[179] Whether Clarke's suspicions were right or not, he became a more avid collector after Ross left the district in 1862.[180] James Lockhart was another important contributor to the Smithsonian. He had joined the HBC in 1849 and assisted one of the HBC's arctic expeditions. By 1860 he was in charge of the company's Fort Yukon.[181] He sent over 1,100 specimens to the Smithsonian, including donations as late as 1868.[182] Other important collectors in the Mackenzie River District included William Lucas Hardisty, a mixed-blood, Red River–educated Chief Trader at Fort Resolution, and Strachan Jones, who, as described by Kennicott, was "a graduate of Toronto College, a gentleman by birth and education and a *brick*, tho' what is called a dry stick."[183] The long list of minor collectors included HBC clerks and officers stationed at posts from Fort Yukon to Labrador, including James Anderson, Robert Campbell, Henry Connolly, Andrew Flett, James Flett, Charles P. Gaudet, Joseph Gladman, Alexander McKenzie, James McKenzie, William McMurray, Governor William Mactavish, Colin Rankin, John Rae, John Reid, Thomas Swanston, Nicol Taylor, and Julian S. Onion (Camsell).[184]

The documents from this period reinforce the impression that labourers cooperated with scientists and officers only if they were paid directly. Debra Lindsay was probably right that "it was difficult for the working class to maintain an interest in activities that differed specifically, but not generally, from the many other time-consuming and physically taxing duties assigned them."[185] However, there were times of the year in which even labourers and tradesmen within the HBC were relatively idle. To Lindsay's explanation we might add that the labourers, many of whom were illiterate and poorly educated, tended to value different forms of leisure than the officers and clerks typically sought out. They also probably placed far less value than officers on the intangible rewards of being acknowledged in the annual report of a scientific institution.

Although evidence is not plentiful, the contributions of aboriginal people to scientific collecting efforts are better documented during this period than during earlier periods in the HBC's history. The evidence supports the impression that the HBC officers, and at least some scientists at the Smithsonian Institution, recognized the value of aboriginal peoples' knowledge of nature.[186] In 1860, the prominent British oologist Alfred Newton wrote an article subsequently republished by the Smithsonian that stated, "the best allies of the collector are the residents in the country, whether aboriginal or settlers, and with them he should always endeavor to cultivate a close intimacy, which may be assisted by the offer of small rewards for the discovery of nests or eggs."[187]

According to William Fitzhugh, Kennicott's collection "utilized native people as collectors and informants to a greater degree than was the case for [Smithsonian] collections made in previous decades."[188] Although it is impossible to say for sure, it certainly is possible that aboriginal people collected a majority of the natural history specimens submitted to scientists from the HBC territories in this period – and in previous periods. The fact that in 1862 MacFarlane noted in passing that upon his arrival at Fort Anderson he "at once set such Indians and Esquimaux as were about the place, to collect; and have since neglected no opportunity of directing their attention to the matter," suggests that reliance on aboriginal collectors was normal and routine.[189]

Scientists and officers in the HBC territories could turn to aboriginal people so easily because of the cooperative relationships that prevailed between HBC traders and Native peoples; relationships that, although unusual in most places in which the Smithsonian was active, were routine in the HBC territories. Officers were not shy about acknowledging that they relied on aboriginal people to collect specimens. Interestingly, traders occasionally identified by name the mixed-blood people with European surnames who collected for them, but very rarely gave the names of Indian people. Donald Gunn, for example, noted that "Indians" collected for him, but did not give their names.[190] Kennicott also frequently noted that he had Native collectors and informants.[191] On one occasion, for example, he wrote "that there is very little chance of my ever killing such things as musk oxen, barren ground bear & reindeer ... I can only hope to get them by hiring the Indians to bring them in from a great distance."[192] Unfortunately, it is impossible to discover what Kennicott or the HBC officers told aboriginal people about the purposes of these collections.

Kennicott's admission that he would have to hire Indians to bring in specimens reinforces the impression that indigenous people, much like HBC labourers – and probably for many of the same reasons – would not collect samples without direct payment in the form of practical trade goods or luxury items. Kennicott wrote Baird in 1860 that "so long as an Indian isn't hungry, or in fact *very* hungry, he is as independent as you please and scorns the idea of working for anything less than very large pay; if he will work at all."[193] In this respect, although aboriginal people needed to preserve specimens destined for scientific research differently than the way they prepared furs for the commercial market, they were paid for both in the same way. As Debra Lindsay has argued, however, officers and scientists routinely belittled their indigenous suppliers. The fact that aboriginal people did not submit specimens gratis probably explains in part why Gunn, Kennicott, and others disparaged Indians as lazy, stupid, or superstitious.[194] Kennicott noted in 1860 that "the officers kindly permit me to take what I want from the stores so I can always offer the Indians good bribes. But they are lazy superstitious Devils and I've got nothing much from them yet."[195]

Kennicott and traders did not rely on indigenous people only to collect specimens; they frequently consulted with aboriginal people about a wide range of matters including questions surrounding the range and abundance of certain animals, their habits and variation, and the times of arrival and departure of migratory animals. HBC traders also surprisingly turned readily to aboriginal expertise on questions upon which a considerable gulf of understanding must have existed between Western and indigenous peoples. Bernard Ross for example, defended his argument that there were three species of snow goose, not one, by appealing to indigenous knowledge. Using the English word *goose* interchangeably with its Cree equivalent, *wavy*, he asserted that "there can be little doubt of the existence of these three species of Snow Geese, (exclusive of the Blue Wavy of Hudson's Bay) as the Slave Lake Indians have a different name for each kind. The first which arrives is the middle-sized species which I believe to be *A. albatus*; next comes the smallest sort, the *A. Rossii*; and lastly the *A. Hyperboreus*, which arrives when the trees are in leaf, and is called the yellow wavy by the Indians."[196]

Ross cited aboriginal expertise on other occasions. Other HBC traders did the same. George Barnston, for example, appealed to aboriginal knowledge to argue that there were two species of lake trout and two species of otter.[197] One might argue, given that this is the kind of argument that a trained scientist was unlikely to find compelling, that HBC traders were more open to aboriginal knowledge than metropolitan scientists were. But it also suggests that they were (ethnocentrically?) blind to how differently aboriginal societies and Western scientists distinguished between one kind of plant or animal and another – or so intent on identifying species new to science that they were willing to resort to any evidence to support their argument. In the case of the snow geese, Spencer Baird accepted Ross's general argument, and scientists continue to distinguish between Ross's goose (*Chen rossii* or *Anser rossii*), and two subspecies of snow goose, the greater snow goose (*Chen caerulescens atlantica*), and the lesser snow goose (*C. c. caerulescens*), which occurs in a white and a blue phase, and often associates with Ross's goose. But more often than not, scientists rejected such arguments. In the case of the snow geese, Western science and "Slave Lake Indians" eventually agreed. On occasions when they did not, it would be facile to suggest that either indigenous knowledge or Western science was incorrect.

The Smithsonian's methods of honouring contributors were developed before the Smithsonian became involved with the HBC. In 1850, during his first year with the Smithsonian, Baird promised potential collectors in the US. Army that "all contributions will, of course, be duly credited to their respective donors in the Museum and in the reports of the Institution."[198] Its published annual reports were extraordinarily effective vehicles for the Smithsonian Institution to publicize and pay tribute to its donors.

A perusal of the Smithsonian Institution's annual reports reveals tributes to individual HBC personnel beginning in the 1850s, and to the company and its directors beginning in 1860, with effusive thanks expressed yearly until 1869. A sample of the tributes published in the annual report of just one year – 1860 – reveals not only the effort that the Smithsonian put into its acknowledgments, but also the lengths the HBC was willing to go to support the institution's efforts. The 1860 annual report of the Smithsonian acknowledged the assistance that the HBC gave to Kennicott by noting that "not only has he been permitted to visit and reside at the different posts, but he has received free transportation of himself and his collections."[199] The report similarly noted that Drexler "was enabled to collect a large number of valuable specimens through the facilities afforded him, and these were sent from Moose factory to London, at the expense of the company; and thence to this country by the Cunard steamers, free of charge; acts of liberality which deserve to be specially noticed."[200] And yet again, the report noted that

> for a most generous cooperation of the Hudson's Bay Company, through Sir George Simpson, and its officers in England and America, the Institution is under the greatest obligations. Every possible facility has been furnished to Mr. Kennicott, not only in permission to visit the different posts, but in the way of free transportation of himself and his collections, quarters at the posts, &c. Wherever he has gone he has found an appreciation of his mission and a readiness to assist, gratifying in the highest degree. Nearly all the gentlemen in charge of different posts have undertaken to make observations in meteorology for the Institution, ... as well as collections of such objects of natural history as he might not succeed in securing himself.[201]

Next, the report listed men individually: "The gentlemen to whom Mr. Kennicott expresses his indebtedness most particularly, after Mr. Ross, are Mr. L. Clarke, Mr. J. Reid, Mr. A. McKenzie, Mr. MacFarlane, and Mr. Hardisty."[202] After detailing some of the contributions of these men, the report reiterated that "without the facilities furnished by the Hudson's Bay Company and its officers, ... the enterprise, in its present extent, would be entirely impracticable."[203] Each year until 1869, the annual reports repeated their tributes to the company, its directors, and a long list of people connected with the company.[204]

Even many years later, the tributes continued. In 1865, company employee J. G. Lockhart submitted a paper on the habits of the moose. Twenty-five years later, the United States National Museum published the paper in its journal. When it published the paper, the editor added a footnote saying that "for more than thirty years the Hudson Bay Company has zealously cooperated with the Smithsonian Institution in increasing the ethnological and natural history collections of the National Museum. The objects thus received from Mr. Robert MacFarlane, Mr. Lockhart and other agents of the company have added greatly to our scientific knowledge of British North America."[205]

As had been the case with David Douglas before him, Kennicott had kept a journal while in the HBC's territories, but did not publish it before his death. But just as excerpts of Douglas's journal were published after his death, segments of Kennicott's were published in 1867. Among the sections published was an endorsement of the HBC that must have pleased its directors:

> The Company pays the Indians very well for their provisions, besides often helping them when they are starving. The popular cry against the Hudson's Bay Company, of injustice to the Indians, is totally unfounded. It is only raised by those who would like to come in here, and with liquor and fancy goods get the furs themselves that the Company now gets, paying useful goods to them. But they would send these tribes to perdition as fast as the southern ones have gone, while the officers of the Hudson's Bay Company take no little pains to secure the well-being of their Indians. And it is their interest to do so.[206]

For their part, the Governor and Committee of the HBC clearly expected that scientists would pay tribute when the company contributed to science. In 1862, William Mactavish informed Spencer Baird that "the Governor and Committee have issued orders that all specimens in Natural History collected for individuals or Societies *in Britain* are to be sent to the Hudsons Bay House and they will be presented by the Governor and Committee."[207] Mactavish noted that the "in Britain" seemed to exclude the Smithsonian, and he would interpret it as such. He continued by explaining, "I am told the reason for the new order is that a Mr. Murray, who applied to the Company for specimens from the Country after getting many, presented them to the Kensington Museum without a word of acknowledgement to the Company. I fancy the unfortunate acted unwittingly, and am not sure that the Company deserved much credit as the officers at their own expense and trouble got the specimens. But still Mr. Murray deserved some punishment for his want of savoir faire."[208]

Mactavish continued his letter by recommending that the Smithsonian send a "complimentary and grateful letter to the Govr. and Committee" – advice that Baird and Henry did not need.[209]

Smithsonian personnel not only remembered to acknowledge, thank, and endorse their patrons, they were very conscious of the need to avoid annoying the HBC directors. Upon learning that American papers had published "d____d nonsense" extracted from his letters home, Kennicott admitted his fear that he would get "into bad odor with Sir Geo Simpson."[210] The directors of the Smithsonian also felt it wise to reassure the HBC that "the Institution will vigorously and carefully respect all the facts bearing on the statistics and internal administration of the company that may come to its knowledge thro' the collection and notes of Mr. Kennicott or other sources, and which the Company might desire not to make public. Nothing bearing in the most remote degree on these subjects will be printed or communicated in any form."[211] The directors at the Smithsonian clearly understood that it was important to ensure that their partnership with the HBC was mutually beneficial.

Documents related to the HBC-Smithsonian cooperation shed considerable light on the many motivations that inspired lay hinterland collectors. For many HBC officers, collecting was at least in part a valued pastime. Recall that in 1859 Kennicott noted, "the officers duty is almost nothing beyond his actual presence. A little less than two months in the year is sufficient for all the writing."[212] Bernard Ross wrote in March 1860 that "there is every wish to aid in forming your collection. This pursuit is an amusement to people whose stores of entertainment are not overly large."[213] Two years later he added, "I only wish that I had taken up some scientific pursuit long ago. You can scarcely fancy the solace that it is to me."[214] Evidently then, collecting was a leisure pursuit for some men.

Like the HBC directors, officers in North America appear to have been motivated to a considerable extent by tributes and thanks they garnered from the prestigious Smithsonian.[215] As George Gibbs noted in 1862, "an acknowledgement from the head of an institution like the Smithsonian is a great inducement to exertion."[216] Bernard Ross confirmed this assessment when he wrote that "it gives me at all times much gratification to advance the cause of Science, as far as the limited sphere in which I move will permit: and in the present case the pleasure is doubly enhanced by affording any assistance to an Institution of such worldwide renown as the Smithsonian."[217] Baird understood that the promise of a credit line in a museum exhibit or in the annual report of a prestigious institution was a significant incentive to collectors. Clearly, being recognized in print was very gratifying for those who contributed. Donald Gunn wrote to Baird in 1857 that "the Volume containing the *Report* of the Institution for 1854, which you had the great kindness to forward me, I received for which I tender you my most sincere and heartfelt thanks for the handsome Manner in which you have been pleased to remember me."[218] Barnston similarly expressed his thanks for the acknowledgments in the 1860 annual report.[219]

If it seems implausible that lay collectors might have expected to become famous, the evidence shows that some were motivated by the hopes of achieving a measure of renown.[220] One person noted that "collectors were repaid with Baird's gratitude and help, with their own sense of accomplishment, and perhaps most importantly, with fame. ... When he named a new species after their collectors, he was conferring on them an enduring fame, the acquisition of which had been a motivation for naturalists for centuries."[221] Men like Baird and Kennicott were well attuned to the personalities of their collectors, and knew how to encourage them. Kennicott for example, explained to Baird that Bernard Ross was "certainly very intelligent and well read" but "like your humble servant he has a very good opinion of himself and is in fact *rather* conceited. ... he likes flattery very well."[222] Ross later corroborated Kennicott's assessment. Not bothering with subtleties, Ross wrote to Baird, "I wish to make myself a name in the scientific world if possible, and I am sure that you will do all in your power to gain it for me."[223] If Ross wanted a name, Baird was quite willing to indulge him. It was in a letter written in November 1859 that Ross first wrote to Baird regarding the snow goose that "there will be noted probably 3 species, instead of only one as now."[224] Convinced, Baird named Ross's goose (*Anser rossii*) after Ross in 1861.[225] Ross was delighted. He wrote, "I am quite flattered at having the snow goose called after me. It is about the pleasantest compliment that could be paid me."[226] Roderick MacFarlane also had several species named after him.[227] Thus did Ross and MacFarlane join a list of HBC officers, starting with Thomas Hutchins, who had species and subspecies named after them. This form of recognition is the sort of "immortality" to which Kennicott alluded when he wrote MacFarlane in 1864. Having the opportunity to name a species after someone else was also a high honour that Ross attempted to grasp. He wrote Baird in 1860 arguing that he might have discovered two new species, saying "I wish them to be called *Bernicla Barnstonii* and *Larus Andersonii* after two esteemed friends."[228]

Failure to acknowledge the assistance of lay collectors could cost a scientist as much as failing to acknowledge the company. Ross was among those who responded in the late 1850s to a circular sent to HBC officers to encourage them to collect specimens for the Scottish naturalist Andrew Murray from the Royal Scottish Museum in Edinburgh. In 1859, Ross wrote Baird that he had previously submitted specimens to

Murray, but his contribution "has not been acknowledged," therefore "every moment that I can spare I will devote to your Institution."[229] Several months later he again told Baird that he had sent Murray a specimen of a rare bird, but remarked that Murray's "omission to acknowledge the receipt of this bird, will be certainly the best way to prevent me from taking the trouble of sending him any more."[230] But Ross's loyalties were fickle. In March of 1860, Kennicott noted that Murray "has won the hearts of Mr. Ross and the others here by getting from the H.B. Co permission to send liquor 'in such quantities as are *actually required* for preserving specimens of Nat. Hist. etc' into Mackenzies River and by telling Mr. Ross to order from York Factory whatever he requires. ... Mr. Ross has ordered 12 gallons, half of which he says he shall drink and use the rest for specimens, considering it '*actually necessary*' to imbibe as much as is used for preservation of specimens."[231]

Within days, a letter penned by Ross himself confirmed the importance of alcohol. Ross wrote, "The greatest present you can confer on the Gentlemen is to send in a good stock of spirits for preserving *one half not medicated* – as we must get liquor in sub rosa – with this *stimulant* there is no doubt but that you will obtain lots of things."[232] That worked. In March 1861 Ross, after thanking Baird for sending him books, added "many thanks also for the whiskey in which I hope to drink your own and Mrs. Baird's health on Xmas day."[233] In the same letter, Ross added that "it is now with me a labor of love, and I am daily becoming more and more attached to the study in particular of ornithology."[234]

Ross's admission that alcohol was a significant inducement to collectors appears to have been well known. The HBC directors understood it, and they were willing to turn a blind eye to the prohibitions against the importation of alcohol into the HBC territories. The case of William L. Hardisty illustrates the fact. In 1867, Mactavish winkingly wrote to Baird that "it is contrary to rule to send spirits of any kind into McKenzie River except for medicinal purposes, so that if as a medical man you consider Hardisty's ailments require something of the kind I may tell you that packages for the Companys officers are never subjected to examination by us."[235]

Gifts of free books encouraged the cooperation of HBC men, from the company's directors to its officers. In 1860 George Barnston lamented, "Fur Traders have to work up amidst a host of difficulties, nothing whatever is at hand for them. No Libraries, no Museums, or collections to refer to, no friend near to consult, so that every step he takes is like an uncertain one, and he finds himself at last so full of mistakes, that he doubts whether he can ever be exact or correct, or sure of any thing, until he actually see the same, upon some other authority."[236] Barnston was exaggerating, perhaps to encourage Baird to send him books. Ross was a much more direct man than Barnston. He wrote, "The greatest favour you can confer on me, if you wish to send me anything, is a few books. I am a *reading* man and have already a well selected library. French and English books are quite of equal interest to me, as I speak the former as fluently as the latter language."[237] Baird already knew that many lay collectors craved books. Baird is known to have rewarded many of his collectors, both in the HBC and elsewhere, by sending books (both scientific and literary), newspapers, and journals.[238] HBC officers expressed their delight at getting them. In 1859 Bernard Ross wrote Baird to thank him for sending "your most interesting works on the fauna of America."[239] The Smithsonian Institution rewarded HBC contributions by sending hundreds of books to many people, from the officers to the directors at the HBC's headquarters in Montreal and London.[240] In early 1860, for example, Joseph Henry sent two boxes of books to George Simpson, together with books for Edward Martin Hopkins (Simpson's secretary and assistant), and James Stewart Clouston, Chief Factor at Lachine.[241] When Simpson thanked Henry, he noted that the books "will form an important addition to a collection of works I have been making, which directly, or indirectly, refer to the discovery, exploration and history of the North American continent and to its aboriginal inhabitants."[242] Other gifts also flattered officers. Ross for example, was thrilled to receive the autograph of US President James Buchanan.[243]

Ross's own letters, and letters written by his contemporaries, suggest that the prestige associated with being an acknowledged contributor to science many have been his primary motivation – a motivation strong enough for him to be underhanded in his methods. Aside from Clarke's insinuation already noted, Kennicott and others also complained that Ross tried to control the conveyance of specimens to the Smithsonian so that he would get the credit for specimens actually gathered by other officers.[244] In 1864, William Hardisty wrote to Kennicott that Roderick MacFarlane's collecting efforts were motivated by a desire "to acquire similar honours as those conferred on Mr. Ross," but Hardisty believed that Ross "gained his distinction by the labors of others while MacFarlane's collections are all his own."[245] In fact, MacFarlane's own letters show that he acquired many (probably most) of his specimens from aboriginal people. He likely used the company's stores to pay for them. But perhaps his transgressions were smaller or less obvious that Ross's, or perhaps his more humble and amiable demeanour led others to overlook his indiscretions.

The lures of fame, flattery, reward, and recognition even led Ross to compromise the interests of the company. In 1859, he sent Baird a list of the numbers of different furs taken out of the Mackenzie River District during the previous decade, but warned Baird, "I wish you to consider the details of this paper AS STRICTLY CONFIDENTIAL, it is what I would not communicate to anyone but a thorough man of Science. The publishing of the details would seriously compromise me; but the inferences of a Scientific nature drawn from it cannot be objected to."[246] Eventually, Ross was punished for indiscretions. In early 1865 Lawrence Clarke wrote Kennicott about Ross, that it "was brot home to him of having misappropriated much of the companys property in obtaining his collections, and has been fined heavily by minutes of council this year; the result is, that people on this side feel an antipathy to meddling with collections of any sort."[247] After learning of Ross's dishonesty, the company's directors apparently decided that all subsequent visitors to its posts had to pay for their expenses.[248] Given the positive image that the company earned through its officers' contributions to science, its directors were probably quite willing to accept that some of the company's trading goods would support their men's efforts, but Ross evidently crossed a line.

The HBC's contributions to science between 1855 and 1870 continued to reflect the priorities of metropolitan scientists, although they naturally swung towards the priorities of scientists at the Smithsonian in particular. If the emphasis on natural history reflected Baird's specialization, the emphasis on climatology and meteorology reflected Joseph Henry's. Because weather and climate were particular interests of Henry, they were also a focus of the Smithsonian's research program from the beginning.[249] Henry, a physicist, was eager to advance the scientific understanding of weather and climate.[250] But meteorological research at the time was utterly dependent on a network of volunteer observers. Already by 1850, the Smithsonian had a corps of 150 such observers, and by 1859 its meteorological research accounted for a third of its budget.[251] According to G. Brown Goode, as a result of the Smithsonian's efforts, "all the essential features for the prediction of meteorological phenomena were in existence in the Smithsonian Institution as early as 1856, having grown up as the result of an extensive series of tabulations of observations recorded by volunteer observers in all parts of the country."[252] In the field of climatology, the Smithsonian published Lorin Blodget's very influential *Climatology of the United States and the Temperate Latitudes of the North American Continent* in 1857, just as the HBC's relationship with the Smithsonian was beginning. However, Blodget's book and the Smithsonian's weather network included little data from British America, and none at all from the HBC territories. Because the HBC posts were not connected to the rest of the world by telegraph, the HBC's weather observations could not be incorporated into the Smithsonian's avant-garde attempt to predict the weather along the east coast of North America, but their meteorological journals were relevant to Henry's efforts to study the typical movements of weather systems, especially storms, across North America and to better understand climate in North America. Furthermore, Baird's natural history program dovetailed well with Henry's meteorology program (often by tapping the same volunteers).[253]

The Smithsonian's approach to natural history in the HBC territories also reflected new intellectual trends in that field. According to historian Morgan Sherwood, the Smithsonian's interest in northwestern North America was linked to the understanding that it "was a bridge between the Old World and the New. It was a fertile, strategic, and virgin ground for testing big questions in biology, geology, and ethnology, an unused laboratory that promised to yield evidence confirming or refuting some of the great theories."[254] The Smithsonian's efforts in the HBC territories were also significant because they came even as Spencer Baird was trying to implement new systematic procedures for collecting, preserving, and labelling specimens.[255] These new procedures were crucial if the specimens were to be useful for better understanding species distribution and abundance, speciation, morphology, and variation. When he reported on the return of Kennicott from the HBC territories, Baird argued that the results of the Smithsonian-HBC cooperation "will serve to fix with precision the relationships of the arctic animals to those of more southern regions, their geographic distribution, their habits and manners."[256] HBC officers had clearly been influenced by the questions that drove Smithsonian scientists. Bernard Ross for example, commented on one of Baird's theories in 1859: "It is extremely probable that your supposition, of the birds that penetrate furthest north being the largest, as well as all migratory animals, is a correct one. The more weakly ones will naturally remain along the road; and the same thing is I think observable respecting fish, regarding the distance that they come up rivers from the Sea. I will with pleasure aid in obtaining a

Series of Measurements to determine the matter. Of the *resident* Northern animals being larger I am doubtful."[257]

One scholar has noted that the Kennicott expedition "produced the largest and most systematically gathered collection of natural history and ethnological materials gathered by the Smithsonian up to that time. It established a method of operating in remote regions using the local native population and the existing infrastructure, in this case, that of the HBC, which was offered to the Smithsonian nearly free of charge."[258] Thanks in part to the systematic way that the collection was amassed, even in 1947 Henry B. Collins could describe the material collected from the HBC territories as "among the most treasured materials in the U.S. National Museum."[259]

The HBC's contributions to ethnology also reflected current trends. The ethnological mission of the Smithsonian dates back to 1847, when it was deemed especially "proper" for the institution to gather evidence pertaining to "the physical history, manners and customs of the various tribes of aborigines on the North American continent."[260] Ethnology however, remained on the fringe of science in 1846, and accordingly, the Smithsonian put little effort into the collection of ethnological materials before the Civil War.[261] But in 1862 geologist and ethnologist George Gibbs was working with the Smithsonian to develop "a comprehensive work upon the ethnology and philology of North America, to be published by the Institution, and to prepare materials for maps showing the location of the Indian tribes at various periods." He suggested to Joseph Henry that "Mr. B.R. Ross would doubtless undertake British North America ... including the Chepewyan family, the Crees, and Knisteneaux, and perhaps the Exquimaux, and prepare a memoir giving the history of the subject and all that is valuable regarding those tribes."[262]

Gibbs was wrong. Hudson's Bay Company officers were considerably less interested in ethnology than Gibbs wished. George Barnston called ethnology "the Science of Amateurs."[263] Bernard Ross wrote, "Ethnology is a dry subject and I return from it with renewed zeal to my natural History Studies."[264] Part of Ross's lack of interest in ethnology may well have stemmed from his belief that indigenous customs and manners had already changed dramatically since the arrival of traders. In a comment similar to that made by Robert Campbell to George Wilson in the 1850s (discussed in chapter 5), Ross argued that even in the 1860s, "the former habits, customs, and traditions of many tribes are completely lost to the world: while even now the aboriginal races, brought into contact in almost every region with the whites, Missionaries and pseudo or real civilization, have imperceptibly lost their ancient ideas, feelings and traditions, and notwithstanding their Asiatic tenacity, insensibly acquired the manners of the dominant race."[265] The salvage anthropologists of the late nineteenth century and early twentieth century would have been struck by Ross's assessment.

Despite their evident lack of enthusiasm for ethnological work, the HBC officers submitted to the Smithsonian Institution some of the first ethnological artefacts collected for any scientific institution in the United States.[266] As early as 1863, Spencer Baird suggested that the Smithsonian's collection of "ethnological peculiarities of the Esquimaux and different tribes of Indians inhabiting the Arctic regions" was likely the pre-eminent collection in the world.[267] HBC officers also wrote some ethnographic articles that were published in the Smithsonian's annual reports. From its founding until the Civil War, "ancient monuments" (Indian mounds) were a major focus of the Smithsonian's ethnological program.[268] The mounds, especially when studied in the empirical, anti-speculative way upon which Joseph Henry insisted, permitted the Smithsonian to pursue an ethnological agenda while steering clear of the racial debates that divided American thinkers.[269] The interest in mounds helps explain why the Smithsonian Institution published Donald Gunn's article on Indian mounds in the vicinity of the Red River Colony in 1867.[270] The Smithsonian's focus on the gathering of material culture similarly helped it avoid divisive racial debates. Thus, much of the Smithsonian's attention was geared towards collecting ethnological material and information relating to the indigenous people of the Mackenzie River region. In its annual report for 1866, the Smithsonian Institution published three articles by HBC officers under the title "Notes on the Tinneh or Chepewyan Indians of British and Russian America."[271] These articles, written by Bernard Ross, William Hardisty, and Stachan Jones are decidedly unromantic portrayals of the Dene. Although the interpretations are mixed, the images portrayed in them approximate more the image of the ignoble savage than the noble savage. All three men disparaged the treatment of women and girls in indigenous societies. Ross and Hardisty in particular also emphasized the physical characteristics of the Indians. According to Debra Lindsay, these "Mackenzie River ethnographies were written in response to one of the earliest attempts to objectify anthropology."[272]

If the HBC's contributions to ethnology were not large, Joseph Henry could boast of them nonetheless. In 1869, he declared, "Thanks to the co-operation of the officers of the Hudson's Bay Company, among whom may be mentioned Messrs. Ross, Gaudet, Hardisty, and Kirkby, but especially Macfarlane, the Institution is in possession of what would appear to be a full representation of the life of the Esquimaux of that region, as illustrated by their dresses, weapons of war, their implements for fishing and the chase, household articles, ornaments, &c."[273]

Scholars have disagreed about the political significance of the work done by Robert Kennicott and the Smithsonian Institution, either with the HBC or subsequently with the Russian-American Division of the Western Union Telegraph Company. Some have argued that the Smithsonian's work with the HBC, combined with Kennicott's experience with the WUT, contributed significantly to the American decisions to purchase Russian America.[274] On the other hand, some have argued that the Smithsonian had little capacity to inform American politicians, because the Smithsonian itself had little knowledge about Russian America before 1867.[275] Debra Lindsay, for example, has written that "the argument that Smithsonian science was the handmaiden of manifest destiny in the 1860s cannot be substantiated on the basis of the empirical data deposited at the Institution before the spring of 1867."[276] A careful assessment of the evidence supports Lindsay's conclusion. Still, although the Smithsonian's cooperation with the HBC was probably unrelated to any anticipated American purchase of Russian America, and although the information that the Smithsonian Institution acquired through that cooperation could have played only a minor role in the actual negotiations for the purchase of the territory, American politicians and Smithsonian Institution directors exaggerated the importance of that information for their own purposes after the negotiations were concluded.

By 1850 few Americans could have been unaware that some expansionists believed that it was the "manifest destiny" of the United States eventually to annex all of North America. So, Henry, Baird, and Kennicott may well have thought that the HBC lands and Russian America were destined to become part of the United States, but it is unlikely that those men undertook their cooperation with the HBC in response to any direct request on the part of American politicians for the Smithsonian to acquire information about the northwestern part of North America. There were internal discussions among Russian officials about the possible sale of Russian America as early as 1850.[277] The Russians first approached the Americans on the issue shortly before the outbreak of the Crimean War (1854–6), but found no interest.[278] Still, the possibility that the United States might purchase Russian America was discussed in New York and London newspapers as early as August 1854.[279] During the Crimean War, French and British forces attacked Petropavlovsk on the Kamchatka Peninsula,[280] but they made no attempt to take Russian America. Although they could easily have seized Russian America, such a move would certainly have angered the United States. Serious talk of selling Russian America to the British or the Americans resumed in Russia in 1857.[281] The first serious discussions regarding a possible sale of the territory between representatives of the American and Russian governments – initiated by Senator William M. Gwin of California – occurred no later than December 1859.[282]

The fact that American politicians and diplomats and the American public could have contemplated the transfer of Russian America to the United States in 1857 raises the possibility that the Smithsonian actually approached the HBC in 1859 about sending Robert Kennicott to the region near Russian America – with the aim of having him also travel to Russian America itself – at least in part in response to requests on the part of politicians for a reconnaissance of the territories. The fact that the president, vice president, and members of the cabinet and Congress were on the Board of Regents and often within walking distance of the Smithsonian meant that government officials could have encouraged this effort.[283] And if they did, Kennicott's failure actually to travel to Russian America when he worked with the HBC, and to learn as much as he wanted when he worked for the Western Union Telegraph would have made his expedition far less successful than they had hoped. But it is unlikely that they did. Given the evidence that Baird was interested in the HBC territories by early 1850, and that the Smithsonian's efforts were heavily influenced (and perhaps driven entirely) by the scientific priorities of the institution and the ambitions of its directors, and the fact that by the mid-1850s American expansionist enthusiasm was at a low ebb, it seems probable that prospects of the territorial expansion of the United States played little part in prompting the Smithsonian's interest in northwestern North America. Only after the Civil War ended could the United States government again seriously consider territorial expansion.

The actual negotiations for the sale of Russian America took place in Washington, DC, in March 1867. Given that the progress and accomplishments of Kennicott's activities in the Mackenzie River District were publicized prominently in the Smithsonian Institution's annual report, beginning with the report for 1859, it would have been a significant oversight on the part of American negotiators if they had neglected to consult with the Smithsonian about the resources, climate, and other characteristics of northwestern North America during those negotiations. Debra Lindsay was right when she argued that the Smithsonian had little knowledge of Russian America before 1867.[284] But Henry B. Collins was also right when he asserted that Kennicott "provided the first substantial and accurate [public] information on the natural history and resources of northwestern Canada and the adjacent part of Alaska."[285] Stated somewhat differently, Morgan Sherwood wrote that "every scrap of information about Alaska was welcome because so few scraps existed."[286]

Despite the fact that the Smithsonian could have given little information about Alaska before 1867, it could, and did, provide information about the Mackenzie River District. Baird and Kennicott certainly knew that the northwestern part of the continent had a considerably milder climate than the same latitudes in northeastern North America. Lorin Blodget's *Climatology of the United States and the Temperate Latitudes of the North American Continent* (1857) said that the climate of northwestern North America should be much warmer than that of the northeast. When he arrived at Fort Simpson during his first summer in the territories, Kennicott saw confirmation of that theory. He remarked that "the timber is much heavier than I had expected to see in latitude 62° N."[287] Indeed, since the tree line in the Mackenzie River District reaches farther north than anywhere else in North America, if Baird had extrapolated based on what he knew of the north-trending isotherms and treeline in the Mackenzie River District and Fort Yukon, he would have provided government officials with optimistic speculation about the potential of Russian America. Still, it is impossible to escape the fact that the Smithsonian possessed only scanty knowledge relating to Russian America in March 1867.

Even if the information that the Smithsonian could provide had only minor significance for American negotiators – a reasonable argument – pro-purchase politicians and the directors of the Smithsonian certainly emphasized the value of that information after the treaty had been concluded. When the news of the treaty to purchase Russian America became public in early April 1867, it sparked immediate controversy, with many critics ridiculing the purchase. Even many who cautiously supported the acquisition did so because they believed that it would be an important step towards the eventual American occupation of all of North America.[288] In order to defend the purchase, Charles Sumner delivered a three-hour speech in the Senate on 9 April 1867.[289] In that speech, he frankly admitted that very little was known about Russian America outside of Russia, and he did not claim that the information acquired thanks to the Smithsonian's cooperation with the HBC was of great importance.[290] Still, Sumner did say that Kennicott "had visited the Youcan country by the way of the Mackenzie river, and contributed to the Smithsonian Institution important information with regard to its geography and natural history, some of which will be found in its reports."[291] So, at the very least, Sumner's reference to Kennicott *was* a small part of his influential speech advocating Senate ratification of the purchase. It was evidently not enough for Joseph Henry. Henry wanted to claim greater credit for the Smithsonian. In his annual report for 1867, Henry claimed that

> it was known that the Institution had for several years been diligently engaged in gathering specimens and collecting information to illustrate the character of the northwest portion of the American continent, and consequently, when the question of the acquisition of Alaska by the United States came under discussion, it was to the Institution that reference was chiefly made by the State Department and the Senate for information in regard to the country. ... Professor Baird ... gave valuable information as to the zoology of the country, from materials which had previously been collected by the Institution.[292]

The statements by Sumner and Henry illustrate the political value of science, regardless of how one assesses the truth of their claims. If the Smithsonian truly did provide the government with very little useful information, the subsequent claims of Henry, Baird, Dall, and Collins reveal how important those men believed it was to convince others that the Smithsonian had been crucial. The fact that the government appropriations for Baird's department more than doubled in 1867 suggests at least that their claim to have assisted the government may have paid dividends.[293]

After 1867, HBC contributions to the Smithsonian Institution declined dramatically. That fact might be understood to support the argument that the Smithsonian had been interested in the region only to learn more about Russian America, or that the HBC might have felt betrayed after the American purchase of Alaska. However, just as the Smithsonian-HBC cooperation may have contributed little valuable information to American negotiators, the decline of cooperation may have had as little to do with international politics. In fact several factors probably explain the decline. One factor is the reorientation of the Smithsonian's research program to Alaska itself, a program facilitated by the Alaska Commercial Company (which had bought the operations of the Russian American Company) and the US government, which was once more able to assist the Smithsonian after the end of the Civil War.[294] After the United States acquired a territory about which Americans knew little, the Smithsonian was well placed to spearhead the explorations, surveys, collecting, and field work intended to make the territory better known.[295] Kennicott's death on 13 May 1866 may also have played a small part. After Kennicott died, MacFarlane was so saddened that he initially indicated his resolve to quit collecting.[296] Although the death of Kennicott may have dampened the enthusiasm of some, MacFarlane did resume collecting. But the deaths of MacFarlane's aboriginal suppliers were probably more important. Epidemics of measles, influenza, and scarlet fever killed so many of MacFarlane's most important Native collectors in 1866 and 1867 that Fort Anderson was closed. The collapse in MacFarlane's donations following those epidemics suggests how crucial aboriginal people were to the supply network.[297] The 1865 discovery that Ross had used company goods to buy specimens, which led to the decision that visitors to HBC posts had to pay their own way, must also have dampened enthusiasm for scientific contributions. Finally, the Smithsonian's immediate research goals for the Mackenzie River District appear to have been largely met by 1867.

The HBC-Smithsonian cooperation is best understood as being driven by a varied motives of scientists, scientific organizations, the HBC as a corporation, and its directors and employees. American expansionist dreams probably played only a minor role. But the cooperation illustrates many of the central arguments presented in this book. Scientific networks were maintained by the self-interest of the many that were involved in their intricate connections, but really flourished when sophisticated and empathetic scientists stirred the scientific enthusiasm of lay collectors. The Smithsonian's particularly well-managed network produced the most effective program of cooperation between metropolitan scientists and lay hinterland practitioners in the history of the HBC's contributions to science up to 1870. HBC men responded enthusiastically to the Smithsonian's research agenda, ensuring that by 1870 the Smithsonian Institution was the most important source of scientific knowledge pertaining to the HBC territories.

Epilogue

When Robert Kennicott suggested to George Simpson in June 1859 that the HBC should establish a museum in Rupert's Land, Simpson dismissed his proposal as unrealistic. It is impossible to know whether or not Kennicott made the suggestion to others, such as William Mactavish, but on 12 February 1862, while Kennicott was still in the HBC territories, a group of people in the Red River settlement established the "Institute of Rupert's Land."[1] Its executive included representatives of the three main groups expected to contribute to its activities: clergy, settlers, and HBC officers. David Anderson, the Lord Bishop of Rupert's Land, was its founding president, and John Christian Schultz, a recent arrival from Canada, and Governor William Mactavish were its secretaries. At its founding meeting held at the HBC's Fort Garry, Anderson tried to inspire the members of the society by recalling the efforts of Kennicott, Drexler, and others. Then Schultz continued with a more detailed description of the contributions of HBC officers, missionaries, and settlers. "It is impossible," he concluded, "to look upon the foundation of such an Institute, without feeling that here, on the remote confines of civilization, we witness the establishment of an outpost of science, from whence we may look for returns of the highest interest and value."[2] Writing optimistically about the new society, the University of Toronto professor Daniel Wilson noted in the Canadian Institute's journal that "the Institute of Rupert's Land, thus happily inaugurated, includes among its members and correspondents educated men both of the resident clergy, and the officers of the Hudson's Bay Company, stationed at many important points over the vast country ranging from the Pacific to Lake Superior and towards the Arctic Sea."[3] Americans too, noted the establishment of the institution. A circular issued by the institute, signed by Mactavish and Schultz, was published in the Smithsonian Institution's Annual Report for 1861.[4] But the society soon collapsed. In May 1863, William Mactavish admitted to Spencer Baird that "the Rupert's Land Institute has blown up like everything else in the Settlement."[5] "Everything else" had blown up in the colony after the Anglican missionary Owen Griffiths Corbett was arrested in December 1862 for attempting to induce an abortion on his servant girl. The crisis unleashed by the arrest and trial badly divided the elite of Red River and damaged most of its institutions.[6]

The fleeting history of the Institute of Rupert's Land contrasts with the long history of the HBC's chartered monopoly. Still, well before the brief existence of the Institute of Rupert's Land, any perceptive observer could foresee the impending end of the HBC's monopoly. It would have been entirely impolitic for the HBC to attempt to sell its rights to the United States without the permission of the British government, although once the American Civil War ended in 1865 it was not inconceivable that the United States might become a willing buyer. In fact, in 1866 one HBC shareholder published a pamphlet that noted, "Rumors are afloat of an intention on the part of the Americans to make an offer for our lands."[7] Some Americans were interested. In March 1866 the United States House of Representatives passed a resolution seeking a report on commercial relations with British North America.[8] Handed responsibility for preparing the report, James Wickes Taylor, leader of the expansionists in Minnesota, prepared instead a draft bill to provide for the annexation of all of British North America, including a proposal to pay the HBC $10 million for its rights to Rupert's Land.[9] The anonymous HBC pamphleteer argued that "there are many who think that we are entitled to look for the highest bidder, whether British, Canadian, American, or Russian."[10]

Taylor's bill failed to pass, but the American threat spurred the British and Canadian governments to push for the prompt and orderly transfer of Rupert's Land to Canada. Urged on by both governments, several British North American colonies (Canada, Nova Scotia, and New Brunswick) federated in 1867 to form the Dominion of Canada, in part to create a union large and stable enough to negotiate the purchase of Rupert's Land. In March 1867 – the very month in which the American purchase of Russian America was negotiated – the British Parliament passed the British North America Act (1867) that set the stage for the creation of the Dominion of Canada on 1 July. The act included a clause providing for the transfer of Rupert's Land to Canada. Even then, some

American expansionists still believed that Rupert's Land would soon be theirs. On 4 March 1868, the *St. Paul Daily Press* argued that "nothing is more evident than that this region must ultimately constitute part of the United States."[11]

Despite the urgency, Canada and the HBC could not agree on terms for the transfer of Rupert's Land. Finally, on 9 March 1869 the British government stepped in and imposed the conditions of sale, which included a sale price of £300,000 ($1.5 million), a bitter disappointment for those who had bought HBC shares on speculation that the company could command a much higher price.[12] On 22 June 1870, the HBC surrendered its rights to Rupert's Land, and on 15 July 1870 those rights were transferred to the Dominion of Canada.[13] The HBC's monopoly rights were ended, Rupert's Land was no more, and the Canadian expansionists' dream of forming a transcontinental bulwark against American manifest destiny was a step closer to realization.

With the end of its chartered monopoly and exclusive licence, the HBC in 1870 became a private joint-stock company with no quasi-state privileges whatsoever. It was the end of an era. For this reason 1870 marks the endpoint of this study, although 1870 may be a less significant milestone than one might assume. The company diversified its operations into retail sales and other enterprises but it continued to use its fur trading operations to support British, Canadian, and American scientific research.

Perhaps no British scientist benefitted more than Charles Elton, whose access to HBC records permitted him to publish in 1942 the first scientific explication of the cyclical population of lynx and snowshoe hare that continues to intrigue biologists to this day.[14] The HBC also cooperated with research conducted by the government of Canada. Robert Bell, on the first page of his 1872 report on the Geological Survey of Canada's (GSC) survey of the region north of Lake Superior, wrote, "As usual, when carrying on our operations in distant parts of the country, we were indebted to the officers of the Hudson Bay Company for much assistance, accommodation and information."[15] In the following years, such acknowledgments were common in the GSC's annual reports. For example, at the beginning of his summary report for 1880, Alfred R.C. Selwyn, the director of the GSC, mentioned the "kind assistance and facilities afforded by the officers of the Company."[16] And near the beginning of his summary report of the activities of the GSC for 1886, he wrote, "Especial thanks are due from the Survey to Mr. Joseph Wrigley, Chief Commissioner of the Hudson's Bay Company, for letters to the officers at the various posts visited by the parties working in the region north of Lake Superior and around Hudson's Bay, and also to the officers themselves for their uniform kindness and valuable assistance in various ways."[17] In the introductory page of his report on his 1887–8 survey of the region around James Bay, Albert Peter Low wrote, "I desire to tender my thanks to the officers of the Hudson Bay Company, met during the two seasons, all of whom extended to myself and the party the greatest hospitality, and to whose kindly assistance the success of my explorations was, in a great measure, due."[18] The company even renewed its cooperation with the Smithsonian Institution in the 1880s. For example, it supported a visit of Lucien M. Turner of the Smithsonian to the Ungava Bay region in 1882, primarily to study the natural history and weather of that region, although the results of his trip are now particularly valuable for their ethnographic observations. Turner's report, originally published in 1894, has been regarded as sufficiently important to be reissued in 2001.[19] The relationship between the HBC and science in northern Canada after 1870 may well be worth another full-length study.

Conclusion

Science is driven by interests. Ironically, much of the prestige connected with supporting, undertaking, and sharing the results of scientific research is derived from the myth of its disinterestedness, objectivity, or even its philanthropy. The HBC may be among the earlier companies to experience the advantages of being perceived to a patron of science, but there have been many others. To be fair, the HBC's (and any company's) sponsorship of science was only a small part of its efforts to maintain a positive image with influential people. After 1818 the HBC's support of Christian missionaries in its territories may have been more important to its corporate image. But the public tributes paid to the company by ostensibly independent and unbiased intellectual elite did much to enhance the company's reputation as a positive corporate citizen. This book shows that the HBC as a corporation, and many of its directors personally, enjoyed other benefits from supporting science, but that the intangible benefit of tribute was probably paramount.

A host of learned institutions benefited from the HBC's patronage, from the Royal Society to the Smithsonian Institution. In an era in which individuals and institutions had few sources of research money, the HBC's assistance was of great importance. The prestige of several institutions was significantly enhanced as a result of HBC assistance. And HBC sponsorship contributed significantly to the prestige of several notable scientists. When scientists understood how to repay the company and its directors, the company's support became enthusiastic.

But science could not flourish simply because it satisfied the interests of the HBC and elite scientists in the metropolis. Science flourished under the auspices of the HBC because scientific networks satisfied many corporate and individual needs, including those of lay practitioners in the HBC territories. The most typical – at least the most productive – relationships between elite metropolitan savants and laymen in the hinterlands were mutually beneficial relationships in which the lay practitioners experienced a wide range of benefits. "Science is meaningless," Leo Tolstoy is often credited as having said, "because it has no answer to the why questions that matter to us: 'what should we do? How shall we live?'"[1] Oddly, although the *results* of science itself might not answer those questions, many people have evidently answered those questions for themselves by *actually doing* science. Many people – from elite scholars to rank amateurs – found considerable joy and meaning in their scientific studies, and some appear to have been truly happy only in the field. Through science, men found companionship with other men with whom they otherwise may have had little on common. But even many lay practitioners who evidently sought less ethereal benefits – direct payment, published acknowledgment, gifts, renown, or promotion – were willing to devote many hours and much energy to efforts that many around them must have thought strange. But these were among the many ingredients that explain why the HBC's contributions to science were as large as they were.

Despite the burgeoning literature on the social history of science and on the history of scientific networks, and despite the existence of many studies focused on particular aspects of the history of science within chartered monopolies, this book represents the first attempt to explore the history of science in a chartered company in its totality. It confirms Mark Harrison's suggestion that such a study can greatly improve our understanding of the history of science, particularly in the British world. It is possible that the HBC was unusual, that its contributions to science were proportionately much larger than (or smaller than, or just very different from) those of other British – and other European – chartered monopolies. If the HBC's support of science was exceptional, its patronage might partially explain why the company was able to retain its exclusive trading rights as long as it did. But it is also possible that the HBC was not unusual, that the HBC was very much like other corporations (chartered or not) in the past and at present that seize upon the advantages of supporting and being seen to support science. In either case, this book demonstrates that a broad study adds significantly to our understanding of the narrative structure of the history of science in the HBC, contributes to our understanding of transatlantic networks, and sheds new light on the history of science in the British World between 1670 and 1870.

This book also shows that "big picture" approaches advocated by John Pickstone allow historians to discern facets of the significance of particular periods and phenomena in broader context. What it does not yet do, except at the most superficial level, is place the history of science in the HBC in the context of the history of science in other corporations, particularly in the context of other chartered monopolies. In that respect, this book merely ploughs the first furrow. Perhaps it will inspire others to undertake similar studies of other companies – perhaps the various East India Companies, the Royal African Company, the Levant Company, the Muscovy Company, or the Russian American Company – or of chartered companies generally. If so, the existence of the present study will allow other historians to cut deeper and straighter furrows than the one modelled here.

Whether typical or atypical, the history of science in the HBC helps other scholars identify some of the possible lines of inquiry. For example, the history of the HBC identifies a wide range of possible motivations on the part of company directors (both corporately and individually), company employees, and indigenous and local people that induced them to assist and participate in scientific study. It also reveals that the company enjoyed a number of benefits because of its support of science. The most important of these appears to have been the external relations value derived from the public tributes offered by prominent and ostensibly disinterested scientists. The documentary evidence consulted for this study offered tantalizing but indirect evidence about the degree to which the directors became intentional about securing tributes from the scientists they assisted. Future research might locate evidence that some companies developed more explicit policies than any that could be located for the HBC.

This book also identifies many of the ingredients of successful (and less successful) long-distance networks that tied metropolitan and elite scientists with lay scientific practitioners in the hinterlands. It identifies some of the trends (from the proliferation of scientific societies to the growth of economic liberalism) and turning points, including the Transits of Venus, in the history of the HBC that must also have influenced the history of science in other companies. But it also reminds us that researchers must be alert to factors unique to each company. Did other companies have directors analogous to Samuel Wegg, George Simpson, J. H. Pelly, and William Mactavish, captains such as Christopher Middleton, or officers such as Thomas Hutchins, Joseph Colen, George Barnston, John McLoughlin, Roderick MacFarlane, and Bernard R.

Ross? Whether they did or did not, our understanding of those men and of the HBC should improve when we are able to answer that question more clearly.

The HBC was only a small part of the scientific networks that supplied men and institutions such as Joseph Banks, W. J. Hooker, Andrew Murray, Spencer Baird, the Royal Society, Kew Gardens, and the Smithsonian Institution with specimens, evidence, and data from afar. We will understand each of them better only when we have a better sense of how they managed their networks. The assistance of the HBC – or at least the scientific activity carried out in HBC territories – was crucially important to the careers and fame of a number of savants: William Wales, Thomas Pennant, David Douglas, John Richardson, and John Henry Lefroy, for example. There is little doubt that the importance of corporate patronage in the careers of many other scientists is yet to be fully understood. A study such as this one can offer only glimpses into the significance of indigenous and resident peoples in the scientific activity carried out in territories ruled by European chartered companies, but considered in the context of additional evidence gathered in other contexts, historians may use the evidence presented in this book to arrive at deeper understandings than were possible here. Was the culture of science and learning in the HBC unusual? An ability to compare it with the intellectual life in other companies would contribute greatly to the kind of big-picture understandings we still lack.

Not only those historians who explore why and when companies *did* support science, but also those interested in why and when they did *not* support science should benefit from illustrations from the history of the HBC. For much of its history the HBC contributed little to science. Furthermore, some of the research carried out under the auspices of the company undermined rather than furthered the company's interests. Decisions to decline to assist scientists may be as revealing as the decisions to embrace opportunities.

The history of science in the HBC offers an important but incomplete understanding of a phenomenon that is obviously hundreds of years old, but continues to this day: company support for scholarly research and publication. It is a practice more complex – both in the past and in the present – than its more vociferous critics or advocates tend to acknowledge. But given its longstanding importance, it is a tradition that certainly warrants continued exploration and analysis.

Notes

Preface

1 See Barbara Belyea, *Dark Storm Moving West* (Calgary: University of Calgary Press, 2007) for Barbara's report on her research.
2 An early report on my research can be found in Ted Binnema, "Theory and Experience: Peter Fidler and the Transatlantic Indian," in *Native Americans and Anglo-American Culture, 1750–1850: The Indian Atlantic*, ed. Tim Fulford and Kevin Hutchings (Cambridge: Cambridge University Press, 2009), 155–70.

Chapter 1

1 This book generally uses terms relating to science and scientific practitioners (like "natural philosophy," "natural philosopher," "natural historian," "naturalist," "virtuoso," and "savant") that the historical actors themselves would have recognized, but following a custom adopted by Michael Hunter and other social historians of science, and to avoid tiresome and repetitious use of such cumbersome terms, this introductory chapter also uses "science" and "scientist" as they are now understood. Although the meaning of the word *science* changed significantly between 1670 and 1870, and the word *scientist* was not coined until 1834, the small risks of using the term *scientist* even anachronistically in a social history of science such as this one are small. See Michael Hunter, *Science and Society in Restoration England* (Cambridge: Cambridge University Press, 1981), passim, esp. 8. Still here and elsewhere in this introduction some discussion of terms is apt. During the early portion of the period under study (as will be evident in some quoted passages), the word *science* was usually used as the word *knowledge*

is used today. The word *science* assumed a meaning much like today's in the early 1700s. See ibid., 8; and Jorge Cañizares-Esguerra, *Nature, Empire, and Nation: Explorations of the History of Science in the Iberian World* (Stanford, CA: Stanford University Press, 2006), 3. The word *scientist* was coined in 1834 in response to the growing differentiation of the scientific disciplines (chemistry, botany, zoology, and physics, for example) already underway by the 1780s. An excellent exploration of the history of words and terms such as *science, scientist, scientific, savant,* and *natural philosopher,* including a discussion of the significance of the emergence of the term *scientist* can be found in Sydney Ross, "Scientist: The Story of Word," *Annals of Science* 18 (1962): 65–86.

2 Jerry Gaston described "disinterestedness" as one of the longstanding central norms of science. See Jerry Gaston, *The Reward System in British and American Science* (New York: John Wiley and Sons, 1978), 6–7. Indeed, one might go so far as to say that during the late seventeenth century, the myth of disinterestedness came to distinguish intellectual activities that were regarded as "scientific" from those that were not. For an important study of a major turning point in this process, see the seminal study by Steven Shapin and Simon Schaffer, *Leviathan and the Air-Pump: Hobbes, Boyle, and the Experimental Life* (Princeton, NJ: Princeton University Press, 1985).

3 The extent of the HBC's territories and the HBC's governance rights over them require some explanation. Specifically, the distinction between "Rupert's Land" and the "Hudson's Bay Company territories" should be understood. In 1670, King Charles II granted the HBC "the sole Trade and Commerce" of a vaguely defined area described as "all those Seas Streightes Bayes Rivers Lakes Creekes and Soundes ... that lye within the entrance of the Streightes commonly called Hudsons Streightes together with all the Landes Cuntryes and Territoryes upon the Coasts and Confynes of the Seas Streightes Bayes Lakes Rivers Creekes and Soundes aforesaid which are not now actually possessed by any of our Subjectes or by the Subjectes of any other Christian Prince or State." E. E. Rich, ed. *Minutes of the Hudson's Bay Company, 1671–1674* (Toronto: Champlain Society, 1942), 131–2. The charter stipulated that this territory was to be considered one of England's "Colonyes in America called *Ruperts Land*"; ibid., 139. In a move that the HBC's directors probably came to regret, the company secured a seven-year parliamentary confirmation of that charter in 1690. The fact that Parliament never renewed its endorsement of the charter meant that from 1697 onwards the company was always vulnerable to the argument that the charter had expired. That helps explain why the HBC always resisted any suggestion that the charter be submitted to a judicial review. See E. E. Rich,

Hudson's Bay Company: 1670–1870. 2 vols. (London: Hudson's Bay Record Society, 1958–9), 1:57–8, 258. In 1821 the British Parliament did tacitly accept the HBC's argument (which it had not always claimed) that Rupert's Land encompassed the entire drainage basin of Hudson Bay and Hudson Strait (1.4 million square miles). In that year Parliament granted the HBC a twenty-one-year renewable licence to vast additional territories outside of Rupert's Land (often referred to as "Indian Territory"). It is important to understand that the HBC Governor and Committee convinced Parliament that the 1821 licence (and subsequent licences) applied only to territories outside Rupert's Land – that the company's rights to Rupert's Land were undiminished by the licence. This meant that the HBC had possessory rights to the area of Rupert's Land (that it claimed could be sold), rights that it did not have to the rest of its territories. In this study, "Rupert's Land" refers strictly to the territory considered to have been covered by the 1670 charter. "HBC territories" refers more broadly to Rupert's Land plus any other territories over which the company had an exclusive trading licence.

4 Brian Schefke, "The Hudson's Bay Company as a Context for Science in the Columbia Department," *Scientia Canadensis* 31, no. 1–2 (2008): 67–84; Brian Schefke, "Imperial Science: A Naturalist in the Pacific Northwest," *Endeavour* 32, no. 3 (2008): 111–6. In these articles Schefke struggles to explain the HBC's support of scientific enterprises that had no potential to contribute directly to the company's profits. For other companies see Harold J. Cook, *Matters of Exchange: Commerce, Medicine, and Science in the Dutch Golden Age* (New Haven and London: Yale University Press, 2007), 225; Deepak Kumar, "The Evolution of Colonial Science in India: Natural History and the East India Company," in *Imperialism and the Natural World*, ed. John M. MacKenzie (Manchester: Manchester University Press, 1990), 51–66. There is no doubt about the significance of the Hudson's Bay Company as an imperial factor. See John S. Galbraith, *Hudson's Bay Company as an Imperial Factor: 1821–1869* (Berkeley and Los Angeles: University of California Press, 1957). However, the argument in the present study is that the connection between HBC's role as patron of science and its role as imperial factor is attenuated.

5 Huib J. Zuidervaart and Rob H. Van Gent, "'A Bare Outpost of Learned European Culture on the Edge of the Jungles of Java': Johan Maurits Mohr (1716–1775) and the Emergence of Instrumental and Institutional Science in Dutch Colonial Indonesia," *Isis* 95, no. 1 (2004): 3.

6 Cook, *Matters of Exchange*, 176.

7 John Gascoigne, *Science in the Service of Empire: Joseph Banks, the British State and the Uses of Science in the Age of Revolution* (Cambridge: Cambridge University Press, 1998), 144, 146.

8 Mark Harrison, "Science and the British Empire," *Isis* 96 (Mar. 2005): 56. Also see Lewis Pyenson and Susan-Sheets Pyenson, *Servants of Nature: A History of Scientific Institutions, Enterprises and Sensibilities* (London: HarperCollins, 1999), 87.
9 Harrison, "Science and the British Empire," 62. Despite the absence of any monographic study of the history of science in any chartered company, the existing literature on aspects of that history is too large to cite here. Not surprisingly, aspects of the history of science in the Dutch East India Company and the English East India Company have been particularly well examined. For readers of Dutch, a worthwhile study can be found in J. Bethlehem and A. C. Meijer, eds., *VOC en Cultuur: Wetenschappelijke en Culturele Relaties tussen Europa en Azië ten Tijde van de Verenigde Oostindische Compagnie* (Amsterdam: Schiphouwer & Brinkman, 1993). Faced with a lack of monographic studies of the English East India Company, readers might begin with Deepak Kumar, *Science and the Raj: A Study of British India* 2nd ed. (New York: Oxford University Press, 2006); and Zaheer Baber, *The Science of Empire: Scientific Knowledge, Civilization, and Colonial Rule in India* (Albany: State University of New York Press, 1996). Some studies that explore the history of science in colonial empires include Lewis Pyenson, *Empire of Reason: Exact Sciences in Indonesia, 1840–1940* (Leiden: Brill, 1989); ibid., *Cultural Imperialism and Exact Sciences: German Expansion Overseas, 1900–1930* (New York: Lang, 1985); and ibid., *Civilizing Mission: Exact Sciences and French Overseas Expansion, 1830–1940* (Baltimore and London: Johns Hopkins University Press, 1993).
10 Despite its theoretical, quasi-sovereign jurisdiction over its territories, the HBC never attempted to treat indigenous people in its territories as "subjects."
11 The first public natural history museum in England was established in 1683 in Oxford. Edward P. Alexander, *Museums in Motion: An Introduction to the History and Functions of Museums* (Nashville: American Association for State and Local History, 1979), 8, 42.
12 For discussions of the history of curiosity, including its connections with Europeans and masculinity, see Susan Scott Parrish, *American Curiosity: Cultures of Natural History in the Colonial British Atlantic World* (Chapel Hill: University of North Carolina Press, 2006), 25–64, esp. 63; Barbara M. Benedict, *Curiosity: A Cultural History of Early Modern Inquiry* (Chicago: University of Chicago Press, 2001), esp. 118–21; and Lorraine Daston and Katherine Park, *Wonders and the Order of Nature, 1150–1750* (New York: Zone Books, 1998). Also see Katie Whitaker, "The Culture of Curiosity," in

Cultures of Natural History, ed. N. Jardine, J.A. Secord and E.C. Spary (Cambridge: Cambridge University Press, 1996), 75–90.
13 Galbraith, *Hudson's Bay Company as an Imperial Factor*, 14.
14 Ibid., 3.
15 Article-length studies of particular aspects of the history of science in the HBC include R.H.G. Leveson Gower, "HBC and the Royal Society," *The Beaver* 256, no. 2 (Sept. 1934): 29–31, 66; R.P. Stearns, "The Royal Society and the Company," *The Beaver* 276, no. 1 (1945): 8–13; James L. Baillie Jr, "Naturalists on Hudson Bay" *Beaver* 277 (Dec. 1946): 36–9; H.C. Diegnan, "The HBC and the Smithsonian," *The Beaver* 278 (June 1947): 3–7; Robert Kerr, "For the Royal Scottish Museum," *Beaver* 284 (June 1953): 32–5; Richard I. Ruggles, "Governor Samuel Wegg, Intelligent Layman of the Royal Society, 1753–1802," *Notes and Records of the Royal Society of London* 32, no. 2 (1978): 181–99; Greg Thomas, "The Smithsonian and Hudson's Bay Company," *Prairie Forum* 10, no. 2 (1985): 283–306; Suzanne Zeller, "The Spirit of Bacon: Science and Self-Perception in the Hudson's Bay Company, 1830–1870," *Scientia Canadensis* 13 (Fall/Winter 1989): 79–101; Rita Griffin-Short, "The Transit of Venus: The Heavens from Hudson Bay, 1769," *The Beaver* 83, no. 2 (Apr./May 2003): 8–11; Schefke, "Hudson's Bay Company as a Context for Science"; and Schefke, "Imperial Science," 111–6. An old and brief survey still worth consulting can be found in R.P. Stearns, *Science in the British Colonies of America* (Urbana: University of Illinois Press, 1970), 247–57. Book-length studies include Debra Lindsay's two fine explorations of the relationship between the HBC and the Smithsonian Institution: Debra Lindsay, *The Modern Beginnings of Subarctic Ornithology: Northern Correspondence to the Smithsonian Institution, 1856–1868* (Winnipeg: Manitoba Record Society, 1991); and Debra Lindsay, *Science in the Subarctic: Trappers, Traders and the Smithsonian Institution* (Washington: Smithsonian Institution Press, 1993). For a book-length study geared towards the HBC's contributions to natural history in the eighteenth century see C. Stuart Houston, Tim Ball, and Mary Houston, *Eighteenth-Century Naturalists of Hudson Bay* (Montreal and Kingston: McGill-Queen's University Press, 2003). Trevor Harvey Levere's examination of the history of science in the North American arctic includes discussion of the HBC's involvement. See his *Science and the Canadian Arctic: A Century of Exploration, 1818–1918* (Cambridge: Cambridge University Press, 1993). Other fur trading companies are less well served. The history of science in the Russian American Company would be a particularly interesting comparison with the HBC because it too was a chartered fur-trading monopoly (albeit one

under much greater government control) operating in northern temperate regions, but as yet has been the subject of only a preliminary survey found in E.A. Oklandnikova, "Science and Education in Russian America," in *Russia's American Colony*, ed. Frederick S. Starr (Durham: Duke University Press, 1987), 218–35.

16 John V. Pickstone, *Ways of Knowing: A New History of Science, Technology and Medicine* (Chicago: University of Chicago Press, 2000), 5.

17 See Parrish, *American Curiosity*; James Delbourgo, *A Most Amazing Scene of Wonders: Electricity and Enlightenment in Early America* (Cambridge, MA: Harvard University Press, 2006); and Andrew J. Lewis, *A Democracy of Facts: Natural History in the Early Republic* (Philadelphia: University of Pennsylvania Press, 2010). An excellent collection of articles on the history of scientific networks in the Atlantic world (although one that includes nothing on the North America north of the present-day United States) can be found in James Delbourgo and Nicholas Dew, eds., *Science and Empire in the Atlantic World* (New York: Routledge, 2008). This book is also intended to revive interest amongst historians about the transatlantic world and the British Empire in the HBC and its territories. A perusal of any history of the transatlantic world or of the British Empire published in the last twenty years will reveal that the HBC, Rupert's Land, and HBC territories, of significant interest to historians until the 1980s, have all but dropped out of their consciousness. The many studies published by Glyndwr Williams, E. E. Rich, and John S. Galbraith before 1980 testify to this past interest. That the prominent historian Jennifer Brown found historians at Oxford University in 2002 unfamiliar with Rupert's Land, and that recently published histories of the British Empire make no meaningful mention of the HBC (including a map of the British Empire in 1830 that is wholly inaccurate in its portrayal of British North America at the time) and which assert that the HBC was "founded in 1670 to monopolize the trade in Canadian furs," is only symptomatic of the general neglect and ignorance of the company and its territories. See Jennifer S.H. Brown, "Rupert's Land, Nituskeenan, Our Land: Cree and English Naming and Claiming around the Dirty Sea," in *New Histories for Old*, ed. Ted Binnema and Susan Neylan (Vancouver: UBC Press, 2007), 18; Sarah Stockwell, *The British Empire: Themes and Perspectives* (Malden, MA: Blackwell, 2008), xviii; Niall Ferguson, *Empire: The Rise and Demise of the British World Order and the Lessons for Global Power* (New York: Basic Books, 2004), 16–17 (quote). One might expect that because the findings of most historians of the Hudson's Bay Company and its territories published since 1975 tend to undermine some of the generalizations found in the historiography of the

Notes to page 8   305

transatlantic world and the British Empire, scholars in those fields would be fascinated by those findings, but the opposite appears to be the case.

18 Mark Harrison, "Science and the British Empire," 56. Recent books that have elaborated our understanding of early American science include Parrish, *American Curiosity*; Lewis, *Democracy of Facts*; Joyce E. Chaplin, *Subject Matter: Technology, the Body, and Science on the Anglo-American Frontier, 1500–1676* (Cambridge, MA: Harvard University Press, 2001); Delbourgo, *A Most Amazing Scene*; and Joyce E. Chaplin, *The First Scientific American: Benjamin Franklin and the Pursuit of Genius* (New York: Basic Books, 2006) Nevertheless, the recent literature suggests that the myth of American exceptionalism is remarkably durable. Older, but still useful studies can be found in John C. Greene, "American Science Comes of Age, 1780–1820," *Journal of American History* 55 (1968): 22–41; and John C. Greene, *American Science in the Age of Jefferson* (Ames: Iowa State University Press, 1983).

19 Greene, *American Science in the Age of Jefferson*, 12. Also see Stearns *Science in the British Colonies*.

20 The number of employees of the company peaked at the time of the HBC's merger with the North West Company in 1821, when the HBC had about 1,983 employees. By 1825, the number had been reduced to 827; John S. Galbraith, *The Little Emperor: Governor Simpson of the Hudson's Bay Company* (Toronto: Macmillan, 1976), 60. In 1861 the number of permanent employees was about 1,400; [A.K. Isbister], "The Hudson's Bay Territories, Their Trade, Productions and Resources; with Suggestions for the Establishment and Economical Administration of a Crown Colony on the Red River and Saskatchewan," *Journal of the Society of Arts* 9 (1 Mar. 1861): 233. Even more remarkable is the small number of officers, surgeons, and clerks (those more likely to contribute to science) employed by the HBC. The hierarchy of officers was complex and changing before 1821, but became more permanent and simple after 1821. The Deed Poll of 1821 created two ranks of commissioned officers: Chief Factors (of whom there would be 25), and Chief Traders (of whom there would be 28); Rich, *Hudson's Bay Company*, 2: 406; Glyndwr Williams, "The Hudson's Bay Company and the Fur Trade: 1670–1870," *Beaver* 314, no. 2 (Autumn, 1983), 50. In 1837 Governor George Simpson indicated that the company had 25 Chief Factors, 27 Chief Traders, 152 clerks, 1200 regular servants, plus "a great number" of "natives" who provided "occasional labour." Great Britain, Parliament, House of Commons, *Copy of the Existing Charter or Grant by the Crown to the Hudson's Bay Company; and Correspondence on the Last Renewal of the Carter, &c.* 1842 (547), 17. It appears that the number of commissioned officers never rose above 53. In 1861, Alexander Kennedy Isbister estimated that, aside from the

commissioned officers, there were 5 surgeons, 87 clerks, and 67 postmasters. He also estimated that the company maintained 152 factories and posts, 65 of them in Rupert's Land. He estimated that the company traded with about 150,000 aboriginal people; [Isbister], "The Hudson's Bay Territories," 233.

21 Nicholas Dew, "Vers la Ligne: Circulating Measurements around the French Atlantic," in *Science and Empire in the Atlantic World*, ed. James Delbourgo and Nicholas Dew, (New York: Routledge, 2008), 60.

22 Kumar, "Evolution of Colonial Science in India," 63.

23 Indeed, after 1812 many retiring officers elected to retire at the small colony at Red River rather than return to Great Britain with their aboriginal wives and mixed-blood families.

24 Zuidervaart and Van Gent, "A Bare Outpost," 25, 31. According to Harold Cook, in the VOC, 10 to 15 percent of all on board a ship travelling between Holland and the Dutch East Indies died on an average trip, and "roughly two of three never made it home"; Cook, *Matters of Exchange*, 178.

25 Zuidervaart and Van Gent, "A Bare Outpost," 30.

26 Ibid., 25.

27 The HBC lost only three ships in its history, none after 1736; Ann M. Carlos and Frank D. Lewis, *Commerce by a Frozen Sea: Native Americans and the European Fur Trade* (Philadelphia: University of Pennsylvania Press, 2010), 45, 210n31. Overland travel was hazardous, but I have found evidence of only one incident (discussed in Chapter 6 of this study) in which two scientists (Robert Wallace and Peter Banks) died along with ten others while travelling with the HBC, apparently because Wallace destabilized a boat after panicking while the brigade was running a rapids.

28 York Factory was closer to London than was Charleston, South Carolina, and not much more than a quarter of the distance from London to India.

29 The description of the HBC as "mean and selfish" can be found in Richard Glover, ed., *David Thompson's Narrative 1784–1812* (Toronto: Champlain Society, 1962), 132. The present study will show that allegations of unreasonable HBC secrecy appeared repeatedly for most of the company's history between the 1730s and 1870s. As late as the 1850s, Alexander Kennedy Isbister argued that the HBC's aim was "to shroud their transactions in the most impenetrable mystery"; Barry Cooper, *Alexander Kennedy Isbister: A Respectable Critic of the Honourable Company* (Ottawa: Carleton University Press, 1988), 181. Officials in the Colonial Office appear to have been unconvinced of Isbister's characterizations at the time, but some scholars have been convinced. Cooper (27) argued that the company was as secretive in the nineteenth century as it had been in the eighteenth.

30 I use the term *tribute* (rather than *thanks*, or *acknowledgment*) because of its dual connotations. *Tribute*, derived from the same Latin word that gave us the word *tribe*, can refer to an actual payment (from tribe to tribe, for example) rooted in some kind of mutually acknowledged obligation. Today, *tribute* more commonly refers to a statement of respect, thanks, and admiration. Even in that context, we still normally describe tribute as being "paid." Thanks might be offered and acknowledgment given, but tribute is paid. It is worth noting that *contribute* (which is what scholars are expected to do), is rooted in the same word, but with a prefix that implies a communal activity.

31 W.O. Hagstrom, "Gift-giving as an Organizing Principle in Science," in *Sociology of Science: Selected Readings*, ed. Barry Barnes (New York: Penguin Books, 1972), 106.

32 Lewis Pyenson has argued that "a focus on the 'pure,' exact sciences offers the possibility of distinguishing economic motivations from cultural ones. ... An astronomical or geophysical observatory extends no promise of immediate financial gain; its establishment relates to general, cultural objectives and to questions of international prestige." Pyenson, *Empire of Reason*, xiv.

33 Pyenson and Sheets-Pyenson, *Servants of Nature*, 321.

34 Nicholas Dew, "Vers la Ligne: Circulating Measurements around the French Atlantic," in Delbourgo and Dew, *Science and Empire*, 58–9.

35 "Life in a Hudson Bay Company's Fort," *Chambers's Journal of Popular Literature, Science, and Art* 828 (8 Nov. 1879): 706.

36 Joseph Henry, Washington, DC, to Thomas Fraser, Secretary, Hudson Bay House, London, England, 29 Oct. 1860, Hudson's Bay Company Archives, Provincial Archives of Manitoba (hereafter HBCA) A.10/48: 459.

37 Galbraith, *Hudson's Bay Company as an Imperial Factor*, 18, 236–8.

38 Glyndwr Williams, ed. *London Correspondence Inward from Sir George Simpson, 1841–2* (London: Hudson's Bay Record Society, 1973), xii. Galbraith added that Simpson "cherished fame." Galbraith, *Little Emperor*, 11.

39 George Simpson, *Narrative of a Journey Round the World, during the Years 1841 and 1842* (London: Henry Colburn, 1847); Williams, *London Correspondence Inward from Sir George Simpson*, 187–9; Zeller, "Spirit of Bacon," 91.

40 "List of Members," *Journal of the Royal Geographical Society* 1 (1831): xviii, xvi. In the second volume of the journal the HBC (as well as Nicholas Garry individually) was listed as one of the society's donors; Ibid. 2 (1832): vi. The Royal Geographical Society was soon a prestigious journal with ties to the elite. Even after the formation of the first anthropological society

in Great Britain in 1843 (the Ethnological Society of London), the Royal Geographical Society included anthropology within its purview.

41 Roy MacLeod has defined "metropolitan science" as "not just the science of Edinburgh or London, or Paris or Berlin, but a way of *doing* science, based on learned societies, small groups of cultivators, certain conventions of discourse, and certain theoretical priorities set in eighteenth-century Western Europe." Colonial science, by contrast, was "fact-gathering," a "derivative science, done by lesser minds working on problems set by savants in Europe." See Roy MacLeod, "On Visiting the 'Moving Metropolis': Reflections on the Architecture of Imperial Science," in *Scientific Colonialism: A Cross-Cultural Comparison*, ed. Nathan Reingold and Marc Rothenberg (Cambridge: Cambridge University Press, 1987), 220. More pithily (on p. 219), MacLeod summarized the distinction this way: "metropolitan science *was* science ... colonial science ... was ... imperial science seen from below."

42 For a discussion of the importance of distinguishing between centrality and peripherality and geographical location, see David Wade Chambers and Richard Gillespie, "Locality in the History of Science: Colonial Science, Technoscience, and Indigenous Knowledge," *Osiris* 15 (2000): 223–9; and David N. Livingstone, *Putting Science in Its Place: Geographies of Scientific Knowledge* (Chicago: University of Chicago Press, 2003).

43 Harold J. Cook, "Physicians and Natural History," in Jardine, Secord, and Spary, *Cultures of Natural History*, 99.

44 Livingstone, *Putting Science in Its Place*, 40–2; Delbourgo, *A Most Amazing Scene*, 18–20, 282–3.

45 Under MacLeod's definitions, some scientists in the colonies would necessarily be categorized among the metropolitan scientists. For example, the American colonial inventor and scientist Benjamin Franklin would have to be categorized as a metropolitan scientist. He won the Royal Society of London's Copley Medal in 1753, was elected a Fellow of the Royal Society in 1756, and conducted experiments of high priority for elite scientists in Europe. The same is true of Thomas Jefferson. No one in the HBC ever matched the scientific credentials of Franklin or Jefferson, but the work of some would have fit MacLeod's definition of metropolitan science. Captain Christopher Middleton, for example, was elected Fellow of the Royal Society on 7 Apr. 1737, and won the Copley Medal in 1742 (Stearns, *Science in the British Colonies*, 251). Thomas Hutchins, another HBC servant who won the Copley Medal in 1783 was arguably another. William Wales and Joseph Dymond were elite scientists carrying out hinterland science when they observed the Transit of Venus at Churchill and

carried out other researches during their stay at Fort Churchill in 1768–9. Conversely, MacLeod's definitions of metropolitan and colonial science would also force us to categorize the vast majority of lay practitioners and collectors who collected specimens and data in Europe and Britain as "colonial scientists." Indeed, during the eighteenth century, a century particularly marked by the contributions of "lay scientists" in England, many people in England participated in science in precisely the way MacLeod describes as "colonial science." This particularly was true of many people who gathered data relating to natural history, meteorology, and phenology.

46 Livingstone, *Putting Science in Its Place*, 43.
47 For example, a biography of the Hookers gives scant attention to the global network of collectors upon which the father and son relied; Mea Allan, *The Hookers of Kew, 1785–1911* (London: Joseph, 1967). A more recent biography of his son, Joseph Hooker, pays more attention to collecting networks, but Joseph Hooker seems to have been unconnected with the HBC; Jim Endersby, *Imperial Nature: Joseph Hooker and the Practices of Victorian Science* (Chicago: University of Chicago Press, 2008).
48 Bernard I. Cohen, "The New World as a Source of Science for Europe," *Actes du IXe Congrès Internationale d'Histoire des Sciences: Barcelona, Madrid, 1–7 Septembre 1959* (Barcelona: Asociación par Historia Ciencia Española, 1960), 96.
49 Zuidervaart and Van Gent, "A Bare Outpost," 32.
50 James Delbourgo and Nicholas Dew, "Introduction: The Far Side of the Ocean," in Delbourgo and Dew, *Science and Empire*, 5.
51 Parrish, *American Curiosity*, 8. Parrish (10) described colonial British American networks as "polycentric." The HBC networks might be described as polycentric insofar as London, Edinburgh, Washington, and Montreal were the major centers of its networks, and insofar as its networks did not exhibit the hierarchical characteristics that the center-periphery image has often implied. Still, it would be misleading to imply that the HBC territories formed a "center" in polycentric networks.
52 Delbourgo and Dew, "Introduction," in Delbourgo and Dew, *Science and Empire*, 12.
53 Ibid., 10. For an introduction see Steven J. Harris, "Networks of Travel, Correspondence, and Exchange," in *The Cambridge History of Science*, ed. Katharine Park and Lorraine Daston, vol. 3, *Early Modern Science* (Cambridge: Cambridge University Press, 2006), 341–62; and Steven J. Harris, "Long-Distance Corporations, Big Sciences, and the Geography of Knowledge," *Configurations* 6 (1998): 269–304.

54 Delbourgo and Dew, "Introduction," in Delbourgo and Dew, *Science and Empire*, 15.

55 Carl Berger makes this argument, although not about the HBC, in *Science, God, and Nature in Victorian Canada* (Toronto: University of Toronto Press, 1983), 3.

56 Margaret C. Jacob, "Afterword: Science, Global Capitalism, and the State," in Delbourgo and Dew, *Science and Empire*, 333. She further argued (333) that "travelers through Europe's extended commercial networks seemed to know instinctively that gathering, collecting, processing, and then theorizing about what was being seen in the Atlantic world could only bring benefits for themselves."

57 Galbraith, *Little Emperor*, 22. Governor George Simpson's "Character Book," in which he described the character of Chief Factors, Chief Traders, clerks, and postmasters in the HBC illustrates the importance Simpson placed on education, learning, and studiousness among officers and prospective officers. See Glyndwr Williams, ed., *Hudson's Bay Miscellany, 1670–1870* (London: Hudson's Bay Record Society, 1975), 151–236. It is important to understand, however, that Simpson's character book reflects the emphasis on education found throughout the HBC's abundant (and under-researched) personnel records which show that men deficient in education were not regarded as worthy of being put in charge of major trading posts. A perusal, for example, of the York Factory Lists of Servants, 1783–1885 (HBCA B.239/f), reveals this emphasis. In the period of rapid expansion (1774–1821) the company sometimes found it necessary to put uneducated or even illiterate men in charge of posts. However, the files in HBCA B.239/f show that the relatively poorly educated men hired for their energy between 1810 and 1821 were especially likely to be laid off after 1821. Men described as "well-educated" were relatively likely to be promoted.

58 The term "Canada" is often applied anachronistically and incorrectly to the HBC territories. This must be because in 1870 the HBC sold most of the former Rupert's Land to Canada, although those small portions of Rupert's Land south of the forty-ninth parallel were ceded to the United States in 1818, and now form parts of four states of the United States. Most of the rest of the HBC territories also eventually became part of Canada, although several other states of the United States now possess other areas of the post-1821 HBC territories. Since Canada did not acquire any portion of the HBC territories before 1870, it would be inappropriate to use the term "Canada" to refer to those territories. In this study, the term "Canada" refers to the territory or political jurisdiction that the historical actors would have recognized as "Canada."

59 Daniel Williams Harmon, *A Journal of Voyages and Travels in the Interiour* [sic] *of North America* (Andover: Flagg and Gould, 1820), 229.
60 Helen E. Ross, *Letters from Rupert's Land, 1826–1840* (Montreal and Kingston: McGill-Queen's University Press, 2009), 99.
61 Robert Kennicott to Spencer Fullerton Baird, 17 Nov. 1859, in Lindsay, *Modern Beginnings of Subarctic Ornithology*, 29. For another account of the typical evening hours "which drag most wearily upon each individual" at trading posts, see "Life in a Hudson's Bay Company's Fort," 706.
62 As the following account suggests, the directors developed an informal policy of encouraging intellectual pursuits and high-minded hobbies among all of its employees, but it not evident that they were encouraged, or that officers pursued scientific activity more than other genteel forms of leisure, except when the directors wished to respond to specific requests from savants. It seems that the company sought to encourage any kind of intellectual (and spiritual) leisure activity that promoted company and personal discipline and a corporate ethos.
63 Quoted in Grace Lee Nute, "Kennicott in the North" *Beaver* (Sept 1943): 30. Geyer refers to Charles Lyell's very influential *Principles of Geology*, 3 vols. (1830–33), and presumably to *The Asiatic Journal and Monthly Register for British India and its Dependencies*. Peter Skene Ogden eventually published *Traits of American-Indian Life and Character* (London: Smith, Elder & Co., 1853). Geyer refers to Fort Vancouver, near the present-day city of Vancouver, Washington, USA.
64 Quoted in Grace Lee Nute, "A Botanist at Fort Colville," *The Beaver* 277 (Sept. 1946): 29.
65 Ibid. For more on McDonald, see Olive and Harold Knox, "Chief Factor Archibald McDonald," *The Beaver* 274 (Mar. 1944): 42–6.
66 Jean-Baptiste-Zacharie Bolduc, "Mission of the Columbia" (trans. Tess E. Jennings, 1937), Jean-Baptiste Bolduc fonds: 1843–1845, British Columbia Archives, MS-0580, pp. 5–6.
67 The literature on these libraries is already large. The most noteworthy studies are Michael R. Angel, "Clio in the Wilderness: or Everyday Reading Habits of the Honourable Company of Merchant Adventurers Trading Into Hudson's Bay," *Manitoba Library Association Bulletin* 19, no. 3 (June 1980): 14–19; Jean Murray Cole, "Keeping the Mind Alive: Literary Leanings in the Fur Trade," *Journal of Canadian Studies* 16, no. 2 (Summer 1981): 87–93; Michael Payne and Gregory Thomas, "Literacy, Literature, and Libraries in the Fur Trade," *The Beaver* 313, no. 4 (Spring 1983): 44–53; Judith Hudson Beattie, "'My Best Friend': Evidence of the Fur Trade Libraries Located in the Hudson's Bay Company Archives," *Épilogue* 8,

no. 1–2 (1993): 1–32; Debra Lindsay, "Peter Fidler's Library: Philosophy and Science in Rupert's Land" in *Readings in Canadian Library History*, ed. Peter F. McNally, (Ottawa: Canadian Library Association, 1986), 209–29; and Leslie D. Castling, "The Red River Library: A Search After Knowledge and Refinement," in McNally, *Readings in Canadian Library History*, 153–66. Also see C.E. L'Ami, "Priceless Books form Old Fur Trade Libraries," *Beaver* 266 (Dec. 1935): 26–9; "The Old Library of Fort Simpson," *Beaver* 5 (Dec. 1924): 20–1; "Books of the North," *Beaver* 288 (Winter 1957): 60–1. To this literature I hope to add contributions of my own in another book on Peter Fidler and the history of science in the HBC.
68 "Life in a Hudson's Bay Company's Fort," 707.
69 John R. Millburn, *Benjamin Martin, Instrument Maker and 'Country Showman'* (Leiden: Noordhoff, 1976), 1, 44. For the early American context, see Delbourgo, *A Most Amazing Scene*, 87–102.
70 "Thomas Hutchins' Manuscript Accompanying Bird and Mammal Specimens Submitted to England from York Factory, 28 Aug. 1772," ed. C. Stuart Houston, Supplementary Document # 2 to Stuart Houston, Tim Ball, and Mary Houston, *Eighteenth-Century Naturalists of Hudson Bay*. More prosaically, James Isham appears to have written his "Observations" to avoid sheer boredom. See E. E. Rich, ed., *James Isham's Observations on Hudsons Bay, 1743 and Notes and Observations on a Book Entitled A Voyage to Hudsons Bay in the Dobbs Galley, 1749* (London: Hudson's Bay Record Society, 1949), lxv.
71 Lorraine Daston, "Afterword: The Ethos of Enlightenment," in *The Sciences in Enlightened Europe*, ed. William Clark, Jan Golinski, and Simon Schaffer (Chicago: University of Chicago Press, 1999), 500. Daston's argument was made about the Enlightenment specifically, but was true both before and after that period.
72 Williams, *Hudson's Bay Miscellany*, 163–4.
73 Circular of George Simpson to the Officers of the Hudson's Bay Company, 31 Mar. 1860, as printed in the *Smithsonian Miscellaneous Collections*, 2 (1862): Article 8, 45.
74 Cohen, "The New World as a Source of Science for Europe," 97.
75 Vladimir Janković, "The Place of Nature and the Nature of Place: The Chorographic Challenge to the History of British Provincial Science," *History of Science* 38 (2000): 93.
76 Benedict, *Curiosity*, 158.
77 HBC posts were stratified. Even in the late nineteenth century, officers ate separately from servants, and followed different routines. For a particularly evocative account, see "Life in a Hudson's Bay Company's Fort."

78 See Lindsay, *Modern Beginnings of Subarctic Ornithology*, viii. The connection between science and standing was not unique in colonial contexts. Individuals in the VOC who did contribute to science, such as Johan Maurits Mohr, found that their work enhanced their reputations in society. See Zuidervaart and Van Gent, "A Bare Outpost," 1–33 (quoted passage is on p. 3).
79 J. Russell Harper, *Paul Kane's Frontier* (Toronto: University of Toronto Press, 1971), 31.
80 Peter Skene Ogden to Paul Kane, 18 Mar. 1851, as published in Harper, *Paul Kane's Frontier*, 335.
81 "Life in a Hudson Bay Company's Fort," 707.
82 Bernard R. Ross to Professor Henry, 30 Nov. 1859, as quoted in *Smithsonian Institution Annual Report* (hereafter *SIAR*), *1859*, 116.
83 An evocative description of the social life in a typical HBC post can be found in "Life in a Hudson Bay Company's Fort." Also see Walter N. Sage, "Life at a Fur Trading Post in British Columbia a Century Ago," *Washington Historical Quarterly* 25, no. 1 (Jan. 1934): 11–22.
84 For a discussion of the role of aboriginal and local peoples in other knowledge networks in British colonies, see Delbourgo, *A Most Amazing Scene*, 183–99.
85 Although the literature has proliferated, the best introduction to the indigenous people's economic relations with the HBC remains Arthur J. Ray, *Indians in the Fur Trade: Their Role as Trappers, Hunters, and Middlemen in the Lands Southwest of Hudson Bay, 1660–1870* (Toronto: University of Toronto Press, 1974); and the best introduction to social relations remains Jennifer S.H. Brown, *Strangers in Blood: Fur Trade Company Families in Indian Country* (Vancouver: UBC Press, 1980). For a discussion of the British aversion to intermarriage and sexual intimacy in early colonial America, see Chaplin, *Subject Matter*, 180–91.
86 George Simpson, Lachine to John F. [sic] Audubon, 24 Jan. 1846, HBCA D.4/67 p. 583 (fo. 300).
87 Letter transcribed into the Edmonton House Post Journal, 22 Nov. 1808; HBCA B.60/a/8. It is not possible to know what the Committee did with the gift, but it seems likely that the Committee would have been willing to show it to learned men, or may have donated it to an institution.
88 Delbourgo and Dew, "Introduction," in Delbourgo and Dew, *Science and Empire*, 6.
89 Ross, "Scientist: The Story of Word." John Pickstone, explains that *natural history* referred to (and can still be used to referred to) the collection, description, inventorying, and classification of natural phenomena. *Natural*

philosophy referred more to the analysis and explanation of scientific evidence. See Pickstone, *Ways of Knowing*, passim, esp. 60.

90 William H. Goetzmann, "Paradigm Lost," in *The Sciences in the American Context*, ed. Nathan Reingold,(Washington, DC: Smithsonian Institution Press, 1979), 25. Also see Ian Jackson, "Exploration as Science: Charles Wilkes and the U.S. Exploring Expedition, 1838–42," *American Scientist* 73 (Sept.–Oct. 1985): 450–61.

91 Cañizares-Esguerra, *Nature, Empire, and Nation*, 4.

92 Delbourgo and Dew, "Introduction," in Delbourgo and Dew, *Science and Empire*, 2, 6.

93 Sverker Sörlin, "Ordering the World for Europe: Science as Intelligence and Information as Seen from the Northern Periphery" *Osiris* 15 (2000): 54. This helps explain why both the Royal Society of London and the HBC can be considered manifestations of Baconianism; Zeller, "Spirit of Bacon," esp. 80.

94 Sörlin, "Ordering the World for Europe," 54.

95 Francisco Hernández's trip to New Spain in 1570 has been credited as the first scientific voyage. Juan Pimentel, "The Iberian Vision: Science and Empire in the Framework of a Universal Monarchy, 1500–1800," *Osiris* 15 (2000): 22. The first British scientific voyage may have been Edmond Halley's voyages in the *Paramore* around the turn of the eighteenth century. Thus James P. Ronda exaggerated the significance of Cook when he wrote that of Cook's time that "voyages of discovery that once had been solely for profit or imperial advantage now became scientific enterprises as well"; James P. Ronda, "'A Chart in His Way': Indian Cartography and the Lewis and Clark Expedition," *Great Plains Quarterly* 4, no. 1 (1984): 4–18.

96 Mary Louise Pratt has described "strategies of representation whereby European bourgeois subjects seek to secure their innocence in the same moment as they assert European hegemony" as "anti-conquest." See Mary Louise Pratt, *Imperial Eyes: Travel Writing and Transculturation* (New York: Routledge, 1992), 7. Given that conquest and imperialism did not carry the stigma then that they do now, I suspect that the strategies were more "pro-humanity" and "pro-scientific" than "anti-conquest."

97 Pickstone, *Ways of Knowing*, 60–82; Nicholas Jardine and Emma Spary, "The Natures of Cultural History," in Jardine, Secord, and Spary, *Cultures of Natural History*, 3; Lewis, *Democracy of Facts*, 11.

98 Cohen, "The New World as a Source of Science for Europe," 98.

99 Karen Ordhal Kupperman, "The Puzzle of the American Climate in the Early Colonial Period" *American Historical Review* 87 (1982): 1262–89; Antonello Gerbi, *The Dispute of the New World: The History of a Polemic*,

1750–1900, rev. and enl. ed. translated by Jeremy Moyle. (Pittsburgh: University of Pittsburgh Press, 1973).

100 The best general history of the HBC remains Rich, *Hudson's Bay Company*. For an excellent brief history see Glyndwr Williams, "The Hudson's Bay Company and the Fur Trade: 1670–1870," *Beaver* 314, no. 2 (Autumn, 1983). One-volume general histories of the region in this period can be found in A. S. Morton, *A History of the Canadian West to 1870–71*, 2nd ed. (Toronto: University of Toronto Press, 1973), and E. E. Rich, *The Fur Trade and the Northwest to 1857*. (Toronto: McClelland and Stewart, 1967); and an excellent general history of the fur trade in Canada, with excellent coverage of the HBC, is Harold A. Innis, *The Fur Trade in Canada: An Introduction to Canadian Economic History*, rev. ed. (Toronto: University of Toronto Press, 1956). A recent economic history of the company that draws on both the pre-1970 corporate histories and the post-1970 studies emphasizing aboriginal peoples is Carlos and Lewis, *Commerce by a Frozen Sea*. A history of labour within the HBC can be found in Edith I. Burley, *Servants of the Honorable Company: Work, Discipline, and Conflict in the Hudson's Bay Company, 1770–1879* (Toronto: Oxford University Press, 1997). None of these general histories contain any substantial reference to the history of science in the HBC.

101 These years are discussed in the first volume of Rich, *Hudson's Bay Company*.

102 Ibid., 2: 1–385; Morton, *History of the Canadian West*, 256–622.

103 Rich, *Hudson's Bay Company*, 2: 112–33; Morton, *History of the Canadian West*, 300–55.

104 Rich, *Hudson's Bay Company*, 2: 134–56, 169–85, 215–31; Morton, *History of the Canadian West*, 508–30.

105 Rich, *Hudson's Bay Company*, 2: 288–352; Morton, *History of the Canadian West*, 531–99.

106 Rich, *Hudson's Bay Company*, 2: 353–400; Morton, *History of the Canadian West*, 600–22.

107 James Cook, *A Voyage to the Pacific Ocean: Undertaken, by the Command of His Majesty, for Making Discoveries in the Northern Hemisphere*. (London: G. Nicol and T. Cadell, 1784), xliv–xlv.

108 Rich, *Fur Trade and the Northwest*, 236–8.

109 The period between 1821 and 1870 is covered in Rich, *Hudson's Bay Company*, 2: 401–937.

110 Galbraith, *Little Emperor*, 87.

111 Williams, *London Correspondence Inward from Sir George Simpson*, xii.

112 Galbraith, *Hudson's Bay Company as an Imperial Factor*, 178.

113 The licence was due to expire in 1842, but the HBC secured an early renewal in 1838. That meant that the second licence expired in 1859.
114 Rich, *Hudson's Bay Company*, 2: 606–748; Morton, *History of the Canadian West*, 710–49. The company's claim arising out of the Oregon settlement was not settled until 1871.
115 An older but excellent biography of Simpson can be found in Galbraith, *Little Emperor*. For a newer biography with new evidence see James Raffan, *Emperor of the North: Sir George Simpson and the Remarkable Story of the Hudson's Bay Company* (Toronto: HarperCollins, 2007).
116 Richard Holmes, *Age of Wonder: How the Romantic Generation Discovered the Beauty and Terror of Science* (London: Harper, 2008), xv–xvi. Also see Delbourgo, *A Most Amazing Scene*, 119. For a recent very welcome addition to the literature see Angela Byrne, *Geographies of the Romantic North* (New York: Palgrave Macmillan, 2013). Holmes dates the second scientific revolution from James Cook's first voyage, begun in 1768, to Charles Darwin's voyage to the Galapagos Islands in 1831. The history of the arctic explorer as hero suggests that Romantic science could be said to have survived into the 1850s.
117 For the Romantic hero as explorer, see Holmes, *Age of Wonder*, 211–34. Holmes's discussion is centered on Mungo Park in Africa, with only a brief mention of arctic exploration (232), but as chapter 5 will show, John Franklin perfectly fit the pattern of the tragic Romantic hero, and those who survived their arctic explorations fit the broader category of Romantic hero.
118 See John Cawood, "Terrestrial Magnetism and the Development of International Collaboration in the Early Nineteenth Century," *Annals of Science* 34 (1977): 551–87; John Cawood, "The Magnetic Crusade: Science and Politics in Early Victorian Britain," *Isis* 70 (1979): 493–518; Deborah Warner, "Terrestrial Magnetism: For the Glory of God and the Benefit of Mankind," *Osiris* 9 (1994): 66–84; and Christopher Carter, "Magnetic Fever: Global Imperialism in the Nineteenth Century," *Transactions of the American Philosophical Society* 99, no. 4 (2009). For a discussion of the origins of the term "Magnetic Crusade," see Carter, "Magnetic Fever," xvi.
119 Carter, "Magnetic Fever," xvi; Cawood, "The Magnetic Crusade," 518.
120 Pratt defined the contact zone as "the space of colonial encounters, the space in which peoples geographically and historically separated come into contact with each other and establish ongoing relations, usually involving conditions of coercion, radical inequality, and intractable conflict." Pratt, *Imperial Eyes*, 6.

Chapter 2

1 The Royal Society was officially named the "Royal Society of London for Improving of Natural Knowledge," in the second charter issued in 1663.
2 For detailed discussions of the charter see Rich, *Hudson's Bay Company*, 1: 52–8; and Morton, *History of the Canadian West*, 54–9.
3 Cañizares-Esguerra, *Nature, Empire, and Nation*, 16, 23.
4 Lisa Jardine and Michael Silverthorne, eds., *The New Organon: Cambridge Texts in the History of Philosophy* (Cambridge and New York: Cambridge University Press, 2000), 100.
5 A good discussion of the Spanish influences on Bacon's thought can be found in Cañizares-Esguerra, *Nature, Empire, and Nation*, 19–22. Also see Delbourgo and Dew "Introduction" in Delbourgo and Dew, *Science and Empire*, 2.
6 Arthur Johnson, ed., *Francis Bacon: The Advancement of Learning and New Atlantis* (Oxford: Clarendon Press, 1974), 221, 229, 239, 245. Bacon noted (229) that Solomon was a natural historian. According to the Bible, Solomon, "described plant life, from the cedar of Lebanon to the hyssop that grows out of walls" (1 Kings 4:33).
7 Henry George Lyons, *The Royal Society, 1660–1940: A History of Its Administration under Its Charters* (Cambridge: Cambridge University Press, 1944), 329.
8 Alison Sandman, "Controlling Knowledge: Navigation, Cartography, and Secrecy in the Early Modern Spanish Atlantic," in Delbourgo and Dew, *Science and Empire*, 31–51. Quoted words are on p. 31.
9 Cañizares-Esguerra, *Nature, Empire, and Nation*, 4.
10 Ibid., 23 ; Sandman, "Controlling Knowledge," in Delbourgo and Dew, *Science and Empire*, 31.
11 E. Dekker, "Early Explorations of the Southern Celestial Sky," *Annals of Science* 44 (1987): 441.
12 Ibid., 469.
13 Ibid., 446.
14 Ibid., 448.
15 Delbourgo and Dew also described Bensalem as "a political ambiguity rather than a transparent utopia"; see Delbourgo and Dew, "Introduction," in Delbourgo and Dew, *Science and Empire*, 2.
16 Lewis Pyenson and Susan-Sheets Pyenson, *Servants of Nature: A History of Scientific Institutions, Enterprises and Sensibilities* (London: HarperCollins, 1999), 74.

17 Royal Society, "Directions for Sea-men, Bound for Far Voyages," *Philosophical Transactions of the Royal Society* 1 (1665–6): 141. The fact that Laurence Rooke died on 27 June 1662 shows that Rooke had completed his task even before the Royal Society was chartered; C. A. Ronan "Laurence Rooke (1622–1662)," *Notes and Records of the Royal Society of London*, 15 (1960): 113–18. Readers will notice idiosyncratic spelling, punctuation, capitalization, grammar, and usage in many quoted passages in this book. In conformity with what has become normal practice, such idiosyncrasies are not normally flagged in this book by the distracting insertion of "[sic]" on such occasions.

18 Robert Boyle, "General Heads for a Natural History of a Countrey [sic], Great or Small," *Philosophical Transactions of the Royal Society* 1 (1665–66): 186–9. Quoted words are on pp. 186 and 189.

19 Stearns, *Science in the British Colonies*, 247; Richard I. Ruggles, *A Country So Interesting: The Hudson's Bay Company and Two Centuries of Mapping, 1670–1870* (Montreal: McGill-Queen's University Press, 1991), 4. Captain James's voyage of 1631–2 itself exemplified Baconian ideals, eliciting interest by the Gresham College professor, Henry Gellibrand, for example, long before the Royal Society was established. See Peter Broughton, "Astronomy in Seventeenth-Century Canada," *Journal of the Royal Astronomical Society of Canada* 75, no. 4 (1981): 185.

20 Stearns, *Science in the British Colonies*, 247, Grace Lee Nute, *Caesars of the Wilderness* (St. Paul: Minnesota Historical Society, 1978), 118.

21 Rich, *Minutes of the Hudson's Bay Company, 1671–1674*, xxvii. Also see Stearns, *Science in the British Colonies*, 247; Zeller, "Spirit of Bacon," 79–101; Ruggles, "Governor Samuel Wegg, Intelligent Layman," 181; E. E. Rich, ed., *Copy-book of Letters Outward &c: Begins 29th May, 1680 Ends 5 July, 1687* (Toronto: Champlain Society, 1948), xxxii.

22 This is according to Stearns, *Science in the British Colonies*, 248. Four were definitely connected with both organizations in 1670. They are Sir Paul Neile (1613–86, founding FRS 1660); Sir Philip Carteret (1642–72, FRS 1665); Anthony Ashley Cooper, 1st Earl of Shaftesbury (1621–83, FRS 1663); and Sir James Hayes (?–1693, FRS 1663); Thomas Birch, *The History of the Royal Society of London, for Improving of Natural Knowledge: From its First Rise* 1 (New York: Johnson Reprint Corp. [1756–7] 1968): xli, xxxvi, xlii, xxxviii. Sir Peter Colleton (1635–94), an Adventurer in 1670, was elected FRS in 1677; Birch, *History of the Royal Society*, 1: xxvi. Sir Christopher Wren (1632–1723) was elected FRS in 1663; Birch, *History of the Royal Society*, 1: xxxiii. Lists of all of the Fellows of the Royal Society can be found on the Royal Society's website: http://royalsociety.org/.

23 Rich, *Minutes of the Hudson's Bay Company, 1671–1674*, 131.
24 Birch, *History of the Royal Society*, 3: 43. The interviews have been published in Birch, *History of the Royal Society*, 3: 43–6, and Stearns, *Science in the British Colonies*, 703–7. There are, however, apparently several versions of the unpublished interview in existence, and the answers to the questions convey knowledge gained from both of Gillam's previous voyages. Grace Lee Nute, *Caesars of the Wilderness*, 118. Bayly and Gillam were in London from Oct. 1671 to the spring of 1672.
25 See Alice M. Johnson, "Charles Bayly," *DCB* 1: 83; and G. Andrews Moriarty, "Zachariah Gillam," *DCB* 1: 337.
26 "A Breviate of Captain Zachariah Gillam's Journal to the North West, in the Nonsuch-Catch, in the Year 1668," in John Seller, *The English Pilot. The Fourth Book. The First Part* (London, circa 1675), 5–9. William Glover has argued that despite the fact that several English mariners had navigated Hudson Strait and Hudson Bay between 1610 and 1632, "given the discrepancies and gaps in information, the voyage was almost akin to a passage into the unknown." Thus, the publication of his information was of some significance in 1675 despite its flaws. See Richard Glover, "The Navigation of the *Nonsuch*, 1668–69," *Northern Mariner* 11, no. 4 (Oct. 2001): 49–66. Quoted passage is on p. 64.
27 Dirick Rembrantz van Nierop, "A Narrative of Some Observations ... Printed at Amsterdam, 1674," *Philosophical Transactions* 9, no. 109 (14 Dec. 1674): 197–208. The discussion of the Northwest Passage is on pp. 207–8.
28 In 1674, Joseph Moxon, Hydrographer to the King, argued that Europeans should seek the Northwest Passage via the North Pole, suggesting that twenty-four hours of daylight during the summer must keep the polar regions ice-free for much of the year. See Joseph Moxon, *A Brief Discourse of a Passage by the North-Pole to Japan, China, &c.* (London: Joseph Moxon, 1674). Moxon's theory influenced European thoughts (and maps) to the time of James Cook.
29 Birch, *History of the Royal Society*, 4: 92. Bernard Cohen reminds us that "we must keep in mind that these eras were credulous to a degree which scholars have largely ignored by concentrating on the skeptical and critical tendencies." Cohen, "The New World as a Source of Science for Europe," 108.
30 Stearns, *Science in the British Colonies*, 249.
31 Ibid., 250. The information about the Indians is cited as being in the Royal Society's *Journal*-Book, X, 212.
32 E. E. Rich, ed., *Minutes of the Hudson's Bay Company, 1679–1684 ; First Part, 1679–82* (Toronto: Champlain Society, 1945), 298. A biography of Nixon can be found at K. G. Davies, "John Nixon," *DCB* 1: 518–20.

33 Rich, *Hudson's Bay Company*, 1: 22, 25, 111.
34 Ibid., 1: 111; The report is published in Appendix A of Rich, *Minutes of the Hudson's Bay Company, 1679–1684 ; First Part, 1679–82*, 239–304.
35 Michael Hunter, *The Boyle Papers: Understanding the Manuscripts of Robert Boyle*, (Burlington, VT: Ashgate, 2007), 131; Davies, "John Nixon," 519. Boyle also owned copies of the interview with Gillam. See Hunter, *Boyle Papers*, 383, 481.
36 See for example Geoffrey Fryer, "John Fryer, F.R.S. and His Scientific Observations, Made Chiefly in India and Persia between 1672 and 1682," *Notes and Records of the Royal Society of London* 33, no. 2 (1979): 175–206.
37 John Gribbin, *The Fellowship: The Story of a Revolution* (London: Allen Lane, 2005), 281; Fryer, "John Fryer," 176.
38 Charles Richard Weld, *A History of the Royal Society* (New York: Arno Press [1848] 1975), 1: 380. A report on the collection was subsequently published as Sam Brown and James Petiver, "An Account of Part of a Collection of Curious Plants and Drugs, Lately Given to the Royal Society by the East India Company," *Philosophical Transactions* 22 (1700/1): 579–94.
39 Weld, *History of the Royal Society*, 328.
40 Harold Hartley, ed., *The Royal Society: Its Origins and Founders* (London: Royal Society, 1960), 164–5; Gribbin, *Fellowship*, 158.
41 Hartley, *Royal Society*, 164–5.
42 Glover, "The Navigation of the *Nonsuch*, 57n31.
43 Davies, "John Nixon," 519.
44 Rich, *Minutes of the Hudson's Bay Company, 1671–1674*, xxvii; Michael Hunter, *The Royal Society and Its Fellows, 1660–1700: The Morphology of an Early Scientific Institution*, 2nd ed. (Oxford: British Society for the History of Science, 1994), 38, 42; Gribben, *Fellowship*, 160.
45 Rich, *Hudson's Bay Company*, 1: 86; Mood, "Adventurers of 1670," 48–53.
46 Hunter, *Royal Society and Its Fellows*, 38–9.
47 Susan Scott Parrish has noted that, in the earlier years, apart from interest generated by a search for the Northwest Passage, British savants were particularly interested in British colonies that were biologically very different from the temperate homeland. Parrish, *American Curiosity*, 19, 114.
48 Hartley, *Royal Society*, 132; Hunter, *Royal Society and Its Fellows*, 40, 130.
49 Hartley, *Royal Society*, 132. The fact that Boyle was also the first president of the missionary organization the New England Company also hints at this priority.
50 H. W. Jones, "Sir Christopher Wren and Natural Philosophy: With a Checklist of His Scientific Activities," *Notes and Records of the Royal Society of London* 13, no. 1 (June 1958): 19–37; Stearns, *Science in the British Colonies*, 248.

51 Rich, *Hudson's Bay Company*, 1: 88.
52 Glyndwr Williams, "The Hudson's Bay Company and Its Critics in the Eighteenth Century," *Transactions of the Royal Historical Society*, 5th series, 20 (1970): 151.
53 Rich, *Copy-book of Letters Outward*, xvii.
54 Ibid., 75; Williams, "The Hudson's Bay Company and Its Critics," 151.
55 For the EIC see Fryer, "John Fryer," and for the Verenigde Oostindische Compagnie (VOC [Dutch East India Company]) see Zuidervaart and Van Gent, "A Bare Outpost," 3.
56 G. Andrews Moriarty, "Zachariah Gillam," 338.
57 Joseph Robson, *An Account of Six Years Residence in Hudson's Bay, From 1733 to 1736 and 1744 to 1747* (London: Printed for J. Payne and J. Bouquet et al., 1752), Appendix 1, pp. 11–13.
58 Williams, "Hudson's Bay Company and the Fur Trade," 10.
59 Nicholas Canny, ed., *The Origins of Empire*, vol. 1 of *Oxford History of the British Empire*, gen. ed. William Roger Lewis (New York: Oxford University Press, 1998), 251; Ann M. Carlos and Jamie Brown Kruse, "The Royal African Company: Fringe Firms and the Role of the Charter," *Economic History Review* 49, no. 2 (1996): 291–313.
60 The HBC managed to endear itself to King William and Parliament very quickly. Rich, *Hudson's Bay Company*, 1: 250–2. The company ensured its short-term survival when it secured a seven-year parliamentary licence in 1690. Still, because the company failed ever to secure a renewal of that licence, the licence turned out in the long term to be a weapon in the arsenal of its opponents. The company was forced after 1697 to base all of its claims on the 1670 Royal Charter. Understandably, the directors did not welcome any scrutiny of its charters and licences after 1697.
61 Joseph Burr Tyrrell, ed., *Documents Relating to the Early History of Hudson Bay* (Toronto: Champlain Society, 1931), xi. Oldmixon's chapter is reprinted in Tyrrell, *Documents Relating to the Early History of Hudson Bay*, 371–411.
62 Given that Robert Hooke, one of the original Fellows of the Royal Society, owned an "extract" of Gorst's 1670–1 journals, copies may have been readily available to those with the requisite connections. See Alice M. Johnson, "Thomas Gorst," *DCB* 1: 343.
63 Tyrrell, *Documents Relating to the Early History of Hudson Bay*, xi.
64 The Copley Medal, named after Sir Godfrey Copley, was first awarded in 1731. It is the Royal Society's most prestigious award.
65 Levere, *Science and the Canadian Arctic*, 98; Glyndwr Williams, "Christopher Middleton," *DCB* 3: 446–50; Rich, *Hudson's Bay Company*, 1: 556–96; C. Middleton, "New and Exact Table Collected from Several Observations, Taken

in Four Voyages to Hudson's Bay ... Shewing the Variation of the Magnetical Needle ... from the Years 1721, to 1725," *Philosophical Transactions of the Royal Society* 34 (1726–7), 73–6. The paper had been presented at a society meeting by Edmond Halley; Stearns, *Science in the British Colonies*, 250.
66 These were published as Christopher Middleton, "Observations on the Weather, in a Voyage to Hudson's Bay in North-America, in the Year 1730," *Philosophical Transactions of the Royal Society* 37 (1731–2): 76–8; Ibid., "Observations of the Variations of the Needle and Weather, Made in a Voyage to Hudson's-Bay, in the Year 1731," *Philosophical Transactions of the Royal Society* 38 (1733–4): 127–33; Ibid., "Observations Made of the Latitude, Variation of the Magnetic Needle, and Weather, by Capt. Christopher Middleton, in a Voyage from London to Hudson's-Bay, Anno 1735," *Philosophical Transactions of the Royal Society* 39 (1735–6): 270–80.
67 *Dictionary of National Biography*, s.v. "Christopher Middleton," 13: 342. Caleb Smith's quadrant was apparently similar to Hadley's; E. G. R. Taylor, *The Mathematical Practitioners of Hanoverian England 1714–1840* (Cambridge: University Press, 1966), 26.
68 Stearns, *Science in the British Colonies*, 251.
69 Christopher Middleton, "An Observation of the Magnetic Needle Being So Affected by Great Cold, That It Would Not Traverse," *Philosophical Transactions of the Royal Society* 40 (1737–8): 310–11; Ibid., "The Use of a New Azimuth Compass for Finding the Variation of the Compass or Magnetic Needle at Sea, with Greater Ease and Exactness Than by Any Ever Yet Contriv'd for That Purpose," *Philosophical Transactions of the Royal Society* 40 (1737–8): 395–8. It is interesting that Zachariah Gillam had reported to the Royal Society almost seventy years earlier that Muscovy glass (mica) prevented the problems that Middleton reported with the compass in cold weather.
70 Ibid., "An Examination of Sea-Water Frozen and Melted Again, to Try What Quantity of Salt is Contained in Such Ice, Made in Hudson's Streights by Capt. Christopher Middleton, F. R. S. at the Request of C. Mortimer, R. S. Secr.," *Philosophical Transactions of the Royal Society* 41 (1739–41): 806–7.
71 Ibid., "The Effects of Cold; Together with Observations of the Longitude, Latitude, and Declination of the Magnetic Needle, at Prince of Wales's Fort, upon Churchill-River in Hudson's Bay, North America," *Philosophical Transactions of the Royal Society* 42 (1742–3): 157–71. According to Charles Weld, Middleton won the Copley Medal for "communication of observations, in the attempt of discovery of N.W. passage to the East Indies

through Hudson's Bay." See Weld, *History of the Royal Society*, 95 or http://royalsociety.org/awards/copley-medal/.
72 Williams, "Christopher Middleton," 449–50.
73 John Richardson indicated in 1836 that "the first collection of Hudson's Bay birds of which I can find any record, are those formed by Mr. Alexander Light, who was sent out, ninety years ago, by the Hudson's Bay Company, on account of his Knowledge of Natural History"; John Richardson, *Fauna Boreali-Americana; or, The Zoology of the Northern Parts of British America* (London, J. Murray, 1829–37), 2: ix. Also see Houston, Ball, and Houston, *Eighteenth-Century Naturalists*, 35. A box appears to have been submitted from Moose Factory in 1738, but was lost. See Rich, *James Isham's Observations*, xxxvi.
74 For John Richardson's suggestion see John Richardson, *Fauna Boreali-Americana* 2: ix. Light clearly did supply specimens to George Edwards from Maryland before Light joined the HBC. See George Edwards, *A Natural History of Birds: Most of Which Have Not Been Figur'd or Describ'd, and Others Very Little Known* (London: For the author, 1743), 1:46. Regarding Light's efforts in Maryland and South Carolina see Baillie, "Naturalists on Hudson Bay," 38. It is unclear where Richardson got the information that Light was hired because of his knowledge of natural history. He may have simply extrapolated from Edwards' *Natural History of Birds*, in which case it was merely an assumption.
75 Richardson, *Fauna Boreali-Americana*, 2: x.
76 The references to Alexander Light are in Edwards, *Natural History of Birds*, 1: 46, 52 (quote), and the references to the gentleman who brought two live birds are on pp. 1: 1 (quote), and 3. For a discussion see C. Stuart Houston, "Birds First Described from Hudson Bay," *The Canadian Field Naturalist* 97 (1983): 87.
77 George Edwards, *Natural History of Birds*, 1: 4.
78 Ibid., 2: 61 ("Great White Owl"), 62 ("Little Hawk Owl"), 72 ("White Partridge").
79 Stearns, *Science in the British Colonies*, 252.
80 Rich, *James Isham's Observations*, xxxvi.
81 Ibid., lxvi.
82 Edwards, *Natural History of Birds*, 3: 107. Isham was in London from Sept. 1745 to the summer of 1746, and from Oct. 1748 to the summer of 1749, Rich, *Hudson's Bay Company*, 1: 577, 582.
83 Edwards, *Natural History of Birds*, 3: 157.
84 Stearns, *Science in the British Colonies*, 522.

85 For example, Light gets credit for supplying the spruce grouse (*Falcipennis canadensis*) that Edwards illustrated, and which Linnaeus then named, although Nicholas Denys had published a description of the spruce grouse and other grouse in 1672. See Elsa G. Allen, "Nicholas Denys, A Forgotten Observer of Birds," *The Auk* 56, no. 3 (1939): 284, 288.
86 For example, he had published Arthur Dobbs, "An Account of a Parhelion [sic] Seen in Ireland: In Two Letters from Arthur Dobbs Esq; Of Castle Dobbs in the County of Antrim, to His Brother Mr. Richard Dobbs of Trinity-College in Dublin; and by This Last Communicated to the Royal Society," *Philosophical Transactions of the Royal Society* 32 (1722–3): 89–92. Details on Dobbs can be found in Morton, *History of the Canadian West*, 206–26. For a biography of this interesting man (which does not discuss his interest in the Northwest Passage, Middleton, or the Hudson Bay Company), see Robert M. Calhoon, "Dobbs, Arthur (1689–1765)," *Oxford Dictionary of National Biography* (Oxford: Oxford University Press, 2004), http://www.oxforddnb.com/index/101007711/ (accessed 29 June 2005).
87 Williams, "Christopher Middleton," 447; Rich, *Hudson's Bay Company*, 1: 562.
88 Morton, *History of the Canadian West*, 208.
89 Ruggles, *Country So Interesting*, 30.
90 Lindsay, *Modern Beginnings of Subarctic Ornithology*, xi.
91 Ruggles, *Country So Interesting*, 31.
92 Rich, *Hudson's Bay Company*, 1: 562.
93 Ibid., 561.
94 Williams, "Christopher Middleton," 447.
95 Morton, *History of the Canadian West*, 209; Ruggles, *Country So Interesting*, 32.
96 Williams, "Christopher Middleton," 447.
97 Rich, *Hudson's Bay Company*, 1: 562.
98 Ibid., 567.
99 Williams, "The Hudson's Bay Company and Its Critics," 153–4.
100 Rich, *Hudson's Bay Company*, 1: 567. The name is rendered "Moore" in the 1749 Parliamentary inquiry's report.
101 Levere, *Science and the Canadian Arctic*, 98–9; Middleton, "The Effects of Cold."
102 The quoted passage comes from John Reinhold Forster, *History of the Voyages and Discoveries Made in the North* (Dublin, Luke White, 1786), 392; Morton, *History of the Canadian West*, 210–11.
103 Morton, *History of the Canadian West*, 211.
104 Arthur Dobbs, *An Account of the Countries Adjoining to Hudson's Bay in the North-west Part of America* (London: printed for J. Robinson, 1744).

105 Morton, *History of the Canadian West*, 213. The HBC managed to convince Parliament to add a provision that explorers were not to infringe on the HBC's trade. Rich, *Hudson's Bay Company*, 1: 566–7.
106 Ironically, because of war the HBC ships had to travel in a convoy with Dobb's ships, escorted by British navy ships commanded by Christopher Middleton. Rich, *Hudson's Bay Company*, 1: 577. For a short biography of Moor, see Glyndwr Williams, "William Moor," *DCB* 3: 471–2.
107 See Morton, *History of the Canadian West*, 213–17.
108 Arthur Dobbs, "A Letter from Arthur Dobbs Esq; of Castle-Dobbs in Ireland, to the Rev. Mr. Charles Wetstein, Chaplain and Secretary to His Royal Highness the Prince of Wales, concerning the Distances between Asia and America," *Philosophical Transactions of the Royal Society* 44 (1746–7): 471–6.
109 Morton, *History of the Canadian West*, 217–18. Dobbs's account of these efforts are described in Arthur Dobbs, *A Short Narrative and Justification of the Proceedings of the Committee Appointed by the Adventurers, to Prosecute the Discovery of the Passage to the Western Ocean of America (1749), Another Attack on the Company's Inactivity and Secretiveness* (London: printed for J. Robinson, 1749).
110 Henry Ellis, *A Voyage to Hudson's-Bay by the "Dobbs Galley" and "California" in the Years 1746 and 1747 for Discovering a North West Passage* (London: H. Whitridge, 1748); [Charles Swaine or Theodorus Swaine Drage?], "The Clerk the *California*," in *An Account of a Voyage for the Discovery of a North-West Passage* (London: Jolliffe 1748, 1749).
111 Raymond Phineas Stearns, "Colonial Fellows of the Royal Society of London, 1661–1788," *Osiris* 8 (1948): 101.
112 Isham's rebuttal is published in Rich, *James Isham's Observations*, 197–240. Notwithstanding its errors, Ellis's book includes natural history observations of some merit. For a discussion see Elsa Guerdrum Allen, "The History of American Ornithology before Audubon," *Transactions of the American Philosophical Society*, new series, 41, no. 3 (1951): 521. Allen's own errors may cause readers to slight the value of Ellis's observations. For example, Allen mistakenly asserts that the range of the pelican that Ellis describes is "a thousand miles farther south" of the Hudson Bay region.
113 Williams, "The Hudson's Bay Company and Its Critics," 167.
114 Rich, *James Isham's Observations*, xciv–xcv; Houston, Ball, and Houston, *Eighteenth-Century Naturalists*, 42.
115 Great Britain, Parliament, House of Commons, *Report from the Committee of the House of Commons Appointed to Inquire into the State and Condition of the Countries Adjoining, Hudson's Bay and of the Trade Carried on There.*

(With a Copy of the Charter of the Hudson's Bay Company) (London: House of Commons, 1749). The list of witnesses included former employees of the company, but none who were still employed by the company.
116 Williams, "The Hudson's Bay Company and Its Critics," 165.
117 Arthur Dobbs "A Letter from Arthur Dobbs Esq; to Charles Stanhope Esq; F. R. S. Concerning Bees, and Their Method of Gathering Wax and Honey" *Philosophical Transactions* 46 (1749–50): 536.
118 Robson, *Account of Six Years Residence*, 6. Despite its official authorship, Arthur Dobbs appears to have had a major role in writing this second book. See Glyndwr Williams, "Joseph Robson," *DCB* 3: 562.
119 Glyndwr Williams, "Arthur Dobbs and Joseph Robson: New Light on the Relationship between Two Early Critics of the Hudson's Bay Company," *Canadian Historical Review* 40 (1959): 133.
120 Robson, *Account of Six Years Residence*, 44.
121 Ibid., 45.
122 Ibid., 7.
123 Ibid., 45.
124 Ibid., 57.
125 Ibid., 46.
126 John Harris, *Navigantium atque Itinerantium Bibliotheca: Or, a Complete Collection of Voyages and Travels Consisting of Above Six Hundred of the Most Authentic Writers*, rev. John Campbell, (London: printed for T. Woodward, A. Ward, S. Birt, et al., 1744–8), 2: 286–93, 434–51. Quoted passage is on p. 2: 437.
127 British Americans based in Philadelphia, Maryland, New York, and Boston mounted their own attempts at the Northwest Passage via Hudson Strait in 1753 and 1754. Chaplin, *First Scientific American*, 147. Their failure even to navigate the strait hints at the value of the HBC's knowledgeable and experienced ship captains.
128 John Douglas's introduction to James Cook, *Voyage to the Pacific Ocean*, 1: xlv.
129 Ruggles, *Country So Interesting*, 42–3. For an analysis of the map drawn in 1716–17 (HBCA G.1/29), another drawn in 1760 (G.2/8), and the map by Matonabbee and Idotlyazee drawn in 1767 (G.2/27) see June Helm, "Matonabbee's Map," *Arctic Anthropology* 26, no. 2 (1989): 28–47.

Chapter 3

1 J.C. Beaglehole, ed., *The Journals of Captain James Cook on His Voyages of Discovery, 1728–1779* (Cambridge: Hakluyt Society, 1955–69), 1: cclxxix. For a history of the Transits of Venus, see Harry Woolf, *The Transits of Venus: A Study of Eighteenth-Century Science* (Princeton, NJ: Princeton University Press, 1959).

2 Churchill was established at the mouth of the Churchill River in 1717. It was replaced by a large stone fort in 1733. Destroyed by the French during their attack of 1782–3, Fort Prince of Wales was replaced by a wooden fort (Churchill) thereafter.
3 John Keill, *An Introduction to the True Astronomy or, Astronomical Lecture, Read in the Astronomical School of the University of Oxford* (London: Bernard Lintot, 1721), 29. Keill here gave the date of the transit according to the Julian calendar. For more on Horrox, see Eli Maor, *June 8, 2004: Venus in Transit* (Princeton: Princeton University Press, 2000), 30–43, and Harry Woolf, *The Transits of Venus: A Study in the Organization and Practice of Eighteenth Century Science* (Princeton, NJ: Princeton University Press, 1959), 4. Notwithstanding Keill's statement, Horrox's friend William Crabtree also observed the 1639 Transit.
4 The relationship between the solar parallax and the solar distance is expressed by the formula p = 206265 r/d, where p is the parallax in arc-seconds, r is Earth's radius and d its distance from the Sun.
5 Woolf, *The Transits of Venus*, 14.
6 Edmond Halley, "Methodus singularis quâ Solis Parallaxis sive distantia à Terra, ope Veneris intra Solem conspiciendæ, tuto determinari poterit [A New Method of Determining the Parallax of the Sun]," *Philosophical Transactions of the Royal Society* 29 (1716): 454–64. Also see Woolf, *The Transits of Venus*, 16.
7 Benjamin Martin, *Venus in the Sun* (London: W. Owen, 1761), xi. John R. Millburn offers a good sense of the man in *Benjamin Martin*.
8 This is explained well in Charles E. Herdendorf, "Captain James Cook and the Transits of Mercury and Venus," *Journal of Pacific History* 21, no. 1–2 (1986): 39–55; and Richard Woolley, "Captain Cook and the Transit of Venus of 1769," *Notes and Records of the Royal Society of London* 24 (1969): 19–32. An account of the observations of the transits geared towards a general audience can be found in J. Donald Fernie, *The Whisper and the Vision: The Voyages of the Astronomers* (Toronto: Clarke, Irwin & Co., 1976), 3–35.
9 Woolf, *The Transits of Venus*, 23.
10 George Costard, *The History of Astronomy, with Its Applications to Geography and Chronology Occasionally Illustrated by the Globes* (London: James Lister, 1767), 184. See Woolf, *The Transits of Venus*, 4.
11 Wayne Orchiston and Derek Howse, "From Transit of Venus to Teaching Navigation: The Work of William Wales," *Journal of Navigation* 53 (2000): 158.
12 Maor, *June 8, 2004*, 77–92. After the observations of 1761, the range of the estimates of the solar parallax narrowed from 8.28 seconds to 10.60 seconds, and estimates of the distance between Earth and the Sun ranged from about 77.8 million miles to 96.2 million miles. Thus, the estimates

ranged as far as 16.8% from the actual distance between Earth and the Sun at the time.
13 Costard, *History of Astronomy*, 185. The astronomical day begins at noon, so the time 10^h 10′ was 10:10 pm – and thus most of the transit was invisible in Paris and London.
14 T. Hornsby, "On the Transit of Venus in 1769," *Philosophical Transactions of the Royal Society* 55 (1765): 344
15 Herdendorf, "Captain James Cook," 43, Orchiston and Howse, "From Transit of Venus," 158. Howse has noted that British planning for the 1761 transit began late. See Derek Howse, *Nevil Maskelyne: The Seaman's Astronomer* (Cambridge, Cambridge University Press, 1989), 21.
16 Woolf, *The Transits of Venus*, 188; Howard T. Fry, *Alexander Dalrymple (1737–1808) and the Expansion of British Trade* (Toronto: University of Toronto Press, 1970), 118.
17 Herdendorf, "Captain James Cook," 45.
18 Levere, *Science and the Canadian Arctic*, 32; Woolf, *The Transits of Venus*, 163. The HBC appears to have received the request on 22 Dec.
19 Woolf, *The Transits of Venus*, 176. The importance of the observations at Fort Prince of Wales was magnified by the fact that weather interfered with all of the observations in Sweden and Russia. See Nevil Maskelyne, "A Letter from Revd. Nevil Maskelyne, B.D.F.R.S. Astronomer Royal, to Rev. William Smith, D.D. Provost of the College of Philadelphia, Giving Some Account of the Hudson's-Bay and Other Northern Observations of the Transit of Venus, June 3d, 1769," *Transactions of the American Philosophical Society* 1 (1769–71): Appendix to the Astronomical and Mathematical Papers, 3.
20 Ruggles, "Governor Samuel Wegg, Intelligent Layman," 186; Leveson Gower, "HBC and the Royal Society," 30.
21 Richard I. Ruggles, "Governor Samuel Wegg: The Winds of Change," *Beaver* 307, no. 2 (1976): 11–12.
22 HBCA A.1/43, Minutes of Governor and Committee, 27 Jan. 1768. The letter also said that the Royal Society observers had to supply their own wood for the observatory, not an unreasonable request given the lack of suitable wood at the site.
23 Ibid., also quoted in Leveson Gower, "HBC and the Royal Society," 30.
24 Griffin-Short, "The Transit of Venus," 10. If Wales envied his brother-in-law Charles Green for getting the assignment to Tahiti, his disappointment must have been lessened when he learned that some of the natural historians, including Green, died of tropical diseases on Cook's first voyage, Herdendorf, "Captain James Cook," 49. Furthermore, if Wales expected winter to be the most miserable season at Fort Prince of Wales, he was surprised

by his experience. Although he claimed to have experienced weather so cold that he heard rocks "bursting," he concluded that thanks to the "muschettos," "I cannot help thinking that the winter is the more agreeable part of the year"; William Wales, "Journal of a Voyage, Made by Order of the Royal Society, to Churchill River, on the North-west Coast of Hudson's Bay; of Thirteen Months Residence in that Country; and of the Voyage Back to England; in the Years 1768 and 1769," *Philosophical Transactions of the Royal Society* 60 (1770): 125, 127.

25 Howse, *Nevil Maskelyne*, 86, 60. This would suggest that Maskelyne, who was on the Royal Society's special committee to plan for the 1769 transit (110), played a significant role in selecting Wales and Dymond for the task. Taylor, *Mathematical Practitioners*, (48) says that Maskelyne recommended Wales for the job.

26 Leveson Gower, "HBC and the Royal Society," 30.

27 According to Francis Lucian Reid, "with the exception of the astronomer Charles Green, who died at sea on the *Endeavour*, Wales's healthy relationship with the *Resolution*'s naval officers was unique among the senior men of science who travelled with Cook during his first two Pacific voyages"; Francis Lucian Reid, "William Wales (ca. 1734–98): Playing the Astronomer," *Studies in the History and Philosophy of Science*, part A, 39, no. 2 (2008): 170–1. The central argument of Reid's article is that Wales was successful because of his deference to authority. I would argue that Wales exhibited an all-round social intelligence that allowed him to earn the respect of those whose patronage he needed, but also the respect and admiration of common men.

28 Hunt and Lamb quoted in Richard Glover, "An Early Visitor to Hudson Bay," *Queen's Quarterly* 55 (1948): 37.

29 In 1789, Dalrymple noted that William Wales had given him "an exact copy made from the *Original*, at *Churchill*, when he was there to observe the Transit of Venus, in 1769." See Alexander Dalrymple, *Plan for Promoting the Fur-Trade and Securing it to This Country, by Uniting the Operations of the East-India and Hudson's-Bay Companys* (London: George Bigg, 1789), 4.

30 Moses Norton as quoted in Leveson Gower, "HBC and the Royal Society," 31.

31 These were published in William Wales and Joseph Dymond, "Astronomical Observations Made by Order of the Royal Society, at Prince of Wales's Fort, on the North-west Coast of Hudson's Bay," *Philosophical Transactions of the Royal Society* 59 (1769): 467–88. Other observations were published in Wales, "Journal of a Voyage," 100–36; and Joseph Dymond and W. Wales, "Observations on the State of the Air, Winds, Weather, etc. Made at the

Prince of Wales's Fort, on the North-west Coast of Hudson's Bay, in the Years 1768 and 1769," *Philosophical Transactions of the Royal Society* 60 (1770): 137–78.
32 Woolf, *The Transits of Venus*, 153.
33 Herdendorf, "Captain James Cook," 50. One of the problems with the observations of the 1761 transit was that there was only one successful observation in the southern hemisphere. See Maor, *June 8, 2004*, 92–110. For a map showing the many locations from which the 1769 transit was viewed see Timothy Ball and David Dyck, "Observations of the Transit of Venus at Prince of Wales's Fort in 1769," *The Beaver* 315, no. 2 (1984): 52.
34 Houston, Ball, and Houston, *Eighteenth-Century Naturalists*, 131.
35 Estimates of the solar parallax were narrowed to between 8.43 and 8.80 seconds; Maor, *June 8, 2004*, 109. This narrowed the range of error to less than 3%, much smaller than the 16.8% following the 1761 transit, but still far wider than the 0.2% Halley had envisioned. Today, scientists estimate the true distance at that time of the year at 93.5 million miles and the solar parallax as 8.794.
36 Beaglehole, *Journals of Captain James Cook*, 1: cclxxii. The "counterpoise theory" that a large continent might lie in the temperate zones west of South America dates back at least to the sixteenth century and included in Cook's time Alexander Dalrymple among its adherents. See Joseph [José] de Acosta, *The Natural & Moral History of the Indies*, trans. Edward Grimston, ed. Clements R. Markham (London: Hakluyt Society, 1880), 1: 170. Also see Andrew S. Cook, "Alexander Dalrymple, (1737–1808)," *Oxford Dictionary of National Biography* (Oxford: Oxford University Press, 2004), http://www.oxforddnb.com/index/101007044/ (accessed 29 June 2005).
37 Jedidiah Morse, *The American Geography; Or a View of the Present Situation of the United States of America*, 2nd ed. (London: John Stockdale, 1792), 12. Morse's argument may have been drawn from William Robertson's *History of America*. See William Robertson, *The History of America*, 3rd ed. (London: W. Strahan, T. Cadell and J. Balfour, 1789), 1: 91–2.
38 Herdendorf, "Captain James Cook," 45.
39 The quoted passage is from Beaglehole, *Journals of Captain James Cook*, 1: cclxxxii–cclxxxiii.
40 Quoted in Leveson Gower, "HBC and the Royal Society," 31.
41 Quoted in Ruggles, "Governor Samuel Wegg, Intelligent Layman," 189. While £250 was a substantial sum of money, the two men were to be allowed to eat at the governor's table each day.
42 HBCA B.239/a/67, York Fort, 6th Nov. 1771 (fo.6d).

43 William B. Ewart, "Thomas Hutchins and the HBC: A Surgeon on the Bay" *Beaver* 75, no. 4 (Aug./Sept. 1995): 38–41.
44 Theodore Sherman Feldman, "The History of Meteorology, 1750–1800: A Study in the Quantification of Experimental Physics," (PhD diss. University of California, Berkeley, 1983), 253. Feldman explains, for example, that the famous French astronomer Joseph-Nicholas Delisle kept weather observations for over twenty years.
45 M. [sic] Lambert, "Exposé de quelques Observations qu'on pouroit faire pour répandre du jour sur la Météorologie," *Nouveaux Memoires de l'Academie Royale des Sciences et Belles-Lettres* (Berlin: 1771): 60. The original French passage reads "Il semble que pour rendre la Météorologie plus scientifique qu'elle ne l'est, il faudroit [sic] imiter les Astronomes, qui, sans s'arrêter d'abord à toutes les minuties, commencent par établir des loix générales & les mouvements moyens. Par là ils se mettent en état de tenir compte des anomalies & de les assujettir pareillement à prédire les phénomenes avec une exactitude qui inspire, même aux plus ignorans, du respect pour l'Astronomie."
46 Ibid.; the map is between pp. 64 and 65.
47 Gilbert Chinard, "The American Philosophical Society and the World of Science (1768–1800)" *Proceedings of the American Philosophical Society*, 87, no. 1 (14 July 1943): 1–11; Pyenson and Sheets-Pyenson, *Servants of Nature*, 351.
48 Chaplin, *First Scientific American*, 204. The second issue of the *Transactions* was not published until 1786.
49 Zuidervaart and Van Gent, "A Bare Outpost," 11, 21, 22–3.
50 Dymond and Wales, "Observations on the State of the Air," 137–78; Ball and Dyck, "Observations of the Transit of Venus," 55.
51 Houston, Ball, and Houston, *Eighteenth-Century Naturalists*, 69.
52 Ibid., 69, 131.
53 A. Burnett Lowe, "Canada's First Weathermen," *Beaver* 292 (Summer 1961): 6.
54 This was almost certainly the first time since Middleton that HBC employees kept meteorological records; Williams, "Hudson's Bay Company and the Fur Trade," 18.
55 Houston, Ball, and Houston, *Eighteenth-Century Naturalists*, 131–2. Ambrometers are rain gauges.
56 Ibid., 133.
57 D.E. Allen, "Natural History in Britain in the Eighteenth Century," *Archives of Natural History* 22 (1993): 342.
58 Ibid., 343.

59 Wales's general observations are published in William Wales, "Journal of a Voyage," 100–36; and a meteorological journal is published in Dymond and Wales, "Observations on the State of the Air," 137–78.
60 Griffin-Short, "The Transit of Venus," 10. William Wales went on to accompany James Cook as astronomer on his second and third voyages, and to train George Vancouver in astronomy and surveying.
61 Glyndwr Williams, ed., *Andrew Graham's Observations on Hudson's Bay, 1767–91* (London: Hudson's Bay Record Society, 1969), xxii. Thus, John Richardson almost certainly erred when he wrote in 1836 that Graham sent specimens to England with William Wales. John Richardson, *Fauna Boreali-Americana* 2: x.
62 David Elliston Allen, *The Naturalist in Britain: A Social History* (Princeton, NJ: Princeton University Press, 1994), 31.
63 Houston, Ball, and Houston, *Eighteenth-Century Naturalists*, 79–80, 107, 57–60.
64 Ibid., 107.
65 Leveson Gower, "HBC and the Royal Society," 31.
66 HBCA A.9/4, 22 Oct. 1771 (fo. 99d), Hudson Bay House, from the Committee to the Commissioners of His Majesty's Customs. Also see Baillie, "Naturalists on Hudson Bay," 38.
67 Ruggles, "Governor Samuel Wegg, Intelligent Layman," 189. The "fossils" were not necessarily petrified remains or impressions of ancient life. At this time the word "fossil" could refer to anything dug out of the earth; Stearns, *Science in the British Colonies*, 11.
68 Leveson Gower, "HBC and the Royal Society," 31.
69 Stearns, "Royal Society and the Company," 12.
70 Houston, "Birds First Described from Hudson Bay," 97. The species are the Eskimo Curlew (*Numenius borealis*), Great Grey Owl (*Strix nebulosa*), Boreal Chickadee (*Parus hudsonicus*), Blackpoll Warbler (*Dendroica striata*), and White-crowned Sparrow (*Zonotrichia leucophrys*). Forster's papers include John Reinhold Forster, "An Account of Some Curious Fishes, Sent from Hudson's Bay," *Philosophical Transactions of the Royal Society* 63, no. 1 (1773): 149–60; Ibid., "An Account of the Birds Sent from Hudson's Bay," *Philosophical Transactions of the Royal Society* 62 (1772): 382–433; and Ibid., "Account of Some Quadrupeds," 370–81.
71 Daines Barrington, "Observations on the Lagopus, or Ptarmigan; In a Letter from the Hon. Daines Barrington, V. P. R. S. to Mathew Maty, M. D. F. R. S." *Philosophical Transactions*, 63 (1773–4): 224–30; and Daines Barrington "Investigation of the Specific Characters Which Distinguish the Rabbit from the Hare: In a Letter to Samuel Wegg, Esq; T. and Vice-President of

the R. S. from the Honourable Daines Barrington, V. P. R. S." *Philosophical Transactions* 62 (1772): 4–14.
72 HBCA A.6/11 Hudson's Bay Company Governor and Committee to Humphrey Marten, Albany, 20 May 1772.
73 Leveson Gower, "HBC and the Royal Society," 31.
74 Houston, Ball, and Houston, *Eighteenth-Century Naturalists*, 161–76.
75 Allen, "History of American Ornithology," 497. Also see Houston, Ball, and Houston, *Eighteenth-Century Naturalists*, 173. Hutchins's original manuscript still resides in the archives of the Royal Society.
76 Allen, "History of American Ornithology," 519–20; Houston, Ball, and Houston, *Eighteenth-Century Naturalists*, 55–60, Samuel Hearne, *A Journey from Prince of Wales's Fort in Hudson's Bay, to the Northern Ocean* (London: A. Strahan and T. Cadell, 1795), 446n.
77 The plants included the common and attractive boreal species, the northern bluebell (*Mertensia paniculata*); James Britten, "An Early Hudson's Bay Collector," *Journal of Botany* 60 (1922): 239. Miller Christy's assertion that Thomas Hutchins sent the plants, while plausible, is unverifiable; Miller Christy, "An Early Hudson Bay Collector" *Journal of Botany* 60 (1922): 337. Given that the Royal Society transmitted specimens that it received from the HBC to the British Museum in the 1770s and early 1780s, it is likely that it also donated samples to other institutions; A.E. Gunther, "The Royal Society and the Foundation of the British Museum, 1753–1781," *Notes and Records of the Royal Society of London* 33, no. 2 (Mar. 1979): 211; Silvio A. Bedini, "The Evolution of the Science Museum," *Technology and Culture* 6, no. 1 (Winter 1965): 19. The donation of specimens to museums was significant because museums were much more linked to research in the nineteenth century than they are today. See Pickstone, *Ways of Knowing*, 73.
78 Houston, Ball, and Houston, *Eighteenth-Century Naturalists*, passim
79 Stearns, *Science in the British Colonies*, 255.
80 Forster, "An Account of Some Curious Fishes," 149–60. Also Forster, "An Account of the Birds Sent from Hudson's Bay," 382–433, and Forster, "Account of Some Quadrupeds," 370–81.
81 Ruggles, "Governor Samuel Wegg, Intelligent Layman," 195; Stearns, "Royal Society and the Company," 12.
82 Ruggles, *Country So Interesting*, 18–19.
83 John Reinhold Forster, "A Letter from Mr. John Reinhold Forster, F. R. S. to William Watson, M. D. Giving Some Account of the Roots Used by the Indians, in the Neighbourhood of Hudson's-Bay, to Dye Porcupine Quills" *Philosophical Transactions of the Royal Society* 62 (1772): 57.

334 Notes to pages 92–100

84 Leveson Gower, "HBC and the Royal Society," 66; Stearns, "Royal Society and the Company," 13.
85 Leveson Gower, "HBC and the Royal Society," 66.
86 Stearns, "Royal Society and the Company," 13; Houston, Ball, and Houston, *Eighteenth-Century Naturalists*, 75.
87 Houston, Ball, and Houston, *Eighteenth-Century Naturalists*, 150.
88 Ibid., 150.
89 Adam Smith, *An Inquiry into the Nature and Causes of the Wealth of Nations* (Oxford: Clarendon Press, 1880), 2: 326–9. See Rich, *Hudson's Bay Company*, 1: 590.

Chapter 4

1 See Chaplin, *First Scientific American*, 116–21.
2 Williams "The Hudson's Bay Company and Its Critics," 171; Ruggles, "Governor Samuel Wegg, Intelligent Layman," 182.
3 Houston, Ball, and Houston, *Eighteenth-Century Naturalists*, 133.
4 William E. Moreau, ed., *Writings of David Thompson* (Toronto: Champlain Society, 2009), 1: 169.
5 Shirlee Anne Smith, "Joseph Colen," *DCB* 5: 195.
6 Ibid.
7 Ibid.
8 Governor and Committee, London to John Ballenden, YF, 31 May 1799, HBCA A.5/4, London Correspondence Book Outwards, General Series, 1796–1808, fo. 51. Angel, "Clio in the Wilderness," 15.
9 H. Christoph Wolfart, "Joseph Howse," *DCB*, 8: 411–14.
10 Houston, Ball, and Houston, *Eighteenth-Century Naturalists*, 133.
11 Royal Society Archives, MA 172. A note on the first page of the second of these journals indicates that this journal was "presented by Joseph Colen Esq. Dec 9, 1802." The fact that most of these two journals are in Fidler's hand does not mean that the readings are his: for a part of the period covered by the journal, neither Fidler nor Colen was present at York Factory.
12 Royal Society Archives M.C. 1(40) Joseph Colen to Sir Joseph Banks, Cirencester, 29 December 1811.
13 Thomas Pennant, *Arctic Zoology*, 2nd ed. (London: Robert Faulder, 1792), Introductory volume (of 3 vols.): ccxxx. Readers are cautioned not to place too much reliance on eighteenth-century meteorological data.
14 Thomas Hutchins, "Experiments On the Dipping Needle, Made by Desire of the Royal Society," *Philosophical Transactions of the Royal Society* 65 (1775): 129–38. The dipping needle was related to important questions regarding

the reliability of the compass at the time. The location of the geographic North Pole can be found by celestial observation, but that place is different than where the magnet points. The difference between the two is "variation" or "declination." The dipping needle measured "inclination" – the tendency of a suspended magnetic needle not to hang perfectly vertically, but slightly in one direction. Compass variation and dip changed over time as the magnetic north pole moved. However, the magnetic north pole was poorly understood until well into the 1800s, and thus the changing declination cast doubt on the reliability of compasses. Edmond Halley's voyage on the *Paramore* in 1698 was related to his theories regarding variation. Research into the dipping needle continued well past the time it was studied in Rupert's Land. John Ross finally located the north magnetic pole (the location of maximum vertical dip – the place where the needle would be perpendicular) in 1831.

15 Thomas Hutchins, "An Account of the Success of Some Attempts to Freeze Quicksilver, at Albany Fort, in Hudson's Bay, in the Year 1775: With Observations on the Dipping-Needle." *Philosophical Transactions of the Royal Society* 66 (1776): 174–81. The experiment is discussed at some length in Hasok Chang, *Inventing Temperature: Measurement and Scientific Progress* (Oxford: Oxford University Press, 2004), 110–15.

16 Although some scientific evidence that mercury could freeze had emerged as early at 1759–60, it was not until the results of experiments conducted in Siberia in 1772 that the evidence was robust. However, the exact freezing point of mercury was still unknown; Chang, *Inventing Temperature*, 105–6; W.E. Knowles Middleton, *A History of the Thermometer and its Uses in Meteorology* (Baltimore: Johns Hopkins Press, 1966), 121.

17 Thomas Hutchins and Joseph Black, "Experiments for Ascertaining the Point of Mercurial Congelation," *Philosophical Transactions of the Royal Society* 73 (1783): 303–70. The assessment of the design is found in Middleton, *History of the Thermometer*, 123. The experiments led to the conclusion that mercury freezes at $-38\frac{2}{3}°$ F ($-39.3°$ C), not far from today's figure of $-38.87°$ C.

18 Houston, Ball, and Houston, *Eighteenth-Century Naturalists*, 70–3.

19 Interestingly, Cavendish had also served on the Royal Society's special committee to plan for the observations of the Transit of Venus. See Simon Schaffer, 'Henry Cavendish, (1731–1810),' *Oxford Dictionary of National Biography* (Oxford: Oxford University Press, 2004), http://www.oxforddnb.com/index/101004937/ (accessed 29 June 2005).

20 John McNab and Henry Cavendish, "An Account of Experiments Made by Mr. John Mc Nab, at Henley House, Hudson's Bay, Relating to Freezing

Mixtures," *Philosophical Transactions of the Royal Society* 76 (1786): 241–72. The quoted passage is on pp. 241–2.

Houston, Ball, and Houston, *Eighteenth-Century Naturalists*, 133; Ruggles, "Governor Samuel Wegg: Intelligent Layman," 190.

21 John McNab and Henry Cavendish, "An Account of Experiments Made by Mr. John McNab, at Albany Fort, Hudson's Bay, Relative to the Freezing of Nitrous and Vitriolic Acids," *Philosophical Transactions of the Royal Society* 78 (1788): 166–81. For a discussion of the significance of Hutchins' and McNab's experiments, see Chang, *Inventing Temperature*, 110–15; and Christa Jungnickel and Russell McCormmach, *Cavendish* (Philadelphia: American Philosophical Society, 1996), 279–82.

22 Pennant, *Arctic Zoology*, 2nd ed., 1: 22–3. Sir Ashton Lever, a correspondent of both Pennant and Latham, collected birds, shells, fossils, and non-European clothing and weaponry, and exhibited them in his personal museum, which, after 1774, was located in London. Allen, "History of American Ornithology," 495.

23 C.S. Mackinnon, "Samuel Hearne," *DCB* 4: 342.

24 Richard Glover, ed., *A Journey from Prince of Wales's Fort in Hudson's Bay to the Northern Ocean, 1769, 1770, 1771, 1772, by Samuel Hearne* (Toronto: Macmillan, 1958), xxix. Also see C. Stuart Houston and Mary I. Houston, "Samuel Hearne, Naturalist," *The Beaver* 67, no. 4 (1987): 23–7.

25 Hearne, *Journey from Prince of Wales's Fort*, vi.

26 Pennant, *Arctic Zoology*, 1: 11.

27 Houston, Ball, and Houston, *Eighteenth-Century Naturalists*, 86; Williams, *Andrew Graham's Observations*, xxii.

28 See, for example, Pennant, *Arctic Zoology*, 2nd ed., introductory volume: ccxcvii, 1: 11, 35.

29 Ibid., introductory volume: cclxxix

30 Ibid.: cclxxvii–cclxxviii.

31 Ibid.: cclxxviii. [This passage appears in the first edition of Thomas Pennant, *Arctic Zoology* (London: Henry Hughs, 1784), vol. 1, *Introduction*: clxxv. Discrepancies between the two surviving manuscript versions of Hearne's journals and the published version have raised questions about Hearne's authorship of sections of the published journals, and those discussions have focused particularly on Hearne's authorship of the account of this "Massacre at Bloody Fall," but Pennant's passage seems to confirm that, whether the story is true or not, and whether it was included in his original manuscript journals or not, the story did originate with Samuel Hearne. For a discussion of the authorship of the journals see I.S. MacLaren, "Samuel Hearne's Accounts of the Massacre at Bloody Fall,

Notes to pages 103–4 337

17 July 1771," *Ariel: A Review of International English Literature* 22 (1991): 25–51; I.S. MacLaren, "Exploration/Travel Literature and the Evolution of the Author," *International Journal of Canadian Studies/Revue internationale d'études canadiennes* 5 (1992): 39–68; and I.S. MacLaren, "Notes on Samuel Hearne's *Journey* from a Bibliographical Perspective," *Papers of the Bibliographical Society of Canada* 31, no. 2 (Fall 1993): 21–45.
32 Cook, *Voyage to the Pacific Ocean*, 1: xlvii. The passage from Hearne's journals is quoted at pp. xlviii–xlix.
33 That Douglas edited Hearne's work is attested to by John Richardson, in George Back, *Narrative of the Arctic Land Expedition to the Mouth of the Great Fish River, and along the Shores of the Arctic Ocean in the Years 1833, 1834, and 1835* (London: J. Murray, 1836), 147. Also see Mackinnon, "Samuel Hearne," 341; Hearne, *Journey from Prince of Wales's Fort*, viii; Joseph Burr Tyrrell, ed., *A Journey from Prince of Wales's Fort in Hudson's Bay to the Northern Ocean in the years 1769, 1770, 1771 and 1772* (Toronto: Champlain Society, 1911), 18.
34 Hearne, *Journey from Prince of Wales's Fort*, viii–ix. It seems quite possible, however, that Hearne had a copy (or perhaps the original) himself. David Thompson noted that when he was stationed at Churchill in 1784–5, "Mr Hearne employed me a few days on his manuscript, entitled 'A Journey to the North.'" See Moreau, *Writings of David Thompson*, 1: 27. John Richardson also later mentioned differences between Hearne's published journals and the "original journal," "a copy of which we saw at Hudson's Bay" while part of John Franklin's first expedition (1819–22). (He mentioned only that the original journals indicated his course and direction more often). See Back, *Narrative of the Arctic Land Expedition*, 147. Clearly, then, several copies of Hearne's journals were created over the years.
35 For an examination of the Romantic adventurer-scientist, see Holmes, *Age of Wonder*, 211–34. For discussions of the importance of masculinity and the broader role of the solitary male explorer in establishing a masculine identity for science, see Mary Terrall, "Gendered Space, Gendered Audience: Inside and Outside the Paris Academy of Sciences," *Configurations* 3, no. 2 (Spring 1995): 217–23; and Thomas Hallock, "Male Pleasure and the Genders of Eighteenth-Century Botanic Exchange: A Garden Tour," *William and Mary Quarterly* 62, no. 4 (October 2005): 715.
36 See R.L. Brett and A.R. Jones, eds., *Wordsworth and Coleridge: Lyrical Ballads*, 2nd ed. (New York: Routledge, 1991), 151–3.
37 Quoted in Ruggles, *Country So Interesting*, 3.
38 Ibid., 28, 15.
39 Henry Epp, ed., *Three Hundred Prairie Years: Henry Kelsey's "Inland Country of Good Report"* (Regina: Canadian Plains Research Center, 1993).

40 Ruggles, *Country So Interesting*, 32.
41 There was no published detailed map of the western coast of Hudson Bay until Christopher Middleton's of 1743; Glyndwr Williams, "Christopher Middleton," 449, 450.
42 For an edition and analysis of Anthony Henday's journals, see Barbara Belyea, ed., *A Year Inland: The Journal of a Hudson's Bay Company Winterer* (Waterloo, Ont.: Wilfrid Laurier University Press, 2000).
43 Quoted in Ruggles, *Country So Interesting*, 10. Henday was the first person to draw a map of his inland travels, although that map has since been lost; Ruggles, *Country So Interesting*, 38.
44 Hearne, *Journey from Prince of Wales's Fort*, 45–6.
45 John Elton's quadrant, invented around 1730, was not a significant improvement over other quadrants available at the time, and was clearly superseded by others by 1740; Taylor, *Mathematical Practitioners*, 175, 123.
46 Hearne, *Journey from Prince of Wales's Fort*, 280.
47 The fact that Hearne did not record the maker of his quadrant suggests that his employers did not prepare him particularly well for this journey. Hearne did indicate the maker of his sextant in the published version of the journals. See ibid., v.
48 Quoted in Ruggles, *Country So Interesting*, 40.
49 Ibid. These journals have been published as Lawrence J. Burpee, ed., "An Adventurer from Hudson Bay: Journal of Matthew Cocking from York Factory to the Blackfeet Country, 1772–1773," *Royal Society of Canada Proceedings and Transactions* series 3, vol. 2 (1908), 89–121.
50 Ruggles, *A Country So Interesting*, 45–6.
51 Houston, Ball, and Houston, *Eighteenth-Century Naturalists*, 63.
52 Because of its much costlier supply lines, the NWC could not compete profitably with the HBC in territories in which both companies operated. It survival was dependent upon its trade in regions where it did not compete with the HBC.
53 Rich, *Fur Trade and the Northwest*, 173–4.
54 Sörlin, "Ordering the World for Europe," 55.
55 Richard I. Ruggles, "Hudson's Bay Company Mapping," in *Old Trails and New Directions: Papers of the Third North American Fur Trade Conference*, ed. Carol M. Judd and Arthur J. Ray (Toronto: University of Toronto Press, 1980), 29.
56 Ruggles, *Country So Interesting*, 18–19.
57 Ibid., 20.
58 Cook, *Voyage to the Pacific Ocean*, lxxx. John Marley was an experienced sailor with the HBC who prepared charts for the company between 1781 and 1784; Ruggles, *Country So Interesting*, 278n1.

59 Fry, *Alexander Dalrymple*, 198–204, 211.
60 Richard Ruggles has noted that the two men are known to have dined together at least 65 times. See Ruggles, "Governor Samuel Wegg: The Winds of Change," 14. Dalrymple himself described Wegg as "My Friend." See A. Dalrymple to Evan Nepean, 11 February 1790, as reprinted in *Report on Canadian Archives, 1889* (Ottawa: Brown Chamberlin, 1890), 35. Dalrymple was elected a Fellow of the Royal Society in 1768. He was hydrographer to the East India Company until he was appointed hydrographer to the British Admiralty in 1795.
61 Pond submitted similar maps to the governments of Britain (1785), the United States (1785), and Russia (1787).
62 *Report on Canadian Archives, 1890* (Ottawa: Brown Chamberlin, 1891), 48–66 (quoted passage on p. 53).
63 "Description of the Country from Lake Superior to Cook's River" *Gentleman's Magazine* 60, no. 3 (March 1790): 197–9. Also see Isaac Ogden, Quebec, to David Ogden, London, 7 November 1789, as printed in Richard H. Dillon, ed., "Peter Pond and the Overland Route to Cook's Inlet," *Pacific Northwest Quarterly* 42, no. 4 (1951): 325, 328. J. Mervin Nooth, FRS, then resident in Quebec, corresponded with Joseph Banks about some of the scientific significance of Pond's explorations. See J. Mervin Nooth to Joseph Banks, 4 November 1789 as printed in Dillon, "Peter Pond," 328–9.
64 *Report on the Canadian Archives, 1889*, 34.
65 E. E. Rich, "Philip Turnor," *DCB* 4: 741. Pond presented a map to Lord Hamilton in April 1785.
66 Joseph Burr Tyrrell, ed., *Journals of Samuel Hearne and Philip Turnor* (Toronto, Champlain Society, 1934), 327. Although David Thompson's account is confused and contradictory, it corroborates Fidler's account. See Glover, *David Thompson's Narrative*, 20, 133.
67 Tyrrell, *Journals of Samuel Hearne and Philip Turnor*, 426.
68 Turnor's journal at Cumberland House from June to September (HBCA B.49/a/22) is published in ibid., 315–24.
69 HBCA B.239/b/49 fo. 10 Joseph Colen to William Walker 24 July 1789.
70 [Aaron Arrowsmith], *Result of Astronomical Observations Made in the Interior Parts of North America* (London: Printed for A. Arrowsmith, 1794); [Aaron Arrowsmith], *Observations made at Slave Lake, Athapascow Lake, ... Cumberland House, Slave Lake, by Philip Turnor, Mr. McKenzie, William Wales and Captain Harwell.* (London: Printed for A. Arrowsmith, 1794).
71 HBCA G 2/32. This map is discussed in Ruggles, *Country So Interesting*, 59–60.
72 Coolie Verner, "The Arrowsmith Firm and the Cartography of Canada," in *Explorations in the History of Canadian Mapping: A Collection of Essays*, ed.

Barbara Farrell and Aileen Desbarats (Ottawa: Association of Canadian Map Libraries and Archives, 1988), 53.
73 HBCA G.4/26.
74 Arrowsmith must also have included information from a 1791 map by Edward Jarvis and Donald McKay; Ruggles, *A Country So Interesting*, 58; Verner, "The Arrowsmith Firm," 53. Evidence that Arrowsmith cooperated with the company before 1794 can be found in the fact that the results of Turnor, Fidler, and Ross's surveys were published as *Result of Astronomical Observations Made in the Interior Parts of North America*, and *Observations Made at Slave Lake*, both of which were "printed for A. Arrowsmith" in 1794, and in the fact that Arrowsmith's first map incorporated the results of those surveys.
75 Verner, "The Arrowsmith Firm," 53.
76 Ruggles, *Country So Interesting*, 60.
77 Tyrrell, *Journals of Samuel Hearne and Philip Turnor*, 89; John Nicks, "David Thompson," *DCB* 8: 879.
78 HBCA E. 3/2 fol. 2–36.
79 Fidler's journal can be found at HBCA E. 3/2 fos. 39d-45d.
80 HBCA E.3/2, fo. 84d, 89.
81 Ruggles, *Country So Interesting*, 62.
82 Ibid., 59–60.
83 HBCA A.6/15 London Correspondence Book Outwards – HBC Official – 1796–1803 Hudson's Bay House, London, 31 May 1797, 15th head.
84 "A map of the Rivers & Lakes above York Factory with communication for Port Nelson River with Churchill River including part of Churchill River by David Thompson 1794 & 1795" is at HBCA G. 2/18, although it is not possible to know when Thompson submitted it, or when it arrived in London.
85 Moreau, *Writings of David Thompson*, 1: xxxi–xxxii.
86 Ruggles, *Country So Interesting*, 62.
87 Theodore Binnema, *Common and Contested Ground: A Human and Environmental History of the Northwestern Plains* (Norman: University of Oklahoma Press, 2001), 166–7.
88 Fidler's journals from Chesterfield House are published in Alice M. Johnson, *Saskatchewan Journals and Correspondence* (London: Hudson's Bay Record Society, 1967), 253–321.
89 These maps are discussed in Theodore Binnema, "How Does a Map Mean?: Old Swan's Map of 1801 and the Blackfoot World," in *From Rupert's Land to Canada: Essays in Honour of John E. Foster*, ed. Theodore Binnema, Gerhard Ens, and Roderick C. Macleod (Edmonton: University of Alberta Press, 2001), 201–24.

90 HBCA E 3/5 fos. 2-22d.
91 HBCA E 3/5, 19 and 29 June 1807. Fidler did not explain why he chose to name the lake after George Hyde Wollaston, who was a influential member of the London Committee from 30 November 1803 to 28 November 1810, but it may well be because Wollaston was instrumental in the London Committee's 1806 decision to accept William Tomison's request to rejoin the company and to lead an expedition to the Athabasca country (an expedition that never actually materialized). See Rich, *Hudson's Bay Company* 2: 283; John Nicks, "William Tomison," *DCB* 6: 775–7. This would suggest that Fidler still wished to do what he could to support the HBC's efforts in the Athabasca country.
92 Ruggles, *Country So Interesting*, 64.
93 Ibid., 64–5.
94 HBCA E3/5, fos. 23d-33.
95 Ruggles, *Country So Interesting*, 66.
96 Ibid., 66. If Arrowsmith kept them, they were probably destroyed when all of Arrowsmith's papers were destroyed during the bombing of London during World War II; Verner, "The Arrowsmith Firm," 47.
97 Fidler wrote many journals. Those that were most scientific in orientation were his district reports, HBCA B.22/e/1 (Red River District Report for 1819), HBCA B.51/e/1(Manitoba District Report for 1820), and HBCA B.51/e/2 (Manitoba District Report for 1821). His private journals (HBCA E.3/2) also contain much of scientific orientation. There is no reason to believe that any of these documents were shared with savants. Angela Byrne makes several very insightful observations about Fidler's scientific endeavours in her *Geographies of the Romantic North*, *passim*, but see especially her case studies, 89–91, 136–44, and 154–7.
98 E.E. Rich, *Colin Robertson's Correspondence Book, September 1817 to September 1822* (London: Hudson's Bay Record Society, 1939), 221–3. Rich reveals that Joseph Colen conveyed some of the information about Howse's explorations to Joseph Banks in a letter of December 1811. Also see Ruggles, *Country So Interesting*, 68.
99 After long delays it was finally published as Joseph Howse, *A Grammar of the Cree Language with which is Combined an Analysis of the Chippeway Dialect*. (London: J.G.F. & J Rivington, 1844). The book is noteworthy for its vehement defence of the intellectual capacities of North American indigenous peoples.
100 Joseph Howse, "Vocabularies of Certain North American Indian Languages" *Proceedings of the Philological Society* 4 (1848–50): 102–22; Ibid., "Vocabularies of Certain North American Languages," *Proceedings of the*

Philological Society 4 (1848–50): 191–206. The first of these offers vocabularies in Blackfoot, Nipissing, and Shawnee, and the second in Chipewyan, Beaver, Sikanni, Kutani Flathead, Okanagan, Shoushwap.
101 Ruggles, *Country So Interesting*, 68–70; K.G. Davies, ed., *Northern Quebec and Labrador Journals and Correspondence, 1819–1835* (London: Hudson's Bay Record Society, 1963).
102 Ruggles, *Country So Interesting*, 71.
103 Moreau, *Writings of David Thompson* 1: 171; Glover, *David Thompson's Narrative*, 132.
104 It is ironic that when he published a very short meteorological journal in 1849, Thompson turned to a journal he kept during his days with the HBC, rather than one of the many he kept during his time with the NWC. David Thompson, "Mean Temperature of Cumberland House and Bedford House, Hudson's Bay Territory, 1789–90, 1795–96" *British American Journal of Medical and Physical Science* 4, no. 11 (March 1849): 302.
105 Susan Faye Cannon, *Science in Culture: The Early Victorian Period* (New York: Science History Publications, 1978), 75. Cannon explained that Alexander von Humboldt considered himself a "scientific traveller," not an explorer.
106 Stearns, *Science in the British Colonies*, 256–7.
107 See Gary E. Moulton, *The Journals of the Lewis and Clark Expedition* (Lincoln: University of Nebraska Press, 1983–2001), 3: 273n3, 275n1, 4: 28n5.
108 See Ruggles, *Country So Interesting*, 65, 280n11.
109 HBCA E.3/2, 31 December 1792. The source referred to is that of today's Red Deer River.
110 Ruggles, *Country So Interesting*, 66, 246; Barbara Belyea, ed., *Columbia Journals: David Thompson* (Montreal and Kingston: McGill-Queen's University Press, 1994), 295–6; Belyea, *Dark Storm Moving West*, 86.
111 Belyea, *Columbia Journals*, 297.
112 Victor G. Hopwood, "David Thompson and His Maps," in Farrell and Desbarats, *Explorations in the History of Canadian Mapping*, 207, 206, 209.
113 Sir Roderick Impey Murchison, "Address to the Royal Society of London," *Proceedings of the Royal Geographical Society of London* 3, no. 1 (1858–9): 319n–320n.
114 Ibid.: 320n.
115 Thomas Pennant, *The Literary Life of the Late Thomas Pennant Esq. by Himself* (London: London: Benjamin and John White, and Robert Faulder, 1793), 3.
116 Ibid., 9.
117 Ibid., 28–9; and the unpaginated advertisement in the introductory volume in Pennant, *Arctic Zoology*, 2nd ed.

118 Richardson, *Fauna Boreali-Americana*, 2: xi.
119 As quoted in Houston, Ball, and Houston, *Eighteenth-Century Naturalists*, 27. Also see p. 28.
120 Pennant, *Arctic Zoology* 2nd ed., in the unpaginated advertisement. In return Hearne acknowledged the help of Pennant's *Arctic Zoology* in his published journals. See Hearne, *Journey from Prince of Wales's Fort*, ix.
121 Hearne, *Journey from Prince of Wales's Fort*, 446n.
122 Cook, *Voyage to the Pacific Ocean*, xliv–xlv.
123 Ibid., lxxxiv.
124 Williams, "The Hudson's Bay Company and its Critics," 170.
125 Ibid., 170.
126 Hearne, *Journey from Prince of Wales's Fort*, xxi–xxii. Hearne refers to the works (already discussed and cited) under the names of Arthur Dobbs and Henry Ellis, and to books by Joseph Robson, *An Account of Six Years Residence in Hudson's Bay, From 1733 to 1736 and 1744 to 1747* (1752), and Mr. Dragge, [often attributed to Theodorus Swaine Drage but probably actually written by Captain Charles Swaine] *The Great Probability of a North-West Passage, Deduced from Observations on the Letter of Admiral de Fonte, Who Sailed from the Callao of Lima on the Discovery of a Communication between the South Sea and the Atlantic Ocean* (1768); and Alexander Cluny, *The American Traveller: Or, Observations on the Present State, Culture and Commerce of the British Colonies in America ... By an Old and Experienced Trader* (1769) On the authorship of *Great Probability* see Howard N. Eavenson, *Map Maker and Indian Traders* (Pittsburgh: University of Pittsburgh Press, 1949).
127 Williams, "The Hudson's Bay Company and its Critics," 171.
128 Morton, *History of the Canadian West*, 421.

Chapter 5

1 Williams, "Hudson's Bay Company and the Fur Trade," 61.
2 Galbraith, *Little Emperor*, 173.
3 Trevor H. Levere, "Science and the Canadian Arctic, 1818–76, from Sir John Ross to Sir George Strong Nares," *Arctic* 41, no. 2 (June 1988): 127.
4 It is no mere coincidence that Mary Shelley's *Frankenstein* was published in 1818.
5 For a broader discussion of science and the search for the Northwest Passage, see Levere, *Science and the Canadian Arctic*, 36–141, 190–238.
6 Ibid., 98.
7 Levere, "Science and the Canadian Arctic, 1818–76," 128.

8 John Gascoigne argued that Barrow became the pre-eminent advocate for science in the British Empire upon the death of Joseph Banks in 1820; Gascoigne, *Science in the Service of Empire*, 196.
9 John Barrow, *A Chronological History of Voyages in the Arctic Regions: Undertaken Chiefly for the Purpose of Discovering a North-East, North-West, or Polar Passage between the Atlantic and Pacific* (London: J. Murray, 1818), 303.
10 Levere, *Science and the Canadian Arctic*, 44.
11 Franklyn Griffiths, "Where Vision and Illusion Meet," in *The Politics of the Northwest Passage*, ed. Franklyn Griffiths (Montreal and Kingston: McGill-Queen's University Press, 1987), 3.
12 Barrow, *Chronological History*, 378–9.
13 Ibid., 364.
14 Otto von Kotzebue, *Voyage of Discovery, into the South Sea and Beering's* [sic] *Straits, for the Purpose of Exploring a North-East Passage Undertaken in the Years 1815–18* (London: Longman, Hurst, Rees, Orme, and Brown, 1821), 1: 91.
15 Barrow, *Chronological History*, 365.
16 Ibid., 280.
17 J. Ross, *A Voyage of Discovery, Made under the Orders of the Admiralty, in His Majesty's Ships Isabella and Alexander, for the Purpose of Exploring Baffin's Bay, and Inquiring into the Probability of a North-West Passage* (London: J. Murray, 1819), 73.
18 Ibid., 74.
19 Lindsay, *Modern Beginnings of Subarctic Ornithology*, xiii; Ross, *Voyage of Discovery*. The publication of Ross's journal unleashed a public battle between Ross and Sabine, in which Sabine accused Ross of plagiarism. See Edward Sabine, *Remarks on the Account of the Late Voyage of Discovery to Baffin's Bay, Published By Captain J. Ross, R.N* (London: John Booth, 1819). Amongst other things, Sabine argued that the duties of naturalist "formed no part of my official engagement." (3, also see 18–19). "I have preferred the Transactions of the Linnean Society as the most appropriate channel for communications on those parts of Natural History which my own knowledge enable me to furnish." (23–4)
20 For more on that topic see Levere, *Science and the Canadian Arctic*, passim.
21 Quote from John Franklin, *Narrative of a Journey to the Shores of the Polar Sea, in the Years 1819–20–21–22*, 2nd ed. (London: J Murray, 1825), xii..
22 C. Stuart Houston, "John Richardson – First Naturalist in the Northwest," *The Beaver* 315, no. 2 (1984): 11.
23 Franklin, *Narrative of a Journey*, ix. Also see Levere, *Science and the Canadian Arctic*, 104.

24 Franklin *Narrative of a Journey*, x.
25 Levere, *Science and the Canadian Arctic*, 103, 104; Peter Steele, *The Man Who Mapped the Arctic: The Intrepid Life of George Back, Franklin's Lieutenant* (Vancouver: Raincoast Books, 2003), 61.
26 E. E. Rich, ed., *Journal of Occurrences in the Athabasca Department by George Simpson, 1820 and 1821, and Report* (London: Hudson's Bay Record Society/Champlain Society, 1938), 65.
27 Circular from William Williams, Governor in Chief to HBC Officers, 17 Jan. 1820, HBCA D.1/3, fo. 15. Williams wrote even more forceful letters to officers who would be most important to Franklin. See William Williams, Cumberland House, to Mr. John Clarke, 12 Jan. 1820, HBCA D.1/3, fo. 6d.
28 William Williams, Cumberland House to Lieut. Franklin,12 Jan. 1820, HBCA D.1/3, 13d.
29 William Williams, Cumberland House to Mr. John Clarke, Isle a la Crosse, 12 Jan. 1820, HBCA D.1/3, fo. 6d. Emphasis in original. The HBC's fear that strategically important information might leak was also reflected in its response to an inquiry from an Alexander Simpson who had worked as a clerk at Moose Factory from 1820 to 1822. Simpson was sternly informed that "the Governor and Committee cannot consent to your publishing any information respecting the affairs of the Company which you may have acquired whilst in their Service." Secretary, HBC, London, to Alexander Simpson, Nairn, 2 Feb. 1825, HBCA A 5/8, fo. 24d. Also see HBCA A 1/54, p. 166.
30 Rich, *Journal of Occurrences in the Athabasca Department*, 261.
31 Ibid., 206. Also see 207.
32 Ibid., 226.
33 Ibid., 243.
34 Ibid., 261.
35 Levere, *Science and the Canadian Arctic*, 103–10.
36 According to Richard Holmes, Edward Parry had established the place of the arctic hero (Parry's *Journal of a Voyage to Discover a North-west Passage* was published in 1821), but the pathos of Franklin's overland expedition outdid that of Parry. For Holmes's discussion of the scientific traveller as Romantic hero, see Holmes, *Age of Wonder*, 211–34. For Parry, see p. 232.
37 Franklin *Narrative of a Journey*, xiii. Franklin offered the same thanks to the NWC on the same page.
38 John Richardson in Back, *Narrative of the Arctic Land Expedition*, 145. Richardson has nevertheless attracted some controversy. See M.A. Macleod and R. Glover "Franklin's First Expedition as Seen by the Fur Traders," *Polar*

Record 15, no. 98 (1971): 669–82; and Janice Cavell, "The Hidden Crime of Dr. Richardson," *Polar Record* 43, no. 225 (2007): 155–64.

39 John Richardson also published a separate volume on the botanical contributions of the expedition, and Joseph Sabine published a paper on the marmots, although Richardson's book did not acknowledge the Hudson's Bay Company, and Sabine's mentioned it only in passing. See John Richardson, *Botanical Appendix to Captain Franklin's Narrative of a Journey to the Shores of the Polar Sea* (London: W. Clowes, 1823); and Joseph Sabine, "Account of Marmots of North America Hitherto Known, with Notices and Descriptions of Three New Species," *Transactions of the Linnean* Society 13 (1822): 579–92. For a discussion of the scientific contributions of Franklin's first expedition see C. Stuart Houston, *Arctic Ordeal: The Journal of John Richardson, Surgeon-Naturalist with Franklin, 1820–1822* (Montreal and Kingston: McGill-Queen's University Press, 1984).

40 John R. Bockstoce, *Furs and Frontiers in the Far North: The Contest among Native and Foreign Nations for the Bering Strait Fur Trade* (New Haven: Yale University Press, 2009), 130. The Anglo-Russian Treaty fixing the boundary between Russian America and British America at the 141st meridian was not signed until 1825.

41 John Franklin, *Narrative of a Second Expedition to the Shores of the Polar Sea in the Years 1825, 1826, and 1827* (London: J. Murray, 1828), x, xi.

42 Ibid., xxiv.

43 Richardson, *Fauna Boreali-Americana*, 1: xviii.

44 Levere, *Science and the Canadian Arctic*, 112.

45 Ibid.

46 Franklin's second expedition is discussed in Levere, *Science and the Canadian Arctic*, 110–25.

47 Ibid., 116–18; Houston, "John Richardson – First Naturalist in the Northwest," 15.

48 John Franklin, Fort Chipewyan, to George Simpson and Chief Factors and Chief Traders, members of Council, Hudson's Bay, York Factory, 20 July 1826, HBCA D.4/119, fo. 15d.

49 Levere, *Science and the Canadian Arctic*, 121, 79–81.

50 Ibid., 124.

51 Richardson, *Fauna Boreali-Americana*, part 1, xvii. Judith F. M. Hoeniger, "Thomas Drummond," *DCB* 6: 221–2. Thomas Drummond, "Sketch of a Journey to the Rocky Mountains and to the Columbia River in North America," *Botanical Miscellany* 1 (1830): 95–6, 178–219.

52 W. J. Hooker, "Account of the Expedition under Captain Franklin, and of the Vegetation of North America, in Extracts of Letters from Dr Richardson, Mr Drummond, and Mr Douglas," *Edinburgh Journal of Science*, 6, no. 1 (Jan. 1827): 107–16.
53 Franklin, *Narrative of a Second Expedition*, xxii–xxiii
54 See for example ibid., 304–5.
55 Ibid., 319.
56 Bockstoce, *Fur and Frontiers*, 179, 204–5.
57 This expedition also produced a publication with notable scientific contributions; J. Ross and J. C. Ross, *Narrative of a Second Voyage in Search of a North-West Passage, and of a Residence in the Arctic Regions during the years 1829, 1830, 1831, 1832, 1833.* (London: A.W. Webster, 1835).
58 George Simpson quoted in Back, *Narrative of the Arctic Land Expedition*, 46.
59 Ibid., 16.
60 Mactavish quoted in William C Wonders, "Introduction," in Back, *Narrative of the Arctic Land Expedition*, xxiii.
61 LAC Hargrave Papers, Thomas Simpson, Fort Confidence, to James Hargrave, 25 Sept. 1838. Previously, Simpson had been more generous. See Steele, *The Man Who Mapped the Arctic*, 218.
62 Historians have generally been critical of Back's leadership skills. Many of Back's contemporaries also commented negatively on his character. Nevertheless, he remained respected and influential in official circles until his death. Later-day scholarly assessments of King's leadership have been much more positive, but King's demeanour alienated himself from almost everyone with any influence in the HBC and the government. See A. H. Beesly, rev. Andrew Lambert, "Sir George Back," *Oxford Dictionary of National Biography*; C. A. Holland, "Sir George Back," *DCB* 10: 26–9; Clements R. Markham, *The Fifty Years' Work of the Royal Geographical Society* (London: John Murray, 1881), 93; Elizabeth Baignet, "Richard King," *Oxford Dictionary of National Biography*; Alan Cooke, "Richard King," *DCB* 10: 406–8.
63 After slavery was abolished in the British Empire in 1833, some humanitarians, led by Thomas Hodgkin, turned their attention to the circumstances of the aboriginal peoples of the empire. The Aborigines Protection Society was established in 1837, and the British Select Committee on Aboriginal People held its investigations in 1836 and 1837. The humanitarians had great influence in the Colonial Office in the late 1830s, although their influence then waned quickly. See Amalie M. Kass and Edward H. Kass, *Perfecting the World: The Life and Times of Dr. Thomas Hodgkin, 1798–1866* (Boston: Harcourt Brace Jovanovich, 1988).

64 Back, *Narrative of the Arctic Land Expedition*, 24.
65 Great Britain, Parliament, *Report from the Select Committee on the Hudson's Bay Company; Together with the Proceedings of the Committee, Minutes of Evidence, Appendix and Index* (London: House of Commons, 1857), 184–9.
66 Back, *Narrative of the Arctic Land Expedition*, 22, 475.
67 Richard King, *Narrative of a Journey to the Shores of the Arctic Ocean in 1833, 1834, and 1835: Under the Command of Capt. Back, R.N.* 2 vols. (London: R. Bentley, 1836), 1: v–ix. Most scholars have accepted the legitimacy of King's criticism, and King's book has been judged the better book. See Baignet, "Richard King." King also separately published "Temperature of Quadrupeds, Birds, Fishes, Plants, Trees, and Earths, as Ascertained at Different Times and Places in Arctic America during Capt. Back's Expedition." *Edinburgh New Philosophical Journal* 21 (1836): 150–1.
68 King, *Narrative of a Journey* 2: 50–5, 62.
69 LAC M19, A21 Hargrave Papers, Thomas Simpson, Fort Confidence to James Hargrave, 17 Jan. 1838.
70 William Barr, ed., *From Barrow to Boothia: The Arctic Journal of Chief Factor Peter Warren Dease, 1836–1839* (Montreal and Kingston: McGill-Queen's University Press, 2002), 6. Quoted passage from Alexander Simpson, *The Life and Travels of Thomas Simpson, the Arctic Discoverer* (London: Richard Bentley, 1845), 176. King did subsequently publish several ethnological articles. See Richard King, "On the Physical Characters of the Esquimaux" *Journal of the Ethnological Society of London* 1 (1848): 45–59; ibid., "On the Industrial Arts of the Esquimaux," *Journal of the Ethnological Society of London* 1 (1848): 277–300; and ibid., "On the Intellectual Character of the Esquimaux," *Journal of the Ethnological Society of London* 1 (1848):127–52. King also published an article related to his efforts to get sponsorship for another expedition. See Richard King, "On the Unexplored Coast of North America," *London, Edinburgh, and Dublin Philosophical Magazine and Journal of Science*, 3rd ser., 20 (1842): 488–94.
71 Patrick Brantlinger, *Dark Vanishings: Discourse on the Extinction of Primitive Races, 1800–1930* (Ithaca, NY: Cornell University Press, 2003), 86–7; George W. Stocking Jr., *Victorian Anthropology* (New York: Free Press, 1987), 242; Kass and Kass, *Perfecting the World*, 269–70.
72 King's damning testimony of 11 July 1836 can be found in Great Britain, Parliament, House of Commons, *Report from the Select Committee on Aborigines (British Settlements)* (Cape Town: C. Struik, 1966), vol. 1, part 1, pp. 639–42. That the committee took King's testimony seriously can be found in vol. 1, part 2, p. 8. Hodgkin paid for his opposition to the HBC in

1837, when he was passed over for promotion at Guy's Hospital. See Kass and Kass, *Perfecting the World*, 57–8, 276–96.

73 Brantlinger, *Dark Vanishings*, 87. The influence of the APS and the humanitarians' movement peaked between 1835 and 1839, when Lord Glenelg served as the Colonial Secretary and James Stephen as his undersecretary. Its influence declined significantly by 1842. See Kass and Kass, *Perfecting the World*, 373–86.

74 Stocking, *Victorian Anthropology*, 240.

75 *The Missionary Register for 1837: Containing the Principal Transactions of the Various Institutions for Propagating the Gospel* (London: L & G Seeley, 1837), 317, 318.

76 Stocking, *Victorian Anthropology*, 242. The APS did include prominent scholars, such as James Cowles Prichard.

77 Kass and Kass, *Perfecting the World*, 393; Stocking, *Victorian Anthropology*, 244; Brantlinger, *Dark Vanishing*, 89; Baignet, "Richard King,"; Cooke, "Richard King.". Many of the members of the Ethnological Society rejected the term "anthropology."

78 Joseph Barnard Davis, "Anthropology and Ethnology," *Anthropological Review* (1868): 396. (394–9). During the early years, members of the APS and the Ethnological Society were very influenced by the works of Prichard and Blumenbach. Richard King's scholarly articles betray an admiration for both of those scholars. For the rise of racialist thought in the society see Kass and Kass, *Perfecting the World*, 507–8.

79 The publication history of King's articles can be explained by the fact that from 1842 to 1848 the Ethnological Society agreed that its proceedings would be published by Robert Jameson's *Edinburgh New Philosophical Journal*. In 1848, when the society inaugurated its own journal, King's articles were reprinted in its first volume. Richard King, "On the Physical Characteristics of the Esquimaux," *Edinburgh New Philosophical Journal* 36 (1844): 296–310; reprinted in *Journal of the Ethnological Society of London* 1 (1848), 45–59;ibid., "On the Industrial Arts of the Esquimaux," *Edinburgh New Philosophical Journal* 42 (1847): 112–35, reprinted in *Journal of the Ethnological Society of London* 1 (1848), 277–300; and ibid., "On the Intellectual Character of the Esquimaux," *Edinburgh New Philosophical Journal* 38 (1845): 303–52, reprinted in *Journal of the Ethnological Society of London*, 1 (1848): 127–52.

80 Baignet, "Richard King,"; Cooke, "Richard King."

81 King remained active in the Ethnological Society throughout its history, and eventually approved of its 1871 merger with the Anthropological

Society, whereupon he became a councillor in the new Society, named the Anthropological Institute of Great Britain and Ireland. King and Hodgkin were among the few members of the APS who actively promoted both the scholarly and the humanitarian goals of the society. Both were convinced that honest inquiry would prove the monogenist theory; Kass and Kass, *Perfecting the World*, 393.

82 Barr, *From Barrow to Boothia*, 7.
83 Quoted in ibid., 20.
84 Simpson, *Life and Travels of Thomas Simpson*, 176–7.
85 Unpublished conference paper by Ken Coates, quoted in Barr, *From Barrow to Boothia*, 6, 7. In its application for renewal of its licence, the HBC, after referring explicitly to the fact that it had "undertaken to fit out an expedition, composed of their own officers and servants, at the sole expense of the Company, to complete the surveys left unfinished by Sir John Franklin, Captain Beechey and Captain Back," noted that "it will be seen what the Company has done in reference to the extension of the British trade on the north-west coast, and the exertions they are making in the causes of discovery and science." Great Britain, Parliament, House of Commons, *Copy of the Existing Charter or Grant*, 14, 15. In another letter, sent on 7 Feb. 1838 to the Lords of the Committee of Privy Council for Trade, while the HBC's application was still being assessed, Pelly argued that Dease and Simpson had been sent "on the grounds of promoting discovery and science ... quite unconnected with any ulterior views towards any pecuniary advantage or benefit arising from trade, but solely for the honour of completing the survey of the northern coast of America, at a cost to the Company." Great Britain, Parliament, House of Commons, *Copy of the Existing Charter or Grant*, 27.
86 Rich, *Hudson's Bay Company*, 2: 647. The dispute was settled to the satisfaction of the HBC in 1838.
87 Barr, *From Barrow to Boothia*, 14.
88 William R. Sampson, "Peter Warren Dease," *DCB* 9: 197.
89 Ibid.
90 Ibid.
91 Rich, *Hudson's Bay Company*, 2: 648.
92 Barr, *From Barrow to Boothia*, 14–17; Sampson, "Peter Warren Dease," 197–8. The contempt for Native people is starkly evident in his correspondence with James Hargrave; see his various letters to Hargrave in LAC Hargrave Papers.
93 Head 14, George Simpson to Peter Dease and Thomas Simpson, 2 July 1836, quoted in Barr, *From Barrow to Boothia*, 25.

94 [Peter Warren] Dease and T. Simpson, "An Account of the Recent Arctic Discoveries by Messrs. Dease and T. Simpson. Communicated by J.H. Pelly, Esq., Governor of the Hudson's Bay Company" *Journal of the Royal Geographical Society of London*: 8 (1838), 213–25, quote from p. 220. According to Simpson the party duly took possession of the land (221). Also see Peter Warren Dease and Thomas Simpson, "An Account of Arctic Discovery on the Northern Shore of America in the Summer of 1838," *Journal of the Royal Geographical Society of London* 9 (1839): 325–30.

95 Barr, *From Barrow to Boothia*, 5.

96 Peter Warren Dease and Thomas Simpson, "Narrative of the Progress of Arctic Discovery on the Northern Shore of America, in the Summer of 1839," *Journal of the Royal Geographical Society of London* 10 (1840): 273, 270. J.H. Pelly and Nicholas Garry (including their status as governor and deputy governor with the HBC) are listed as two of 535 members of the Royal Geographical Society in the *Journal of the Royal Geographical Society* 1 (1831): xviii, xvi.

97 Dease and Simpson, "Narrative of the Progress of Arctic Discovery," 274. Simpson again indicated that he performed the ritual to take possession of the land for Great Britain (330). The claim is understandable if one considers that Simpson appears to have considered the Northwest Passage to extend from the Bering Sea to midway along the Arctic coast, and the Northeast Passage to extend from there to the Atlantic; Barr, *From Barrow to Boothia*, 280.

98 Sampson, "Peter Warren Dease," 198.

99 Dease and Simpson, "An Account of the Recent Arctic Discoveries," 224.

100 Thomas Simpson, *Narrative of the Discoveries on the North Coast of America; Effected by the Officers of the Hudson's Bay Company during the Years 1836–39* (London, Richard Bentley, 1843), 409–81; Sampson, "Peter Warren Dease," 197.

101 Simpson, *Narrative of the Discoveries*, xii–xiii. In 1844, J. C. Platt described the Simpson-Dease explorations as "well known"; J. C. Platt, "Old Trading Companies," in *London*, ed. Charles Knight (London: Charles Knight, 1844), 6: 49.

102 Simpson, *Narrative of the Discoveries*. Quoted passages are on p. 1.

103 Rich, *Hudson's Bay Company*, 2: 650.

104 Sampson, "Peter Warren Dease," 198; Simpson, *Narrative of the Discoveries*, xiii–xix, Barr, *From Barrow to Boothia*, 273; Rich, *Hudson's Bay Company*, 2: 650; "The President's Address on Presenting Medals," *Journal of the Royal Geographical Society*, 9 (1839): ix–xii. Nicholas Garry, the Deputy Governor, responded to the President's address by saying that "it is a great

satisfaction that the Hudson's Bay Company, as a commercial Company, have been able to extend their discoveries not only *within*, but *beyond* Her Majesty's dominions; and that you, Sir, on the part of the Geographical Society, should have expressed your approbation that they have not limited or restricted their endeavours, in time or expense, whenever they could aid the great cause of the advancement of geographical science and discovery" (xii).

105 Rich, *Hudson's Bay Company*, 2: 649.
106 Barr, *From Barrow to Boothia*, 273–93.
107 John Craig, "John Bell," *DCB* 9: 42.
108 Sampson, "Peter Warren Dease," 198.
109 John Richardson, "On the Frozen Soil of North America" *Edinburgh New Philosophical Journal* 30 (1840–1): 110–11. Also see Schefke, "Hudson's Bay Company as a Context for Science," 74.
110 Richardson, "On the Frozen Soil of North America," 111.
111 Ibid.; John Richardson, *Arctic Searching Expedition: Journal of a Boat-Voyage through Rupert's Land and the Arctic Sea* (London: Longman, Brown, Green, and Longmans, 1851), 1: 165–6.
112 John Richardson, "Note on the Best Points in British North America for Making Observations on the Temperature of the Air; and also for the Height of the Station above the Level of the Sea," *Journal of the Royal Geographical Society* 9 (1839): 121–5; Murdoch McPherson and John Richardson, "Register of the Temperature of the Atmosphere, Kept at Fort Simpson, North America, in the Years 1837, 1838, 1839, and 1840," *Edinburgh New Philosophical Journal* 30 (1840–1): 124–7.
113 Peter Warren Dease and John Richardson, "On the Cultivation of the Cerealea in the High Latitudes of North America." *Edinburgh New Philosophical Journal*, 30 (1840–1): 123. Richardson, *Arctic Searching Expedition*, 1: 153, 165–6, 214.
114 George Simpson to Governor and Committee, London, 21 June 1843, HBCA A.12/2, fo. 171d.
115 Rich, *Hudson's Bay Company*, 2: 723–6.
116 Ibid., 723, 725.
117 R. L. Richards, "John Rae," *DCB* 12: 876.
118 E. E. Rich and A.M. Johnson, eds., *Rae's Arctic Correspondence 1844–45* (London: Hudson's Bay Record Society, 1943), xiv. Rae also described himself as "addicted to field-sports" since childhood; John Rae, *Narrative of an Expedition to the Shores of the Arctic Sea in 1846 and 1847* (London: T. & W. Boone, 1850), 73.

119 Quoted words by R. M. Ballantyne, 7 Sept. 1845, as found in R. M. Ballantyne, *Hudson's Bay*, 2nd ed. (Edinburgh and London: 1848), 225–6.
120 Richards, "John Rae," 876. This opinion must have been widespread, for in 1860, while at Fort Simpson, Robert Kennicott wrote that "Dr. Rae is said to have been the best snow shoe walker ever known in these parts – beating the natives"; Kennicott to Baird, 23 Mar. 1860, in Lindsay, *Modern Beginnings of Subarctic Ornithology*, 43.
121 George Simpson, London, to Thomas Simpson, 4 June 1840, in Barr, *From Barrow to Boothia*, 289–91.
122 Rae, *Narrative of an Expedition*, 1–2.
123 LAC Hargrave Papers, George Simpson, Hudson Bay House, to James Hargrave, 2 Dec. 1844.
124 LAC Hargrave Papers, H.H. Berens, London, 31 Mar. 1845 to James Hargrave, York Factory
125 Rich, *Hudson's Bay Company*, 2: 461, 465. See the short biography of Ross in R. Harvey Fleming, ed., *Minutes of Council Northern Department of Rupert [sic] Land, 1821–31* (Toronto: Champlain Society, 1940), 453.
126 LAC Hargrave Papers, Donald Ross, Norway House to James Hargrave, York Factory, 29 Apr. 1845.
127 Major Davenzac, quoted in Michael Paul Rogin, *Fathers and Children: Andrew Jackson and the Subjugation of the American Indian* (New Brunswick, NJ: Transaction, 1991), 107.
128 John O'Sullivan, "Annexation," *Democratic Review* 17 (July 1845): 5. (article from 5–10)
129 Rich, *Hudson's Bay Company*, 2: 540, 724.
130 Ibid., 723–6. The troops were actually used to enforce the HBC's trading monopoly.
131 Ibid., 541–3, 790; Galbraith, *Little Emperor*, 157.
132 LAC Hargrave Papers, James Douglas, Fort Vancouver, to James Hargrave, York Factory, 24 Mar. 1847.
133 Galbraith, *Hudson's Bay Company as an Imperial Factor*, 251–82.
134 Rae, *Narrative of an Expedition*, 17.
135 Ibid., 15; Lindsay, *Modern Beginnings of Subarctic Ornithology*, xiii.
136 Rich and Johnson, *Rae's Arctic Correspondence 1844–45*, 32. Rae also admitted at the time that "my stock of scientific knowledge is small"; Rae, Sault Ste. Marie to Simpson, 29 July 1845, in Rich and Johnson, *Rae's Arctic Correspondence*, 13.
137 Rae, *Narrative of an Expedition*, 14; Lindsay, *Modern Beginnings of Subarctic Ornithology*, xiii.

138 Rae, *Narrative of an Expedition*, 18.
139 Ibid., 16.
140 Richardson, *Arctic Searching Expedition*, 1: 24.
141 Rae, *Narrative of an Expedition*, 125.
142 Levere, *Science and the Canadian Arctic*, 196. Rae noted that he submitted specimens to the British Museum in 1847 in his "Notes on Some of the Birds and Mammals of the Hudson's Bay Company's Territory, and of the Arctic Coast of America," *Journal of the Linnean Society of London, Zoology*, 20, no. 119 (Nov. 1888): 140.
143 Richardson, *Arctic Searching Expedition*, 2: 425.
144 Lindsay, *Modern Beginnings of Subarctic Ornithology*, xiii.
145 Ibid.
146 J. W. Dawson, "Gleanings in the Natural History of the Hudson's Bay Territories, by the Arctic Voyagers," *Canadian Naturalist and Geologist* 2, no. 3 (July 1857): 170.
147 Richardson, *Arctic Searching Expedition* 1: 48.
148 Ibid., 1: 48. The instructions are at 1: 27–31.
149 Ibid., 1: 49.
150 Ibid., 2: 59.
151 LAC M19, A21 Hargrave Papers, John Rae (Chicago, Ill.) to James Hargrave (Sault Ste. Marie), 26 Feb. 1852.
152 Richards, "John Rae," 877.
153 Ibid., 878; Vilhjalmur Stefansson, "Rae's Arctic Correspondence," *Beaver* 284 (Mar. 1954): 36.
154 R. C. Wallace, "Rae of the Arctic," *Beaver* 284 (Mar. 1954): 29.
155 Rich, *Hudson's Bay Company*, 2: 649.
156 Peter Lund Simmonds, *Sir John Franklin and the Arctic Regions* (Buffalo: George H. Derby, 1852), 195.
157 Wallace, "Rae of the Arctic," 33.
158 Richards, "John Rae," 878–9; Robert L. Richards, *Dr. John Rae* (Whitby, Eng.: Caedmon, 1985), 77.
159 Richards, "John Rae," 878.
160 William Mactavish (Red River) to James Hargrave (Hudson Bay House, Lachine) in LAC James Hargrave papers, 26 Oct. 1859.
161 John Rae, "On the Esquimaux," *Transactions of the Ethnological Society of London* 4 (1865): 139.
162 Rae, "On the Esquimaux," 138–53; John Rae, "On the Condition and Characteristics of Some of the Native Tribes of the Hudson's Bay Company's Territories," *Journal of the Society of Arts* 30 (1882): 483–96.
163 Rae, "On the Esquimaux," 138.

164 Ibid., 153.
165 Ibid., 143.
166 Rae, "On the Condition and Characteristics of Some of the Native Tribes," 496.
167 Cooper, *Alexander Kennedy Isbister*, 6. Cooper provides the only full-length biography of A. K. Isbister. Also see Sylva Van Kirk, "Alexander Kennedy Isbister," *DCB* 11: 445–6.
168 Cooper, *Alexander Kennedy Isbister*, 7.
169 A. K. Isbister, "Some Account of Peel River, N. America," *Journal of the Royal Geographical Society of London* 15 (1845): 335.
170 Isbister's early appointment to "apprentice postmaster" rather than "clerk" seemed to signal that he was not in line for eventual promotion to Chief Trader. Native-born employees and "servants" often found themselves relegated to that position while European men (particularly former employees of the NWC) were more likely to be promoted. Cooper, *Alexander Kennedy Isbister*, 5–9. Also see Denise Fuchs, "Embattled Notions: Constructions of Rupert's Land's Native Sons, 1760 to 1861," *Manitoba History* 44 (2002–3): 10–17; Carol M. Judd, "Native Labour and Social Stratification in the Hudson's Bay Company's Northern Department 1770–1870," *The Canadian Review of Sociology and Anthropology* 17, no. 1 (1980): 305–14; ibid., "'Mixt Bands of Many Nations,' 1821–70," in *Old Trails and New Directions: Papers of the Third North American Fur Trade Conference*, ed. Carol M. Judd and Arthur J. Ray (Toronto: University of Toronto Press, 1980), 127–46; Brown, *Strangers in Blood*, 114, 119, 184–5, 195–6.
171 Cooper, *Alexander Kennedy Isbister*, 33.
172 Isbister, "Some Account of Peel River, N. America," 334. The article is discussed in Cooper, *Alexander Kennedy Isbister*, 30–2.
173 Cooper, *Alexander Kennedy Isbister*, 43. Isbister appears also to have submitted a journal to the Royal Geographical Society in 1845, but that the journal was not published. Ibid., 287.
174 Cooper, *Alexander Kennedy Isbister*, 77.
175 A. K. Isbister, *A Few Words on the Hudson's Bay Company; With a Statement of the Grievances of the Native and Half-Caste Indians, Addressed to the British Government through Their Delegates now in London* (London: C. Gilpin, 1846), quoted passage on p. 1; Rich, *Hudson's Bay Company*, 2: 545, 791.
176 Galbraith, *Hudson's Bay Company as an Imperial Factor*, 288, 318–23 (quoted passage on p. 319); Rich, *Hudson's Bay Company*, 2: 791.
177 Great Britain, Parliament, House of Commons, *Hansard*, 3d ser., vol. 100 (13 July 1848), cols. 470–80; 3d ser. vol. 106 (5 July 1849), cols. 1344–63.

178 Great Britain, Parliament, House of Commons, *Papers Relating to the Legality of the Powers in Respect to Territory, Trade, Taxation and Government Claimed or Exercised by the Hudson's Bay Company*. 1850 (542), quoted passage on p. 7.
179 Cooper, *Alexander Kennedy Isbister*, passim but see esp. 121–39.
180 Ibid., 241.
181 Isbister's many publications are listed, and many are discussed in ibid., 306–9 (although the true authorship of some of the unsigned articles attributed to him is open to question.) The discussion here is focused on his signed scientific articles dealing with the HBC territories.
182 A. K. Isbister, "On the Chippewyan Indians," *Transactions of the Sections* (Ethnology Section) in *Report of the Seventeenth Meeting of the British Association for the Advancement of Science* (London: John Murray, 1848), 119–21; ibid., "On the Nehanni Tribe of a Koloochian Class of American Indians," ibid., 121; and ibid., "On the Loucheux Indians," ibid., 121–2. These articles are discussed in Cooper, *Alexander Kennedy Isbister*, 35–9.
183 Isbister, "On the Chippewyan Indians," 120.
184 Ibid.
185 Ibid.
186 A. K. Isbister, "On the Geology of the Hudson's Bay Territories, and of Portions of the Arctic and North-Western Regions of America; with a Coloured Geological Map," *Quarterly Journal of the Geological Society* 11 (1855): 497–520. References to Kennedy and Barnston are on 505 and 506. That article was reprinted in A. K. Isbister, "On the Geology of the Hudson's Bay Territories, and of Portions of the Arctic and Northwestern Regions of America," *American Journal of Science and Arts* 21 (May 1856): 313–38. These articles are discussed in Cooper, *Alexander Kennedy Isbister*, 39–43.
187 Levere, *Science and the Canadian Arctic*, 177.
188 Richardson, *Fauna Boreali-Americana*, 1: ix.
189 Ibid., 3: xiii.
190 Ibid., 1: xxxvi.
191 Ibid., 1: xix. Richardson later specifically mentioned that the Zoological Society had received a "long-tailed star-nose" from Moose Factory, (ibid., 1: 13) and a mountain goat (ibid., 1: 269).
192 Ibid., 2: xii. The very informal way in which specimens were requested may be seen from a letter George Simpson wrote to James Hargrave in 1840. Simpson wrote, "In passing through the [Hudson] Straits I wish you would get hold of two or three handsome large, thorough bred

Esquimaux pup dogs, with prick ears, and magnificent curling tails; the colour uniform, either Black or Brindled; they are intended as present to the Zoological, Mr. Halkett, and Lord Selkirk. If there be any Buffaloes, or any Bears, Foxes &c, or Birds, let them all be addressed to me; and be good enough to direct Mr. Herd's attention to them very particularly"; LAC Hargrave Papers, George Simpson, London, to James Hargrave, 8 June 1840. There are many examples of mention of specimens getting to naturalists without information about how. For example, in 1829, David Douglas mentioned that a female white tailed deer had been presented to the Zoological Society by the HBC; David Douglas, "Observations on Two Undescribed Species of North American Mammalia," *Zoological Journal* 4 (1829): 332; Douglas also mentioned that mountain goat skins had been collected for Nicholas Garry; Jack Nisbet, *The Collector: David Douglas and the Natural History of the Northwest* (Seattle: Sasquatch Books, 2009), 146.
193 Richardson, *Fauna Boreali-Americana*, 1: 233.
194 Ibid., 1: xviii–xix.
195 Ibid., 1: xviii–xix, xxxvi, 3: x, 158.
196 Ibid., 2: ix.
197 Ibid., 1: 108–9.

Chapter 6

1 Lynn Barber, *Heyday of Natural History: 1820–1870* (London: Jonathan Cape, 1980), 23. Also see Allen, *Naturalist in Britain*, as well as Delbourgo, *A Most Amazing Scene*, 102–9.
2 Explicit references to the link between observation of nature and religious faith are few, but several naturalists discussed in this chapter are known to have been devout Christians, including John Scouler and William Fraser Tolmie. See Malcolm Nicolson, "John Scouler (1804–1871)," *Oxford Dictionary of National Biography* (Oxford: Oxford University Press, 2004–10); W. Kaye Lamb, "William Fraser Tolmie," *DCB* 11: 885–7; and Dorothy O. Johansen, "William Fraser Tolmie of the Hudson's Bay Company" *Beaver* 268, no. 2 (Sept. 1937): 32. For literature on the significance of male bonding within scientific networks, see Hallock, "Male Pleasure."
3 See James R. Gibson, *Lifeline of the Oregon Country* (Vancouver: UBC Press, 1998); Richard Mackie *Trading beyond the Mountains: The British Fur Trade on the Pacific, 1793–1843* (Vancouver: UBC Press, 1996).
4 Fleming, *Minutes of Council of the Northern Department, 1821–1831*, 85.
5 Rich, *Hudson's Bay Company*, 2: 441.

6 Frederick Merk, ed., *Fur Trade and Empire: George Simpson's Journal Entitled Remarks Connected with the Fur Trade in the Course of a Voyage from York Factory to Fort George and Back to York Factory 1824–5*, rev. ed. (Cambridge, MA: Harvard University Press, 1968), 111–12. John Work did note receiving a letter dated 16 Apr. 1825 requesting that Work "do me the favour to collect all the seeds plants Birds and quadrupids [sic] & mice & rats you can and let them be forwarded by the ship of next season to ... H. B. Cmy., London." See T. C. Elliott, ed., "Journal of John Work," *Washington Historical Quarterly* 5, no. 2 (1914): 99. In 1829, Work submitted a lengthy document (as his "Report on District") that "Answers to Queries on Natural History." Most of the document provides information about the aboriginal peoples and geography of the district. See HBCA B.45/e/2. I am indebted to Jack Nisbet for calling my attention to these documents.
7 Governor and Committee to McLoughlin, 20 Sept. 1826, as quoted in Schefke, "Hudson's Bay Company as a Context for Science," 74.
8 Richardson, *Fauna Boreali-Americana*, 2: xii.
9 John Richardson, "On Aplodontia, a New Genus of the Order Rodentia, Constituted for the Reception of the Sewellel, a Burrowing Animal which Inhabits the North Western Coast of America," *Zoological Journal* 4 (1829): 336. Also see Richardson, *Fauna Boreali-Americana*, 1: xix, xxxvi.
10 Athelstan George Harvey, *Douglas of the Fir, A Biography of David Douglas, Botanist* (Cambridge: Harvard University Press, 1947), 17; William Morwood, *Traveler in a Vanished Landscape: The Life and Times of David Douglas* (Newton Abbot, Eng.: Readers Union, 1974), 12. The Horticultural Society of London sought plants for practical or ornamental interest, but did not seek plants that were of only botanical interest; Harold R. Fletcher, *The Story of the Royal Horticultural Society: 1804–1968* (London: Oxford University Press, 1969), 81. For some context, see Andrew Cunningham, "The Culture of Gardens," in Jardine, Secord, and Spary, *Cultures of Natural History*, 38–56.
11 Cohen, "The New World as a Source of Science for Europe," 111; Fletcher, *Story of the Royal Horticultural Society*, 81.
12 Fletcher, *Story of the Royal Horticultural Society*, 79.
13 A. G. Harvey, "John Jeffrey: Botanical Explorer," *British Columbia Historical Quarterly* 10 (Oct. 1946): 281.
14 Nisbet, *The Collector*, 7; William Keddie, "Biographical Notice of the Late John Scouler," *Transactions of the Geological Society of Glasgow* 4 (1874): 194. Aside from Nisbet's biography, David Douglas is the subject of several biographies, including Harvey, *Douglas of the Fir*; and Morwood, *Traveler in a Vanished Landscape*. Also see Jack Nisbet, *David Douglas, A Naturalist at*

Notes to pages 171–2 359

Work: An Illustrated Exploration across Two Centuries in the Pacific Northwest (Seattle: Sasquatch Books, 2012). Douglas was preparing journals for publication before he died. They were published in their entirety for the first time in David Douglas, *Journal Kept by David Douglas During His Travels in North America, 1823–1827 … with Appendices Containing a List of the Plants Introduced by Douglas and an Account of his Death in 1834* (London: William Wesley & Son, 1914), and subsequently as David Lavender, ed., *The Oregon Journals of David Douglas, of His Travels and Adventures among the Traders and Indians in the Columbia, Willamette and Snake River Regions During the Years 1825, 1826 and 1827*, 2 vols. (Ashland, OR: Oregon Book Society, 1972), and John Davies, ed., *Douglas of the Forests: The North American Journals of David Douglas* (Seattle: University of Washington Press, 1980).

15 William Jackson Hooker, "A Brief Memoir of the Life of Mr. David Douglas, with Extracts from His Letters" *Companion to the Botanical Magazine* 2 (1836): 82.
16 Davies, *Douglas of the Forests*, 25.
17 Governor and Committee to Chief Factors in the Columbia District, E. E. Rich, ed., *The Letters of John McLoughlin, from Fort Vancouver to the Governor and Committee: First Series, 1825–38* (London: Hudson's Bay Record Society, 1941), 15. (Letter of Instruction, 22 July 1824, HBCA A.6/21, fol 11d)
18 John Scouler, "Journal of a Voyage to N.W. America [Part I]," *Quarterly of the Oregon Historical Society* 6 (1905): 55.
19 John Scouler, "Account of a Voyage to Madeira, Brazil, Juan Fernandez, and the Gallapagos [sic] Islands, Performed in 1824 and 1825, with a View of Examining Their Natural History, &c." *Edinburgh Journal of Science* 5, no. 2 (Oct. 1826): 196. Scouler described his opportunity to go to the Columbia District as one that "unexpectedly presented itself," Scouler, "Journal of a Voyage to N.W. America [Part I]," 54. For evidence about how Scouler was chosen, see William Smith, Secretary, HBC to John Scowler, 3 July 1826, HBCA A.5/8, fo. 76. That letter indicates that the HBC had sought a surgeon "not only skilful in his profession but well qualified for scientific pursuits."
20 Hooker quoted in Keddie, "Biographical Notice," 195.
21 Scouler, "Account of a Voyage," 196.
22 Scouler, "Journal of a Voyage to N.W. America [Part I]," 54.
23 John Scouler, "Journal of a Voyage to N.W. America, Part II." *The Quarterly of the Oregon Historical Society* 6 (1905): 164.
24 The books range from George Vancouver's journals to Alexander Hamilton's *A Treatise on the Management of Female Complaints*, and include works in natural philosophy (John Playfair's *Outlines of Natural Philosophy*, for

example), and works on natural history by William Bingley, James Edward Smith, and W. J. Hooker. See Fort George, Columbia, Account Book, 1821 [Library] Inventory, 1821, B.76/d/2, fo. 30d-31.
25 John Scouler, "Journal of a Voyage to N.W. America, Part III" *The Quarterly of the Oregon Historical Society* 6 (1905): 279. That this roused "the indignation of the natives," see Keddie, "Biographical Notice," 199. Scouler later published an article that depicted skulls. See John Scouler, "Remarks on the Form of the Skull of a North American Indian" *Zoological Journal* 4 (1829): 304–9. (The skulls, including one still carrying desiccated skin, are depicted on plates 10 and 11 at the end of the volume).
26 David Douglas to W. J. Hooker, 24 Mar. 1826 in Hooker, "Account of the Expedition under Captain Franklin," 113.
27 Scouler, "Account of a Voyage," 196.
28 Keddie, "Biographical Notice," 200. Anderson University remains as the John Anderson campus of the University of Strathclyde. Aside from other articles published from his connections with the HBC, elsewhere cited in this study, can be added John Scouler, "On the Temperature of the North West Coast of America," *Edinburgh Journal of Science* 6 (1827): 251–3.
29 Hooker, "A Brief Memoir of the Life of Mr. David Douglas," 83. Nicolson, "John Scouler (1804–1871)," *Oxford Dictionary of National Biography*. Keddie, "Biographical Notice of the Late John Scouler," 201; Blodwen Lloyd, "John Scouler, M.D. Ll.D., F.L.S. (1804–1871)," *Glasgow Naturalist* 18, no. 4 (Feb. 1962): 210–12.
30 John Scouler, "On the Indian Tribes Inhabiting the North-West Coast of America," *Journal of the Ethnological Society of London* 1 (1848): 228–52.
31 Reuben Hafen Le Roy, *Mountain Men and Fur Traders of the Far West: Eighteen Biographical Sketches* (Lincoln: University of Nebraska Press, 1982), 112.
32 F. Arago, "Sur le Climat de la Côte Orientale de l'Amerique du Nord," *Comptes Rendus Hebdomadaires des Séances de l'Académie des Sciences* 1 (13 July 1835): 266–8; F. Arago, "Observations météorologiques faites sur lá côte occidentale de l'Amerique du Nord, au fort Vancouver, rivière Columbia," *Comptes Rendus Hebdomadaires des Séances de l'Académie des Sciences* 6 (Jan.–June 1838): 120–1. Arago may have acquired the first of McLoughlin's meteorological journals during his visit to Edinburgh in 1834, to attend meetings of the British Association for the Advancement of Science. See the obituary of Arago in the *Monthly Notices of the Royal Astronomical Society*, 14 (1854): 106 (obituary runs from p. 102 to p. 107).
33 Scouler, "Journal of a Voyage to N.W. America, Part II," 168.
34 Quoted in Erwin F. Lange, "Dr. John McLoughlin and the Botany of the Pacific Northwest," *Madroño* 14 (Oct. 1958): 269.

35 David Douglas, Spokan House to George Simpson, 14 Apr. 1826, HBCA D.4/119, fo. 20.
36 Lange, "Dr. John McLoughlin," 270; Harvey, *Douglas of the Fir*, 61.
37 Nisbet, *The Collector*, 79.
38 Hoeniger, "Thomas Drummond," 222; M. L. Tyrwhitt-Drake, "David Douglas," *DCB* 6: 218. Douglas was not at Carlton House long enough to see interesting flowers and collect their seeds, so Drummond's gesture was a generous one.
39 Drummond, "Sketch of a Journey to the Rocky Mountains," 216–18; Richardson, *Fauna Boreali-Americana*, 1: xvii–xviii. Douglas knew of and had planned to rendezvous with the Franklin party. Douglas had consulted with John Richardson in England before they went their separate ways; Nisbet, *The Collector*, 26. Douglas may not have known then that Drummond was going to accompany the Franklin expedition, but he wrote to Hooker in 1826 with apparent interest that "I learn that a Mr Drummond, probably the botanist of that name who has lived at Forfar, accompanies the [Franklin] expedition as a naturalist"; David Douglas to W. J. Hooker, 24 Mar. 1826 in Hooker, "Account of the Expedition under Captain Franklin," 114. They almost certainly expected to meet along the Saskatchewan River; Nisbet, *The Collector*, 152–3.
40 Harvey, *Douglas of the Fir*, 132.
41 William Beattie Booth, quoted in Hooker, "A Brief Memoir of the Life of Mr. David Douglas," 142.
42 Schefke, "Hudson's Bay Company as a Context for Science," 77; Nisbet, *The Collector*, 173.
43 John Lindley, *Edward's Botanical Register* 16 (1830): Item 1349.
44 Hooker, "A Brief Memoir of the Life of Mr. David Douglas," 140–2.
45 Morwood, *Traveler in a Vanished Landscape*, 129.
46 Harvey, *Douglas of the Fir*, 155–6; Morwood, *Traveler in a Vanished Landscape*, 129–30; Nisbet, *The Collector*, 180.
47 Hooker, "A Brief Memoir of the Life of Mr. David Douglas," 142.
48 Ibid. One contemporary also noted that he became "quite a sauvage in his appearance and manners"; Harvey, *Douglas of the Fir*, 151, also see 157–60.
49 Hooker, "A Brief Memoir of the Life of Mr. David Douglas," 143. The Horticultural Society and Zoological Societies also supported this trip; Harvey, *Douglas of the Fir*, 164.
50 Nisbet, *The Collector*, 193.
51 Douglas noted being impressed by the Spanish missionaries' love of science; Ibid., 205.
52 Harvey, *Douglas of the Fir*, 207.

53 Nisbet, *The Collector*, 241–2.
54 Tyrwhitt-Drake, "David Douglas," 218.
55 Hooker, "A Brief Memoir of the Life of Mr. David Douglas," 140.
56 Fletcher, *Story of the Royal Horticultural Society*, 104; Ann Lindsay Mitchell and Syd House, *David Douglas: Explorer and Botanist* (London: Aurum Press, 1999), 180–96.
57 Douglas quoted in Hooker, "A Brief Memoir of the Life of Mr. David Douglas," 82.
58 Davies, *Douglas of the Forests*, 161.
59 Nisbet, *The Collector*, 254.
60 Scouler, "Journal of a Voyage to N.W. America, Part II," 193:
61 Douglas quoted in Nisbet, *The Collector*, 196.
62 Harvey, *Douglas of the Fir*, 56.
63 Lamb, "William Fraser Tolmie," 885. Johansen, "William Fraser Tolmie," 29–32; HBC Committee Minutes, 6 June 1832, HBCA A.1/58, fo. 3d; Governor and Committee, London to J. McLoughlin, Fort Vancouver, 12 Sept. 1832, HBCA A.6/22 fo. 139d.
64 Lamb, "William Fraser Tolmie," 886, 887; *SIAR, 1855* (Washington: A. O. P Nicholson, 1856), 60.
65 John Scouler, "Observations on the Indigenous Tribes of the N.W. Coast of America," *Journal of the Royal Geographic Society* 11 (1841): 216–50.
66 W. Fraser Tolmie and George M. Dawson, *Comparative Vocabularies of the Indian Tribes of British Columbia with a Map Illustrating Distribution* (Montreal: Dawson Brothers, 1884), 3.
67 For more on Jameson see Suzanne Zeller, "The Colonial World as Geological Metaphor: Strata(gems) of Empire in Victorian Canada." *Osiris* 15 (2000): 88.
68 Scouler, "Journal of a Voyage to N.W. America [Part I]," 56–7.
69 Meredith Gairdner, MD, "Notes on the Geography of the Columbia River" *Journal of the Royal Geographic Society* 11 (1841): 250, 256; HBCA A.10/3 17 May 1836, fos. 99-99d,
70 By his own account, the exertion required to retrieve the skull caused Gairdner to cough up blood from his tubercular lungs. A. G. Harvey, "Chief Concomly's Skull," *Oregon Historical Quarterly* 40 (1939): 164–6.
71 M. Gairdner, "Letter from Dr. M. Gairdner, 19th Mar. 1834, Fort Vancouver" *Edinburgh New Philosophical Journal,* (Jan. 1836): 205–7; Meredith Gairdner, "Meteorological Observations Made at Fort Vancouver, from June 7, 1833 to May 31, 1834" *Edinburgh New Philosophical Journal* 20 (1836): 67; Meredith Gairdner, "General Table of Meteorological Observations at

Fort Vancouver, from June 1, 1834 to May 13, 1835" *Edinburgh New Philosophical Journal* 21 (1836): 152–3.
72 Anonymous, "A List of the Collectors Whose Plants Are in the Herbarium of the Royal Botanic Gardens, Kew, to 31st December, 1899," *Bulletin of Miscellaneous Information (Royal Gardens, Kew)*, 169/171 (Jan.–Mar. 1901): 25.
73 Richardson, *Fauna Boreali-Americana*, 3: 158. The species was renamed *Oncorhynchus mykiss* in the 1990s.
74 George Simpson to J. McLoughlin, Norway House, 28 June 1836 (Confidential), HBCA D.4/22, fo. 38d.
75 Kate Colquhoun, *"The Busiest Man in England": Life of Joseph Paxton, Garden Architect, and Victorian Visionary* (Boston: David R. Godine, 2006), 83, 93; Violet R. Markham, *Paxton and the Bachelor Duke* (London: Hodder and Stoughton, 1935), 63–72. George Simpson was apparently Peter Bank's father-in-law; Gibson, *Lifeline of the Oregon Country*, 42.
76 Colquhoun, *"Busiest Man in England"*, 35; Markham, *Paxton and the Bachelor Duke*, 65.
77 Markham, *Paxton and the Bachelor Duke*, 68.
78 Circular from G. Simpson, London, to James Keith and the Gentlemen in Charge of the District Posts in the Hudson's Bay Company Service, 12 Mar. 1838, as reproduced in Joseph Paxton, "Botanical Expedition to Columbia, and Melancholy Loss of Messrs. Wallace and Banks" *The Gardeners Magazine and Register of Rural & Domestic Improvement* 17 (Aug. 1839): 136.
79 Paxton, "Botanical Expedition to Columbia," 136.
80 Ibid., 135. J. A. Stevenson, "Disaster in the Dalles," *Beaver* 273 (Sept. 1942): 19–21.
81 Allan, *Hookers of Kew*, 78–9.
82 Ibid., 89.
83 Ibid., 15; Colquhoun, *"Busiest Man in England"*, 82; Harvey, *Douglas of the Fir*, 156.
84 Archibald Barclay, Secretary of HBC, London to John McLoughlin, Fort Vancouver, 28 Sept. 1843, HBCA A.6/26, fo. 86. Barclay's response to Hooker is found at Archibald Barclay, Secretary of HBC, London to Hooker, 5 Oct. 1843, HBCA 5/14, p 188.
85 Susan Delano McKelvey, *Botanical Exploration of the Trans-Mississippi West, 1790–1850* (Corvallis, OR: Oregon State University Press, [1955] 1991), 811.
86 Richard Glover, "The Man Who Did Not Go to California," Canadian Historical Association *Historical Papers* 10 (1975): 101.
87 William Jackson Hooker, "Figure and Description of Castanea Chrysophylla" *London Journal of Botany* 2 (1843): 496.

88 Ibid., 496–7.
89 Glover, "The Man Who Did Not Go to California," 98–9, 101; Margaret Arnett MacLeod, ed., *Letters of Letitia Hargrave* (Toronto: Champlain Society, 1947), 175; E. E. Rich, ed., *The Letters of John McLoughlin, from Fort Vancouver to the Governor and Committee: Third Series, 1844–46* (London: Hudson's Bay Record Society, 1944), 59n3. For more on Burke, see Clifford M. Drury, "Botanist in Oregon in 1843–44 for Kew Gardens, London," *Oregon Historical Quarterly* 41, no. 2 (June 1940). For more on the earl of Derby's collection (although nothing on Burke), see S. J. Woolfall, "History of the 13th Earl of Derby's Menagerie and Aviary at Knowsley Hall, Liverpool (1806–1851)," *Archives of Natural History* 17 (1990): 1–47.
90 Governor and Committee to George Simpson, 1 Apr. 1843, HBCA A.6/26, fo. 59.
91 MacLeod, *Letters of Letitia Hargrave*, 176.
92 Glover, "The Man Who Did Not Go to California," 99–101; Rich, *The Letters of John McLoughlin, from Fort Vancouver to the Governor and Committee: Third Series*
93 McKelvey, *Botanical Exploration*, 796.
94 Rich, *The Letters of John McLoughlin, from Fort Vancouver to the Governor and Committee: Third Series*, 59.
95 Glover, "The Man Who Did Not Go to California," 103.
96 Ibid., 104–5.
97 McKelvey, *Botanical Exploration*, 814.
98 Glover, "The Man Who Did Not Go to California," 105.
99 Ibid., 106.
100 Perhaps Hooker later realized that he had been too harsh with Burke. He did name a species after Burke; R. K. Beattie, "Joseph Burke up to 1853," *Madroño* 13, no. 8 (Oct. 1956): 260.
101 Glover, "The Man Who Did Not Go to California," 106. McKelvey also defended Burke. See McKelvey, *Botanical Exploration*, 792–817.
102 Beattie, "Joseph Burke up to 1853," 259–61.
103 McKelvey, *Botanical Exploration*, 788.
104 McKelvey, *Botanical Exploration*, 660.
105 Keir B. Sterling et al., eds., *Biographical Dictionary of American and Canadian Naturalists and Environmentalists* (Westport, CT: Greenwood Press, 1997), 253; McKelvey, *Botanical Exploration*, 774–5; Mae Reed Porter and Odessa Davenport, *Scotsman in Buckskin: Sir William Drummond Stewart and the Rocky Mountain Fur Trade* (New York: Hastings House, 1963), 216. That Hooker and Gray were friends is explained in Bruce Sinclair, "Americans Abroad: Science and Cultural Nationalism in the Early Nineteenth

Century," in *The Sciences in the American Context*, ed. Nathan Reingold (Washington, DC: Smithsonian Institution Press, 1979), 41, 47 (article goes from pp. 35–53).
106 McKelvey, *Botanical Exploration*, 772–3.
107 Ibid., 772; Drury, "Botanist in Oregon," 182. Although Geyer did not know Hooker personally, he later wrote to Hooker, noting that he was the botanist who "set out at your desire, as expressed in a letter to ... Dr. Geo. Engelmann at St. Louis, ... to explore the upper Missouri country," quoted in Drury, "Botanist in Oregon," 185; Frederick V. Coville, "Added Botanical Notes on Carl A. Geyer," *Oregon Historical Quarterly*, 42 (Dec. 1941): 323.
108 Drury, "Botanist in Oregon," 186; McKelvey, *Botanical Exploration*, 778. McKelvey suggests that the AFC officials may have disliked Geyer.
109 Porter and Davenport, *Scotsman in Buckskin*, 216.
110 Lisa Strong, "American Indians and Scottish Identity in Sir William Drummond Stewart's Collection," *Winterthur Portfolio* 35, no. 2/3 (Summer-Autumn 2000): passim, esp. 143.
111 John McLoughlin, Oregon City, to Governor and Committee, 1 July 1846, in Rich, *The Letters of John McLoughlin, from Fort Vancouver to the Governor and Committee: Third Series*, 161; Nute, "Botanist at Fort Colville," 28.
112 Nute, "Botanist at Fort Colville," 29; Rich, *The Letters of John McLoughlin, from Fort Vancouver to the Governor and Committee: Third Series*, 160.
113 Nute, "Botanist at Fort Colville," 29. Stewart had apparently been told that Geyer would have to pay £100 for his passage to London on the company's ship; Rich, *The Letters of John McLoughlin, from Fort Vancouver to the Governor and Committee: Third Series*, 53.
114 Drury, "Botanist in Oregon," 186. An overturned canoe destroyed three years' worth of Lüders's collections. Lüders noted that "the kindest assistance was offered me by the gentlemanly officers at Fort Vancouver," but the accident ended his career as a travelling botanist. Clara T. Runge, "Frederick George Jacob Lueders: Naturalist and Philosopher, 1818–1904" *The Wisconsin Magazine of History* 15, no. 3 (Mar. 1932): 353–4.
115 McKelvey, *Botanical Exploration*, 774–6.
116 Ibid., 788–90; Porter and Davenport, *Scotsman in Buckskin*, 251; Strong, "American Indians and Scottish Identity," 143.
117 Charles A. Geyer, "Notes on the Vegetation and General Character of the Missouri and Oregon Territories" *London Journal of Botany* 4 (1845): 482.
118 J. McNab, "On the Discoveries of Mr. John Jeffrey and Mr. Robert Brown, Collectors to the Botanical Expeditions to British Columbia between the years 1850 and 1866," *Transactions of the Botanical Society Edinburgh* 11

(1872): 322–3; Fletcher, *Story of the Royal Horticultural Society*, 179; Harvey, "John Jeffrey," 282.
119 Quoted in Strong, "American Indians and Scottish Identity," 149.
120 Harvey, "John Jeffrey," 282.
121 Ibid., 290.
122 Frank A. Lang, "John Jeffrey in the Wild West: Speculations on his Life and Times (1828–1854?)" *Kalmiopsis*, 13 (2006): 5; Harvey, "John Jeffrey," 283–7.
123 McNab, "On the Discoveries of Mr. John Jeffrey and Mr. Robert Brown," 325; Harvey, "John Jeffrey," 287–8.
124 Harvey, "John Jeffrey," 288.
125 See F. V. Coville, "The Itinerary of John Jeffrey, an Early Botanical Explorer of Western North America." *Proceedings of the Biological Society of Washington* 11 (1897): 57–60; Erwin F. Lange, "John Jeffrey and the Oregon Botanical Expedition." *Oregon Historical Quarterly* 69 (1967): 111–24; McNab, "On the Discoveries of Mr. John Jeffrey and Mr. Robert Brown," 322–38.
126 Fletcher, *Story of the Royal Horticultural Society*, 179.
127 In 1858, Murray wrote that he had been secretary of "an association" that sought to "to procure seeds of new and valuable hardy trees and plants from Oregon and the neighbouring districts"; Andrew Murray, "Contributions to the Natural History of the Hudson's Bay Company's Territories: Part I – Rein-Deer" *Edinburgh New Philosophical Journal* 7, no. 2 (Apr. 1858): 190. For a suggestion that this initiated him into the world of science, see J. F. M. Clark, "Andrew Murray" *Oxford Dictionary of National Biography Online* (2004).
128 Anonymous, "Andrew Murray, F. L. S.," *Entomologist's Monthly Magazine* 3, no. 14 (1877–8): 215–16; Clark, "Andrew Murray." Formerly the London Horticultural Society, the name of the society was changed in 1861 after receiving a royal charter.
129 Clark, "Andrew Murray."
130 Murray, "Contributions to the Natural History of the Hudson's Bay Company's Territories: Part I,"190–1.
131 Ibid., 191.
132 Ibid., 191.
133 Andrew Murray, "Contributions to the Natural History of the Hudson's Bay Company's Territories, Part II – Mammalia (continued)," *Edinburgh New Philosophical Journal* (Apr. 1859): 210.
134 Ibid., 216.

135 Murray, "Contributions to the Natural History of the Hudson's Bay Company's Territories: Part I,": 191.
136 Debra Lindsay, "The Hudson's Bay Company-Smithsonian Connection and Fur Trade Intellectual Life: Bernard Rogan Ross, a Case Study," in *"Le Castor Fait Tout": Selected Papers of the Fifth North American Fur Trade Conference, 1985*, ed. Bruce G. Trigger, Toby Morantz, and Louise Dechêne, (Montreal: Lake St. Louis Historical Society, 1987), 600; Murray, "Contributions to the Natural History of the Hudson's Bay Company's Territories, Part II," 229; Portions of the letter are quoted in Thomas Blakiston, "On the Birds of the Interior of British North America," *Ibis* 5 (1863):145.
137 Despite the title of the second part ("Mammalia"), this article includes sections on birds and fish.
138 Murray, "Contributions to the Natural History of the Hudson's Bay Company's Territories: Part I," 189–90.
139 Ibid., 190.
140 Ibid., 190.
141 McNab, "On the Discoveries of Mr. John Jeffrey and Mr. Robert Brown," 333. For more on Brown, including his botanical surveys of Vancouver Island (when no longer under HBC jurisdiction), see John Hayman, "Robert Brown," *DCB* 12: 129–30.
142 Kerr, "For the Royal Scottish Museum," 32.
143 P. J. Hartog, rev. by R. G. W. Anderson, "Wilson, George (1818–1859)," *Oxford Dictionary of National Biography* (Oxford, Oxford University Press, 2004).
144 George Simpson, Hudson's Bay House, Lachine to George Barnston, Norway House, 4 May 1857, HBCA B.239/c/10, fo. 237.
145 Dale Idiens, "Chipewyan Artefacts Collected by Robert Campbell and Others in the National Museums of Scotland," in *Proceedings of the Fort Chipewyan and Fort Vermilion Bicentennial Conference: September 23–25, 1988, Provincial Museum of Alberta, Edmonton, Alberta*, ed. Patricia Alice McCormack, and R. G. Ironside (Edmonton: Alberta Culture and Multiculturalism, Boreal Institute for Northern Studies, 1990), 278 and Kerr, "For the Royal Scottish Museum," 33. For Barnston see Lindsay, *Science in the Subarctic*, 50.
146 Kerr, "For the Royal Scottish Museum," 35.
147 Idiens, "Chipewyan Artefacts," 280.
148 Nisbet, *The Collector*, 138; Jennifer S. H. Brown and Sylvia M. Van Kirk, "George Barnston," *DCB* 11: 52; Also a biography in Fleming, *Minutes of Council of the Northern Department, 1821–1831*, 427 (where Simpson is

quoted as saying that Barnston had been "well educated"); G. A. Dunlop and C. P. Wilson, "George Barnston," *Beaver* 272 (Dec. 1941): 16–17.
149 LAC Hargrave Papers, George Barnston, Fort Langley to Hargrave, YF, 17 Feb. 1828.
150 Williams, *Hudson's Bay Miscellany*, 231.
151 Ibid.
152 LAC Hargrave Papers, George Barnston, Martins Falls, to James Hargrave, YF, 17 July 1834.
153 For a similar argument see Suzanne Zeller, "Humboldt and the Habitability of Canada's Great Northwest" *Geographical Review* 96, no. 3 (2006), 388.
154 Nisbet, *The Collector*, 138.
155 Barnston quoted in Nisbet, *The Collector*, 193.
156 LAC Hargrave papers, George Barnston, Norway House, to James Hargrave, 22 June 1831.
157 LAC Hargrave Papers, George Barnston to Hargrave, 1 June 1832.
158 LAC Hargrave Papers, George Barnston, Martins Falls, to James Hargrave, YF, 17 July 1834.
159 LAC Hargrave Papers, George Barnston, Martin's Fall to James Hargrave, 2 and 5 Feb. 1836. Many years later, Barnston published "Abridged Sketch of the Life of David Douglas, Botanist, with a Few Details of His Travels and Discoveries," *Canadian Naturalist and Geologist*, 5 (1860): 120–32, 200–8, 267–78.
160 LAC Hargrave Papers, George Barnston, Martin's Fall to Hargrave, 1 Feb. 1837
161 Brown and Van Kirk, "George Barnston," 52.
162 G.P. de T. Glazebrook, ed., *The Hargrave Correspondence, 1821–1843* (Toronto: Champlain Society, 1938), 293.
163 Barnston to Baird, 26 July 1860, in Lindsay, *Modern Beginnings of Subarctic Ornithology*, 72.
164 George Barnston, "Observations on the Progress of the Seasons as Affecting Animals and Vegetables at Martin's Falls, Albany River, Hudson's Bay," *Edinburgh New Philosophical Journal* 30 (1840–1): 252–6. That Barnston sent this to Richardson, and that Richardson had it published, see Richardson, *Arctic Searching Expedition*, 2: 241.
165 Barnston to Baird, 26 July 1860, in Lindsay, *Modern Beginnings of Subarctic Ornithology*, 72.
166 Thomas, "The Smithsonian and the Hudson's Bay Company," 287, citing Barnston to Simpson, 3 Jan. 1844, HBCA D.5/10, fo. 28
167 Barnston to Baird, 26 July 1860, in Lindsay, *Modern Beginnings of Subarctic Ornithology*, 72. John Edward Gray (1800–75) was keeper of the British

Museum at the time, and E. Doubleday and Adam White worked with the insect collection of that museum; Lindsay, *Modern Beginnings of Subarctic Ornithology*, 71.
168 Adam White in Richardson, *Arctic Searching Expedition*, 2: 356.
169 Ibid. The several catalogues of insects in the collection of the British Museum subsequently published by John Edward Gray and others bear out White's prediction.
170 P. P. Harper and W. E. Ricker, "Distribution of Ontario Stoneflies (Plecoptera) *Proceedings of the Entomological Society of Ontario* 125 (1994): 45. To Barnston, explain Harper and Ricker (45), goes the distinction of being "the first entomologist to describe the peculiar drumming habit that is widespread among stoneflies."
171 Suzanne E. Zeller and John H. Noble, "James Barnston," *DCB* 8: 61–2 (quote on p. 62).
172 A. N. Rennie, "Obituary: James Barnston, M.D.," *Canadian Naturalist and Geologist*, 3 (1858): 224–6 (quoted passage on p. 225); Zeller and Noble, "James Barnston," 61.
173 Henry J. Morgan, *Bibliotheca Canadensis or a Manual of Canadian Literature* (Ottawa: G.E. Desbarats, 1867), 19.
174 Rennie, "Obituary: James Barnston, M.D.," 225.
175 W. S. M. D'Urban, "A Systematic List of Lepidoptera Collected in the Vicinity of Montreal," *The Canadian Naturalist and Geologist* 5, no. 4 (1860): 242.
176 Lindsay, *Science in the Subarctic*, 50.
177 Ibid.
178 Ibid.
179 W. S. M. D'Urban, "Catalogue of Coleoptera Collected by George Barnston, Esq., of the Hon. Hudson's Bay Company, in the Hudson's Bay Territories" *The Canadian Naturalist and Geologist* 5, no. 3 (June 1860): 227.
180 George Barnston, "Remarks upon the Geographical Distribution of the Order Ranunculaceae, throughout the British Possessions of North America," *Canadian Naturalist and Geologist* 2 (Mar. 1857): 12–19; George Barnston, "Remarks upon the Geographical Distribution of Plants in the British Possessions of North America" *Canadian Naturalist and Geologist* 3, no. 1 (Feb. 1858): 26–32; George Barnston, "Remarks on the Geographical Distribution of the Cruciferae throughout the British Possessions in North America," *The Canadian Naturalist and Geologist* 4 (Feb. 1859): 1–12; George Barnston, "Geographical Distribution of the Genus Allium in British North America *The Canadian Naturalist and Geologist* 4 (Feb. 1859): 116–21; George Barnston, "Recollections of the Swans and Geese of Hudson's Bay," *Canadian Naturalist* 6 (Feb. 1861): 337–44; George Barnston,

"Remarks on the Genus Lutra and on the Species inhabiting North America," *The Canadian Naturalist and Geologist* 8 (Feb. 1863): 147–60; George Barnston, "On a Collection of Plants from British Columbia, Made by Mr. James Richardson in the Summer of 1874," *Canadian Naturalist and Quarterly Journal of Science*, new ser., 8 (Apr. 1876): 90–4.

181 George Barnston, "Recollections of the Swans and Geese of Hudson's Bay," *Ibis* 2, no. 3 (July 1860): 254. The paper of the same title subsequently published in the *Canadian Naturalist* is a slightly revised version of this paper.

Chapter 7

1 Cawood, "Terrestrial Magnetism"; Cawood, "The Magnetic Crusade"; Warner, "Terrestrial Magnetism"; Carter, "Magnetic Fever."
2 This is a point made by Carter in "Magnetic Fever," xxvi. For Carter's discussion (and critique) of the literature's emphasis on science "as agent of imperial power, not a self-motivated force," especially in the context of the British science and geophysical science in the nineteenth century, see xviii–xix.
3 An excellent recent assessment of the "fur desert" policy can be found in John Phillip Reid, *Contested Empire: Peter Skene Ogden and the Snake River Expeditions* (Norman: University of Oklahoma Press, 2002).
4 Rich, *Hudson's Bay Company*, 2: 563–748; Morton, *History of the Canadian West*, 710–49.
5 Galbraith, *Little Emperor*, 143.
6 Ibid., 30–1.
7 In Nov. 1845, when John Ballenden wrote George Simpson to recommend that he grant Paul Kane permission to travel to the HBC territories, Simpson replied by writing, "You do not say whether Mr. Kane the artist is a British subject or an American. If a British subject I shall have much pleasure in facilitating his projected visit to the interior." George Simpson, Lachine to John Ballenden, Sault Ste. Marie, 17 Nov. 1845, HBCA D. 4/33, fo. 107.
8 Harvey, *Douglas of the Fir*, 194; McKelvey, *Botanical Exploration*, 510–11.
9 McKelvey, *Botanical Exploration*, 512.
10 Thomas Nuttall, "A Catalogue of a Collection of Plants Made Chiefly in the Valleys of the Rocky Mountains of Northern Andes, towards the Sources of the Columbia River, by Mr. Nathaniel B. Wyeth, and Described by T. Nuttall," *Journal of the Academy of Natural Science of Philadelphia* 7 (1834): 5–60.

11 Porter and Davenport, *Scotsman in Buckskin*, 77, 90.
12 Anonymous, "Biographical Sketch of the Late Thomas Nuttall," *The Gardener's Monthly and Horticultural Advertiser* 2 (1860): 23.
13 Rich, *Letters of John McLoughlin, First Series*, 126; Rich, *The Letters of John McLoughlin, from Fort Vancouver to the Governor and Committee: Third Series*, 161.
14 John K. Townsend, *Narrative of a Journey across the Rocky Mountains to the Columbia River* (Philadelphia: Henry Perkins, 1839), 169.
15 Ibid.
16 Lange, "Dr. John McLoughlin," 271; W. Kaye Lamb, "John McLoughlin," *DCB* 8: 578; McKelvey, *Botanical Exploration*, 587. Gairdner and Nuttall left Fort Vancouver on the same ship, on 1 Oct. 1835; McKelvey, *Botanical Exploration*, 613.
17 Townsend, *Narrative of a Journey across the Rocky Mountains*, 349.
18 Ibid., 263.
19 Sinclair, "Americans Abroad," 41.
20 Sinclair, "Americans Abroad," 41, 47, 42.
21 Ibid., 44.
22 Lamb, "John McLoughlin," 578.
23 Ibid., 579.
24 Ibid., 580.
25 Ibid., 580.
26 Peter Skene Ogden to John McLeod, 25 Feb. 1837, in "Documents: Oregon Missionaries" *Washington Historical Quarterly* 2, no. 2 (1908): 260. Given Ogden's double negative, it is difficult to know what he actually meant.
27 Quoted in Lange, "Dr. John McLoughlin," 271.
28 Quoted in Ibid., 271–2. According to Karl A. Geyer, for their part the HBC men were unimpressed by Frémont; Nute, "Botanist at Fort Colville," 30.
29 Rich, *The Letters of John McLoughlin, from Fort Vancouver to the Governor and Committee: Third Series*, cxxii. The criticism of McLoughlin came from outside the company as well. In 1856, the Aborigines Protection Society attributed the loss of the southern portion of Columbia District to the United States in 1846 to McLoughlin "having uniformly encouraged the emigration of settlers from the United States, and of having discouraged that of British subjects"; Aborigines Protection Society, *Canada West and the Hudson's Bay Company: A Political and Humane Question of Vital Importance to the Honour of Great Britain, to the Prosperity of Canada, and the Existence of the Native Tribes* (London: William Tweedie, 1856), 12.
30 Nute, "Botanist at Fort Colville," 31.

31 John J. Audubon, New York, to George Simpson, 29 Apr. 1845, HBCA D.5/13, fo. 470.
32 Ibid., fo. 470d. Through George Back, Audubon had evidently in 1836 sought permission from Governor Pelly to travel to the HBC territories "for the sole purpose of collecting Birds," but it is unclear whether the Governor ever responded to the request. See George Back to Governor Pelly, 5 May 1836, HBCA A.10/3. At the time (as now) the classification of species and subspecies of bears in northwestern North America was unclear. At the time the "barren ground bear" *Ursus arctos* was regarded as an inhabitant of the "the barren-grounds and Arctic coasts. Distinguished from the U. horibilis by its smaller size and reddish coloration"; Bernard R. Ross, "A List of Mammals, Birds, and Eggs Observed in the McKenzie's River District, with Notices," *Natural History Review* 7 (July 1862): 273. Thus, the barren ground bear was a subspecies of the brown bear, but perhaps not a subspecies that is still recognized today.
33 John H. Scott, Montreal, to George Simpson, Lachine, 18 Nov. 1845, HBCA D.5/15, fo. 415.
34 George Simpson, Lachine, to John F. [sic] Audubon, New York, 24 Jan. 1846, HBCA D.4/67 p. 584.
35 Ibid., p. 583–4. Simpson's letter to William Sinclair instructing him to obtain specimens for Audubon can be found at George Simpson, Lachine, to William Sinclair, Churchill, 22 Nov. 1845, HBCA D.4/67 (fo. 235d). Even though he was cautious, Simpson's words were probably overly optimistic about the chances of procuring a musk ox from Churchill. The normal range of the musk ox is far to the northwest of Churchill. Perhaps Simpson hoped that Rae would procure a specimen during his expedition.
36 John James Audubon and John Bachman, *The Quadrupeds of North America*, (New York: V.G. Audubon, 1846–54).
37 Ibid., 2: 211–12. The National Institute for the Promotion of Science was founded in 1840. After 1842 it was called the National Institute; Curtis M. Hinsley, *The Smithsonian and the American Indian: Making a Moral Anthropology in Victorian America* (Washington: Smithsonian Institution Press, [1981] 1994), 17.
38 Audubon and Bachman, *Quadrupeds of North America*, 3: 49–50.
39 The specimen he saw was probably one collected by William Edward Parry and presented to the museum by the Admiralty. See John Edward Gray, *List of the Specimens of Mammalia in the Collection of the British Museum* (London: The Trustees, 1843), 153.
40 Audubon and Bachman, *Quadrupeds of North America*, 3: 49.

41 Ibid., 72.
42 Ibid., 72. The answer to Audubon's question is 1854 (The Philadelphia Zoological Garden). The London Zoological Garden had opened in 1828. See Alexander, *Museums in Motion*, 112, 103.
43 Galbraith, *Little Emperor*, 188.
44 Galbraith, *Hudson's Bay Company as an Imperial Factor*, 336, 356.
45 See Rich, *Hudson's Bay Company*, 2: 813; and Alvin C. Gluek Jr., *Minnesota and the Manifest Destiny of the Canadian Northwest* (Toronto: University of Toronto Press, 1965).
46 Galbraith, *Hudson's Bay Company as an Imperial Factor*, 305.
47 For a discussion of this event from the point of view of the HBC see Rich, *Hudson's Bay Company*, 2: 795.
48 It is beyond the scope of the present volume to examine the history of Canadian science before 1870, or even the history of Canada's expansionist movement in more than a cursory fashion. Interested readers should consult Doug Owram, *Promise of Eden: The Canadian Expansionist Movement and the Idea of the West, 1856–1900* (Toronto: University of Toronto Press, 1980), 3–58; Suzanne Zeller, *Inventing Canada: Early Victorian Science and the Idea of a Transcontinental Nation* (Toronto: University of Toronto Press, 1987), 118–44; Suzanne Zeller, *Land of Promise, Promised Land: The Culture of Victorian Science in Canada* (Ottawa: Canadian Historical Association, 1996), and A. B. McKillop, *A Disciplined Intelligence: Critical Inquiry and Canadian Thought in the Victorian Era* (Montreal: McGill-Queen's University Press, 1979). The purpose in the following section is to argue that Lefroy, Kane, and the Canadian Institute were more important in the history of Canada perceptions of the HBC territories than the literature suggests. A summary of the Canadian expansionists in relation to the HBC can be found in Rich, *Hudson's Bay Company*, 2: 794–9.
49 Galbraith, *Little Emperor*, 89. In 1833 the company bought Simpson a substantial home at Lachine; Galbraith, *Little Emperor*, 122. Then in 1853 Simpson bought Isle de Dorval in the St. Lawrence River near Montreal, where he established a respectable private garden; Galbraith, *Little Emperor*, 128; Nute, "Kennicott in the North," 28.
50 Galbraith, *Little Emperor*, 90.
51 Zeller, "Spirit of Bacon," 88; Thomas, "The Smithsonian and Hudson's Bay Company," 286; Berger, *Science, God, and Nature*, 4–5; Pyenson and Sheets-Pyenson, *Servants of Nature*, 330.
52 Zeller, "Spirit of Bacon," 88; Thomas, "The Smithsonian and Hudson's Bay Company," 286.

53 Pyenson and Sheets-Pyenson, *Servants of Nature*, 330, Susan Sheets-Pyenson, *John William Dawson: Faith, Hope, and Science* (Montreal and Kingston: McGill-Queen's University Press, 1996), 167–9.
54 Sheets-Pyenson, *John William Dawson*, 171, 168.
55 Pyenson and Sheets-Pyenson, *Servants of Nature*, 334.
56 Carter, "Magnetic Fever," 153; Cawood, "The Magnetic Crusade," 518.
57 Carter, "Magnetic Fever," 152; Cawood, "The Magnetic Crusade," 495.
58 Cawood, "The Magnetic Crusade," 496.
59 These factors ranged from the 1820 discovery that magnetism and electricity were linked, to the shipwrecks caused by flawed compass readings linked to the increasing use of metal in ships.
60 Cawood, "The Magnetic Crusade," 498.
61 Ibid., 502.
62 Carter, "Magnetic Fever," 128–9.
63 Ibid., 144.
64 Ibid., 143–5
65 William Whewell, *History of the Inductive Sciences from the Earliest to the Present Time*, 2nd ed. (London: John W. Parker, 1847), 3: 75.
66 Edward Sabine, "Observations on the Magnetism of the Earth, Especially of the Arctic Regions; in a Letter from Capt. Edward Sabine, to Professor Renwick," *American Journal of Science and Art* 17, no. 1 (1830): 151.
67 Zeller, *Inventing Canada*, 116; A. D. Thiessen, "The Founding of the Toronto Magnetic Observatory and the Canadian Meteorological Service," *Journal of the Royal Astronomical Society of Canada* 40 (1940): 312–15.
68 John Henry Lefroy, *Magnetical and Meteorological Observations at Lake Athabaska and Fort Simpson, by Captain J.H. Lefroy, Royal Artillery; and at Fort Confidence, in Great Bear Lake, by Sir John Richardson C.B., M.D* (London: Longman, Brown, Green, and Longmans, 1855), ix. These were supplemented by two observatories in the British Isles (Dublin and Greenwich) and three others in the British Empire funded by the Artillery and Admiralty (on St. Helena Island, at Cape Town, and Hobart, Van Diemen's Land), and four in India (funded by the EIC). Other countries supported many other observatories in their mother countries and colonies. For a list see Cawood, "The Magnetic Crusade," 513.
69 Edward Sabine, "Contributions to Terrestrial Magnetism. No. VII," *Philosophical Transactions of the Royal Society* 136 (1846): 238.
70 Zeller, *Inventing Canada*, 127–9.
71 Carol M. Whitfield and Richard A. Jarrell, "John Henry Lefroy," *DCB* 11: 508; Zeller, *Inventing Canada*, xiv; Zeller, "Humboldt and the Habitability of Canada's Great Northwest," 389–90.

72 Lefroy, *Magnetical and Meteorological Observations*, x; Zeller, *Inventing Canada*, 127–30. Lefroy published his personal diary of the expedition in 1883; J. H. Lefroy, *Diary of a Magnetic Survey of a Portion of the Dominion of Canada, Chiefly in the North-Western Territories, Executed in the Years 1842–44* (London: Longmans, 1883). His wife ensured the publication of his autobiography after his death; Sir J. H. Lefroy, *Autobiography*, ed. Lady Lefroy (London: Pardon and Sons, 1895).
73 John McLean, *Notes of a Twenty-Five Years' Service in the Hudson's Bay Territory* (London: Richard Bentley, 1849), 2: 233.
74 Ibid., 229.
75 Ibid., 228–9. The term *mangeur de lard* (literally "pork-eater"), analogous to "greenhorn," was used to denote newcomers to the territories. Winterers (*hivernants*) were experienced veterans.
76 J. H. Lefroy, *On the Probable Number of the Native Indian Population of British America from the Proceedings of the Canadian Institute* (Toronto: Printed by Hugh Scobie, ca. 1853). This was reprinted as J. H. Lefroy, "On the Probable Number of the Native Indian Population of British America" *The American Journal of Science and Arts*, 2nd series, 16 (Nov. 1853): 189–203.
77 J. H. Lefroy, "Second Report on Observations of the Aurora Borealis, 1850–51,"*London, Edinburgh and Dublin Philosophical Magazine and Journal of Science*, 4th series, 4 (1852): 59–68. Lefroy had previously published another paper to which was appended the instructions for lay keepers of auroral registers that obviously guided the HBC employees. See J. H. Lefroy, "Preliminary Report on the Observations of the Aurora Borealis, Made by the Non-commissioned Officers of the Royal Artillery, at the Various Guard-rooms in Canada," *Edinburgh and Dublin Philosophical Magazine and Journal of Science*, 3rd series, 36 (1850): 457–66.
78 Circular from James Anderson to officers in charge of Mackenzie River Posts, 16 Mar. 1852, in William McMurray Correspondence Inward, 1852–3, HBCA E.61/2, fo. 1.
79 Ibid., fo. 1.
80 Ibid., fo. 1.
81 Ibid., fo. 1d.
82 F. H. Schofield, *A Brief Sketch of the Life and Services of Retired Chief Factor, R. MacFarlane*. (Winnipeg: S.J. Clarke, 1913), 3.
83 Harper, *Paul Kane's Frontier*, 14.
84 Paul Kane, "Incidents of Travel on the North-West Coast, Vancouver's Island, Oregon, &c., &c, The Chinook Indians," *Canadian Journal* 3, no. 12 (July 1855), 278. In *Wanderings of an Artist*, he later described his goal as "to sketch pictures of the principal chiefs, and their original customs,

to illustrate their manners and customs, and to represent the scenery of an almost unknown country." Paul Kane, *Wanderings of an Artist among the Indians of North America from Canada to Vancouver's Island and Oregon through the Hudson's Bay Company's Territory and Back Again* (London: Longman, Brown, Green, Longmans and Roberts, 1859), viii.

85 John Ballenden, Sault Ste. Marie to George Simpson, Lachine, 29 Oct. 1845, HBCA D. 5/15, fo. 3.

86 J. H. Lefroy, Toronto, to G. Simpson, Lachine, 16 Mar. 1846, HBCA D 5/16, fo. 408. Understandably, given the status of the Oregon Crisis at the time, Simpson wanted to confirm that Kane was a British subject before approving his entry into the HBC territories; Harper, *Paul Kane's Frontier*, 16.

87 Simpson as quoted in Harper, *Paul Kane's Frontier*, 327. Harper judged that "no artist ever travelled and lived for three years more cheaply than Kane had done on his 1846–8 wanderings. He had had free hospitality and travelling privileges with the Hudson's Bay Company men"; ibid., 27–8.

88 George Simpson to Paul Kane, n.d., quoted in ibid., 330.

89 Ibid., 28; J. Russell Harper, "Paul Kane," *DCB* 10: 391.

90 Harper, *Paul Kane's Frontier*, 328 (Douglas and McBean's letters) and 329 (Lewes's letter). Kane's finished canvasses are more obviously subjective than his preliminary sketches; ibid., 36.

91 See Owram, *Promise of Eden*, 26–36; Zeller, *Inventing Canada*, 118–44; and Zeller, *Land of Promise,*, 9–12.

92 David Bain, "George Allan and the Horticultural Gardens," *Ontario History*, 87, no. 3 (1995): 233.

93 Zeller, *Inventing Canada*, 132; "William Allan," *DCB* 8: 10. Bain, "George Allan and the Horticultural Gardens," 233.

94 Zeller, *Inventing Canada*, 132.

95 See Kenneth R. Lister, *Paul Kane, The Artist: Wilderness to Studio* (Toronto: Royal Ontario Museum, 2010), 20.

96 Harper, *Paul Kane's Frontier*, 28; Lister, *Paul Kane*, 21, 389; Ian S. MacLaren, "On the Trail of Paul Kane's *Wanderings of an Artist*," *Prairie Fire* 10, no. 3 (Autumn 1989), 241. Kane's claims as faithful recorder of reality had significant staying power. In 1949, the prominent Canadian anthropologist T. F. McIlwraith argued that "Kane may well be criticized as an artist, but the records of his brush are still of value as objective descriptions"; T. F. McIlwraith, "Anthropology," in *Royal Canadian Institute Centennial Volume*, ed. W. Stewart Wallace (Toronto: Royal Canadian Institute, 1949), 4–5.

97 Harper, *Paul Kane's Frontier*, 34.

98 Ibid., 35; Harper, "Paul Kane," 391–2. All of the hundred paintings were evidently delivered by 1856; Harper, *Paul Kane's Frontier*, 320–1. Allan was abroad for most of the time between 1848 and 1852.

99 Harper surmised as much in *Paul Kane's Frontier*, 29.
100 Ibid., 43, 337.
101 Owram, *Promise of Eden*, 36. Owram's discussion of that period can be found on pp. 26–36.
102 See ibid.; Zeller, *Invention Canada*; and Zeller, *Land of Promise*.
103 Owram, *Promise of Eden*, 39.
104 Owram argued that the expansionist movement was stimulated by Canadian prosperity, railway expansion, the fear of an impending shortage of agricultural land within the province of Canada, and the fear of commercial stagnation. Owram, *Promise of Eden*, 41–4.
105 Richard A. Jarrell, *The Cold Light of Dawn: A History of Canadian Astronomy* (Toronto: University of Toronto Press, 1988), 74. Also see W. Stewart Wallace, "A Sketch of the History of the Royal Canadian Institute, 1849–1949," in Wallace, *Royal Canadian Institute Centennial Volume*, 123.
106 Zeller, *Inventing Canada*, 116. The Geological Survey of Canada may have been of relatively little significance for Canada's relationship with the HBC territories, but it did produce William E. Logan's analysis of Canada's coal that predicted that no significant coal deposits would ever be found within the borders of the colony. That prediction, coupled with the knowledge that the HBC territories had abundant coal resources, made the annexation of Rupert's Land attractive to Canadian expansionists.
107 S. Stewart Wallace, "A Sketch of the History of the Royal Canadian Institute, 1849–1949," in Wallace, *Royal Canadian Institute Centennial Volume*, 131; Zeller, *Inventing Canada*, 154.
108 George W. Allan, "Presidential Address," *Canadian Journal*, new series, 1, no. 2 (Feb. 1856): 98. [98–104] The journal was edited until 1856 by Henry Youle Hind. The journal's full name was *The Canadian Journal: A Repertory of Industry, Science, and Art, and a Record of the Proceedings of the Canadian Institute*.
109 *Canadian Journal* 3 (1855): 378. Allan cannot have been very active during the first year because he spent much of it abroad; Bain, "George Allan and the Horticultural Gardens," 237.
110 Wallace, *Royal Canadian Institute Centennial Volume*, 171. Allan was elected president of the Canadian Institute again in 1858.
111 William Allan moved to Toronto in the mid-1790s, when York (Toronto) was a small village. Allan bought park lot 5 (running north from today's Queen Street to Bloor Street between George/Huntley Streets and Sherbourne Street) in 1819. This is the property that George Allan inherited in 1853. See "William Allan," *DCB*, 8: 11–12. George Allan supported many causes in subsequent years. For example, in 1856 he offered land to the

Toronto Horticultural Society (land now known as Allan Horticultural Gardens). See Bain, "George Allan and the Horticultural Gardens."
112 *Canadian Journal* 3 (1855): 20; *Canadian Journal* 3 (1855): 137. Financial difficulties prevented the institute from actually ever building on the site; Wallace, "A Sketch of the History of the Royal Canadian Institute," 140.
113 Wallace, "A Sketch of the History of the Royal Canadian Institute," 140. The new series was issued in a new format. The name of the journal was also changed slightly to *The Canadian Journal of Industry, Science, and Art*.
114 See the institute's list of members in *Canadian Journal* 3, no. 17 (1855): 402–5. Sir George Simpson was elected a member in June 1855; *Proceedings of the Canadian Institute (Canadian Journal)* 3 , no. 17 (Dec. 1855): 316, 393.
115 Kane, "Incidents of Travel on the North-West Coast," 273–9. That the paper was read at the Institute 14 Mar. 1855 can be found on p. 273, and that the article was republished in Toronto's *Daily Colonist* in Aug. 1855 can be found in Harper, *Paul Kane's Frontier*, 39.
116 *Canadian Journal* 3 (1855): 243.
117 Paul Kane, "The Chinook Indians," *Canadian Journal*, new series, 2, no. 7 (Jan. 1857), 11–30. This article appeared almost unchanged in *Wanderings of an Artist*.
118 Paul Kane, "Notes of a Sojourn among the Half-breeds, Hudson's Bay Company's Territory, Red River," *Canadian Journal*, new series, 1 (1856): 128–38.
119 It was published as Paul Kane, "Notes of Travel among the Walla-Walla Indians," *Canadian Journal*, new series, 1 (1856), 417–24. See p. 417 for evidence that it was read on 5 Apr. 1856.
120 George Simpson, Lachine, to William G. Smith, London, 25 Feb. 1858, HBCA D.4/54, fos. 73d-74.
121 Kane, *Wanderings of an Artist*.
122 Harper, *Paul Kane's Frontier*, 44.
123 The paper is listed among those presented but not published in *Proceedings of the American Association for the Advancement of Science* 11 (1857), part 2: 159.
124 Daniel Wilson, *Prehistoric Man: Researches into the Origin of Civilisation in the Old and the New World* (Cambridge: Macmillan, 1862). For his thanks (including the quoted words), see 1: xiv–xv. For Kane's illustrations, see 1: ii and 2: ii. For reference to him see, 1: 6, 131, 156, 192, 198, 199 and 2: 14, 17–18, 27–8, 83, 85, 259, 316, 318–20, 351, 429–30. Wilson's use of Kane's illustrations mirrored James Cowles Prichard's use of George Catlin's work in his *Natural History of Man* (1843).

125 I. S. MacLaren, "Paul Kane and the Authorship of *Wanderings of an Artist*," in *From Rupert's Land to Canada: Essays in Honour of John E. Foster*, ed. Theodore Binnema, Gerhard J. Ens, and R. C. Macleod (Edmonton: University of Alberta Press, 2001), 239.

126 MacLaren, "On the Trail of Paul Kane's *Wanderings of an Artist*," 40. Also see MacLaren, "Paul Kane and the Authorship of *Wanderings of an Artist*," 239.

127 MacLaren, "On the Trail of Paul Kane's *Wanderings of an Artist*," 339, 242.

128 Kane, "Notes of a Sojourn among the Half-breeds," 138, repeated in Kane, *Wanderings of an Artist*, 97–8.

129 Quoted in Beckles Willson, *The Life of Lord Strathcona and Mount Royal, G.C.M.G., G.C.V.O. (1820–1914)*. (London: Cassell, 1915), 109.

130 Kane, *Wanderings of an Artist*, 43–5.

131 Harper, *Paul Kane's Frontier*, 38–9, 332.

132 Russell Harper argued that strains were evident in the relationship at this time; ibid., 30.

133 Ibid., 29, 332.

134 Ibid., 320.

135 Willson, *Life of Lord Strathcona and Mount Royal*, 109.

136 Kane, "Notes of a Sojourn among the Half-breeds," 137, repeated in Kane, *Wanderings of an Artist*, 96.

137 Willson, *Life of Lord Strathcona and Mount Royal*, 109. Perhaps this assertion sheds light on a passage in Kane, *Wanderings of an Artist* (on p. 434), in which Kane referred to "the pleasure of again meeting with Sir George Simpson, and several gentlemen to whose kindness I had before been deeply indebted." The wording of the sentence seems to suggest that, although he was pleased to meet Simpson again, Kane could not bring himself to write that he was deeply indebted to Simpson.

138 Bain, "George Allan and the Horticultural Gardens," 240.

139 Reference is to Lefroy's *Magnetical and Meteorological Observations*.

140 George Simpson, Lachine, to Paul Kane, Toronto, 25 Feb. 1858, HBCA D.4/54, fo. 73.

141 George Simpson, Lachine, to William G. Smith, London, 25 Feb. 1858, HBCA D.4/54, fos. 73d-74.

142 Kane, *Wanderings of an Artist*, dedication page.

143 In *Wanderings of an Artist*, Kane implied that he was born in Canada, and in the 1861 census, he claimed to have been born in Toronto. In fact, he arrived as a boy with his family in 1819. Harper, *Paul Kane's Frontier*, 6. The degree to which Kane succeeded in developing a myth of his childhood in Canada can be see in Daniel Wilson's obituary for him. See

      Daniel Wilson, "Paul Kane, the Canadian Artist," *Canadian Journal of Science, Literature, and History* 13, no. 1 (1871): 66.
144 Wilson, "Paul Kane," 41–2.
145 Willson, *Life of Lord Strathcona and Mount Royal*, 108–9.
146 Donald Swainson, "Allan Macdonell (McDonell)," *DCB* 11: 552–5.
147 Owram, *Promise of Eden*, 3.
148 *Montreal Gazette*, 26 Apr. 1856, as reprinted in Aborigines Protection Society, *Canada West and the Hudson's Bay Company*, 12.
149 The letters appear on p. 2 of the Toronto *Globe*, on 19, 27, and 30 Aug., and 2, 6, 15, 30 Sept. That "Huron" was Macdonell is explained in Owram, *Promise of Eden*, 39.
150 Owram, *Promise of Eden*, 26–36; Zeller, *Inventing Canada*, 97. Zeller's *Inventing Canada* explores the history of the connection between science and a British American transcontinental nation, as does her shorter updated booklet, Zeller, *Land of Promise*. Also see Suzanne Zeller, "Classical Codes: Biogeographical Assessments of Environment in Victorian Canada," *Journal of Historical Geography* 24, no. 1 (1998): 20–35.
151 Owram, *Promise of Eden*, 50–4. The quoted words of Alfred Roche are found on p. 54. In a letter published in the Toronto *Globe*, "Huron" (Allan Macdonell) described the HBC as "a company which is to all intent and purposes foreign and opposed to the true interests of our country"; Toronto *Globe*, 6 Sept. 1856, 2.
152 Great Britain, Parliament, House of Commons, *Copy of a Despatch from the Secretary of State for the Colonies to the Governor-General of Canada, Together with the Reply of the Governor General* 1857, Session 1 (113), 2–3; Great Britain, Parliament, House of Commons, *Copy of the Letter Addressed by Mr. Chief Justice Draper to Her Majesty's Secretary of State for the Colonies* ... 1857, Session 2 (104); Toronto *Globe*, 10 Dec. 1856, 2.
153 VanKoughtnet quoted in Galbraith, *Little Emperor*, 194
154 Owram, *Promise of Eden*, 59.
155 Ibid., 3, 56. Quoted passage on p. 56.
156 Suzanne Zeller argued that Isbister's "On the Geology of the Hudson's Bay Territories," "gave Canadian expansionists a splendid reference source during the second half of the nineteenth century"; Zeller, *Inventing Canada*, 98.
157 Toronto *Globe*, 14 Sept. 1855, 2.
158 Ibid., 2.
159 Joseph Henry, "Meteorology in Its Connection with Agriculture," in *Report of the Commissioner of Patents for the Year 1856: Agriculture* (Washington: Cornelius Wendell, 1857), 457. Henry elaborated on this

assessment in a passage on p. 481, Henry described the whole region between the 98th meridian and the Rocky Mountains as "a barren waste," and most of the region all the way to the Pacific Ocean as "a wilderness unfitted for the use of the husbandman." Alexander Isbister seized on those words in 1861. See Isbister, "On the Geology of the Hudson's Bay Territories," 236.

160 Lorin Boldget, *Climatology of the United States, and of the Temperate Latitudes of the North American Continent* (Philadelphia: J.B. Lippencott, 1857), 532–4, quoted passage on p. 533.
161 Ibid., 432–4; Owram, *Promise of Eden*, 65–7, . Zeller, "Humboldt and the Habitability of Canada's Great Northwest," 391–4.
162 James Cooper, "On the Distribution of the Forests and Trees of North America, with Notes on its Physical Geography," *SIAR, 1858*, 246–80.
163 Galbraith, *Little Emperor*, 188.
164 Great Britain, Parliament, *Report from the Select Committee on the Hudson's Bay Company; Together with the Proceedings of the Committee, Minutes of Evidence, Appendix and Index* (London: House of Commons, 1857), iii.
165 Morton, *History of the Canadian West*, 818.
166 Richard King was in fact hostile to the company, but his assertion that there was a farm of between 1500 and 2000 acres near Cumberland House must have badly damaged his credibility. On the other hand, John Rae's independence from the company (Rae was then retired) was emphasized. See Great Britain, Parliament, *Report from the Select Committee on the Hudson's Bay Company*, 312–19 (King), and 26–44 (Rae).
167 See Great Britain, Parliament, *Report from the Select Committee on the Hudson's Bay Company*, 12–26 (Lefroy), 150–68 (Richardson), and 184–9 (Back).
168 Rich, *Hudson's Bay Company*, 2: 780; Galbraith, *Hudson's Bay Company as an Imperial Factor*, 344.
169 Great Britain, Parliament, *Report from the Select Committee on the Hudson's Bay Company*, iii. The colony of British Columbia was created in 1858.
170 Ibid., iii.
171 Ibid., iv. The members of the Canadian Legislative Council and Legislative Assembly continued to argue that "Canada should not be called upon to compensate the said Company of any portion of such territory from which they may withdraw or be compelled to withdraw"; Great Britain, Parliament, House of Commons, *Papers Relative to the Hudson's Bay Company's Charter and Licence of Trade*, 3.
172 Great Britain, Parliament, *Report from the Select Committee on the Hudson's Bay Company*, iv.
173 Owram, *Promise of Eden*, 38.

174 Irene M. Spry, ed, *The Papers of the Palliser Expedition, 1857–1860* (Toronto: Champlain Society, 1968), xxii, xxiv.
175 John Palliser, *Papers Relative to the Exploration by Captain Palliser of that Portion of British North America Which Lies between the Northern Branch of the River Saskatchewan and the Frontier of the United States; and between the Red River and Rocky Mountains* (London: G. E. Eyre and W. Spottiswoode, 1859), 3.
176 Ibid., 4. For a scholarly study of the expedition see Irene M. Spry, *The Palliser Expedition: An Account of John Palliser's British North American Exploring Expedition* (Toronto: Macmillan, 1973), and Spry, *Papers of the Palliser Expedition.*
177 Spry, *Papers of the Palliser Expedition,* xix, xxvi–xxxviii.
178 John Palliser, *The Journals, Detailed Reports, and Observations Relative to the Exploration, by Captain Palliser, of that Portion of British North America, Which, in Latitude, Lies between the British Boundary Line and the Height of Land or Watershed of the Northern or Frozen Ocean Respectively, and in Longitude, between the Western Shore of Lake Superior and the Pacific Ocean during the Years 1857, 1858, 1859, and 1860* (London: G. E. Eyre and W. Spottiswoode, 1863), 20.
179 W. L. Morton, *Henry Youle Hind: 1823–1908* (Toronto: University of Toronto Press, 1980), 29–30, 79.
180 Ibid., 79.
181 Recall the letter from Donald Ross to James Hargrave of Apr. 1845, quoted in chapter 5. Also see Galbraith, *Hudson's Bay Company as an Imperial Factor,* 331.
182 Donald Ross to George Simpson, 21 Aug. 1848, quoted in Galbraith, *Little Emperor,* 190.
183 Simpson quoted in Cooper, *Alexander Kennedy Isbister,* 246. John Galbraith summarized the conclusion of the HBC this way: "why not make a virtue of necessity by selling what was worthless before the buyer became aware of the fact?"; Galbraith, *Little Emperor,* 190. Also see Galbraith, *Hudson's Bay Company as an Imperial Factor,* 332, 338, 374; and Rich, *Hudson's Bay Company,* 2: 796.
184 Rich, *Hudson's Bay Company,* 2: 799.
185 S. J. Dawson, H. Y. Hind, et al., *Papers Relative to the Exploration of the Country between Lake Superior and the Red River Settlement* (London: George Edward Eyre and William Spottiswoode, 1859), 12.
186 Lewis H. Thomas, "The Hind and Dawson Expeditions: 1857–58," *Beaver* 289 (Winter 1958): 40–1.
187 Ibid., 42.

188 Ibid., 44.
189 Morton, *Henry Youle Hind*, 59.
190 Henry Youle Hind, *Narrative of the Canadian Red River Exploring Expedition of 1857 and of the Assiniboine and Saskatchewan Exploring Expedition of 1858*. (London: Longman, Green, Longman, and Roberts, 1860), 2: 234.
191 Ibid.
192 Morton, *Henry Youle Hind*, 54. Morton also wrote (81–2) that Hind "used his scientific skills and his explorations to point to new resources and invite men to develop them. He was not, in fact, the restrained professional scientist," and (82) "he was first and foremost the frontier publicist, the promoter of new ventures. As such, he would be distrusted by scientists."
193 Zeller, *Land of Promise*, 14. Also see Morton, *Henry Youle Hind*, 81–2.
194 Hind, *Narrative of the Canadian Red River Exploring Expedition*, 2: 272.
195 Palliser, *Papers Relative to the Exploration by Captain Palliser*, 6.
196 Ibid., 7.
197 Morton, *Henry Youle Hind*, 41; Dawson, Hind, et al., *Papers Relative to the Exploration of the Country between Lake Superior and the Red River Settlement*, 24.
198 Extract from William H. Seward, *A Cruise to Labrador, Log of the Schooner Emerence, Correspondence of the Albany Evening Journal* as reprinted in Appendix 2 in Henry Youle Hind, *Explorations in the Interior of the Labrador Peninsula the Country of the Montagnais and Nasquapee Indians* (London: Longman, Green, Longman, Roberts, & Green, 1863), 2: 252–3.
199 Ibid.

Chapter 8

1 Lindsay, *Science in the Subarctic*, 7.
2 Lindsay, *Modern Beginnings of Subarctic Ornithology*, ix.
3 Ibid., xiv. Lindsay then overstated her nuanced argument by writing that the Smithsonian's cooperation with the HBC was "the first large-scale, apolitical, and noncommercial scientific study of Rupert's Land." Her analysis actually explicitly acknowledges the importance of this cooperation for the Smithsonian's position in internal US politics.
4 Rich, *Hudson's Bay Company*, 3: 827.
5 Ibid., 851.
6 Barnston to Baird, 28 Jan. 1862, in Lindsay, *Modern Beginnings of Subarctic Ornithology*, 128.

7 Rich, *Hudson's Bay Company*, 3: 851.
8 Annual Report of the Smithsonian for the year 1853, 126.
9 Hinsley, *Smithsonian and the American Indian*, 27.
10 E. F. Rivinus and E. M. Youssef, *Spencer Baird of the Smithsonian* (Washington: Smithsonian Institution Press, 1992), 109.
11 Ross to Baird, 20 June 1860 in Lindsay, *Modern Beginnings of Subarctic Ornithology*, 53. Hartwell Bowsfield, "Bernard Rogan Ross," *DCB* 10: 629. The fact that Ross was a founding Fellow of the Anthropological Society of London reinforces the idea that he was a polygenist. The society was formed in 1863 by anti-Darwinian polygenist secessionists from the Ethnological Society of London. See J. W. Burrow, "Evolution and Anthropology in the 1860s: The Anthropological Society of London, 1863–71" *Victorian Studies* 7, no. 2 (Dec. 1963): 137–54; Ronald Rainger, "Race, Politics, and Science: The Anthropological Society of London in the 1860s," *Victorian Studies* 22, no. 1 (Autumn 1978): 51–70; and Kass and Kass, *Perfecting the World*, 507–8. For a broader contextual study see George W. Stocking Jr., "The Persistence of Polygenist Thought in Post-Darwinian Anthropology," in *Race, Culture, and Evolution: Essays in the History of Anthropology* (Chicago: University of Chicago Press, 1982), 42–68. The evidence concerning Ross's beliefs is ambiguous. In his notebook, Ross wrote that "as to the theories of some eminent Ethnologists that a distinct and seperate [sic] pair was required to people America, I cannot perceive the necessity"; Lindsay, "The Hudson's Bay Company-Smithsonian Connection," 606. Ross certainly believed that the distant ancestors of the Dene peoples had come to North America from Asia.
12 Lindsay, *Modern Beginnings of Subarctic Ornithology*, 9.
13 Barnston to Baird, 26 July 1860, in Lindsay, *Modern Beginnings of Subarctic Ornithology*, 73.
14 *SIAR, 1855*, 51–2, 59; *SIAR, 1860*, 116.
15 Williams, *London Correspondence Inward from Sir George Simpson*, 186.
16 Joseph James Hargrave, *Red River* (Montreal: Printed for the author by John Lovell, 1871), 246.
17 Hargrave, *Red River*, 246.
18 Barnston to Baird, 26 July 1860, in Lindsay, *Modern Beginnings of Subarctic Ornithology*, 72. John Edward Gray (1800–75) was keeper of the British Museum at the time, and E. Doubleday and Adam White worked with the insect collection of that museum; ibid., 71. Barnston did tell Baird, however, that he felt a particular obligation to McGill College and the Natural History Society of Montreal because they had much smaller collections than American institutions. See ibid., 162n36.

19 Barnston to Baird, 26 July 1860, in Lindsay, *Modern Beginnings of Subarctic Ornithology*, 74.
20 Rivinus and Youssef, *Spencer Baird*, 8–9.
21 Ibid., 6.
22 Ibid., 10.
23 Robert V. Bruce, *Launching of Modern American Science: 1846–76* (New York: Knopf, 1987), 188. That scientist was Alexander Dallas Bache.
24 *SIAR*, 1897, part 2, 27.
25 Lindsay, *Science in the Subarctic*, 124.
26 Bruce, *Launching of Modern American Science*, 198.
27 Annual Report of the Board of Regents of the Smithsonian Institution, 1860, 48
28 William W. Fitzhugh, "Origins of Museum Anthropology at the Smithsonian Institution and Beyond," *Smithsonian Contributions to Anthropology* 44 (2002): 180.
29 Rivinus and Youssef, *Spencer Baird*, 56. Rivinus and Youssef argued that there is no reason to believe that the different interests and priorities of Baird and Spencer brought the two men into conflict (68).
30 Ibid., 2.
31 Ibid., 82.
32 Goetzmann, "Paradigm Lost," 26.
33 Rivinus and Yousef, *Spencer Baird*, 64.
34 Bruce, *Launching of American Science*, 300; Lindsay, *Science in the Subarctic*, 35.
35 Lindsay, *Science in the Subarctic*, 7. While the Civil War diverted attention and money away from science, George Daniels also argued that the war was a great watershed in the history of scientific institutions in the United States because it gave scientists such as Henry and Bache greater influence over the shaping of scientific policy in the United States. See George H. Daniels, *Science in American Society: A Social History* (New York: Alfred A. Knopf, 1971), 268.
36 Rich, *Hudson's Bay Company*, 3: 815–6.
37 Lindsay, *Science in the Subarctic*, 3.
38 Rivinus and Yousef, *Spencer Baird*, 81.
39 John Pickstone expertly distinguishes between "natural history" and "analysis" as "ways of knowing" in his *Ways of Knowing*, passim, but see especially pp. 2–8, 60–105.
40 These were identified as interests that motivated the Smithsonian's interest in the HBC territories. See Committee of the Chicago Academy of Sciences, "Biography of Robert Kennicott," *Transactions of the Chicago Academy of*

Sciences, 1, no. 2 (1869): 141-2; Fitzhugh, "Origins of Museum Anthropology," 186.
41 Rivinus and Yousef, *Spencer Baird*, 85.
42 *SIAR, 1855*, 60.
43 A letter from Donald Gunn (Red River) to Spencer Baird, dated 20 Feb. 1856, alludes to earlier correspondence between those two men; Lindsay, *Modern Beginnings of Subarctic Ornithology*, 1.
44 Gunn to Baird, 20 Feb. 1856, in Ibid., 1-3.
45 In 1862, Baird provided Gunn with a fifty-dollar expense account to underwrite his efforts; Sterling et al., *Biographical Dictionary*, 332.
46 *SIAR, 1868*, 22.
47 Donald Gunn, "Notes of an Egging Expedition to Shoal Lake, West of Lake Winnipeg," *SIAR, 1867*, 427-32; and Donald Gunn, "Indian Remains near Red River Settlement, Hudson's Bay Territory," *SIAR, 1867*, 399-400.
48 Joseph Henry, Secretary, Smithsonian Institution, Washington, to Sir George Simpson, 10 Nov. 1857, HBCA D.5/45 fo. 265.
49 Ibid., fo. 265d.
50 Ibid., fo. 266d.
51 George Simpson, Lachine, to Joseph Henry 10 Dec. 1857, HBCA D.4/77, p. 575.
52 Bernard R. Ross to John Richardson, 10 Nov. 1861, in Bernard H.[sic] Ross, "On the Mammals, Birds, &c., of the Mackenzie River District," *Natural History Review* 7 (July 1862): 270.
53 F. H. Schofield, *Life and Services of Retired Chief Factor, R. MacFarlane*, 3.
54 Ross to Joseph Henry, 28 Nov. 1858, in Lindsay, *Modern Beginnings of Subarctic Ornithology*, 11.
55 Ibid.
56 Debra Lindsay has suggested that Ross might have been inspired to make his first contacts with the Smithsonian through George Gibbs, who was a friend of the Smithsonian, and with whom Ross had had previous contact while Gibbs served as a geologist and botanist with the North West Boundary Survey Commission (1857-60); Lindsay, *Science in the Subarctic*, 52.
57 Bernard R. Ross to Professor Henry, 30 Nov. 1859, as quoted in *SIAR, 1859*, 116.
58 Ibid.
59 Ibid. Flett was an Orcadian assistant trader.
60 Francis Napier, Washington, to George Simpson, 19 Mar. 1859, HBCA D.5/48 fo. 403; Lindsay, *Science in the Subarctic*, 42.

61 George Simpson to Joseph Henry, 28 Mar. 1859, HBCA D.4/78, p. 175.
62 George Simpson, (31 Mar. 1860) "Circular to the Officers of the Hudson's Bay Company," Smithsonian Miscellaneous Collections, 2 (1862): article 8, 45. Evidence that the letter was actually drafted at the Smithsonian can be found at HBCA D.5/51, fos. 397-397d, Draft of cover letter to circular Henry to Simpson (Circular) 16 Mar. 1860.
63 Thomas, "The Smithsonian and the Hudson's Bay Company," 291.
64 Joseph Henry, (20 Apr. 1860) "Circular to Officers of the Hudson's Bay Company," Smithsonian Miscellaneous Collections 2 (1862): article 8, 41.
65 Ibid.: article 8, 41–4.
66 Lindsay, *Science in the Subarctic*, 46, Committee of the Chicago Academy of Sciences, "Biography of Robert Kennicott," 133.
67 Committee of the Chicago Academy of Sciences, "Biography of Robert Kennicott," 134–5.
68 Committee of the Chicago Academy of Sciences, "Biography of Robert Kennicott," 134; Ronald S. Vasile, "The Early Career of Robert Kennicott, Illinois's Pioneering Naturalist," *Illinois Historical Journal* 87 (1994): 156n; Rivinus and Yousef, *Spencer Baird*, 85; Fitzhugh, "Origins of Museum Anthropology," 187; Thomas, "The Smithsonian and the Hudson's Bay Company," 290. Kennicott is acknowledged as a donor of specimens to the Smithsonian in the Annual Report of the Smithsonian Institution 1853, 56. Also see Lindsay, *Science in the Subarctic*, 47.
69 Committee of the Chicago Academy of Sciences, "Biography of Robert Kennicott," 137–8. During the nineteenth century, university students who fell ill were commonly advised to abandon their studies and find work outdoors. Going west where men might spend their time "sketching, botanizing, geologizing, fishing or hunting," were especially recommended. See Sheila M. Rothman, *Living in the Shadow of Death: Tuberculosis and the Social Experience of Illness in American History* (New York: Basic Books, 1994), 45, 135, 152 (quoted passage on p. 152).
70 Thomas, "The Smithsonian and the Hudson's Bay Company," 290. For evidence that the directors of the Smithsonian were aware of this trip and received specimens, see: see *SIAR, 1857*, 48; and Committee of the Chicago Academy of Sciences, "Biography of Robert Kennicott," 139.
71 Thomas, "The Smithsonian and the Hudson's Bay Company," 290.
72 Committee of the Chicago Academy of Sciences, "Biography of Robert Kennicott," 142; James Alton James, *The First Scientific Exploration of Russian America and the Purchase of Alaska* (Evanston and Chicago: Northwestern University, 1942), 6.

73 Vasile, "Early Career of Robert Kennicott," 168.
74 Sheets-Pyenson, *John William Dawson*, 103–4, 169–71. The meeting was marked by warm feelings by Americans for the Britain and Canada. See *Proceedings of the American Association for the Advancement of Science* 11 (1857): part 2, 163. The proceedings also reveal that the meeting had a significant impact on the scientific community in Canada. Although relatively few Canadians were members of the association before the Montreal meeting, more than 120 new members in 1857 reported a Canadian address, bringing the number of members that did so to more than 10 percent. *Proceedings of the American Association for the Advancement of Science* 11 (1857): xxiii–lv. The American Association for the Advancement of Science (est. 1848) was modelled after its British namesake. In the early 1850s it had lobbied the British government for the continued funding of the Toronto Observatory.
75 Committee of the Chicago Academy of Sciences, "Biography of Robert Kennicott," 142–3; James, *First Scientific Exploration*, 46; Quoted passage from Nute, "Kennicott in the North," 28. That the two men expected Kennicott to travel via St. Paul, see George Simpson, Lachine, to Robert Kennicott, Chicago, 15 Apr. 1859, HBCA D.4/78, 109. Aside from the Smithsonian, Kennicott's trip was supported by the University of Michigan, the Audubon Club of Chicago, the Chicago Academy of Sciences, and a number of private benefactors. See *SIAR, 1859*, 51.
76 Committee of the Chicago Academy of Sciences, "Biography of Robert Kennicott," 143; *SIAR, 1859*, 66; Thomas, "The Smithsonian and the Hudson's Bay Company," 291. John Rae had submitted specimens to the Smithsonian; *SIAR, 1859*, 66. Dawson must have shared some of the results of his surveys. See James, *First Scientific Exploration*, 154.
77 Committee of the Chicago Academy of Sciences, "Biography of Robert Kennicott," 144–5. Kennicott had owned Simpson's *Narrative of a Journey Round the World, during the Years 1841 and 1842 (1847)* by around 1853; James, *First Scientific Exploration*, 1.
78 Committee of the Chicago Academy of Sciences, "Biography of Robert Kennicott," 145.
79 The Annual Report of the Smithsonian for 1860 indicates that George Barnston donated samples from Lake Superior; Annual Report of the Board of Regents of the Smithsonian Institution, 1860, 75. The quote passage is on p. 70. Also see Thomas, "The Smithsonian and the Hudson's Bay Company," 292; Lindsay, *Modern Beginnings of Subarctic Ornithology*, xvii.
80 Kennicott to Baird, 15 June 1859, in Lindsay, *Modern Beginnings of Subarctic Ornithology*, 12–16.
81 Ibid., 13.

82 Thomas, "The Smithsonian and the Hudson's Bay Company," 292; Palliser, *Papers Relative to the Exploration by Captain Palliser*, 35.
83 For evidence that Kennicott referred to Ross as "Barny," see Kennicott to Baird, 15 June 1859, in Lindsay, *Modern Beginnings of Subarctic Ornithology*, 14. That the HBC traders referred to Kennicott as "Kenny," see Nute, "Kennicott in the North," 28.
84 Reference to the letter, and Simpson's approval of Kennicott's request can be found in George Simpson to Joseph Henry, 13 Jan. 1860, HBCA D.4/79 (pp. 402–1 are numbered in reverse). The London Governor and Committee then endorsed Simpson's approval. See George Simpson to Joseph Henry, 8 Feb. 1860, HBCA D.4/79. Henry's effusive thanks can be found at Henry to Simpson, 21 Mar. 1860, HBCA D.5/51, fo. 421.
85 Bernard R. Ross to Professor Henry, 30 Nov. 1859, as quoted in *SIAR, 1859*, 116.
86 Robert V. Bruce named Joseph Henry and Bache as the two most important pioneers of American science; Bruce, *Launching of Modern American Science*, 15.
87 A. D. Bache, (Coast Survey Office) to George Simpson, 3 May 1860, HBCA D.5/52, fo. 14–15. Given the timing of the eclipse and location of the line of the shadow (at the very northern tip of Labrador), the HBC was not in a position to assist the expedition led by S. Alexander that traveled to Labrador.
88 C. H. Davis, Commander, USN, Superin Naut. Alm., Cambridge, Massachusetts, to George Simpson, Lachine, 4 May 1860, HBCA D.5/52, fo. 16.
89 George Simpson to C.H. Davis, 7 May 1860, HBCA D.4/79, pp. 139–8 [the book is numbered backwards].
90 Samuel Hubbard Scudder, *Winnipeg Country: Or Roughing It with an Eclipse Party* (Boston: Cupples, Upham, & Company, 1886), 71; Daniel Wilson, "Science in Rupert's Land," *The Canadian Journal of Industry, Science, and Art* 7 (1862): 337. William Mactavish was attentive to instructions from George Simpson "to provide whatever assistance the eclipse party required in carrying out their scientific pursuits." See Simon Newcomb's diary entry for 25 June 1860 in J.E. Kennedy and S.D. Hanson, "Excerpts from Simon Newcomb's Diary of 1860," *Journal of the Royal Astronomical Society of Canada* 90, no.5/6(1996): 298.
91 Jules Verne, *The Fur Country: Or, Seventy Degrees North Latitude*, trans. N. D'Anvers (Boston: James R. Osgood and Company, 1874). In this novel, the HBC assists the fictional astronomer Thomas Black in his efforts to observe the eclipse of 18 July 1860. Black's experiences are, not surprisingly, more bizarre than those of the members of the historical party, although the weather in his case is more cooperative.

92 Kennicott to Baird, 17 Nov. 1859, in Lindsay, *Modern Beginnings of Subarctic Ornithology*, 22.
93 Bockstoce, *Furs and Frontiers in the Far North*, 220–1.
94 Lindsay, *Science in the Subarctic*, 41. William L. Hardisty assisted Kennicott's collecting activities at Fort Resolution in the spring of 1860; Nute, "Kennicott in the North," 30.
95 Thomas, "The Smithsonian and the Hudson's Bay Company," 294. The Smithsonian's annual report updated readers on Kennicott's travels (and paid effusive tribute to the HBC) each year; *SIAR, 1859,* 51, 66; *SIAR, 1860,* 69–70; *SIAR, 1861,* 59–61; *SIAR, 1862,* 39–40; *SIAR, 1863,* 36, 52–3.
96 Committee of the Chicago Academy of Sciences, "Biography of Robert Kennicott," 214, 215; Nute, "Kennicott in the North," 31.
97 Lindsay, *Science in the Subarctic*, 107–8; Committee of the Chicago Academy of Sciences, "Biography of Robert Kennicott," 216.
98 Lindsay, *Science in the Subarctic*, 106–7.
99 Committee of the Chicago Academy of Sciences, "Biography of Robert Kennicott," 219.
100 Lindsay, *Science in the Subarctic*, 112.
101 Fitzhugh, "Origins of Museum Anthropology," 188.
102 Those who have argued that Kennicott committed suicide include Lindsay, *Science in the Subarctic*, 118–19, 166. The note is mentioned in Committee of the Chicago Academy of Sciences, "Biography of Robert Kennicott," 223. William Healey Dall's explanation of Kennicott's death also implies suicide. He wrote that Kennicott "died last May of disease of the heart on a desolate northern beach, alone! He was murdered; not by the merciful knife but by slow torture of the mind. By ungrateful subordinates, by an egotistic and selfish commander, by anxiety to fulfil his commands, while those that gave them were lining their pockets in San Francisco. I am so nervous from rage and grief that I can hardly write." See Morgan B. Sherwood, *Exploration of Alaska, 1865–1900* (New Haven and London: Yale University Press, 1965), 24. For accounts of Kennicott's death that do not attempt to explain the cause of his death, see Fitzhugh, "Origins of Museum Anthropology," 188; James, *First Scientific Exploration*, 17.
103 Gunn to Baird, 1 Mar. 1867, in Lindsay, *Modern Beginnings of Subarctic Ornithology*, 202.
104 Vasile, "Early Career of Robert Kennicott," 150.
105 Committee of the Chicago Academy of Sciences, "Biography of Robert Kennicott," 218–9.
106 Vasile, "Early Career of Robert Kennicott," 153–4.

107 John A. Kennicott, West Northfield, Illinois, to George Simpson, 26 Apr. 1860, D.5/51, fol 630d. Also see Committee of the Chicago Academy of Sciences, "Biography of Robert Kennicott," 133.
108 Lindsay, *Science in the Subarctic*, 118. His memorialists expressed it this way: "At an early period in his life, his health was extremely delicate, and it was doubtful whether he would long survive; but at the age of thirteen he began to show signs of physical improvement, and by the help of athletic exercises he ultimately became hardy enough to endure the rigors of the severest Arctic winters"; Committee of the Chicago Academy of Sciences, "Biography of Robert Kennicott," 133.
109 Lindsay, *Science in the Subarctic*, 118. In the late nineteenth century, many assumed that life in large urban centers bred mental illness, not just physical illness. See Rothman, *Living in the Shadow of Death*, 135.
110 Vasile, "Early Career of Robert Kennicott," 169.
111 Lindsay, *Science in the Subarctic*, 44.
112 Thomas, "The Smithsonian and the Hudson's Bay Company," 300.
113 Mactavish to Baird, 21 May 1863, in Lindsay, *Modern Beginnings of Subarctic Ornithology*, 147.
114 MacFarlane to Baird, 10 May 1864, in Lindsay, *Modern Beginnings of Subarctic Ornithology*, 158. More evidence of Dallas's unenthusiastic aid to the Smithsonian can be found in Macfarlane to Kennicott, 9 Sept. 1864, in Lindsay, *Modern Beginnings of Subarctic Ornithology*, 165.
115 Lindsay, *Modern Beginnings of Subarctic Ornithology*, xx.
116 Joseph Henry, Secretary, Smithsonian Institution, Washington to Sir George Simpson, [Governor, Hudson's Bay Company, Lachine] 10 Nov. 1857, HBCA D.5/45 fo. 265.
117 George Simpson, "Circular to the Officers of the Hudson's Bay Company," article 8, 45.
118 Kennicott to Baird, 15 June 1859, in Lindsay, *Modern Beginnings of Subarctic Ornithology*, 15.
119 Mactavish to Baird, 23 Nov. 1860 in ibid., 92. Mactavish seems to be alluding to James Barnston, but Barnston had died in May 1858. Daniel Wilson, at the University of Toronto, was also from Edinburgh, but he was not a botanist.
120 C. Stuart Houston, "Bernard Rogan Ross," in Sterling et al., *Biographical Dictionary*, 686.
121 Ross to Baird, 20 June 1860, in Lindsay, *Modern Beginnings of Subarctic Ornithology*, 54. Sir William Edmond Logan was the director of the Canadian Geological Survey.

392 Notes to pages 261–4

122 Ross to Baird, 20 Nov. 1861 in ibid., 126.
123 Barnston to Baird, 6 Nov. 1861, in ibid., 124.
124 Lindsay, *Science in the Subarctic*, 95.
125 Schofield, *Life and Services of Retired Chief Factor, R. MacFarlane*, 5. The Royal Colonial Institute survives today as the Royal Commonwealth Society.
126 H. C. Deignan, "The HBC and the Smithsonian," *The Beaver* 278 (June 1947): 5.
127 Ibid., 6.
128 Ibid., 6.
129 Ibid., 6.
130 Rivinus and Youssef, *Spencer Baird*, 68–80.
131 Major R. Lachlan, "In Behalf of the Establishment of a Provincial System of Meteorological Observations," *Canadian Journal* 3 (1855): 408.
132 Ibid.
133 That is the title of the eighth chapter of Rivinus and Youssef, *Spencer Baird*. See p. 81.
134 Thomas, "The Smithsonian and the Hudson's Bay Company," 300. An excellent examination of Baird's network of collectors can be found in William A. Deiss, "Spencer F. Baird and His Collectors," *Journal of the Society for the Bibliography of Natural History* 9 (1980): 635–45.
135 Deiss, "Spencer F. Baird and His Collectors," 636. For a discussion of the significance of Baird's life as boyhood collector, including his involvement with Audubon, see Marcia Bonta, "Baird of the Smithsonian," *Pennsylvania Heritage* 6, no. 3 (Summer 1980): 19–20.
136 Annual Report of the Board of Regents of the Smithsonian Institution, 1897, Part 2: A Memorial of George Brown Goode Together with a Selection of His Papers on Museums and on the History of Science in America, 55th Congress, 2nd Session, Doc. No. 575, 19.
137 Rivinus and Youssef, *Spencer Baird*, 60.
138 Kennicott to Baird, 23 June 1861, in Lindsay, *Modern Beginnings of Subarctic Ornithology*, 112.
139 Roderick Ross MacFarlane, "Land and Sea Birds Nesting within the Arctic Circle in the Lower Mackenzie River District," The Historical and Scientific Society of Manitoba, *Transactions*, series 1, no. 39 (June 1889), accessed online, http://www.mhs.mb.ca/docs/transactions/1/mackenzie birds.shtml.
140 George Simpson to Joseph Henry, 13 Jan. 1860, HBCA D.4/79, p. 402.
141 Thomas, "The Smithsonian and the Hudson's Bay Company," 302, citing James, *First Scientific Exploration*, 11.

142 Ibid.
143 MacFarlane, "Land and Sea Birds Nesting within the Arctic Circle," http://www.mhs.mb.ca/docs/transactions/1/mackenziebirds.shtml.
144 Clarke to Baird, 21 June 1861, as quoted in Lindsay, *Science in the Subarctic*, 49.
145 Nute, "Kennicott in the North," 30.
146 Hargrave, *Red River*, 245.
147 Lindsay, *Modern Beginnings of Subarctic Ornithology*, xxii.
148 Kennicott to MacFarlane, 15 Apr. 1864, as quoted in James, *First Scientific Exploration*, 11. There is reason to doubt the accuracy of James's transcription. The letter was written before Kennicott arrived in the Russian District. I have never seen the original, but I suspect that the portion of the letter James transcribed as "R. District" should probably be "M.R. District."
149 Joseph Henry, Washington, to George Simpson, Lachine, dated 23 Jan. 1859 [actually 1860], HBCA D.5/51, fos. 126d-127.
150 Spencer F. Baird, Washington, to George Simpson, 21 Apr. 1860, HBCA D.5/51, fo. 605.
151 Lindsay, *Modern Beginnings of Subarctic Ornithology*, xx. Admittedly, the trip to Moose Factory, although much shorter than the trip to Fort Simpson, was an arduous one. George Simpson had warned Henry that the trip would be difficult, "requiring that the person you send should be accustomed to such travelling & to the fatigue, exposure, plain fare &c. incidental to it"; George Simpson to Joseph Henry, 11 Feb. 1860, HBCA D.4/79, p. 329.
152 Lindsay, *Modern Beginnings of Subarctic Ornithology*, xx.
153 Joseph Henry, Washington, to Thomas Fraser, Secretary, Hudson Bay House, London, 29 Oct. 1860, HBCA A.10/48, p. 459.
154 Thomas Fraser, Secretary, Hudson Bay House, London to Joseph Henry, 23 Nov. 1860, HBCA A.5/24, p. 121.
155 Ibid., pp. 121-2.
156 Annual Report of the Board of Regents of the Smithsonian Institution, 1860, 70-1.
157 Bowsfield, "Bernard Rogan Ross," 629; Lindsay, *Modern Beginnings of Subarctic Ornithology*, xviii.
158 Ross to Baird, 10 Nov. 1860, in Lindsay, *Modern Beginnings of Subarctic Ornithology*, 89.
159 Bernard R. Ross to Professor Henry, 30 Nov. 1859, as quoted in *SIAR, 1859*, 116.
160 Bowsfield, "Bernard Rogan Ross," 629; Thomas, "The Smithsonian and the Hudson's Bay Company," 293; Bernard H. [sic] Ross, "On the

Mammals, Birds, &c., of the Mackenzie River District," *Natural History Review* 7 (July 1862): 269–70; Lindsay, "The Hudson's Bay Company-Smithsonian Connection," 600. He contributed at least "two Kootchin skulls" to the Academy of Natural Sciences; Lindsay, "The Hudson's Bay Company-Smithsonian Connection," 611.

161 Bernard R. Ross, "On the Indian Tribes of the McKenzie River District and the Arctic Coast," *Canadian Naturalist and Geologist* 4 (1859): 190–5; Bernard R. Ross, "A Popular Treatise on the Fur-Bearing Animals of the Mackenzie River District," *Canadian Naturalist and Geologist* 6 (1861): 5–36; Bernard R. Ross, "List of Species of Mammals and Birds – Collected in McKenzie's River District during 1860–61," *Canadian Naturalist and Geologist* 6 (1861): 441–4; Ross, "On the Mammals, Birds, &c., of the Mackenzie River District"; Bernard R. Ross, "An Account of the Botanical and Mineral Products, Useful to the Chipewyan Tribes of Indians, Inhabiting the McKenzie River District," *Canadian Naturalist & Geologist* 7 (1862): 133–7; Ross, "List of Mammals, Birds, and Eggs, Observed in the McKenzie's River District, with Notices," 137–55. This last article includes a reprinting, under the same title, of a portion (pages 171–90) of Ross's article in the *Natural History Review* (1862).

162 Bowsfield, "Bernard Rogan Ross," 629. The fact that Ross was a founding member of the Anthropological Society of London reinforces the impression that he was a polygenist. The society was established in 1863 by anti-Darwinian polygenists. Although not all members of the society were of that persuasion, many were. See Rainger, "Race, Politics, and Science: The Anthropological Society of London in the 1860s," 51–70. Also see Stocking, *Victorian Anthropology*, 245–57.

163 Lindsay, *Modern Beginnings of Subarctic Ornithology*, xviii.

164 Illustrated in Fitzhugh, "Origins of Museum Anthropology," 189.

165 Annual Report of the Board of Regents of the Smithsonian Institution, 1860, 36.

166 Ross to Baird, 15 Apr. 1861, in Lindsay, *Modern Beginnings of Subarctic Ornithology*, 111; Bowsfield, "Bernard Rogan Ross," 629.

167 *SIAR, 1860*, 80

168 Thomas, "The Smithsonian and Hudson's Bay Company," 283; William F. Butler penned a very laudatory account of Macfarlane in *Wild North Land* (Philadelphia: Porter & Coates, 1874), 112–16.

169 Lindsay, *Modern Beginnings of Subarctic Ornithology*, xviii.

170 Kennicott to Baird, 23 Mar. 1860, in ibid., 44.

171 Thomas, "The Smithsonian and Hudson's Bay Company," 301.

172 Lindsay, *Modern Beginnings of Subarctic Ornithology*, 133.

173 E. A. Preble, "Roderick Ross MacFarlane, 1833–1920," *The Auk* 39 (1922): 208–9.
174 Thomas, "The Smithsonian and the Hudson's Bay Company," 302.
175 Lindsay, *Modern Beginnings of Subarctic Ornithology*, xix.
176 R. MacFarlane, "On an Expedition down the Beghula or Anderson River," *Canadian Record of Science* 4, no. 1 (Jan. 1890): 28–53; MacFarlane, "Land and Sea Birds Nesting within the Arctic Circle," http://www.mhs.mb.ca/docs/transactions/1/mackenziebirds.shtml; Robert MacFarlane, "Notes on and List of Birds and Eggs Collected in Arctic America, 1861–1866," *Proceedings of the United States National Museum* 14 (1891): 413–46; R. MacFarlane, "List of Birds and Eggs Observed and Collected in the North-West Territories of Canada, Between 1880 and 1894," in *Through the Mackenzie Basin*, ed. Charles Mair (Toronto: William Briggs, 1908), 286–470; R. MacFarlane, "Notes on Mammals Collected and Observed in the Northern Mackenzie River District, North-West Territories of Canada with Remarks on Explorers and Explorations of the Far North," in Mair, *Through the Mackenzie Basin*, 150–283; Roderick Ross MacFarlane, "Notes on the Mammals Collected and Observed in the Northern Mackenzie River District, North-West Territories of Canada, with Remarks on Explorers and Explorations of the Far North," *Proceedings of the United States National Museum* 28 (1905): 673–764.
177 *SIAR, 1868* (Washington: Government Printing Office, 1869), 57.
178 Ibid., 22.
179 Kennicott to Baird, 29 June 1860, in Lindsay, *Modern Beginnings of Subarctic Ornithology*, 58. Ross was long known for his tactlessness. Donald Ross described him in 1843 as "one of the greatest blunderers this country ... ever produced"; Debra Lindsay, *Science in the Subarctic*, 92.
180 Nute, "Kennicott in the North," 30. For more on Clarke, see Lindsay, *Science in the Subarctic*, 148n32 and 152n66.
181 Lindsay, *Modern Beginnings of Subarctic Ornithology*, xix.
182 Ibid.; *SIAR, 1868*, 57. On Lockhart, also see Lindsay, *Science in the Subarctic*, 152n68.
183 Kennicott to Baird, 23 June 1861, in Lindsay, *Modern Beginnings of Subarctic Ornithology*, 113. On Hardisty, see Lindsay, *Science in the Subarctic*, 151n64, and Jennifer Brown, "William Lucas Hardisty," *DCB* 11: 384–5. Strachan Jones was the son of Thomas Mercer Jones (commissioner of the Canada Company, a land and colonization company in Upper Canada) and grandson of John Strachan, the prominent Anglican bishop of Toronto and head of the Family Compact. See Lindsay, *Science in the Subarctic*, 152n71.

184 Each year's annual report of the Smithsonian Institution included a section entitled "List of Donations," and another entitled "Lost of Meteorological Stations and Observers." The names listed are gleaned from those sections in the *SIAR* between 1859 and 1869. Short biographies of many of these men can be found in the notes in Lindsay, *Science in the Subarctic*, 150–2.
185 Ibid., 56.
186 Ibid., 63–75.
187 Alfred Newton, "Suggestions for Forming Collections of Birds' Eggs," in *The Zoologist*, ed. Edward Newman, 18 (1860): 7199–200 (article from pp. 7189–201). This is reprinted from the circular written for the Smithsonian Institution. The article can also be found in the Smithsonian Miscellaneous Collections 2 (1862): 10–22.
188 Fitzhugh, "Origins of Museum Anthropology," 188.
189 Lindsay, *Modern Beginnings of Subarctic Ornithology*, 133.
190 Gunn to Baird, 2 June 1857, in ibid., 4.
191 See for example, extracts from Kennicott's journals in Committee of the Chicago Academy of Sciences, "Biography of Robert Kennicott," 171–6.
192 Thomas, "The Smithsonian and the Hudson's Bay Company," 296.
193 Kennicott to Baird, 29 June 1860, in Lindsay, *Modern Beginnings of Subarctic Ornithology*, 56 (emphasis in original). Also see Ross to Henry, 25 July 1859, in ibid., 16. For a discussion see Lindsay, *Science in the Subarctic*, 64.
194 Lindsay, *Modern Beginnings of Subarctic Ornithology*, 16.
195 Kennicott to Baird, 23 Mar. 1860, in ibid., 42.
196 Ross, "List of Mammals, Birds, and Eggs," 151–2.
197 Lindsay, *Science in the Subarctic*, 66–7.
198 Rivinus and Youssef, *Spencer Baird*, 83.
199 *SIAR, 1860*, 48.
200 Ibid.
201 *SIAR, 1860*, 69.
202 Ibid., 69.
203 Ibid., 70.
204 *SIAR, 1861*, 39–40, 49, 59–61, 64–7, 69, 85; *SIAR, 1862*, 39–40, 55, 57–8, 62; *SIAR, 1863*, 36, 44, 52–3, 58–61, 64, 91; *SIAR, 1864*, 30, 49, 74, 81–2, 101, 108; *SIAR, 1865*, 50, 86–7, 92, 101; *SIAR, 1866*, 26, 47–8, 51; *SIAR, 1867*, 44, 74; *SIAR, 1868*, 21, 22, 56–8, 68; *SIAR, 1869*, 31–2, 55, 69.
205 J. G. Lockhart, "Notes on the Habits of the Moose in the Far North of British North America in 1865," Proceedings of the National Museum," 13, no. 827 (1890): 305.
206 Committee of the Chicago Academy of Sciences, "Biography of Robert Kennicott," 202.

207 Mactavish to Baird, 26 Feb. 1862, in Lindsay, *Modern Beginnings of Subarctic Ornithology*, 129. Emphasis in original.
208 Ibid., 129–30. If Andrew Murray felt chastised, he was apparently not repentant. His *The Geographical Distribution of Mammals* (London: Day and Son, 1866), makes no mention of the HBC, its directors, or its officers. The reference is probably to the museum known from 1857 to 1899 as the South Kensington Museum (London). See "The South Kensington Museum; Its Educational Resources," *The Museum: A Quarterly Magazine of Education, Literature, and Science* 1, no. 1 (Apr. 1861): 66–72; Christopher Whitehead, *The Public Museum in Nineteenth Century Britain: The Development of the National Gallery* (Burlington, VT: Ashgate, 2005), 78.
209 Mactavish to Baird, 26 Feb. 1862, in Lindsay, *Modern Beginnings of Subarctic Ornithology*, 130.
210 Kennicott to Baird, 23 Mar. 1860, in ibid.., 47.
211 Joseph Henry, Washington, to Dallas, Fort Garry, 16 Apr. 1863, HBCA D.8/1, fo. 234-234d.
212 Kennicott to Baird, 17 Nov. 1859, in Lindsay, *Modern Beginnings of Subarctic Ornithology*, 29.
213 Ross to Baird, 25 Mar. 1860, in ibid., 48.
214 Ross to Baird, 1 June 1862, in Lindsay, *Modern Beginnings of Subarctic Ornithology*, 132.
215 Thomas, "The Smithsonian and the Hudson's Bay Company," 297.
216 George Gibbs to Joseph Henry, 7 Nov. 1862, *SIAR, 1863*, 88.
217 Ross to Joseph Henry, 28 Nov. 1858, in Lindsay, *Modern Beginnings of Subarctic Ornithology*, 11.
218 Gunn to Baird, 2 June 1857, in ibid., 4.
219 Barnston to Baird, 28 Jan. 1862, in ibid., 128.
220 See Lindsay's chapter entitled "Recognition and Reward," in *Science in the Subarctic*, 89–101.
221 Rivinus and Youssef, *Spencer Baird*, 91.
222 Kennicott to Baird, 17 Nov. 1859, in Lindsay, *Modern Beginnings of Subarctic Ornithology*, 30.
223 Ross to Baird, 10 July 1861, as quoted in Lindsay, *Science in the Subarctic*, 100.
224 Lindsay, *Modern Beginnings of Subarctic Ornithology*, 34.
225 Ibid., xviii.
226 Ross to Baird, 20 Nov. 1861, in ibid., 126.
227 Sterling et al., *Biographical Dictionary*, 488.
228 Baird to Ross, 20 June 1860, in Lindsay, *Modern Beginnings of Subarctic Ornithology*, 51. Ross's wish to have the species named after friends was prevented by the fact that naturalists determined that these two specimens did not represent new species. Even Ross obviously suspected

that his identification of *Bernicla barnstonii* as a separate species would not withstand scrutiny of scientists. See Ross, "List of Mammals, Birds, and Eggs," 152. In that same publication, Ross attempted, admitting his doubts, to name a rodent *Arctomys kennicotti* (140).

229 Ross to Baird, 29 July 1859, in Lindsay, *Modern Beginnings of Subarctic Ornithology*, 17.
230 Ross to Baird, 26 Nov. 1859, in ibid., 35.
231 Kennicott to Baird, 23 Mar. 1860, in Lindsay, *Modern Beginnings of Subarctic Ornithology*, 46. The evidence is scant but it does suggest that Murray was generally less able to forge friendly relations with traders than were Kennicott or Baird. According to one account, Murray had "a countenance that rarely relaxed into a smile, he yet had a kindness of manner that made him respected by all, and a fund of dry humour that told irresistibly upon his hearers, although uttered with what almost amounted to an appearance of unconsciousness on his part"; Anonymous, "Andrew Murray," 216. Perhaps it was difficult for Murray to convey in correspondence the more congenial aspects of his personality. At any rate, I have been able to find no expressions of affection for Murray written by HBC officers.
232 Ross to Baird, 25 Mar. 1860, in Lindsay, *Modern Beginnings of Subarctic Ornithology*, 48.
233 Ross to Baird, 18 Mar. 1861, in Lindsay, *Modern Beginnings of Subarctic Ornithology* , 105.
234 Ibid.
235 Mactavish to Baird, 11 Feb. 1867, quoted in Lindsay, *Science in the Subarctic*, 95.
236 Barnston to Baird, 26 July 1860, in Lindsay, *Modern Beginnings of Subarctic Ornithology*, 76.
237 Ross to Baird, 25 Mar. 1860, in ibid., 48.
238 Rivinus and Yousef, *Spencer Baird*, 86.
239 Ross to Baird, 29 July 1859, in Lindsay, *Modern Beginnings of Subarctic Ornithology*, 17.
240 Lindsay, *Science in the Subarctic*, 94.
241 Joseph Henry, Washington, to George Simpson, Lachine, dated 23 Jan. 1859 [actually 1860], HBCA D.5/51, fos. 126d and 127d. Henry subsequently offered books to Governor Dallas; Joseph Henry, Washington, to Dallas, Fort Garry, 16 Apr. 1863, HBCA D.8/1fol 233d.
242 George Simpson to Joseph Henry, 11 Feb. 1860, HBCA D.4/79, p. 329.
243 Ross to Baird, 15 Apr. 1861, in Lindsay, *Modern Beginnings of Subarctic Ornithology*, 110.

244 Kennicott to Baird, 29 June 1860, in ibid., 58.
245 Hardisty to Kennicott, 30 Nov. 1864, in Lindsay, *Science in the Subarctic*, 60.
246 Ross to Baird, 26 Nov. 1859, in Lindsay, *Modern Beginnings of Subarctic Ornithology*, 33.
247 Clarke to Kennicott, 16 Jan. 1865, as quoted in Lindsay, *Science in the Subarctic*, 53.
248 Ibid.
249 James R. Fleming, "Meteorology at the Smithsonian Institution, 1847–74: The Natural History Connection," *Archives of Natural History* 16, no. 3 (1989): 277.
250 Bruce, *Launching of Modern American Science*, 194–5.
251 Fleming, "Meteorology at the Smithsonian Institution," 277; Frank Rives Millikan, "Joseph Henry: Father of Weather Service," in "Joseph Henry American Physicist," Smithsonian Institution Archives, http://siarchives.si.edu/history/jhp/joseph03.htm. Also see Frank Millikan, "Joseph Henry's Grand Meteorological Crusade," Weatherwise 50, no. 5 (Oct./Nov. 1997): 14–18.
252 George Brown Goode, "The Origin of the National Scientific and Educational Institutions of the United States," *Annual Report of the American Historical Association for the Year 1889*, 96.
253 Goode, "The Origin of the National Scientific and Educational Institutions," 96–8; Fleming, "Meteorology at the Smithsonian Institution," 275–84.
254 Sherwood, *Exploration of Alaska*, 37–8.
255 Lindsay, *Science in the Subarctic*, 30, 34. Baird's "Directions for Collecting, Preserving, and Transporting Specimens of Natural History," was published in the *SIAR 1856*, 235–53.
256 *SIAR, 1863*, 53.
257 Ross to Baird, 26 Nov. 1859, in Lindsay, *Modern Beginnings of Subarctic Ornithology*, 34.
258 See Fitzhugh, "Origins of Museum Anthropology," 187.
259 Henry B. Collins, "Wilderness Exploration and Alaska's Purchase," *Living Wilderness*, 11, no. 19 (1947), 17.
260 Fitzhugh, "Origins of Museum Anthropology," 179. Also see Lindsay, *Science in the Subarctic*, 77–88.
261 Hinsley, *Smithsonian and the American Indian*, 35, 67.
262 George Gibbs to Joseph Henry, 18 Nov. 1862, as published in the *SIAR, 1862*, 90.
263 Barnston to Baird, 6 Nov. 1861, in Lindsay, *Modern Beginnings of Subarctic Ornithology*, 124.

264 Ross to Baird, 1 June 1862, in Lindsay, *Modern Beginnings of Subarctic Ornithology*, 132.
265 Quoted in Lindsay, *Science in the Subarctic*, 82.
266 Lindsay, *Science in the Subarctic*, 77.
267 *SIAR, 1863*, 53.
268 Fitzhugh, "Origins of Museum Anthropology," 179–80; Robert E. Bieder, *Science Encounters the Indian* (Norman: University of Oklahoma Press, 1986), 115–18; Hinsley, *Smithsonian and the American Indian*, 23–4, 35; Lewis, *Democracy of Facts*, 78–106. The Smithsonian's first publication was E. G. Squier and E. H. Davis, *Ancient Monuments of the Mississippi Valley* (Washington: Smithsonian Institution, 1848).
269 Hinsley, *Smithsonian and the American Indian*, 35.
270 Donald Gunn, "Indian Remains," 399–400. For more on the scholarship on mounds in Rupert's Land, see Gwen Rempel, "The Manitoba Mound Builders: The Making of an Archaeological Myth, 1857–1900," *Manitoba History* 28 (1994): 12–18.
271 Bernard R. Ross, "The Eastern Tinneh," *SIAR 1866*, 304–11; W. L. Hardisty, "The Loucheux Indians," *SIAR 1866*, 311–20; Strachan Jones, "The Kutchin Tribes," *SIAR 1866*, 320–7. Ross's paper is evidently a condensed version of the paper he submitted. The full paper can be found in the Smithsonian Institution Archives, RU 72221; Lindsay, *Modern Beginnings of Subarctic Ornithology*, 131.
272 Lindsay, *Science in the Subarctic*, 87.
273 Annual Report of the Board of Regents of the Smithsonian Institution, 1868, 30.
274 Archibald Williamson Shiels, *Purchase of Alaska* (College, AK: University of Alaska Press, 1967), esp. xii–xiii; James, *First Scientific Exploration*, 29, 45.
275 Lindsay, *Science in the Subarctic*, 123–4; Sherwood, *Exploration of Alaska*, 32–3.
276 Lindsay, *Science in the Subarctic*, 123.
277 Bockstoce, *Furs and Frontiers in the Far North*, 288.
278 James, *First Scientific Exploration*, 20.
279 "Russian Possessions in America," from the *New York Herald*, as reprinted in *London Times*, 10 Aug. 1854; *New York Times*, 25 Aug. 1854. In the *London Times* of 10 Aug. 1854, A. K. Isbister argued that Russia could not sell its American possessions to the United States because the HBC's 1840 lease of portions of Russian America meant that Great Britain was in actual possession of the most valuable portions of Russian America. In his speech on the cession of Russian America to the United States

in Apr. 1867, Charles Sumner mentioned that this lease was due to expire in June 1867; Shiels, *Purchase of Alaska*, 37, 65.
280 Bockstoce, *Furs and Frontiers in the Far North*, 160.
281 Ibid., 289.
282 Shiels, *Purchase of Alaska*, 6, 34.
283 There was little reason for the Smithsonian to hide evidence of correspondence to that effect, if it had received it, although much of the institution's official correspondence was destroyed by fire in 1865; James, *First Scientific Exploration*, 32.
284 Lindsay, *Science in the Subarctic*, 123–4.
285 Collins, "Wilderness Exploration," 17. Collins emphasized the significance of Kennicott's work with the telegraph, but included the work he did with the HBC. Shiels identified Collins as acting chair of the Smithsonian in *Purchase of Alaska*, xi.
286 Sherwood, *Exploration of Alaska*, 33.
287 Committee of the Chicago Academy of Sciences, "Biography of Robert Kennicott," 166.
288 James, *First Scientific Exploration*, 22.
289 The text of the speech can be found in Shiels, *Purchase of Alaska*, 22–126.
290 Ibid., 54–9.
291 Ibid., 59.
292 *SIAR, 1867*, 43.
293 Lindsay, *Science in the Subarctic*, 124, 168n11. Scientists at the Smithsonian Institution were not alone in exaggerating their knowledge of Russian America. In 1854 A. K. Isbister implied that his explorations "some distance into Russian America" (less far than Kennicott's) enabled him to comment on the value of the Russian possessions in America; *London Times*, 10 Aug. 1854.
294 Fitzhugh, "Origins of Museum Anthropology," 190.
295 Ibid., 188.
296 Thomas, "The Smithsonian and the Hudson's Bay Company," 302.
297 Lindsay, *Science in the Subarctic*, 60–1.

Epilogue

1 Wilson, "Science in Rupert's Land," 336.
2 Ibid., 339.
3 Ibid., 340.
4 *SIAR, 1861*, 105–6.

5 Mactavish to Baird, 21 May 1863, in Lindsay, *Modern Beginnings of Subarctic Ornithology*, 146. This resolves some of the questions about the fate of the institute unanswered in T. C. B. Boon, "The Institute of Rupert's Land and Bishop David Anderson," Historical and Scientific Society of Manitoba, *Transactions* series 3, no.18 (1961–2): 92–114. Scholarly publications associated with the institute do include James Hunter, *A Lecture on the Grammatical Construction of the Cree Language* (London: Society for Promoting Christian Knowledge, 1875), which was based on a presentation to the institute on 2 Apr. 1862, and W. W. Kirkby, "The Indians of the Youcon," Appendix 3 in Henry Youle Hind, *Explorations in the Interior of the Labrador Peninsula the Country of the Montagnais and Nasquapee Indians* (London: Longman, Green, Longman, Roberts, & Green, 1863), 2: 254–7, which was also originally transmitted to the Institute in 1862.
6 See Thomas G. Boreskie, "Owen Griffiths Corbett," *DCB* 13: 215–17.
7 Anonymous, *The Hudson Bay Company. "A Million": Shall We Take It? Addressed to the Shareholders of the Company by One of Themselves* (London: A. H. Baily, 1866), 4.
8 Angry with unofficial British support for the Confederacy during the Civil War, the United States government announced in 1866 that it was abrogating the 1854 Reciprocity Treaty (Elgin-Marcy Treaty) that had permitted a measure of free trade between British North American and the United States.
9 Galbraith, *Hudson's Bay Company as an Imperial Factor*, 408–9. Also see Rich, *Hudson's Bay Company* 2: 867.
10 Anonymous, *The Hudson Bay Company. "A Million": Shall We Take It?*, 5.
11 Russell W. Fridley, "When Minnesota Coveted Canada," *Minnesota History* 41, no. 2 (Summer 1968): 79.
12 Galbraith, *Hudson's Bay Company as an Imperial Factor*, 423; Rich, *Hudson's Bay Company* 2: 890. Other major stipulations of the agreement were that the company was also allowed ownership of parcels of land upon which its posts were situated, and up to 1/20th of the land in the fertile belt of the prairies.
13 Chester Martin, "The Royal Charter," *Beaver* 276 (June 1945): 26.
14 Charles Sutherland Elton and Mary Nicholson, "The Ten-year Cycle in Numbers of the Lynx in Canada." *Journal of Animal Ecology* 11 (1942): 215–44. For a discussion of the significance of Elton's discussion of the population cycle, see Charles J. Krebs, Rudy Boonstra, Stan Boutin, and A. R. E. Sinclair, "What Drives the 10-year Cycle of Snowshoe Hares?" *BioScience* 51 (2001): 25–35. Elton is also discussed in Tina Loo, *States of Nature* (Vancouver: UBC Press, 2007), 93–120.

15 *Report of Progress, Geological Survey of Canada, 1870–71*, (Ottawa: I.B. Taylor, 1872), 322. The information included "keeping for us a regular register of the readings of the barometer and thermometer throughout the season, as well as for many other favors" (322). Another acknowledgment of the assistance of the HBC is found in James Richardson's report of his survey north of Lac St. Jean, 283.
16 Geological and Natural History Survey of Canada, *Reports of Progress for 1879–80* (Montreal: Dawson Brothers, 1881), 2.
17 Alfred R.C. Selwyn, "Summary Report," *Annual Report of the Geological Survey of Canada for 1886*, (Montreal: Dawson Brothers, 1887), 1.
18 Annual Report, Geological Survey of Canada Annual Report, 1887–8, (Montreal: William Foster Brown & Co. 1889), vol. 3, part 2, 139.
19 Lucien M. Turner, *Ethnology of the Ungava District, Hudson Bay Territory* (Washington, DC: Smithsonian Institution Press, 2001).

Conclusion

1 Quoted in Max Weber, "Science as Vocation," in *The Vocation Lectures*, ed. David S. Owne and Tracy B Stong, trans. Rodney Livingstone (Indianapolis: Hackett, 2004), 17.

Bibliography

Abbreviations

DCB: *Dictionary of Canadian Biography* (Printed and online versions were identical at time of writing)
HBCA: Hudson's Bay Company Archives
ODNB: *Oxford Dictionary of National Biography* (online)
SIAR: *Smithsonian Institution Annual Report*

Archival Sources

British Columbia Archives, Victoria, British Columbia, Canada
  MS-0580 Jean-Baptiste Bolduc fonds: 1843–5.
Hudson's Bay Company Archives, Provincial Archives of Manitoba, Winnipeg, Manitoba, Canada (HBCA)
  Documents cited include files from each of the following sections of the HBCA:
    Section A: Records of the London Governor and Committee
    Section B: Post Records
    Section C: Ship's Records
    Section D: (North American) Governor's Papers and Commissioner's Office
    Section E: Private Records
    Section G: Cartographic Records
Library and Archives Canada, Ottawa, Ontario, Canada (LAC)
  MG 19-A 21, Series 1, James Hargrave and family fonds. "Hargrave Correspondence" series, Volume 1–20, Microfilm Reels C-73 to C-79.

Royal Society Archives, London, England (RSA)
MA 69 Peter Fidler, Meteorological Journal at Isle à la Crosse, at Clapham House, Deer Lake, Isle à la Crosse, 1809–1811"
MC 1 (40) Joseph Colen papers

Published Sources

Aborigines Protection Society. *Canada West and the Hudson's Bay Company: A Political and Humane Question of Vital Importance to the Honour of Great Britain, to the Prosperity of Canada, and the Existence of the Native Tribes.* London: William Tweedie, 1856.
Acosta, Joseph [José] de. *The Natural & Moral History of the Indies.* ed. Clements R. Markham, trans. Edward Grimston. 2 vols. London: Hakluyt Society, 1880.
Alexander, Edward P. *Museums in Motion: An Introduction to the History and Functions of Museums.* Nashville: American Association for State and Local History, 1979.
Allan, Mea. *The Hookers of Kew, 1785–1911.* London: Joseph, 1967.
Allen, D.E. "Natural History in Britain in the Eighteenth Century." *Archives of Natural History* 22 (1993): 333–47.
–. *The Naturalist in Britain: A Social History.* Princeton, NJ: Princeton University Press, 1994.
Allen, Elsa G. "Nicholas Denys: A Forgotten Observer of Birds." *The Auk* 56, no. 3 (1939): 283–90.
–. "The History of American Ornithology before Audubon." *Transactions of the American Philosophical Society.* New Series. 41, no. 3 (1951): 387–591.
Angel, Michael R. "Clio in the Wilderness: or Everyday Reading Habits of the Honourable Company of Merchant Adventurers Trading Into Hudson's Bay." *Manitoba Library Association Bulletin* 19, no. 3 (June 1980): 14–19.
Anonymous. "Andrew Murray, F.L.S." *Entomologist's Monthly Magazine* 14, no. 3 (1877–78): 215–16.
–. "Biographical Sketch of the Late Thomas Nuttall." *The Gardener's Monthly and Horticultural Advertiser* 2 (1860): 21–3.
–. "Books of the North." *Beaver* 288 (Winter 1957): 60–1.
–. "A List of the Collectors Whose Plants Are in the Herbarium of the Royal Botanic Gardens, Kew, to 31st December, 1899." *Bulletin of Miscellaneous Information (Royal Gardens, Kew)* 169/171 (Jan.–Mar. 1901): 1–80.
–. Obituary of Jean François Arago. *Monthly Notices of the Royal Astronomical Society* 14 (1854): 102–7.
–. "The Old Library of Fort Simpson." *Beaver* 5 (December 1924): 20–1.

Arago, F. "Observations météorologiques faites sur la côte occidentale de l'Amérique du Nord, au fort Vancouver, rivière Columbia." *Comptes Rendus Hebdomadaires des Séances de l'Académie des Sciences* 6 (Jan.–June 1838): 120–1.
–. "Sur le Climat de la Côte Orientale de l'Amérique du Nord." *Comptes Rendus Hebdomadaires des Séances de l'Académie des Sciences* 1 (13 July 1835): 266–8.
Arrowsmith, A. *Observations made at Slave Lake, Athapascow Lake, ... Cumberland House, Slave Lake, by Philip Turnor, Mr. McKenzie, William Wales and Captain Harwell.* London: printed by author, 1794.
–. *Result of Astronomical Observations made in the Interior Part of North America.* London: printed by author, 1794.
Audubon, John James, and John Bachman. *The Quadrupeds of North America.* 3 vols. New York: V.G. Audubon, 1846–54.
Baber, Zaheer. *The Science of Empire: Scientific Knowledge, Civilization, and Colonial Rule in India.* Albany: State University of New York Press, 1996.
Back, George. *Arctic Artist: The Journal and Paintings of George Back, Midshipman with Franklin, 1819–1822.* Edited by C. Stuart Houston. Montreal: McGill-Queen's University Press, 1994.
–. *Narrative of the Arctic Land Expedition to the Mouth of the Great Fish River, and along the Shores of the Arctic Ocean in the Years 1833, 1834, and 1835.* London: J. Murray, 1836.
Baillie, James L. Jr. "Naturalists on Hudson Bay." *Beaver* 277 (December 1946): 36–9.
Bain, David. "George Allan and Horticultural Gardens." *Ontario History.* 87, no. 3 (1995): 231–51.
Ball, Timothy, and David Dyck. "Observations of the Transit of Venus at Prince of Wales's Fort in 1769." *The Beaver* 315, no. 2 (1984): 51–6.
Ballantyne, R. M. *Hudson's Bay.* 2nd ed. Edinburgh and London: 1848.
Barber, Lynn. *Heyday of Natural History: 1820–1870.* London: Jonathan Cape, 1980.
Barnston, George. "Abridged Sketch of the Life of David Douglas, Botanist, with a Few Details of His Travels and Discoveries." *The Canadian Naturalist and Geologist* 5 (1860): 120–32, 200–8, 267–78.
–. "Geographical Distribution of the Genus Allium in British North America." *The Canadian Naturalist and Geologist* 4 (Feb. 1859): 116–21.
–. "Life in a Hudson's Bay Company's Fort." *Chambers's Journal of Popular Literature, Science, and Art* 828 (8 November 1879): 705–7.
–. "Observations on the Progress of the Seasons as Affecting Animals and Vegetables at Martin's Falls, Albany River, Hudson's Bay." *Edinburgh New Philosophical Journal* 30 (1840–1): 252–6.

–. "On a Collection of Plants from British Columbia, Made by Mr. James Richardson in the Summer of 1874." *Canadian Naturalist and Quarterly Journal of Science*, new series 8 (April 1876): 90–4.
–. "Recollections of the Swans and Geese of Hudson's Bay." *Canadian Naturalist* 6 (Feb. 1861): 337–44.
–. "Recollections of the Swans and Geese of Hudson's Bay." *Ibis* 2, no. 3 (July 1860): 253–9.
–. "Remarks on the Genus Lutra and on the Species Inhabiting North America." *The Canadian Naturalist and Geologist* 8 (Feb. 1863): 147–60.
–. "Remarks on the Geographical Distribution of the Cruciferae throughout the British Possessions in North America." *The Canadian Naturalist and Geologist* 4 (Feb. 1859): 1–12.
–. "Remarks upon the Geographical Distribution of Plants in the British Possessions of North America." *Canadian Naturalist and Geologist* 3, no. 1 (Feb. 1858): 26–32.
–. "Remarks upon the Geographical Distribution of the Order Ranunculaceae, throughout the British Possessions of North America." *Canadian Naturalist and Geologist* (1857): 12–20.
Barr, William, ed. *From Barrow to Boothia: The Arctic Journal of Chief Factor Peter Warren Dease, 1836–1839*. Montreal and Kingston: McGill-Queen's University Press, 2002.
Barrington, Daines. "Investigation of the Specific Characters Which Distinguish the Rabbit from the Hare: In a Letter to Samuel Wegg, Esq; T. and Vice-President of the R. S. from the Honourable Daines Barrington, V. P. R. S." *Philosophical Transactions* 62 (1772): 4–14.
–. "Observations on the Lagopus, or Ptarmigan; In a Letter from the Hon. Daines Barrington, V. P. R. S. to Mathew Maty, M. D. F. R. S." *Philosophical Transactions*, 63 (1773–4): 224–30.
Barrow, John. *A Chronological History of Voyages in the Arctic Regions: Undertaken Chiefly for the Purpose of Discovering a North-East, North-West, or Polar Passage between the Atlantic and Pacific*. London: J. Murray, 1818.
Beaglehole, J.C., ed. *The Journals of Captain James Cook on His Voyages of Discovery, 1728–1779*. 4 vols. Cambridge: Hakluyt Society, 1955–69.
Beattie, Judith Hudson. "'My Best Friend': Evidence of the Fur Trade Libraries Located in the Hudson's Bay Company Archives." *Épilogue* 8, no. 1–2, (1993): 1–32.
Beattie, R. K. "Joseph Burke up to 1853." *Madroño* 13, no. 8 (Oct. 1956): 259–61.
Bedini, Silvio A. "The Evolution of the Science Museum." *Technology and Culture* 6, no. 1, (Winter 1965): 1–29.

Bell, Thomas. "Description of a New Species of Agama, Brought from the Columbia River by Mr. Douglass." *Transactions of the Linnean Society of London* 16 (1833): 105–7.
Belyea, Barbara, ed. *Columbia Journals: David Thompson*. Montreal and Kingston: McGill-Queen's University Press, 1994.
–. *Dark Storm Moving West*. Calgary: University of Calgary Press, 2007.
–, ed. *A Year Inland: The Journal of a Hudson's Bay Company Winterer*. Waterloo, ON: Wilfrid Laurier University Press, 2000.
Benedict, Barbara M. *Curiosity: A Cultural History of Early Modern Inquiry*. Chicago: University of Chicago Press, 2001.
Berger, Carl. *Science, God and Nature in Victorian Canada*. Toronto: University of Toronto Press, 1983.
Bethlehem, J., and A. C. Meijer, eds. *VOC en Cultuur: Wetenschappelijke en Culturele Relaties tussen Europa en Azië ten Tijde van de Verenigde Oostindische Compagnie*. Amsterdam: Schiphouwer & Brinkman, 1993.
Bieder, Robert. *Science Encounters the Indian: The Early Years of American Ethnology*. Norman: University of Oklahoma Press, 1986.
Binnema, Ted (Theodore). *Common and Contested Ground: A Human and Environmental History of the Northwestern Plains*. Norman: University of Oklahoma Press, 2001.
–. "How Does a Map Mean?: Old Swan's Map of 1801 and the Blackfoot World." In *From Rupert's Land to Canada: Essays in Honour of John E. Foster*, ed. Theodore Binnema, Gerhard Ens, and Roderick C. Macleod. Edmonton: University of Alberta Press, 2001, 201–24.
–. "Theory and Experience: Peter Fidler and the Transatlantic Indian." In *Native Americans and Anglo-American Culture, 1750–1850: The Indian Atlantic*, ed. Tim Fulford and Kevin Hutchings. Cambridge: Cambridge University Press, 2009, 155–70.
Birch, Thomas. *The History of the Royal Society of London, for Improving of Natural Knowledge: From Its First Rise*. 4 vols. New York: Johnson Reprint Corp., [1756–7] 1968.
Blakiston, Thomas. "On Birds Collected and Observed in the Interior of British North America." *Ibis* 3 (1861): 314–320; 4 (Jan. 1862): 3–10.
–. "On the Birds of the Interior of British North America." *Ibis* 5 (1863): 39–87, 121–55.
Bockstoce, John R. *Furs and Frontiers in the Far North: The Contest among Native and Foreign Nations for the Bering Strait Fur Trade*. New Haven: Yale University Press, 2009.
Boldget, Lorin. *Climatology of the United States, and of the Temperate Latitudes of the North American Continent*. Philadelphia: J. B. Lippencott, 1857.

Bonta, Marcia. "Baird of the Smithsonian." *Pennsylvania Heritage* 6, no. 3 (Summer 1980): 19–23.

Boon, T. C. B. "The Institute of Rupert's Land and Bishop David Anderson." Historical and Scientific Society of Manitoba. *Transactions* 3rd ser., no. 18 (1961–2): 92–114.

Boyle, Robert. "General Heads for a Natural History of a Countrey [sic], Great or Small." *Philosophical Transactions of the Royal Society* 1 (1665–6): 186–9.

Brantlinger, Patrick. *Dark Vanishings: Discourse on the Extinction of Primitive Races, 1800–1930*. Ithaca, NY: Cornell University Press, 2003.

Brett, R. L., and A. R. Jones, eds. *Wordsworth and Coleridge: Lyrical Ballads*. 2nd ed. New York: Routledge, 1991.

Britten, James. "An Early Hudson's Bay Collector." *Journal of Botany* 60 (1922): 239.

Broughton, Peter. "Astronomy in Seventeenth-Century Canada." *Journal of the Royal Astronomical Society of Canada* 75, no. 4 (1981): 175–208.

Brown, Jennifer S.H. "Rupert's Land, *Nituskeenan*, Our Land: Cree and English Naming and Claiming around the Dirty Sea." In *New Histories for Old*, ed. Ted Binnema and Susan Neylan. Vancouver: UBC Press, 2007, 18–40.

–. *Strangers in Blood: Fur Trade Company Families in Indian Country* Toronto: University of British Columbia Press, 1980.

Brown, Sam, and James Petiver. "An Account of Part of a Collection of Curious Plants and Drugs, Lately Given to the Royal Society by the East India Company." *Philosophical Transactions* 22 (1700–1): 579–594.

Bruce, Robert V. *The Launching of Modern American Science: 1846–1876*. New York: Knopf, 1987.

Burley, Edith I. *Servants of the Honorable Company: Work, Discipline, and Conflict in the Hudson's Bay Company, 1770–1879*. Toronto: Oxford University Press, 1997.

Burpee, Lawrence J., ed. "An Adventurer from Hudson Bay: Journal of Matthew Cocking from York Factory to the Blackfeet Country, 1772–1773." *Royal Society of Canada Proceedings and Transactions* ser. 3, vol. 2 (1908): 89–121.

Burrow, J. W. "Evolution and Anthropology in the 1860s: The Anthropological Society of London, 1863–71." *Victorian Studies* 7, no. 2 (Dec. 1963): 137–54.

Butler, William F. *Wild North Land*. Philadelphia: Porter & Coates, 1874.

Byrne, Angela. *Geographies of the Romantic North*. New York: Palgrave Macmillan, 2013.

Cañizares-Esguerra, Jorge. *Nature, Empire, and Nation: Explorations of the History of Science in the Iberian World*. Stanford, CA: Stanford University Press, 2006.

Cannon, Susan Faye. *Science in Culture: The Early Victorian Period*. New York: Science History Publications, 1978.

Carlos, Ann M., and Frank D. Lewis. *Commerce by a Frozen Sea: Native Americans and the European Fur Trade*. Philadelphia: University of Pennsylvania Press, 2010.

Carlos, Ann M., and Jamie Brown Kruse. "The Royal African Company: Fringe Firms and the Role of the Charter." *Economic History Review* 49, no. 2 (1996): 291–313.

Carter, Christopher. "Magnetic Fever: Global Imperialism in the Nineteenth Century," *Transactions of the American Philosophical Society* 99, no. 4 (2009).

Castling, Leslie D. "The Red River Library: A Search after Knowledge and Refinement." In *Readings in Canadian Library History*, ed. Peter F. McNally. Ottawa: Canadian Library Association, 1986, 153–66.

Cavell, Janice. "The Hidden Crime of Dr. Richardson." *Polar Record* 43, no. 225 (2007): 155–164.

Cawood, John. "Terrestrial Magnetism and the Development of International Collaboration in the Early Nineteenth Century" *Annals of Science* 34 (1977): 551–87.

–. "The Magnetic Crusade: Science and Politics in Early Victorian Britain," *Isis* 70 (1979): 493–518.

Chambers, David Wade, and Richard Gillespie. "Locality in the History of Science: Colonial Science, Technoscience, and Indigenous Knowledge." *Osiris* 15 (2000): 221–40.

Chang, Hasok. *Inventing Temperature: Measurement and Scientific Progress*. Oxford: Oxford University Press, 2004.

Chaplin, Joyce E. *Subject Matter: Technology, the Body, and Science on the Anglo-American Frontier, 1500–1676*. Cambridge, MA: Harvard University Press, 2001.

–. *The First Scientific American: Benjamin Franklin and the Pursuit of Genius*. New York: Basic Books, 2006.

Chinard, Gilbert. "The American Philosophical Society and the World of Science (1768–1800)." *Proceedings of the American Philosophical Society* 87, no. 1 (14 July 1943): 1–11.

Christy, Miller. "An Early Hudson Bay Collector." *Journal of Botany* 60 (1922): 337.

Clark, William, Jan Golinski, and Simon Schafffer, eds. *The Sciences in Enlightened Europe*. Chicago: University of Chicago Press, 1999.

Clerk, the California. [probably Charles Swaine or Theodorus Swaine Drage]. *An Account of a Voyage for the Discovery of a North-West Passage*. London: Jolliffe, 1748, 1749.

Cohen, I. Bernard. "The New World as a Source of Science for Europe." *Actes du IXe Congrès Internationale d'Histoire des Sciences: Barcelona Madrid, 1–7 Septembre 1959*. Barcelona: Asociación para la Historia de la Ciencia Española, 1960, 96–130.

Cole, Jean Murray. "Keeping the Mind Alive: Literary Leanings in the Fur Trade." *Journal of Canadian Studies* 16, no. 2 (Summer 1981): 87–93.

Collins, Henry B. "Wilderness Exploration and Alaska's Purchase." *Living Wilderness* 11, no. 19 (1947): 17–18.

Colquhoun, Kate. *"The Busiest Man in England": Life of Joseph Paxton, Garden Architect, and Victorian Visionary*. Boston: David R. Godine, 2006.

Committee of the Chicago Academy of Sciences. "Biography of Robert Kennicott." *Transactions of the Chicago Academy of Sciences* 1, no. 2 (1869): 133–226.

Cook, Harold J. *Matters of Exchange: Commerce, Medicine, and Science in the Dutch Golden Age*. New Haven and London: Yale University Press, 2007.

Cook, James. *A Voyage to the Pacific Ocean: Undertaken, by the Command of His Majesty, for Making Discoveries in the Northern Hemisphere: To Determine the Position and Extent of the West Side of North America, Its Distance from Asia, and the Practicability of a Northern Passage to Europe; Performed under the Direction of Captains Cook, Clerke, and Gore, in His Majesty's Ships the Resolution and Discovery, in the Years 1776, 1777, 1778, 1779, and 1780*. London: G. Nicol and T. Cadell, 1784.

Cooper, Barry. *Alexander Kennedy Isbister: A Respectable Critic of the Honourable Company*. Ottawa: Carleton University Press, 1988.

Cooper, James. "On the Distribution of the Forests and Trees of North America, with Notes on its Physical Geography." *SIAR* (1858): 246–280.

Costard, George. *The History of Astronomy, with its Applications to Geography and Chronology Occasionally Illustrated by the Globes*. London: James Lister, 1767.

Coville, Frederick V. "Added Botanical Notes on Carl A. Geyer." *Oregon Historical Quarterly* 42 (Dec. 1941): 323–4.

–. "The Itinerary of John Jeffrey, an Early Botanical Explorer of Western North America." *Proceedings of the Biological Society of Washington* 11 (1897): 57–60.

Dalrymple, Alexander. *Memoir of a Map of the Lands around the North-Pole*. London: George Bigg 1789.

–. *Plan for Promoting the Fur-Trade and Securing It to This Country, by Uniting the Operations of the East-India and Hudson's-Bay Companys*. London: George Bigg, 1789.

Daniels, George H. *Science in American Society: A Social History*. New York: Alfred A. Knopf, 1971.

Daston, Lorraine. "Afterword: The Ethos of Enlightenment." In *The Sciences in Enlightened Europe*, ed. William Clark, Jan Golinski, and Simon Schaffer. Chicago: University of Chicago Press, 1999, 495–504.

Daston, Lorraine, and Katherine Park. *Wonders and the Order of Nature, 1150–1750*. New York: Zone Books, 1998.

Davies, John, ed. *Douglas of the Forests: The North American Journals of David Douglas*. Seattle: University of Washington Press, 1980.

Davies, K. G., ed., *Northern Quebec and Labrador Journals and Correspondence, 1819–1835*. London: Hudson's Bay Record Society, 1963.

Dawson, G. M. "The Progress and Trend of Scientific Investigation in Canada." *Transactions of the Royal Society of Canada* ser. 1, vol. 12 (1894): 3–17.

Dawson, J. W. "Gleanings in the Natural History of the Hudson's Bay Territories, by the Arctic Voyagers." *Canadian Naturalist and Geologist* 2, no. 3 (July 1857): 170–195.

Dawson, S. J., H. Y. Hind, et al., *Papers Relative to the Exploration of the Country between Lake Superior and the Red River Settlement*. London: George Edward Eyre and William Spottiswoode, 1859.

Dease, Peter Warren, and John Richardson. "On the Cultivation of the Cerealea in the High Latitudes of North America." *Edinburgh New Philosophical Journal*, 30 (1840–1): 123–24.

Dease [Peter Warren], and T. Simpson. "An Account of the Recent Arctic Discoveries by Messrs. Dease and T. Simpson. Communicated by J.H. Pelly, Esq., Governor of the Hudson's Bay Company." *Journal of the Royal Geographical Society of London*: 8 (1838): 213–25.

Dease, Peter Warren, and Thomas Simpson. "An Account of Arctic Discovery on the Northern Shore of America in the Summer of 1838." *Journal of the Royal Geographical Society of London* 9 (1839): 325–30.

–. "Narrative of the Progress of Arctic Discovery on the Northern Shore of America, in the Summer of 1839." *Journal of the Royal Geographical Society of London* 10 (1840): 268–74.

Deiss, William A. "Spencer F. Baird and His Collectors." *Journal of the Society for the Bibliography of Natural History* 9 (1980): 635–45.

Dekker, E. "Early Explorations of the Southern Celestial Sky." *Annals of Science* 44 (1987): 439–70.

Delbourgo, James. *A Most Amazing Scene of Wonders: Electricity and Enlightenment in Early America*. Cambridge, MA: Harvard University Press, 2006.

Delbourgo, James, and Nicholas Dew, eds. *Science and Empire in the Atlantic World*. New York: Routledge, 2008.

Dictionary of Canadian Biography.vols 1–13. Toronto: University of Toronto Press, 1966–1994.

Diegnan, H. C. "The HBC and the Smithsonian." *The Beaver* 278 (June 1947): 3–7.

Dillon, Richard H., ed. "Peter Pond and the Overland Route to Cook's Inlet." *Pacific Northwest Quarterly* 42, no. 4 (1951): 324–9.

Dobbs, Arthur. "A Letter from Arthur Dobbs Esq; of Castle-Dobbs in Ireland, to the Rev. Mr. Charles Wetstein, Chaplain and Secretary to His Royal

Highness the Prince of Wales, Concerning the Distances between Asia and America." *Philosophical Transactions of the Royal Society* 44 (1746–7): 471–76.

—. "A Letter from Arthur Dobbs Esq; to Charles Stanhope Esq; F. R. S. concerning Bees, and Their Method of Gathering Wax and Honey." *Philosophical Transactions of the Royal Society* 46 (1749–50): 536–49.

—. "An Account of a Parhelion [sic] Seen in Ireland: In Two Letters from Arthur Dobbs Esq; Of Castle Dobbs in the County of Antrim, to His Brother Mr. Richard Dobbs of Trinity-College in Dublin; and by This Last Communicated to the Royal Society." *Philosophical Transactions of the Royal Society* 32 (1722–3): 89–92.

—. *An Account of the Countries Adjoining to Hudson's Bay in the North-west Part of America.* London: Printed for J. Robinson, 1744.

"Documents: Oregon Missionaries." *Washington Historical Quarterly* 2, no. 2 (1908): 254–64.

Douglas, David. *Journal Kept by David Douglas During His Travels in North America, 1823–1827 ... With Appendices Containing a List of the Plants Introduced by Douglas and an Account of His Death in 1834.* London: William Wesley & Son, 1914.

—. "Observations on Two Undescribed Species of North American Mammalia." *Zoological Journal* 4 (1829): 330–2.

Drummond, Thomas. "Sketch of a Journey to the Rocky Mountains and to the Columbia River in North America." *Botanical Miscellany* 1 (1830): 95–96, 178–219.

Drury, Clifford M. "Botanist in Oregon in 1843–44 for Kew Gardens, London." *Oregon Historical Quarterly* 41, no. 2 (June 1940): 182–8.

Dunlop, G. A., and C.P. Wilson. "George Barnston." *Beaver* (Dec. 1941): 16–17.

D'Urban, W. S. M. "A Systematic List of Lepidoptera Collected in the Vicinity of Montreal." *The Canadian Naturalist and Geologist* 5, no. 4 (1860): 242–66.

Dyck, David. "Observations of the Transit of Venus at Prince of Wales's Fort in 1769." *Beaver* 315, no. 2 (1984): 51–6.

Dymond, Joseph, and William Wales. "Observations On the State of the Air, Winds, Weather, &c. Made at the Prince of Wales's Fort, on the North-West Coast of Hudson's Bay, in the Years 1768 and 1769." *Philosophical Transactions of the Royal Society* 60 (1770): 137–78.

Eavenson, Howard N. *Map Maker and Indian Traders.* Pittsburgh: University of Pittsburgh Press, 1949.

Edwards, George. *A Natural History of Birds: Most of Which Have Not Been Figur'd or Describ'd, and Others Very Little Known.* 4 vols. London: For the author, 1743–51.

Elliott, T. C., ed. "Journal of John Work." *Washington Historical Quarterly* 5, no. 2 (1914): 83–115.

Ellis, Henry. *A Voyage to Hudson's-Bay by the "Dobbs Galley" and "California" in the Years 1746 and 1747 for Discovering a North West Passage.* London: H. Whitridge, 1748.

Endersby, Jim. *Imperial Nature: Joseph Hooker and the Practices of Victorian Science.* Chicago: University of Chicago Press, 2008.

Epp, Henry, ed. *Three Hundred Prairie Years: Henry Kelsey's "Inland Country of Good Report."* Regina: Canadian Plains Research Center, 1993.

Ewart, William B. "Thomas Hutchins and the HBC: A Surgeon on the Bay." *Beaver* 75, no. 4 (Aug./Sept. 1995): 39–41.

Feldman, Theodore Sherman. "The History of Meteorology, 1750–1800: A Study in the Quantification of Experimental Physics." PhD diss., University of California, Berkeley, 1983.

Ferguson, Niall. *Empire: The Rise and Demise of the British World Order and the Lessons for Global Power.* New York: Basic Books, 2004.

Fernie, J. Donald. *The Whisper and the Vision: The Voyages of the Astronomers.* Toronto: Clark Irwin, 1976.

Fitzhugh, William W. "Origins of Museum Anthropology at the Smithsonian Institution and Beyond." *Smithsonian Contributions to Anthropology* 44 (2002): 179–200.

Fleming, James R. "Meteorology at the Smithsonian Institution, 1847–1874: The Natural History Connection." *Archives of Natural History* 16, no. 3 (1989): 275–84.

Fleming, R. Harvey, ed. *Minutes of Council Northern Department of Rupert [sic] Land, 1821–31.* Toronto: Champlain Society, 1940.

Fletcher, Harold R. *The Story of the Royal Horticultural Society 1804–1968.* London: Oxford University Press for The Royal Horticultural Society, 1970.

Forster, J. R. "An Account of the Birds Sent from Hudson's Bay." *Philosophical Transactions of the Royal Society* 62 (1772): 382–433.

–. "A Letter from Mr. John Reinhold Forster, F. R. S. to William Watson, M. D. Giving Some Account of the Roots Used by the Indians, in the Neighbourhood of Hudson's-Bay, to Dye Porcupine Quills." *Philosophical Transactions of the Royal Society* 62 (1772): 54–9.

–. "Account of Some Quadrupeds from Hudson's Bay." *Philosophical Transactions of the Royal Society* 62 (1772): 370–81.

–. "An Account of Some Curious Fishes, Sent from Hudson's Bay." *Philosophical Transactions of the Royal Society* 63, no. 1 (1773): 149–60.

–. *History of the Voyages and Discoveries Made in the North.* Dublin: Luke White, 1786.

Franklin, John. *Narrative of a Journey to the Shores of the Polar Sea, in the Years 1819–20–21–22*. 2nd ed. London: J. Murray, 1825.
–. *Narrative of a Second Expedition to the Shores of the Polar Sea in the Years 1825, 1826, and 1827*. London: J. Murray, 1828.
Fridley, Russell W. "When Minnesota Coveted Canada." *Minnesota History* 41, no. 2 (Summer 1968): 76–9.
Fry, Howard Tyrrell. *Alexander Dalrymple (1737–1808) and the Expansion of British Trade*. Toronto: University of Toronto Press, 1970.
Fryer, Geoffrey. "John Fryer, F.R.S. and His Scientific Observations, Made Chiefly in Indian and Persia between 1672 and 1682." *Notes and Records of the Royal Society of London* 33, no. 2 (1979): 175–206.
Fuchs, Denise. "Embattled Notions: Constructions of Rupert's Land's Native Sons, 1760 to 1861." *Manitoba History* 44 (2002–3): 10–17.
Galbraith, John S. *The Hudson's Bay Company as an Imperial Factor: 1821–1869*. Berkeley and Los Angeles: University of California Press, 1957.
–. *The Little Emperor: Governor Simpson of the Hudson's Bay Company* Toronto: Macmillan, 1976.
Gairdner, Meredith. "General Table of Meteorological Observations at Fort Vancouver, from June 1, 1834 to May 13, 1835." *Edinburgh New Philosophical Journal* 21 (1836): 152–3.
–. "Letter from Dr. M. Gairdner." [March 1834?] *Edinburgh New Philosophical Journal*, (Jan. 1836): 206–7.
–. "Letter from Dr. M. Gairdner, 19th March 1834, Fort Vancouver." *Edinburgh New Philosophical Journal*, (Jan.1836): 205–6.
–. "Meteorological Observations Made at Fort Vancouver, from June 7, 1833 to May 31, 1834." *Edinburgh New Philosophical Journal* 20 (1836): 67.
–. "Notes on the Geography of the Columbia River." *Journal of the Royal Geographic Society* 11 (1841): 250–7.
Gascoigne, John. *Science in the Service of Empire: Joseph Banks, the British State and the Uses of Science in the Age of Revolution*, Cambridge: Cambridge University Press, 1998.
Gaston, Jerry. *The Reward System in British and American Science*. New York: John Wiley and Sons, 1978.
Gerbi, Antonello. The *Dispute of the New World: The History of a Polemic, 1750–1900*. Rev. and enl. ed. translated by Jeremy Moyle. Pittsburgh: University of Pittsburgh Press, 1973.
Geyer, Charles A. [Carl Augustus]. "Notes on the Vegetation and General Character of the Missouri and Oregon Territories." *London Journal of Botany* 4 (1845): 479–492, 653–662; 5 (1846): 22–41, 198–208, 285–310, 509–524.

Gibson, James R. *Lifeline of the Oregon Country.* Vancouver: UBC Press, 1998.
Gillispie, Charles Coulston, gen. ed. *Dictionary of Scientific Biography.* 16 vols. New York: Charles Scribner's Sons, 1970–80.
Glazebrook, G. P. de T., ed. *The Hargrave Correspondence, 1821–1843.* Toronto: Champlain Society, 1938.
Glover, Richard, ed. *A Journey from Prince of Wales's Fort in Hudson's Bay to the Northern Ocean, 1769, 1770, 1771, 1772, by Samuel Hearne.* Toronto: Macmillan, 1958.
–. "An Early Visitor to Hudson Bay." *Queen's Quarterly* 55, no. 1 (1948): 37–45.
–, ed. *David Thompson's Narrative 1784–1812.* Toronto: Champlain Society, 1962.
–. "The Man Who Did Not Go to California." Canadian Historical Association. *Historical Papers* 10 (1975): 95–112.
Glover, William. "The Navigation of the *Nonsuch*, 1668–69." *Northern Mariner* 11, no. 4 (Oct. 2001): 49–66.
Gluek, Alvin C., Jr. *Minnesota and the Manifest Destiny of the Canadian Northwest.* Toronto: University of Toronto Press, 1965.
Goetzmann, William H. "Paradigm Lost." In *The Sciences in the American Context*, ed. Nathan Reingold. Washington, DC: Smithsonian Institution Press, 1979, 21–34.
Goode, George Brown. "The Origin of the National Scientific and Educational Institutions of the United States." *Annual Report of the American Historical Association for the Year 1889*, 53–161.
Gray, John Edward. *List of the Specimens of Mammalia in the Collection of the British Museum.* London: The Trustees, 1843.
Great Britain, Parliament, House of Commons, *Copy of a Despatch from the Secretary of State for the Colonies to the Governor-General of Canada, Together with the Reply of the Governor General.* 1857, session 1 (113).
–. *Copy of the Existing Charter or Grant by the Crown to the Hudson's Bay Company; and Correspondence on the Last Renewal of the Charter, &c.* 1842, (547).
–. *Copy of the Letter Addressed by Mr. Chief Justice Draper to Her Majesty's Secretary of State for the Colonies, Bearing the Date of May 1857, Together with a Copy of the Memorandum.* 1857, session 2 (104).
–. *Hansard*, ser., vol. 100 (13 July 1848), cols. 470–80 and 3d ser. vol. 106 (5 July 1849), cols. 1344–63.
–. *Papers Relative to the Hudson's Bay Company's Charter and Licence of Trade.* 1859, session 1 (2507), 3.
–. *Report from the Committee of the House of Commons Appointed to Inquire into the State and Condition of the Countries Adjoining, Hudson's Bay and of the Trade Carried on There. (With a copy of the Charter of the Hudson's Bay Company).* London, House of Commons, 1749.

–. *Report from the Select Committee on Aborigines (British Settlements)*. Cape Town: C. Struik, 1966.

–. *Report from the Select Committee on the Hudson's Bay Company; Together with the Proceedings of the Committee, Minutes of Evidence, Appendix and Index*. London: House of Commons, 1857.

Greene, John C. *American Science in the Age of Jefferson*. Ames: Iowa State University Press, 1983.

Gribbin, John. *The Fellowship: The Story of a Revolution*. London: Allen Lane, 2005.

Griffin-Short, Rita. "The Transit of Venus: The Heavens from Hudson Bay, 1769." *The Beaver* 83, no. 2 (April/May 2003): 8–11.

Griffiths, Franklyn, ed. *The Politics of the Northwest Passage*. Montreal and Kingston: McGill-Queen's University Press, 1987.

Gunn, Donald. "Indian Remains near Red River Settlement, Hudson's Bay Territory." *SIAR, 1867*, 399–400.

–. "Notes of an Egging Expedition to Shoal Lake, West of Lake Winnipeg." *SIAR 1867*, 427–32.

Gunther, A. E. "The Royal Society and the Foundation of the British Museum, 1753–1781." *Notes and Records of the Royal Society of London* 33, no. 2 (March 1979): 207–16.

Hagstrom, W. O. "Gift-giving as an Organizing Principle in Science." In *Sociology of Science: Selected Readings*, ed. Barry Barnes. New York: Penguin Books, 1972, 105–20.

Halley, Edmond. "Methodus singularis quâ Solis Parallaxis sive distantia à Terra, ope Veneris intra Solem conspiciendæ, tuto determinari poterit [A New Method of Determining the Parallax of the Sun]." *Philosophical Transactions of the Royal Society* 29 (1716): 454–64.

Hallock, Thomas. "Male Pleasure and the Genders of Eighteenth-Century Botanic Exchange: A Garden Tour." *William and Mary Quarterly* 62, no. 4 (Oct. 2005): 697–718.

Hardisty, W. L. "The Loucheux Indians." *SIAR 1866*, 311–20.

Hargrave, Joseph James. *Red River*. Montreal: Printed for the author by John Lovell, 1871.

Harmon, Daniel Williams. *A Journal of Voyages and Travels in the Interiour [sic] of North America* Andover: Flagg and Gould, 1820.

Harper, J. Russell. *Paul Kane's Frontier*. Toronto: University of Toronto Press, 1971.

Harper, P. P., and W. E. Ricker. "Distribution of Ontario Stoneflies (Plecoptera)." *Proceedings of the Entomological Society of Ontario* 125 (1994): 43–66.

Harris, John. *Navigantium atque Itinerantium Bibliotheca: Or, a Complete Collection of Voyages and Travels Consisting of Above Six Hundred of the Most Authentic Writers.* Revised by John Campbell. 2 vols. London, printed for T. Woodward, A. Ward, S. Birt, et al., 1744–8.

Harris, Steven J. "Long-Distance Corporations, Big Sciences, and the Geography of Knowledge." *Configurations* 6 (1998) 269–304.

–. "Networks of Travel, Correspondence, and Exchange." In *The Cambridge History of Science: Early Modern Science*, ed. Katharine Park and Lorraine Daston. Vol. 3. Cambridge: Cambridge University Press, 2006, 341–62.

Harrison, Mark. "Science and the British Empire." *Isis* 96, no. 1 (2005): 56–63.

Hartley, Harold, ed. *The Royal Society: Its Origins and Founders*. London: Royal Society, 1960.

Harvey, A. G. "Chief Concomly's Skull." *Oregon Historical Quarterly* 40 (1939): 161–7.

–. "John Jeffrey: Botanical Explorer." *British Columbia Historical Quarterly* 10 (Oct. 1946): 281–90.

–. *Douglas of the Fir: A Biography of David Douglas Botanist.* Cambridge, MA: Harvard University Press, 1947.

Hearne, Samuel. *A Journey from Prince of Wales's Fort in Hudson's Bay, to the Northern Ocean.* London: A. Strahan and T. Cadell, 1795.

Helm, June. "Matonabbee's Map." *Arctic Anthropology* 26, no. 2 (1989): 28–47.

Henry, Joseph. "Meteorology in Its Connection with Agriculture." In *Report of the Commissioner of Patents for the Year 1856: Agriculture.* Washington: Cornelius Wendell, 1857, 455–95.

Herdendorf, Charles E. "Captain James Cook and the Transits of Mercury and Venus." *Journal of Pacific History* 21, no. 1–2 (1986): 39–55.

Hind, Henry Youle. *Narrative of the Canadian Red River Exploring Expedition of 1857 and of the Assiniboine and Saskatchewan Exploring Expedition of 1858.* 2 vols. London: Longman, Green, Longman, and Roberts, 1860.

Hinsley, Curtis M. *The Smithsonian and the American Indian: Making a Moral Anthropology in Victorian America.* Washington: Smithsonian Institution Press, [1981] 1994.

Holmes, Richard, *Age of Wonder: How the Romantic Generation Discovered the Beauty and Terror of Science.* London: Harper, 2008.

Hooker, W. J. "Account of the Expedition under Captain Franklin, and of the Vegetation of North America, in Extracts of Letters from Dr Richardson, Mr Drummond, and Mr Douglas." *Edinburgh Journal of Science*, 6, no. 1 (Jan. 1827): 107–16.

–. "A Brief Memoir of the Life of Mr. David Douglas, with Extracts from His Letters." *Companion to the Botanical Magazine* 2 (1836): 79–182.

–. "Figure and Description of Castanea Chrysophylla." *London Journal of Botany* 2 (1843): 495–7.

Hopwood, Victor G. "David Thompson and His Maps." In *Explorations in the History of Canadian Mapping: A Collection of Essays*, ed. Barbara Farrell and Aileen Desbarats. Ottawa: Association of Canadian Map Libraries and Archives, 1988, 205–10.

Hornsby, T. "On the Transit of Venus in 1769." *Philosophical Transactions of the Royal Society* 55 (1765): 326–44.

Houston, C. Stuart. *Arctic Ordeal: The Journal of John Richardson, Surgeon-Naturalist with Franklin, 1820–1822*. Montreal and Kingston: McGill-Queen's University Press, 1984.

–. "Birds First Described from Hudson Bay." *The Canadian Field Naturalist* 97 (1983): 95–8.

–. "John Richardson – First Naturalist in the Northwest." *The Beaver* 315, no. 2 (1984): 10–15.

Houston, C. Stuart, and Mary I. Houston. "Samuel Hearne, Naturalist." *The Beaver* 67, no. 4 (1987): 23–7.

Houston, C. Stuart, Tim Ball, and Mary Houston. *Eighteenth-Century Naturalists of Hudson Bay*. Montreal and Kingston: McGill-Queen's University Press, 2003.

Howse, Derek. *Nevil Maskelyne: The Seaman's Astronomer*. Cambridge: Cambridge University Press, 1989.

Howse, Joseph. *A Grammar of the Cree Language with Which Is Combined an Analysis of the Chippeway Dialect*. London: J. G. F. & J. Rivington, 1844.

–. "Vocabularies of Certain North American Indian Languages." *Proceedings of the Philological Society* 4 (1848–50): 102–22.

–. "Vocabularies of Certain North American Languages." *Proceedings of the Philological Society* 4 (1848–50): 191–206.

Hunter, Michael. *Science and Society in Restoration England*. Cambridge: Cambridge University Press, 1981.

–. *The Boyle Papers: Understanding the Manuscripts of Robert Boyle*. Burlington, VT: Ashgate, 2007.

–. *The Royal Society and Its Fellows, 1660–1700: The Morphology of an Early Scientific Institution*. 2nd ed. Oxford: British Society for the History of Science, 1994.

Hutchins, Thomas, and Joseph Black. "Experiments for Ascertaining the Point of Mercurial Congelation." *Philosophical Transactions of the Royal Society of London* 73 (1783): 303–70.

Hutchins, Thomas. "An Account of the Success of Some Attempts to Freeze Quicksilver, at Albany Fort, in Hudson's Bay, in the Year 1775: With

Observations on the Dipping-Needle." *Philosophical Transactions of the Royal Society* 66 (1776): 174–81.
–. "Experiments On the Dipping Needle, Made by Desire of the Royal Society." *Philosophical Transactions of the Royal Society* 65 (1775) 129–38.
Idiens, Dale. "Chipewyan Artefacts Collected by Robert Campbell and Others in the National Museums of Scotland." In *Proceedings of the Fort Chipewyan and Fort Vermilion Bicentennial Conference: September 23–25, 1988, Provincial Museum of Alberta, Edmonton, Alberta*, ed. Patricia Alice McCormack, R. G. Ironside. Edmonton: Alberta Culture and Multiculturalism, Boreal Institute for Northern Studies Published by Boreal Institute for Northern Studies, 1990, 278–80.
Innis, Harold A. *The Fur Trade in Canada: An Introduction to Canadian Economic History*. Rev. ed. Toronto: University of Toronto Press, 1956.
[Isbister, A. K.] "The Hudson's Bay Territories, Their Trade, Productions and Resources; with Suggestions for the Establishment and Economical Administration of a Crown Colony on the Red River and Saskatchewan." *Journal of the Society of Arts* 9 (1 March 1861): 230–46.
Isbister, A. K. *A Few Words On the Hudson's Bay Company; With a Statement of the Grievances of the Native and Half-Caste Indians, Addressed to the British Government through Their Delegates now in London*. London: C. Gilpin, 1846.
–. "On the Chippewyan Indians," *Transactions of the Sections* in *Report of the Seventeenth Meeting of the British Association for the Advancement of Science*. London: John Murray, 1848, 119–21.
–. "On the Geology of the Hudson's Bay Territories, and of Portions of the Arctic and North-Western Regions of America; with a Coloured Geological Map." *Quarterly Journal of the Geological Society* 11 (1855): 497–520. Reprinted as A. K. Isbister. "On the Geology of the Hudson's Bay Territories, and of Portions of the Arctic and Northwestern Regions of America." *American Journal of Science and Arts* 21 (May 1856): 313–38.
–. "On the Loucheux Indians." *Transactions of the Sections* in *Report of the Seventeenth Meeting of the British Association for the Advancement of Science*. London: John Murray, 1848, 121–2.
–. "On the Nehanni Tribe of a Koloochian Class of American Indians," *Transactions of the Sections* in *Report of the Seventeenth Meeting of the British Association for the Advancement of Science*. London: John Murray, 1848, 121.
–. "Some Account of Peel River, N. America." *Journal of the Royal Geographical Society of London* 15 (1845): 332–45.
Jackson, Ian. "Exploration as Science: Charles Wilkes and the U.S. Exploring Expedition, 1838–42." *American Scientist* 73 (Sept.–Oct. 1985): 450–61.

James, James Alton. *The First Scientific Exploration of Russian America and the Purchase of Alaska*. Evanston and Chicago: Northwestern University, 1942.

Janković, Vladimir. "The Place of Nature and the Nature of Place: The Chorographic Challenge to the History of British Provincial Science." *History of Science* 38 (2000): 79–113.

Jardine, Lisa, and Michael Silverthorne, eds. *The New Organon: Cambridge Texts in the History of Philosophy*. Cambridge and New York: Cambridge University Press, 2000.

Jardine, N., J. A. Secord, and E. C. Spary, eds. *Cultures of Natural History*. Cambridge: Cambridge University Press, 1996.

Jarrell, Richard A. *The Cold Light of Dawn: A History of Canadian Astronomy*. Toronto: University of Toronto Press, 1988.

Johansen, Dorothy O. "William Fraser Tolmie of the Hudson's Bay Company." *Beaver* 268, no. 2 (Sept. 1937): 29–32.

Johnson, Alice M. *Saskatchewan Journals and Correspondence*. London: Hudson's Bay Record Society, 1967.

Johnson, Arthur, ed., *Francis Bacon: The Advancement of Learning and New Atlantis*. Oxford: Clarendon Press, 1974.

Jones, H. W. "Sir Christopher Wren and Natural Philosophy: With a Checklist of His Scientific Activities." *Notes and Records of the Royal Society of London* 13, no. 1 (June 1958): 19–37.

Jones, Strachan. "The Kutchin Tribes." *SIAR 1866*, 320–7.

Judd, Carol M. "'Mixt Bands of Many Nations,' 1821–70." In *Old Trails and New Directions: Papers of the Third North American Fur Trade Conference*, ed. Carol M. Judd and Arthur J. Ray. Toronto: University of Toronto Press, 1980, 127–146.

–. "Native Labour and Social Stratification in the Hudson's Bay Company's Northern Department 1770–1870." *The Canadian Review of Sociology and Anthropology* 17, no. 1 (1980): 305–14.

Jungnickel, Christa, and Russell McCormmach. *Cavendish*. Philadelphia: American Philosophical Society, 1996.

Kane, Paul. "Incidents of Travel on the North-West Coast, Vancouver's Island, Oregon, &c., &c, The Chinook Indians." *Canadian Journal* 3, no. 12 (July 1855), 273–9.

Kass, Amalie M., and Edward H. Kass. *Perfecting the World: The Life and Times of Dr. Thomas Hodgkin, 1798–1866*. Boston: Harcourt Brace Jovanovich, 1988.

Keddie, William. "Biographical Notice of the Late John Scouler, MD, LLD, FLS." *Transactions of the Geological Society of Glasgow* 4 (1874): 194–205.

Keill, John. *An Introduction to the True Astronomy or, Astronomical Lecture, Read in the Astronomical School of the University of Oxford*. London: Bernard

Lintot, 1721.Kennedy, J.E. and S.D. Hanson, "Excerpts from Simon Newcomb's Diary of 1860." *Journal of the Royal Astronomical Society of Canada* 90, no. 5/6(1996): 292–303.
Kerr, Robert. "For the Royal Scottish Museum." *Beaver* 284 (June 1953): 32–5.
King, Richard. *Narrative of a Journey to the Shores of the Arctic Ocean in 1833, 1834, and 1835: Under the Command of Capt. Back, R.N.* 2 vols. London: R. Bentley, 1836.
–. "On the Industrial Arts of the Esquimaux." *Edinburgh New Philosophical Journal* 42 (1847): 112–35. Reprinted in *Journal of the Ethnological Society of London* 1 (1848): 277–300.
–. "On the Intellectual Character of the Esquimaux." *Edinburgh New Philosophical Journal* 38 (1845): 303–52. Reprinted in *Journal of the Ethnological Society of London* 1 (1848):127–52.
–. "On the Physical Characters of the Esquimaux." *Edinburgh New Philosophical Journal* 36 (1844): 296–310. Reprinted in *Journal of the Ethnological Society of London* 1 (1848): 45–59.
–. "On the Unexplored Coast of North America." *London, Edinburgh, and Dublin Philosophical Magazine and Journal of Science* 3rd ser., 20 (1842): 488–94.
–. "Temperature of Quadrupeds, Birds, Fishes, Plants, Trees, and Earths, as Ascertained at Different Times and Places in Arctic America during Capt. Back's Expedition." *Edinburgh New Philosophical Journal* 21 (1836): 150–1.
Knox, Olive and Harold. "Chief Factor Archibald McDonald." *The Beaver* 274 (March 1944): 42–6.
Kotzebue, Otto von. *Voyage of Discovery, into the South Sea and Beering's [sic] Straits, for the Purpose of Exploring a North-East Passage Undertaken in the Years 1815–18.* 3 vols. London: Longman, Hurst, Rees, Orme, and Brown, 1821.
Kumar, Deepak. "The Evolution of Colonial Science in India: Natural History and the East India Company." In *Imperialism and the Natural World*, ed. John M. MacKenzie. Manchester: Manchester University Press, 1990, 51–66.
–. *Science and the Raj: A Study of British India*. 2nd ed. New York: Oxford University Press, 2006.
Kupperman, Karen Ordhal. "The Puzzle of the American Climate in the Early Colonial Period." *American Historical Review* 87 (1982): 1262–89.
Lachlan, Major R. "In Behalf of the Establishment of a Provincial System of Meteorological Observations." *Canadian Journal* 3 (1855): 406–9.
Lambert, M. "Exposé de quelques Observations qu'on pouroit faire pour répandre du jour sur la Météorologie." *Nouveaux Mémoires de l'Académie Royale des Sciences et Belles-Lettres* (1771): 60–5.
L'Ami, C.E. "Priceless Books form Old Fur Trade Libraries." *Beaver* 266 (Dec. 1935): 26–9.

Lang, Frank A. "John Jeffrey in the Wild West: Speculations on His Life and Times (1828–1854?)" *Kalmiopsis*, 13 (2006): 1–12.
Lange, Erwin F. "Dr. John McLoughlin and the Botany of the Pacific Northwest." *Madroño* 14 (Oct. 1958): 268–72.
–. "John Jeffrey and the Oregon Botanical Expedition." *Oregon Historical Quarterly* 69 (1967): 111–24.
Lavender, David, ed. *The Oregon Journals of David Douglas, of His Travels and Adventures among the Traders and Indians in the Columbia, Willamette and Snake River Regions during the Years 1825, 1826 and 1827*. 2 vols. Ashland, OR: Oregon Book Society, 1972.
Le Roy, Reuben Hafen. *Mountain Men and Fur Traders of the Far West: Eighteen Biographical Sketches*. Lincoln: University of Nebraska Press, 1982.
Lefroy, J. H. *Diary of a Magnetic Survey of a Portion of the Dominion of Canada, Chiefly in the North-Western Territories, Executed in the Years 1842–44*. London: Longmans, 1883.
–. *Magnetical and Meteorological Observations at Lake Athabaska and Fort Simpson, by Captain J.H. Lefroy, Royal Artillery; and at Fort Confidence, in Great Bear Lake, by Sir John Richardson C.B., M.D.* London: Longman, Brown, Green, and Longmans, 1855.
–. "On the Probable Number of the Native Indian Population of British America." *The American Journal of Science and Arts* 2nd series, 16 (Nov. 1853): 189–203.
–. *On the Probable Number of the Native Indian Population of British America: From the Proceedings of the Canadian Institute*. Toronto: Printed by Hugh Scobie, ca. 1853.
–. "Preliminary Report on the Observations of the Aurora Borealis, Made by the Non-commissioned Officers of the Royal Artillery, at the Various Guardrooms in Canada." *Edinburgh and Dublin Philosophical Magazine and Journal of Science* 3rd series, 36 (1850): 457–66.
–. "Second Report on Observations of the Aurora Borealis, 1850–51." *London, Edinburgh and Dublin Philosophical Magazine and Journal of Science* 4th series, 4 (1852): 59–68.
Levere, Trevor H. "Science and the Canadian Arctic, 1818–76, from Sir John Ross to Sir George Strong Nares." *Arctic* 41, no. 2 (June 1988): 127–37.
–. *Science and the Canadian Arctic: A Century of Exploration, 1818–1918*. Cambridge: Cambridge University Press, 1993.
Leveson Gower, R. H. G. "HBC and the Royal Society." *The Beaver* 256, no. 2 (Sept. 1934): 29–31, 66.
Lewis, William Roger, gen. ed. *Oxford History of the British Empire: The Origins of Empire*, ed. Nicholas Canny. vol. 1. New York: Oxford University Press, 1998.

Lindsay, Debra. "Peter Fidler's Library: Philosophy and Science in Rupert's Land." In *Readings in Canadian Library History*, ed. Peter F. McNally. Ottawa: Canadian Library Association, 1986, 209–29.

–. *Science in the Subarctic: Trappers, Traders and the Smithsonian Institution*. Washington: Smithsonian Institution Press, 1993.

–. "The Hudson's Bay Company-Smithsonian Connection and Fur Trade Intellectual Life: Bernard Rogan Ross, a Case Study." *"Le Castor Fait Tout": Selected Papers of the Fifth North American Fur Trade Conference, 1985*, ed. Bruce G. Trigger, Toby Morantz and Louise Dechêne. Montreal: Lake St. Louis Historical Society, 1987, 587–617.

–. *The Modern Beginnings of Subarctic Ornithology: Northern Correspondence to the Smithsonian Institution, 1856–1868*. Winnipeg: Manitoba Record Society, 1991.

Lister, Kenneth R. *Paul Kane, The Artist: Wilderness to Studio*. Toronto: Royal Ontario Museum, 2010.

Livingstone, David N. *Putting Science in Its Place: Geographies of Scientific Knowledge*. Chicago: University of Chicago Press, 2003.

Lloyd, Blodwen. "John Scouler, M.D. Ll.D., F.L.S. (1804–1871)." *Glasgow Naturalist* 18, no. 4 (Feb. 1962): 210–12.

Lockhart, J. G. "Notes on the Habits of the Moose in the Far North of British North America in 1865." *Proceedings of the National Museum* 13, no. 827 (1890): 305–8.

Lowe, A. Burnett. "Canada's First Weathermen." *Beaver* 292 (Summer 1961): 4–7.

Lyons, Henry George. *The Royal Society, 1660–1940: A History of Its Administration under Its Charters*. Cambridge: Cambridge University Press, 1944.

MacFarlane, R.R. "Land and Sea Birds Nesting within the Arctic Circle in the Lower Mackenzie River District." The Historical and Scientific Society of Manitoba. *Transactions* ser. 1, no. 39 (June 1889), http://www.mhs.mb.ca/docs/transactions/1/mackenziebirds.shtml.

–. "List of Birds and Eggs Observed and Collected in the North-West Territories of Canada, Between 1880 and 1894." In Charles Mair, *Through the Mackenzie Basin*. Toronto: William Briggs, 1908, 286–470.

–. "Notes on and List of Birds and Eggs Collected in Arctic America, 1861–1866." *Proceedings of the United States National Museum* 14 (1891): 413–46.

–. "Notes on Mammals Collected and Observed in the Northern Mackenzie River District, North-West Territories of Canada with Remarks on Explorers and Explorations of the Far North." In Charles Mair, *Through the Mackenzie Basin*. Toronto: William Briggs, 1908, 150–283.

–. "Notes on the Mammals Collected and Observed in the Northern Mackenzie River District, North-West Territories of Canada, with Remarks on

Explorers and Explorations of the Far North." *Proceedings of the United States National Museum* 28 (1905): 673–764.

–. "On an Expedition down the Beghula or Anderson River." *Canadian Record of Science* 4, no. 1 (Jan. 1890): 28–53.

Mackie, Richard. *Trading beyond the Mountains: The British Fur Trade on the Pacific, 1793–1843*. Vancouver: UBC Press, 1996.

MacLaren, I. S. "Exploration/Travel Literature and the Evolution of the Author." *International Journal of Canadian Studies/Revue internationale d'études canadiennes* 5 (1992): 39–68.

–. "Notes on Samuel Hearne's *Journey* from a Bibliographical Perspective." *Papers of the Bibliographical Society of Canada* 31, no. 2 (Fall 1993): 21–45.

–. "On the Trail of Paul Kane's *Wanderings of an Artist*," *Prairie Fire* 10, no. 3 (Autumn 1989): 28–41.

–. "Paul Kane and the Authorship of *Wanderings of an Artist*." In *From Rupert's Land to Canada: Essays in Honour of John E. Foster*, ed. Theodore Binnema, Gerhard J. Ens, and R. C. Macleod. Edmonton: University of Alberta Press, 2001, 225–47.

–. "Samuel Hearne's Accounts of the Massacre at Bloody Fall, 17 July 1771." *Ariel: A Review of International English Literature* 22 (1991): 25–51.

MacLeod, Margaret Arnett, ed. *Letters of Letitia Hargrave*. Toronto: Champlain Society, 1947.

Macleod, M. A., and R. Glover "Franklin's First Expedition as Seen by the Fur Traders." *Polar Record* 15, no. 98 (1971): 669–82.

MacLeod, Roy. "On Visiting the 'Moving Metropolis': Reflections on the Architecture of Imperial Science." In *Scientific Colonialism: A Cross-Cultural Comparison*, ed. Nathan Reingold and Marc Rothenberg. Cambridge: Cambridge University Press, 1987, 217–49.

Maor, Eli. *June 8, 2004: Venus in Transit*. Princeton: Princeton University Press, 2000.

Markham, Violet R. *Paxton and the Bachelor Duke*. London: Hodder and Stoughton, 1935.

Martin, Benjamin. *Venus in the Sun*. London: W. Owen, 1761.

Maskelyne, Nevil. "A Letter from Revd. Nevil Maskelyne, B.D.F.R.S. Astronomer Royal, to Rev. William Smith, D.D. Provost of the College of Philadelphia, Giving Some Account of the Hudson's-Bay and Other Northern Observations of the Transit of Venus, June 3d, 1769." *Transactions of the American Philosophical Society* 1 (1769–71): Appendix to the Astronomical and Mathematical Papers, 1–4.

McKelvey, Susan Delano. *Botanical Exploration of the Trans-Mississippi West, 1790–1850*. Corvallis, OR: Oregon State University Press, [1955] 1991.

McLean, John. *Notes of a Twenty-Five Years' Service in the Hudson's Bay Territory.* 2 vols. London: Richard Bentley, 1849.
McNab, James. "On the Discoveries of Mr. John Jeffrey and Mr. Robert Brown, Collectors to the Botanical Expeditions to British Columbia between the Years 1850 and 1866." *Transactions of the Botanical Society Edinburgh* 11 (1872): 322–38.
McNab, John, and Henry Cavendish. "An Account of Experiments Made by Mr. John McNab, at Albany Fort, Hudson's Bay, Relative to the Freezing of Nitrous and Vitriolic Acids." *Philosophical Transactions of the Royal Society* 78 (1788): 166–81.
–. "An Account of Experiments Made by Mr. John McNab, at Henley House, Hudson's Bay, Relating to Freezing Mixtures." *Philosophical Transactions of the Royal Society* 76 (1786): 241–72.
McPherson, Murdoch, and John Richardson. "Register of the Temperature of the Atmosphere, Kept at Fort Simpson, North America, in the Years 1837, 1838, 1839, and 1840." *Edinburgh New Philosophical Journal* 30 (1840–1): 124–7.
Merk, Frederick, ed. *Fur Trade and Empire: George Simpson's Journal Entitled Remarks Connected with the Fur Trade in the Course of a Voyage from York Factory to Fort George and Back to York Factory 1824–5.* Rev. ed. Cambridge, MA: Harvard University Press, 1968.
Middleton, C. "New and Exact Table Collected from Several Observations, Taken in Four Voyages to Hudson's Bay ... Shewing the Variation of the Magnetical Needle ... from the Years 1721, to 1725." *Philosophical Transactions of the Royal Society* 34 (1726–7): 73–6.
–. "An Examination of Sea-Water Frozen and Melted Again, to Try What Quantity of Salt is Contained in Such Ice, Made in Hudson's Streights by Capt. Christopher Middleton, F. R. S. at the Request of C. Mortimer, R. S. Secr." *Philosophical Transactions of the Royal Society* 41 (1739–41): 806–7.
–. "An Observation of the Magnetic Needle Being So Affected by Great Cold, That It Would Not Traverse." *Philosophical Transactions of the Royal Society* 40 (1737–8): 310–1.
–. "Observations Made of the Latitude, Variation of the Magnetic Needle, and Weather, by Capt. Christopher Middleton, in a Voyage from London to Hudson's-Bay, Anno 1735." *Philosophical Transactions of the Royal Society* 39 (1735–6): 270–80.
–. "Observations of the Variations of the Needle and Weather, Made in a Voyage to Hudson's-Bay, in the Year 1731." *Philosophical Transactions of the Royal Society* 38 (1733–4): 127–33.
–. "Observations on the Weather, in a Voyage to Hudson's Bay in North-America, in the Year 1730." *Philosophical Transactions of the Royal Society* 37 (1731–2): 76–8.

–. "The Effects of Cold; Together with Observations of the Longitude, Latitude, and Declination of the Magnetic Needle, at Prince of Wales's Fort, upon Churchill-River in Hudson's Bay, North America." *Philosophical Transactions of the Royal Society* 42 (1742–3): 157–71.

–. "The Use of a New Azimuth Compass for Finding the Variation of the Compass or Magnetic Needle at Sea, with Greater Ease and Exactness Than by Any Ever Yet Contriv'd for That Purpose." *Philosophical Transactions of the Royal Society* 40 (1737–8): 395–8.

Middleton, W. E. Knowles. *A History of the Thermometer and Its Uses in Meteorology*. Baltimore: Johns Hopkins Press, 1966.

Millburn, John R. *Benjamin Martin, Instrument Maker and 'Country Showman.'* Leiden: Noordhoff, 1976.

Mitchell, Ann Lindsay, and Syd House. *David Douglas: Explorer and Botanist*. London: Aurum Press, 1999.

Mood, Fulmer. "Adventurers of 1670." *The Beaver* 276 (June 1956): 48–53.

Moreau, William E., ed. *The Writings of David Thompson*. Toronto: Champlain Society, 2009.

Morgan, Henry J. *Bibliotheca Canadensis: Or a Manual of Canadian Literature*. Ottawa: G.E. Desbarats, 1867.

Morse, Jedidiah. *The American Geography; Or a View of the Present Situation of the United States of America*. 2nd ed. London: John Stockdale, 1792.

Morton, A. S. *History of the Canadian West to 1870–71*. 2nd ed. Toronto: University of Toronto Press, 1973.

Morton, W. L. *Henry Youle Hind, 1823–1908*. Toronto: University of Toronto Press, 1980.

Morwood, William. *Traveler in a Vanished Landscape: The Life and Times of David Douglas*. Newton Abbot, [England]: Readers Union, 1974.

Moulton, Gary E. *The Journals of the Lewis and Clark Expedition*. 13 vols. Lincoln: University of Nebraska Press, 1983–2001.

Moxon, Joseph. *A Brief Discourse of a Passage by the North-Pole to Japan, China, &c*. London: Joseph Moxon, 1674.

Murchison, Sir Roderick Impey. "Address to the Royal Society of London." *Proceedings of the Royal Geographical Society of London* 3, no. 1 (1858–9): 224–346.

Murray, Andrew. "Contributions to the Natural History of the Hudson's Bay Company's Territories: Part I – Rein-Deer." *Edinburgh New Philosophical Journal* 7, no. 2 (April 1858): 189–210.

–. "Contributions to the Natural History of the Hudson's Bay Company's Territories, Part II – Mammalia (continued)" *Edinburgh New Philosophical Journal* 9, no. 2 (April 1859): 210–20.

–. "Contributions to the Natural History of the Hudson's Bay Company's Territories, Part III – Aves" *Edinburgh New Philosophical Journal* 9, no. 2 (April 1859): 221–31.
–. *The Geographical Distribution of Mammals*. London: Day and Son, 1866.
Nicolson, Malcolm. "Alexander von Humboldt, Humboldtian Science and the Origin of the Study of Vegetation." *History of Science* 25 (1987): 167–94.
–. "Scouler, John (1804–1871)." *Oxford Dictionary of National Biography*.In *Oxford Dictionary of National Biography*, edited by H.C.G. Matthew and Brian Harrison. Oxford: OUP, 2004. Online ed., edited by Lawrence Goldman, May 2010. http://www.oxforddnb.com/view/article/24944.
Nierop, Dirck Rembrantz van. "A Narrative of Some Observations Made upon Several Voyages, Undertaken to Find a Way for Sailing about the North to the East-Indies, and for Returning the Same Way from Hence Hither: Together with Instructions Given by the Dutch East-India Company For the Discovery of the Famous Land of Jesso Near Japan. To Which is Added a Relation of Sailing through the Northern America to the East-Indies. Englished by the Publisher out of Dutch, Which Had Been Compos'd by Dirick Rembrantz van Nierop, and Printed at Amsterdam. 1674." *Philosophical Transactions* 9, no. 109 (14 Dec. 1674): 197–208.
Nisbet, Jack. *The Collector: David Douglas and the Natural History of the Northwest*. Seattle: Sasquatch Books, 2009.
–. *David Douglas, A Naturalist at Work: An Illustrated Exploration across Two Centuries in the Pacific Northwest*. Seattle: Sasquatch Books, 2012.
Nute, Grace Lee. "A Botanist at Fort Colville." The Beaver 277 (Sept. 1946): 28–31.
–. *Caesars of the Wilderness*. St. Paul: Minnesota Historical Society, 1978.
–. "Kennicott in the North." *Beaver* (Sept. 1943): 28–33.
Nuttall, Thomas. "A Catalogue of a Collection of Plants Made Chiefly in the Valleys of the Rocky Mountains of Northern Andes, towards the Sources of the Columbia River, by Mr. Nathaniel B. Wyeth, and Described by T. Nuttall." *Journal of the Academy of Natural Science of Philadelphia* 7 (1834): 5–60.
Oklandnikova, E. A. "Science and Education in Russian America." In *Russia's American Colony*, ed. Frederick S. Starr. Durham: Duke University Press, 1987, 218–48.
Orchiston, Wayne, and Derek Howse. "From Transit of Venus to Teaching Navigation: The Work of William Wales." *Journal of Navigation* 53 (2000): 156–66.
Owram, Doug. *Promise of Eden: The Canadian Expansionist Movement and the Idea of the West, 1856–1900*. Toronto: University of Toronto Press, 1980.

Palliser, John. *Further Papers Relative to the Exploration by the Expedition under Captain Palliser of that Portion of British North America which Lies between the Northern Branch of the River Saskatchewan and the Frontier of the United States; and between the Red River and the Rocky Mountains, and Thence to the Pacific Ocean.* London: G. Eyre and W. Spottiswoode for Her Majesty's Stationery Office, 1860.

–. *Papers Relative to the Exploration by Captain Palliser of that Portion of British North America which Lies between the Northern Branch of the River Saskatchewan and the Frontier of the United States; and between the Red River and Rocky Mountains.* London: G. E. Eyre and W. Spottiswoode, 1859.

–. *The Journals, Detailed Reports, and Observations Relative to the Exploration, by Captain Palliser, of that Portion of British North America, which, in Latitude, Lies between the British Boundary Line and the Height of Land or Watershed of the Northern or Frozen Ocean Respectively, and in Longitude, between the Western Shore of Lake Superior and the Pacific Ocean during the Years 1857, 1858, 1859, and 1860.* London: G. E. Eyre and W. Spottiswoode, 1863.

Parrish, Susan Scott. *American Curiosity: Cultures of Natural History in the Colonial British Atlantic World.* Chapel Hill: University of North Carolina Press, 2006.

Paxton, Joseph. "Botanical Expedition to North America: Melancholy Loss of the Collectors, Messrs. Wallace and Banks." *Paxton's Magazine of Botany, and Register of Flowering Plants* 6 (1839): 135–7.

Payne, Michael, and Gregory Thomas. "Literacy, Literature, and Libraries in the Fur Trade." *The Beaver* Outfit 313, no. 4 (Spring 1983): 44–53.

Pennant, Thomas. *Arctic Zoology.* 2nd ed. 3 vols. London: Robert Faulder, 1792.

–. *Arctic Zoology.* 2 vols. London: Henry Hughes, 1784.

–. *The Literary Life of the Late Thomas Pennant Esq. by Himself.* London: Benjamin and John White, and Robert Faulder, 1793.

Pickstone, John V. *Ways of Knowing: A New History of Science, Technology and Medicine.* Chicago: University of Chicago Press, 2000.

Pimentel, Juan. "The Iberian Vision: Science and Empire in the Framework of a Universal Monarchy, 1500–1800." *Osiris* 15 (2000): 17–30.

Platt, J. C. "Old Trading Companies." In *London,* ed. Charles Knight. London: Charles Knight, 1844, 6: 49–64.

Porter, Mae Reed, and Odessa Davenport. *Scotsman in Buckskin: Sir William Drummond Stewart and the Rocky Mountain Fur Trade.* New York: Hastings House, 1963.

Pratt, Mary Louise. *Imperial Eyes: Travel Writing and Transculturation.* New York: Routledge, 1992.

Preble, E. A. "Roderick Ross MacFarlane, 1833–1920." *The Auk* 39 (1922): 203–9.

Pyenson, Lewis. *Civilizing Mission: Exact Sciences and French Overseas Expansion, 1830–1940*. Baltimore and London: Johns Hopkins University Press, 1993.
–. *Cultural Imperialism and Exact Sciences: German Expansion Overseas, 1900–1930*. New York: Lang, 1985.
–. *Empire of Reason: Exact Sciences in Indonesia, 1840–1940*. Leiden: Brill, 1989.
Pyenson, Lewis, and Susan-Sheets Pyenson. *Servants of Nature: A History of Scientific Institutions, Enterprises and Sensibilities*. London: HarperCollins, 1999.
Rae, John. *Narrative of an Expedition to the Shores of the Arctic Sea in 1846 and 1847*. London: T. & W. Boone, 1850.
–. "Notes on Some of the Birds and Mammals of the Hudson's Bay Company's Territory, and of the Arctic Coast of America." *Journal of the Linnean Society of London*, Zoology, 20, no. 119 (Nov. 1888): 136–145.
–. "On the Condition and Characteristics of Some of the Native Tribes of the Hudson's Bay Company's Territories." *Journal of the Society of Arts* 30 (1882): 483–96.
–. "On the Esquimaux." *Transactions of the Ethnological Society of London* 4 (1865): 138–53.
Raffan, James. *Emperor of the North: Sir George Simpson and the Remarkable Story of the Hudson's Bay Company*. Toronto: HarperCollins, 2007.
Rainger, Ronald. "Race, Politics, and Science: The Anthropological Society of London in the 1860s." *Victorian Studies* 22, no. 1 (Autumn 1978): 51–70.
Ray, Arthur J. *Indians in the Fur Trade: Their Role as Trappers, Hunters, and Middlemen in the Lands Southwest of Hudson Bay, 1660–1870*. Toronto: University of Toronto Press, 1974.
Reid, Francis Lucian. "William Wales (ca. 1734–1798): Playing the Astronomer." *Studies in the History and Philosophy of Science*, part A, 39, no. 2 (2008): 170–5.
Reingold, Nahan, and Marc Rothenberg, eds. *Scientific Colonialism: A Cross-Cultural Comparison*. Cambridge: Cambridge University Press, 1987.
Rempel, Gwen. "The Manitoba Mound Builders: The Making of an Archaeological Myth, 1857–1900." *Manitoba History* 28 (1994): 12–18.
Rennie, A. N. "Obituary: James Barnston, M.D." *Canadian Naturalist and Geologist* 3 (1858): 224–6.
Report on Canadian Archives, 1889. Ottawa: Brown Chamberlin, 1890.
Report on Canadian Archives, 1890. Ottawa: Brown Chamberlin, 1891.
Rich, E. E. *Colin Robertson's Correspondence Book, September 1817 to September 1822*. London: Hudson's Bay Record Society, 1939.
–, ed. *Copy-book of Letters outward &c: Begins 29th May, 1680 Ends 5 July, 1687*. Toronto: Champlain Society, 1948.
–. *Hudson's Bay Company: 1670–1870*. 2 vols. London: Hudson's Bay Record Society, 1958–59.

–, ed. *James Isham's Observations on Hudsons Bay, 1743 and Notes and Observations on a Book Entitled A Voyage to Hudsons Bay in the Dobbs Galley, 1749*. London: Hudson's Bay Record Society, 1949.

–, ed. *Journal of Occurrences in the Athabasca Department by George Simpson, 1820 and 1821, and Report*. London: Hudson's Bay Record Society/Champlain Society, 1938.

–, ed. *Minutes of the Hudson's Bay Company, 1671–1674*. Toronto: Champlain Society, 1942.

–, ed. *Minutes of the Hudson's Bay Company, 1679–1684 ; First Part, 1679–82*. Toronto: Champlain Society, 1945.

–. *The Fur Trade and the Northwest to 1857*. Toronto: McClelland and Stewart, 1967.

–, ed. *The Letters of John McLoughlin, from Fort Vancouver to the Governor and Committee: First series, 1825–38*. London: Hudson's Bay Record Society, 1941.

–, ed. *The Letters of John McLoughlin, from Fort Vancouver to the Governor and Committee: Third Series, 1844–46*. London: Hudson's Bay Record Society, 1944.

Rich, E. E., and A. M. Johnson, eds. *Rae's Arctic Correspondence 1844–45*. London: Hudson's Bay Record Society, 1943.

Richards, Robert L. *Dr. John Rae*. Whitby, Eng.: Caedmon, 1985.

Richardson, John. *Arctic Searching Expedition: Journal of a Boat-Voyage through Rupert's Land and the Arctic Sea*. 2 vols. London: Longman, Brown, Green, and Longmans, 1851.

–. *Botanical Appendix to Captain Franklin's Narrative of a Journey to the Shores of the Polar Sea*. London: W. Clowes, 1823.

–. *Fauna Boreali-Americana; or, The Zoology of the Northern Parts of British America*. 4 vols. London: J. Murray, 1829–37.

–. "Note on the Best Points in British North America for Making Observations on the Temperature of the Air; and also for the Height of the Station above the Level of the Sea." *Journal of the Royal Geographical Society* 9 (1839): 121–5.

–. "On Aplodontia, a New Genus of the Order Rodentia, Constituted for the Reception of the Sewellel, a Burrowing Animal Which Inhabits the North Western Coast of America." *Zoological Journal* 4 (1829): 333–7.

–. "On the Frozen Soil of North America." *Edinburgh New Philosophical Journal* 30 (1840–1): 110–23.

Rivinus, E. F., and E. M. Youssef, *Spencer Baird of the Smithsonian*. Washington: Smithsonian Institution Press, 1992.

Robertson, William. *The History of America*. 3rd ed. 3 vols. London: W. Strahan, T. Cadell, and J. Balfour, 1780.

Robson, Joseph. *An Account of Six Years Residence in Hudson's Bay, From 1733 to 1736 and 1744 to 1747*. London: Printed for J. Payne and J. Bouquet et al., 1752.

Ronan, C. A. "Laurence Rooke (1622–1662)." *Notes and Records of the Royal Society of London* 15 (1960): 113–18.

Ronda, James P. "'A Knowledge of Distant Parts': The Shaping of the Lewis and Clark Expedition." *Montana* 41 (1991): 4–18.

Ross, Bernard H. [sic] "On the Mammals, Birds, &c., of the Mackenzie River District." *Natural History Review* 7 (July 1862): 269–90.

Ross, Bernard R. "An Account of the Botanical and Mineral Products, Useful to the Chipewyan Tribes of Indians, Inhabiting the McKenzie River District." *Canadian Naturalist & Geologist* 7 (1862): 133–7.

–. "The Eastern Tinneh." *SIAR 1866*, 304–11.

–. "A List of Mammals, Birds, and Eggs Observed in the McKenzie's River District, with Notices." *Natural History Review* 7 (July 1862): 271–90.

–. "List of Mammals, Birds, and Eggs, Observed in the McKenzie's River District, with Notices." *Canadian Naturalist and Geologist* 7, no. 2 (1862), 137–55.

–. "List of Species of Mammals and Birds – Collected in McKenzie's River District during 1860–61." *Canadian Naturalist and Geologist* 6 (1861): 441–44.

–. "On the Indian Tribes of the McKenzie River District and the Arctic Coast." *Canadian Naturalist and Geologist* 4 (1859): 190–5.

–. "A Popular Treatise on the Fur-Bearing Animals of the Mackenzie River District." *Canadian Naturalist & Geologist* 6 (1861): 5–36.

Ross, Helen E. *Letters from Rupert's Land, 1826–1840*. Montreal and Kingston: McGill-Queen's University Press, 2009.

Ross, J. A. *A Voyage of Discovery, Made under the Orders of the Admiralty, in His Majesty's Ships Isabella and Alexander, for the Purpose of Exploring Baffin's Bay, and Inquiring into the Probability of a North-west passage*. London: J. Murray, 1819.

Ross, Sydney. "Scientist: The Story of Word." *Annals of Science* 18 (1962): 65–86.

Rothman, Sheila M. *Living in the Shadow of Death: Tuberculosis and the Social Experience of Illness in American History*. New York: Basic Books, 1994.

Royal Society. "Directions for Sea-men, Bound for Far Voyages." *Philosophical Transactions of the Royal Society* 1 (1665–6): 140–3.

Ruggles, Richard I. *A Country So Interesting: The Hudson's Bay Company and Two Centuries of Mapping, 1670–1870*. Montreal: McGill-Queen's University Press, 1991.

–. "Governor Samuel Wegg, Intelligent Layman of the Royal Society, 1753–1802." *Notes and Records of the Royal Society of London* 32, no. 2 (1978): 181–99.

–. "Governor Samuel Wegg: The Winds of Change." *The Beaver* 307, no. 2 (1976): 10–20.

–. "Hudson's Bay Company Mapping." In *Old Trails and New Directions: Papers of the Third North American Fur Trade Conference*, ed. C. M. Judd and A. J. Ray. Toronto: University of Toronto Press, 1980, 24–36.

Runge, Clara T. "Frederick George Jacob Lueders: Naturalist and Philosopher, 1818–1904." *The Wisconsin Magazine of History*, 15, no. 3 (March 1932): 350–5.

Sabine, Edward. "Account of Marmots of North America Hitherto Known, with Notices and Descriptions of Three New Species." *Transactions of the Linnean Society* 13 (1822): 579–92.

–. "Contributions to Terrestrial Magnetism. No. VII." *Philosophical Transactions of the Royal Society* 136 (1846): 237–336.

–. "Observations on the Magnetism of the Earth, Especially of the Arctic Regions; in a Letter from Capt. Edward Sabine, to Professor Renwick." *American Journal of Science and Arts* 17, no. 1 (Jan. 1830): 145–56.

–. *Remarks on the Account of the Late Voyage of Discovery to Baffin's Bay, Published By Captain J. Ross, R. N.* London: John Booth, 1819.

Sage, Walter N. "Life at a Fur Trading Post in British Columbia a Century Ago." *Washington Historical Quarterly* 25, no. 1 (Jan. 1934): 11–22.

Schefke, Brian. "Imperial Science: A Naturalist in the Pacific Northwest." *Endeavour* 32, no. 3 (2008): 111–6.

–. "The Hudson's Bay Company as a Context for Science in the Columbia Department." *Scientia Canadensis* 31, no. 1–2 (2008): 67–84.

Schofield, F. H. *A Brief Sketch of the Life and Services of Retired Chief Factor, R. MacFarlane*. Winnipeg: S. J. Clarke, 1913.

Scouler, John. "Account of a Voyage to Madeira, Brazil, Juan Fernandez, and the Gallapagos [sic] Islands, Performed in 1824 and 1825, with a View of Examining their Natural History, &c." *Edinburgh Journal of Science* 5, no. 2 (Oct. 1826): 195–214.

–. "Journal of a Voyage to N.W. America." (Part I) *The Quarterly of the Oregon Historical Society* 6 (1905): 54–75.

–. "Journal of a Voyage to N.W. America, Part II." *The Quarterly of the Oregon Historical Society* 6 (1905): 159–205.

–. "Journal of a Voyage to N.W. America, Part III." *The Quarterly of the Oregon Historical Society* 6 (1905): 276–87.

–. "Observations on the Indigenous Tribes of the N.W. Coast of America." *Journal of the Royal Geographic Society* 11 (1841): 216–50.

–. "On the Indian Tribes Inhabiting the North-West Coast of America." *Journal of the Ethnological Society of London* 1 (1848): 228–52.

–. "On the Temperature of the North West Coast of America." *Edinburgh Journal of Science* 6 (1827): 251–3.

–. "Remarks on the Form of the Skull of a North American Indian." *Zoological Journal* 4 (1829): 304–9.
Scudder, Samuel Hubbard. *Winnipeg Country: Or Roughing it with an Eclipse Party*. Boston: Cupples, Upham, & Company, 1886.
Seller, John. *The English Pilot: The Fourth Book; The First Part*. London: J. Mount, T. Page, and W. Mount, circa 1675.
Shapin, Steven, and Simon Schaffer. *Leviathan and the Air-Pump: Hobbes, Boyle, and the Experimental Life*. Princeton, NJ: Princeton University Press, 1985.
Sheets-Pyenson, Susan. *John William Dawson: Faith, Hope, and Science*. Montreal and Kingston: McGill-Queen's University Press, 1996.
Sherwood, Morgan B. *Exploration of Alaska, 1865–1900*. New Haven and London: Yale University Press, 1965.
Shiels, Archibald Williamson. *The Purchase of Alaska*. College, AK: University of Alaska Press, distributed by the University of Washington Press, 1967.
Simmonds, Peter Lund. *Sir John Franklin and the Arctic Regions*. Buffalo: George H. Derby, 1852.
Simpson, Alexander. *The Life and Travels of Thomas Simpson, the Arctic Discoverer*. London: Richard Bentley, 1845.
Simpson, George. *Narrative of a Journey Round the World, during the Years 1841 and 1842*. London: Henry Colburn, 1847.
Simpson, Thomas. *Narrative of the Discoveries on the North Coast of America; Effected by the Officers of the Hudson's Bay Company during the Years 1836–39*. London, Richard Bentley, 1843.
Sinclair, Bruce. "Americans Abroad: Science and Cultural Nationalism in the Early Nineteenth Century." In *The Sciences in the American Context*, ed. Nathan Reingold. Washington, DC: Smithsonian Institution Press, 1979, 35–53.
Smith, Adam. *An Inquiry into the Nature and Causes of the Wealth of Nations*. 2 vols. Oxford: Clarendon Press, [1776] 1880.
Sörlin, Sverker. "Ordering the World for Europe: Science as Intelligence and Information as Seen from the Northern Periphery." *Osiris* 15 (2000): 51–69.
"The South Kensington Museum; Its Educational Resources." *The Museum: A Quarterly Magazine of Education, Literature, and Science* 1, no. 1 (April 1861): 66–72.
Spry. Irene M. *The Palliser Expedition: An Account of John Palliser's British North American Exploring Expedition*. Toronto: Macmillan, 1973.
–, ed. *The Papers of the Palliser Expedition, 1857–1860*. Toronto: Champlain Society, 1968.
Squier, E. G., and E. H. Davis, *Ancient Monuments of the Mississippi Valley*. Washington: Smithsonian Institution, 1848.
Stearns, R. P. "Colonial Fellows of the Royal Society of London, 1661–1788." *Osiris* 8 (1948): 73–121.

–. "The Royal Society and the Company." *The Beaver* 276, no. 1 (1945): 8–13.
–. *Science in the British Colonies of America*. Urbana: University of Illinois Press, 1970.
Steele, Peter. *The Man Who Mapped the Arctic: The Intrepid Life of George Back, Franklin's Lieutenant*. Vancouver: Raincoast Books, 2003.
Stefansson, Vilhjalmur. "Rae's Arctic Correspondence." *Beaver* 284 (March 1954): 36–7.
Sterling, Keir B., George A. Cevasco and Richard A. Harmond, eds. *Biographical Dictionary of American and Canadian Naturalists and Environmentalists*. Westport, CT: Greenwood Press, 1997.
Stevenson, J. A. "Disaster in the Dalles." *Beaver* 273 (Sept. 1942): 19–21.
Stewart, Susan J. "Paul Kane Paintings Rediscovered." *Journal of Canadian Art History* 5, no. 2 (1981): 85–93.
Stocking, George W., Jr. "The Persistence of Polygenist Thought in Post-Darwinian Anthropology." In George W. Stocking, *Race, Culture, and Evolution: Essays in the History of Anthropology*. Chicago: University of Chicago Press, 1982.
–. *Victorian Anthropology*. New York: Free Press 1987.
Strong, Lisa. "American Indians and Scottish Identity in Sir William Drummond Stewart's Collection." *Winterthur Portfolio* 35, no. 2–3 (Summer–Autumn 2000): 127–55.
Taylor, E. G. R. *The Mathematical Practitioners of Hanoverian England: 1714–1840*. Cambridge: Cambridge University Press for the Institute of Navigation, 1966.
Terrall, Mary. "Gendered Space, Gendered Audience: Inside and Outside the Paris Academy of Sciences." *Configurations* 3, no. 2 (Spring 1995): 207–32.
Thiessen, A. D. "The Founding of the Toronto Magnetic Observatory and the Canadian Meteorological Service." *Journal of the Royal Astronomical Society of Canada* 40 (1940): 308–48.
Thomas, Greg. "The Smithsonian and Hudson's Bay Company." *Prairie Forum* 10, no. 2 (1985): 283–306.
Thomas, Lewis H. "The Hind and Dawson Expeditions: 1857–58." *Beaver* 289 (Winter 1958): 39–45.
Thompson, David. "Mean Temperature of Cumberland House and Bedford House, Hudson's Bay Territory, 1789–90, 1795–96." *British American Journal of Medical and Physical Science* 4, no. 11 (March 1849): 302.
Tolmie, W. Fraser, and George M. Dawson. *Comparative Vocabularies of the Indian Tribes of British Columbia with a Map Illustrating Distribution*. Montreal: Dawson Brothers, 1884.
Townsend, John K. *Narrative of a Journey across the Rocky Mountains to the Columbia River*. Philadelphia: Henry Perkins, 1839.

Turner, Lucien M. *Ethnology of the Ungava District, Hudson Bay Territory.* Washington, DC: Smithsonian Institution Press, 2001.

Tyrrell, Joseph Burr, ed. *A Journey from Prince of Wales's Fort in Hudson's Bay to the Northern Ocean in the Years 1769, 1770, 1771 and 1772.* Toronto: Champlain Society, 1911.

–, ed. *Documents Relating to the Early History of Hudson Bay.* Toronto: Champlain Society, 1931.

–, ed. *Journals of Samuel Hearne and Philip Turnor.* Toronto, Champlain Society, 1934.

Vasile, Ronald S. "The Early Career of Robert Kennicott, Illinois's Pioneering Naturalist." *Illinois Historical Journal* 87 (1994): 150–70.

Verne, Jules. *The Fur Country: Or, Seventy Degrees North Latitude,* trans. N. D'Anvers. Boston: James R. Osgood and Company, 1874.

Verner, Coolie. "The Arrowsmith Firm and the Cartography of Canada." In *Explorations in the History of Canadian Mapping: A Collection of Essays,* ed. Barbara Farrell and Aileen Desbarats. Ottawa: Association of Canadian Map Libraries and Archives, 1988, 47–54.

Wales, W. "Journal of a Voyage, made by Order of the Royal Society, to Churchill River, on the north-west Coast of Hudson's Bay; of Thirteen Months Residence in that Country; and of the Voyage back to England; in the years 1768 and 1769." *Philosophical Transactions of the Royal Society* 60 (1770): 100–36.

Wales, W., and Dymond, J. "Astronomical Observations made by Order of the Royal Society, at Prince of Wales's Fort, on the North-west Coast of Hudson's Bay." *Philosophical Transactions of the Royal Society* 59 (1769): 467–88.

Wallace, R. C. "Rae of the Arctic." *Beaver* 284 (March 1954): 28–33.

Warner, Deborah. "Terrestrial Magnetism: For the Glory of God and the Benefit of Mankind."*Osiris* 9 (1994): 66–84.

Weld, Charles Richard. *A History of the Royal Society.* 2 vols. New York: Arno Press [1848] 1975.

Whewell, William. *History of the Inductive Sciences from the Earliest to the Present Time.* 2nd ed. 3 vols. London: John W. Parker, 1847.

Whitehead, Christopher. *The Public Museum in Nineteenth Century Britain: The Development of the National Gallery.* Burlington, VT: Ashgate, 2005.

Williams, Glyndwr, ed. *Andrew Graham's Observations on Hudson's Bay, 1767–91.* London: Hudson's Bay Record Society, 1969.

–. "Arthur Dobbs and Joseph Robson: New Light on the Relationship between Two Early Critics of the Hudson's Bay Company." *Canadian Historical Review* 40 (1959): 132–6.

–. "The Hudson's Bay Company and Its Critics in the Eighteenth Century." *Transactions of the Royal Historical Society* ser. 5, 20 (1970): 149–71.

–. "The Hudson's Bay Company and the Fur Trade: 1670–1870." *Beaver* 314, no. 2 (Autumn, 1983).
–. *Hudson's Bay Miscellany, 1670–1870.* London: Hudson's Bay Record Society, 1975.
–, ed. *London Correspondence Inward from Sir George Simpson, 1841–42.* London: Hudson's Bay Record Society, 1973.
Willson, Beckles. *The Life of Lord Strathcona and Mount Royal, G.C.M.G., G.C.V.O. (1820–1914).* London: Cassell, 1915.
Wilson, Daniel. "Paul Kane, the Canadian Artist." *Canadian Journal of Science, Literature, and History* 13, no. 1 (1871): 66–72.
–. "Science in Rupert's Land." *The Canadian Journal of Industry, Science, and Art* 7 (1862): 336–47.
Wonders, William C. "Introduction." In George Back, *Narrative of the Arctic Exploring Expedition to the Mouth of the Great Fish River and along the Shores of the Arctic Ocean, in the Years 1833, 1834, and 1835.* Edmonton: Hurtig, 1970, i–xxviii.
Woolf, Harry. *The Transits of Venus: A Study in the Organization and Practice of Eighteenth Century Science.* Princeton, NJ: Princeton University Press, 1959.
Woolfall, S. J. "History of the 13th Earl of Derby's Menagerie and Aviary at Knowsley Hall, Liverpool (1806–1851)." *Archives of Natural History* 17 (1990): 1–47.
Woolley, Richard. "Captain Cook and the Transit of Venus of 1769." *Notes and Records of the Royal Society of London* 24, no. 1 (1969): 19–32.
Zeller, Suzanne. "Classical Codes: Biogeographical Assessments of Environment in Victorian Canada." *Journal of Historical Geography* 24, no. 1 (1998): 20–35.
–. "The Colonial World as Geological Metaphor: Strata(gems) of Empire in Victorian Canada." *Osiris* 15 (2000): 85–107.
–. "Humboldt and the Habitability of Canada's Great Northwest." *Geographical Review* 96, no. 3 (2006): 382–98.
–. *Inventing Canada: Early Victorian Science and the Idea of a Transcontinental Nation.* Toronto: University of Toronto Press, 1987.
–. *Land of Promise, Promised Land: The Culture of Victorian Science in Canada.* Ottawa: Canadian Historical Association, 1996.
–. "The Spirit of Bacon: Science and Self-Perception in the Hudson's Bay Company, 1830–1870." *Scientia Canadensis* 13 (Fall/Winter 1989): 79–101.
Zuidervaart, Huib J., and Rob H. van Gent. "'A Bare Outpost of Learned European Culture on the Edge of the Jungles of Java': Johan Maurits Mohr (1716–1775): and the Emergence of Instrumental and Institutional Science in Dutch Colonial Indonesia." *Isis* 95, no. 1 (2004): 1–33.

Index

Bold numbers indicate references to figures.

aboriginal people: Chinook, 173, 180; as contributors to scientific knowledge, 21, 31, 44, 67, 88, 91, 106, **107**, 116, 179, 192, 269, 272–4, 280, 289, 297, 313n84; Dakota (Sioux), 241; Inuit (Esquimaux, Eskimos), 148, 158, 161, 163, 285; Peigan, 113; perceptions of scientific instruments, 25; relations of with HBC traders, 12, 31, 44, 144, 147, 159, 163, 166, 223, 231, 272, 276, 302n10; responses to government exploring expeditions, 236; as subjects of study, 53–4, 74, 92, 103, 145, 179, 192–3, 283–5; treatment of women, 163, 284. *See also* ethnology

Aborigines Protection Society (1837), 145–6, 165–6, 347n63, 349n72, 350n81

Academy of Natural Sciences (Philadelphia 1812), 268

Account of Six Years Residence (1752), 72

Acosta, José de (1539–1600), 33, 53

Acton House (Rocky Mountain House), 37, 122

Admiralty (British), 13, 41, 51, 130–2, 139, 143–4, 15–61

Alaska, 244, 287, 288. *See also* Russian America

Alaska Commercial Company (1868), 289

Albany House. *See* Fort Albany (Albany House)

alcohol, 24, 90, 187, 223, 248, 263, 276, 279

Allan, George William (1822–1901), 218–22, 225–6

Allan, William (1770–1853), 218

Allan (Robinson), Louisa, 218, 221

ambrometers (rain gauges), 86

American Association for the Advancement of Science (1848), 211, 222, 252–3, 388n74

American Fur Company, 24, 187, 208

American Ornithological Society, 261

American Philosophical Society (1743), 85, 123

Americans, 200–8, 238–89, 291–3

American Traveller, 125

Anderson, David (1814–85), 290

Anderson, James (1812–67), 193, 215, 249, 271

Anderson University Museum (Glasgow), 173, 256
Anglo-Russian Convention (1825), **4**, 152, 346n40
animals, as subjects of study, 53, 66, 71–2, 89, 124, 134, 184–5, 205–7, 246–50, 273–4, 282–3. *See also* birds; mammals; zoology
Anthropological Institute of Great Britain and Ireland, 350n81
Anthropological Society of London (1863), 268, 349n47, 384n11
anthropology 284, 307n40, 349n77. *See also* ethnology
Arago, François Jean Dominique (1786–1853), 173
Arctic Searching Expedition (1851) (Richardson), 152, 160
Arctic Zoology (1784–5) (Pennant), 99, 102, 123–4
Aristotle (384–22 BC), 48
Arrowsmith, Aaron (1750–1823), **96**, 112, 114–19, **115**, 121–2, 125, 133, 149
Artillery (British), 13, 42, 134, 200, 213
Asiatic Journal, 24
Assiniboine and Saskatchewan Expedition (1858), 234
Assiniboine River, 117
astronomy, 25, 32, 36, 55, 75–83, 85, 95, 118, 120, 134, 172
Athabasca Country, **4**, 37, 39, 110–12, 117, 135, 137, 232
Athabasca River, 116
Audubon, John James (1785–1851), 31, 42, 205–7, 238, 259, 263, 372n32
Audubon, John Woodhouse, 206, 207
aurora borealis, 134, 139, 141, 144, 158, 214–15

Bache, Alexander Dallas (1806–67), 203, 253, 255, 385n35
Bachman, John (1790–1874), 206–7
Back, George (1796–1878), 134, 137, 142–4, 150, 231
Back River (Great Fish River), **136**, 143, 152
Bacon, Francis (1561–1626), 6, 32, 48, 244
Baffin Bay, 134
Baird, Spencer Fullerton (1823–87), 239, 241–9, **247**, 252, 260–5, 269, 273–4, 276–82, 284–8, 291, 297
Baldwin, Robert (1804–58), 221
Baldwin, William Warren (1775–1844), 221
Balfour, John Hutton (1808–84), 189
Ballenden, John (*c.* 1812–56), 216
Banks, Joseph (1743–1820), 82, 91, 99, 194, 297
Banks, Peter (d. 1838), 181, 183, 185, 198, 306n27
Barclay, Archibald (1785–1855), 183
Barnston, George (1800–83), 39, 166, 169, 179, 181, 192–8, 241–3, 253, 261, 274, 278–9, 283, 296
Barnston, James (1831–58), 179, 196, 197
barometers, 34, 85–6
Barrington, Daines (1727–1800), 90, 124
Barrow, John (1764–1848), 132, 142, 344n8
Batavia (Jakarta) (Dutch East Indies), 9, 85
Bayly, Charles (*c.* 1630–80), 53
Beale, John, 52
bears, 205, 249, 260, 273, 372n32
beavers, 54, 168, 191

Beechey, Frederick William (1796–1856), 136, 140–1
Bell, John (1799–1868), 152, 164
Bell, Robert (1841–1917), 292
Bellot Strait, 161
Berens, Henry Hulse (1804–83), 155
Berens, Joseph (1773–1825), 138
Berens River, 119
Berger, Carl, 22
Bering, Vitus (1681–1741), 70
Bering Strait, 130, **136,** 140, 258
Bird, James (*c.* 1773–1856), 31
birds, as subjects of study, 54, 62–7, 72, 87–92, 123, 141, 150, 164, 167–8, 191, 197, 204, 248, 265, 273–4, 282, 323n73, 324n85
Black, Joseph (1728–99), 100–1
Black, Samuel (1780–1841), 170
Blakiston, Thomas Wright (1832–91), 232
Blodget, Lorin (1823–1901), 229
Boat Encampment, 141
Bolduc, Jean-Baptiste-Zacharie (1818–89), 25
books, 9; in the HBC territories, 9, 10, 25, 98, 172, 194, 215, 249, 359n24; as reward for scientific contributions, 15, 30, 279, 280. *See also* libraries
Boothia Peninsula, 150, 153–4, 161
boredom, 24, 193, 277
Botanical Society of London (1836), 24
botany, 41, 141, 144, 157, 164, 167, 171–2, 176, 183, 188–9, 192, 197. *See also* plants, as subjects of study
Bourgeau (Bourgeaux), Eugène (1813–77), 232, 255
Boyle, Robert (1627–91), 51, 54, 56
Brackenridge, William D. (1810–93), 205

British Association for the Advancement of Science (1831), 163, 203, 211
British Columbia, 192, 208, 232, 258
British Museum, The, 13, 24, 28, 159–60, 168, 196–7, 207, 243, 261, 268, 333n77, 354n142
British North America Act (1867), 291
British parliamentary committees: Committee on the Hudson's Bay Company (1749), 71; Select Committee on Aboriginal Peoples (1836), 11, 145; Select Committee on the Hudson's Bay Company (1857), 12, 43, 143, 145, 163, 231
British Zoology (1761–6) (Pennant), 88, 123
Brown, George (1818–80), 221
Brown, Robert (1773–1858), 174
Brown, Robert (1842–95), 192
Buchan, Alexander (d. 1769), 82
Buchanan, James (1791–1868), 280
Buckingham House, 112–14
Burke, Joseph (1812–73), 184–6, 189, 198

California: Alta California (Mexico), 177, 183, 194; Baja (Lower) California (Mexico), 244
California (USA), 186
Cambridge (USA), 212
Campbell, Colin (*c.* 1787–1853), 39, 213–14
Campbell, John (1708–75), 73
Campbell, Robert (1808–94), 193, 271, 283
Canada, 207, 209, 218, 220–1, 226–37; acquisition of Rupert's Land (1870), 292; defined, 310n58; expansionist movement in, 43, 199, 200,

209, 226–37, 292; federation (1867), 291; science in, 13, 42, 209, 211–26, 253
Canada (grey) goose (*Branta canadensis*), 197–8
Canadian Geological Museum (Montreal), 197
Canadian Institute (Royal Canadian Institute) (1849), 43, 163, 220–2, 225–7, 233, 262, 290
Canadian Journal, 220–2, 233
Canadian Naturalist and Geologist, 210
Canadians, 42–3, 107, 113, 117, 137, 147, 160, 166, 199–200, 208–12, 218–37
cannibalism, 138, 161
caribou (reindeer), 150, 191, 242, 249, 253, 268, 273
Carlton House, 174
Carteret, Sir Philip (1641–72), 47, 56, 318n22
Cartier, George-Étienne (1814–73), 221
cartography, 32, 38, 50, 95–7, 104–23, 126, 135
Castanopsis chrysophylla, 183–6
Catalogue of the Animals of North America (1771) (Forster), 89
Catlin, George (1796–1872), 215
Cavendish, Henry (1731–1810), 100, 335n19
Cavendish, William George Spencer, 6th Duke of Devonshire (1790–1858), 181
Chamisso, Adelbert von (1781–1838), 133
Charles Fort (Rupert House), 58, 60
Charles II (King), 47
Charlton Island, 52, 54
chartered monopolies: characterizations of, 4–5, 10–11, 93, 146
Chatsworth Gardens, 181

Chesterfield House, 116
Chesterfield Inlet, 71, 73, 109
Chicago, Illinois, 228, 253, 256, 258
Children, John George (1777–1852), 144
Chiswick Gardens (London), 171, 176
Christie, Alexander (1792–1872), 224
Christopher, William (1729–97), 73, 90, 109, 124
Churchill (Fort Churchill) (Fort Prince of Wales), 68–9, 72–3, 75, 79–80, 82, 84–6, 88, 97, 102, 105, 113–14, 116–17, 155, 158–59, 163, 168, 206, 327n2
Churchill River, 116–17
Church Missionary Society, 118
Civil War (USA) (1861–65), 209, 240–2, 245–6, 283–4, 286, 289, 291, 385n35, 402n8
Clarke, Lawrence (Laurence) (1832–90), 254, 264, 271, 275, 281
Cleveland, Ohio, 253
climate, 8, 9, 34, 41, 72, 74, 99, 160, 171, 189, 228–30, 252, 281, 287; isotherms, 228–9
Climatology of the United States (1857) (Blodget), 229, 282, 287
clothing, 29, 193, 216, 256, 268
Clouston, James (fl. 1808–27), 118
Clouston, James Stewart (*c.* 1826–74), 214, 280
coal, 166, 228, 232, 377n106
Cocking, Matthew (1743–99), 105–7
Cohen, Bernard, 20, 28, 319n29
Colen, Joseph (*c.* 1751–1818), 86, 97, 99, 111, 113, 296
Coleridge, Samuel Taylor (1772–1834), 41
Colleton, Peter (1635–94), 56, 318n22
Colonial Office, 12, 39, 146, 164–5, 177, 306n29, 347n63

Columbia District (Oregon Country), **4**, 24–5, 39–40, 153, 156, 164, 169–88, **175**, 199, 201, 204, 207–8, 237–88, 246
Columbia River, 116, 141, 170, 174–5, 179–80, 183, 185, 203
Columbia River Fishing and Trading Company, 203
Columbus, Christopher (1451–1506), 32, 81
Colville, Andrew (1779–1856), 215
Committee Bay, 158
Comparative Vocabularies of the Indian Tribes of British Columbia (1884) (Tolmie and Dawson), 179
compasses, 62. *See also* geomagnetism
"Complaint of a Forsaken Indian Woman" (Wordsworth) (1798), 103
Concomly, Chief (d. 1830), 180
Connolly, Henry (d. 1910), 271
Cook, James (1728–79), 33, 75, 79–82, 88, 103, 109–10, 116, 124, 136
Cook's Inlet, 110
Cooper, Barry, 164
Cooper, James G. (1830–1902), 229, **230**
Copley, Sir Godfrey (1653–1709), 321n64
Copley Medal (Royal Society), 26, 61–3, 67, 70, 100, 101, 308n45, 321n64
Coppermine River, **69**, 109, 131, 135–6, 139, 140, 154; mouth of, **69**, 102–3, 110, **115**, 130–6, **136**, 139, 148
Corbett, Owen Griffiths (1823–1909), 291
Costard, George (1710–82), 77
Crimean War (1854–56), 285
Cumberland House, 37, 98, 106, 108–9, 111, 113, 117, 141, 164, 168, 174, 189, 213, 255, 381n166

curiosity: as motive for scientific study, 26, 250; perceptions of, 6, 29, 52; as a positive trait, 66–7, 102

Dall, William Healey (1845–1927), 244, 288
Dallas, Alexander Grant (1816–82), 260
Dalles des Morts (Death Rapids), 183
Dalrymple, Alexander (1737–1808), 79, 102, 109–11, 125, 330n36, 339n60
Davis, Charles Henry (1807–77), 255
Davis, John (c. 1550–1605), 131
Davis Strait, 131, 134
Dawson, George Mercer (1849–1901), 179
Dawson, John William (1820–99), 210
Dawson, Simon James (1818–1902), 253
Dease, Peter Warren (1788–1863), 39, 142–3, 147–53, 157, 168, 170
Dease-Simpson expeditions (1837–9), **136**, 147–53, **149**
Delbourgo, James, 21, 51
Dew, Nicholas, 21, 51
dipping needles, 51, 100. *See also* geomagnetism
Dobbs, Arthur (1689–1765), 67–8, 70–1, 73, 93, 125
Doubleday, Edward (1811–49), 196, 384n18
Douglas, David (1799–1834), 41, 169, 171–9, **175**, 181, 183, 186, 189, 193–8, 201–2, 259, 263, 276, 297
Douglas, James (1803–77), 157, 189, 195, 216
Douglas, John, Bishop of Salisbury (1721–1807), 103, 124
Drage (Dragge), Theodorus Swaine (c. 1712–74), 125
Drexler, Constantin, 254, 265–6, 275, 290

Drummond, Thomas (c. 1790–1835), 139, 141, 167, 174, 186
Dutch East Indies, 9, 20, 85, 108
Dymond, Joseph, 79–82, 85–6, 88, 105

East India Company (Dutch), 5, 9
East India Company (English), 5–7, 9, 55–6, 60, 91, 109, 296; surgeons in, 9
Eastmain River, 118
eclipses (solar): of 1771, 83; of 1780, 83; of 1860, 255–6
Edinburgh, Scotland: links with HBC, 23, 30, 32, 100, 180, 189–90, 192, 197, 261, 268
Edinburgh Journal of Science, 141
Edinburgh New Philosophical Journal, 180, 191, 349n79
Edmonton House (Fort Edmonton), 31, 114, 141, 185, 216, 258
Edwards, George (1694–1773), 61–2, 65–7, 123, 167
Edward's Botanical Register, 176
elite science, defined, 17–18
Ellice, Edward (1783–1863), 123, 190, 202
Ellis, Henry (1721–1806), 71, 125
Elson, Thomas, 141, 150
Elton, Charles Sutherland (1900–91), 292
Elton, John (fl. 1730–32), 61
Engelmann, George (1809–84), 187–8, 191
entomology, 133–4, 190, 195–6. *See also* insects and spiders
Eschscholtz, Johann Friedrich von (1793–1831), 133
Essay on the History of Civil Society (1767) (Ferguson), 26

ethnography, 102–3, 158, 160, 164, 166, 173, 193, 222, 283–4, 293
Ethnological Society of London (1843), 145
ethnology, 31–2, 50, 54, 66, 82, 88, 133, 145, 147, 158, 166, 193, 239, 245, 261, 268–9, 271, 275, 282–4. *See also* anthropology
exploration (geographical), 32, 38, 68–9, 71, 105, 116, 119, 126, 130, 134, 142, 146, 150, 152, 252; in the arctic, 24, 103, 129–30, 133, **136**, 146, 148, 151, 155, 159; HBC's approaches to, 104, 106, 108, 120; NWC's approaches to, 119; as science, 32–3, 49, 314n95

Falconer, William (1739–86), 86
fame: as motive or result of scientific study, 8, 18, 20, 26, 41, 103, 176, 206, 278, 281, 297
Fauna Boreali-Americana (1829–37) (Richardson), 167–8, 178
Fenchurch Street, 4–5, 11, **128**, 242
Fendler, Augustus (1813–83), 187
Ferrel, William (1817–91), 255
fertile belt, 232
Few Words on the Hudson's Bay Company (1846) (Isbister), 165
Fidler, Peter (1769–1822), **96**, 97–8, 111–18, 120–3
Finlay, Jacques-Raphael (Jaco) (1768–1828), 121
Finlay River, 170
fish: as subjects of study, 66, 82, 87–8, 90, 92, 167–8, 191, 282; lake trout, 274; steelhead trout (*Oncorhynchus mykiss, Salmo gairdnerii*), 180
Fitton, William Henry (1780–1861), 144

Fitzhugh, William, 272
Flathead Lake, 118
Fleming, Sandford (1827–1915), 221
Flett, Andrew (fl. 1846–81), 30, 250, 271
Flett, James (c. 1825–99), 271
Flora Americana Septentrionalis (1771) (Forster), 89
Flora Boreali-Americana (1833–40) (Hooker), 167, 178
Forster, John (Johann) Reinhold (1729–98), 89, 91–2, 125
Fort Albany (Albany House), 58, 83, 88, 100, 106
Fort Anderson, 254, 269, 272, 289
Fort Assiniboine, 141
Fort Chipewyan, 135, 147, 213
Fort Colville (Colvile), 24–5, **175**, 189
Fort Confidence, 151
Fort Churchill. *See* Churchill (Fort Churchill) (Fort Prince of Wales)
Fort Dunvegan, 213
Fort Franklin, 140–1
Fort Garry, 155, 236, 290
Fort Good Hope, 142, 147, 213, 254
Fort Liard, 256
Fort McLoughlin, 179
Fort McPherson (Peel River House), 152, 164, 254, 256
Fort Nelson, 168
Fort Nisqually, 179
Fort Prince of Wales (Churchill). *See* Churchill (Fort Churchill) (Fort Prince of Wales)
Fort Providence, 135
Fort Rae, 254, 256, 271
Fort Reliance, 152, 155
Fort Resolution, 254, 256, 271

Fort Simpson (Mackenzie River), 164, 213, 254–6, 258, 268–9, 287
Fort Simpson (Pacific Ocean), 177
Fort Snelling, 156
Fort St. James, 177
Fort Union, 209
Fort Vancouver, 25, 170, 172–3, 177, 179–80, 185, 187, 189, 194, 202–5, 216
Fort Walla Walla, 25, 179–80, 185
Fort William, 122, **182**, 213, 234, 253
Fort Yukon, 152, 254, 256, 271, 287
fossils, 89, 166, 249, 261, 336n22
foxes, 31, 150, 205, 207
Frankenstein (1818) (Shelley), 132
Franklin, John (1786–1847), 134–5, **136**, 137–43, 146–52, **154**, 159–61, 174
Fraser, Simon (1776–1862), 119
Fraser, Thomas, 266
Fraser River, 112, 116, 177
Fraser River gold rush (1858), 208
Frémont, John C. (1813–90), 187, 205
Frobisher, Benjamin (1742–87), 110
Frobisher, Joseph (1740–1810), 110
Frobisher, Martin (c. 1539–1594), 131
Frog Portage, 117
Fury and Hecla Strait, **136**, 140, 153–4, 159

Gairdner, Dr. Meredith (1809–37), 168, 177, 179–80, 186, 203
Galbraith, John, 15, 39, 165, 234, 304n17
gardens, 41–2, 72, 91, 171, 176–7, 181, 183–4, 186–9, 202, 205, 252, 297
Garry, Nicholas (c. 1782–1856), 17, 150, 168, 178
Gascoigne, John, 5
Gaudet, Charles P. (1827–1917), 254, 271, 285

gender, 6, 8, 30, 42, 103, 153, 169
General Synopsis of Birds (1781–1801) (Latham), 90
Gentleman's Magazine, 72, 110, 125
geography, 81–2, 95, 103–4, 112, 120, 133–4, 158, 160, 209, 288
Geological Survey of Canada (1842), 179, 220, 268, 292
geology, 134, 141, 144, 157, 159, 166, 195, 282
Geomagnetical and Meteorological Observatory (Toronto) (1841), 42, 200, 211–12, 220
geomagnetism, 42, 132, 134, 141, 144, 159, 210–13, 335n14; Magnetic Crusade (geomagnetic survey of the 1840s), 42, 200, 210–13; magnetic dip (inclination), 51, 158; magnetic variation (declination), 51, 53–4, 61, 132, 158; north magnetic pole, 130, 132, 211
Geyer, Karl (Charles or Carl) Andreas (1809–53), 24–5, 186–9, 205
giant chinkapin (*Castanopsis chrysophylla*), 183–6
Gibbs, George, (1815–73), 277, 283
Gillam, Zachariah (1636–82), 52, 53, 58
Gladman, George (1800–63), 233
Gladman, Joseph (1796–1875), 271
Gladstone, William E. (1809–98), 165
Glasgow Botanic Garden (1817), 171, 183
Globe (Toronto), 226, 228
Glorious Revolution (1688), 60
Gloucester House, 108
golden chestnut (*Castanopsis chrysophylla*), 183–6
Goode, George Brown (1851–96), 281
Gordon, Alexander (1813–71), 187

Gorst, Thomas (fl. *c.* 1668–87), 60
governments: their expectations of chartered companies, 6, 10, 36; their role in funding scientific activity, 6, 19, 28, 211
Graham, Alexander (1733–1815), 83, 87–90, 100, 102, 105–7, **107**,124
Grammar of the Cree Language (1844) (Howse), 118, 166
Gray, Asa (1810–88), 33, 187, 203
Gray, John Edward (1800–75), 159, 196, 261, 268
Great American Desert, 232
Great Bear Lake, 140
Great Slave Lake, 105, 111, 135, 137, 142, 155, 212, 271
Green, Charles (1735–71), 80, 81, 82
Greenwich House, 116
Grey Coat Hospital School (Westminster, London), 106, 111
Groseillers, Médard Chouart, Sieur des (1618–96), 52, 58
grouse (and ptarmigans), 54, 90, 150, 324n85
Gulf of Boothia, 152
Gunn, Donald (1797–1878), 242, 248, 252, 259, 272–3, 277, 284
Gwin, William M. (1805–85), 286

Hadley, John (1682–1744), 61
Hagstrom, Warren O., 11
Haida Gwaii (the Queen Charlotte Islands), 172
Haldane, John (fl. 1798–1857), 168
Halley, Edmond (1656–1742), 54–5, 61, 76–7
Hardisty, Mrs. William Lucas, 269
Hardisty, William Lucas (c. 1822–81), 214, 254, 271, 275, 279–80, 284–5
hares, 90

Hargrave, James (1798–1865), 24, 143, 155–56, 191, 193
Hargrave, Joseph James (1841–94), 193, 243, 264
Hargrave, Letitia (1813–54), 185
Harmon, Daniel Williams (1778–1843), 23
Harrison, Mark, 6, 7, 295
Harvard University, 187, 202
Hawaii, 177, 180, 203
Hayes, James (1637–94), 47, 54, 56
Hayes River, 58
Hearne, Samuel (1745–92), 38, 41, **69**, 74, 91, 102–3, 105, 108–12, **115**, 119, 124–5, 131–2, 136, 139, 146, 148, 158, 168
Hector, James (1834–1907), 232
Henday, Anthony, **69**, 105
Henley House, 59, 100, 104, 108
Henry, Joseph (1797–1878), 15, 203, 210–11, 214, 228, 239, 241, 244–6, 249, 250, 252–3, 260, 262, 265, 276, 280–5, 288
Hincks, Francis (1807–85), 221
Hind, Henry Youle (1823–1908), 221, 233
Hind Expeditions (1857, 1858), 43, **182**, 253
hinterland science, 17, 20, 33, 239; defined, 17
History of Astronomy (1767) (Costard), 77
Hodges, William (1744–97), 82
Hodgkin, Thomas (1798–1866), 144
Hodgson, John, 106
Hood, Robert (1797–1821), 134
Hooker, Joseph Dalton (1817–1911), 203
Hooker, William Jackson (1785–1865), 24–5, 141, 144, 151, 159, 167, 169, 171–3, 176–9, 183–9, 203

Hopkins, Edward Martin (1820–93), 264, 280
Hornsby, Thomas (1733–1810), 78
Horrox (Horrocks), Jeremiah (1619–41), 76–7
horticultural organizations, 13, 41, 171, 174, 176–7, 190. *See also* gardens
Horticultural Society of London (1804) (Royal Horticultural Society), 171, 177, 358n10
horticulture, 41, 171, 189
Howland, William P. (1811–1907), 221
Howse, Joseph (1774–1852), 98, 118, 122, 166
Howse Pass, 121
Hoy, Philo Romayne (1816–92), 252
Hudson, George (c. 1761–90), 111
Hudson, Henry (d. 1611), 53, 131
Hudson Bay coast: scientific knowledge of, 49, 53
Hudson House, 108, 109
Hudson's Bay Company: expansion after 1774, 106; Governor and Committee, 57; Mackenzie River District, 142; no official government representation on London Committee, 6; transportation routes, **4**, 9
Hudson's Bay Company charters and licences, 3, 12, 39, 165, 178, 245; 1670 charter, 6, 35, 47, 49, 52, 156, 191; controversies over, 43, 68, 70–1, 93, 160, 226–7, 234; end of, 292; licence of 1821, **4**, 39, 41, 144, 146, 170; licence of 1838, 151, 192, 223, 229, 231; licence of 1859, 232, 246
Hudson's Bay Company employees: clerks, 29–30, 214, 269, 271, 272; non-officers, 29–30, 271; number of, 22; officers, 9–10, 12, 21–30, 34,

39, 67, 88, 90, 139, 141, 149, 162, 168–9, 184, 186, 192, 201–2, 213, 216, 234, 240, 242, 250, 253, 256, 258, 266, 272, 276–8, 280, 282, 284–5, 290; ship captains, 29, 36, 49, 54, 57, 61, 73, 296; significance of furloughs and retirement, 23, 36, 49, 62, 66, 88, 100, 118, 159, 196, 197; surgeons, 9, 29, 36, 100, 106; surveyors, 38, 92, **96**, 108, 113–14, 116, 120

Hudson's Bay Company territories, **4**, 6–9, 15, 38–40, **96**, 157, **175**, **182**, 201, 207–9, 218–22, 226–9, **230**, 231–37, 240–1, 285–6, 291–2, 300n3, 310n58, 377m106; agricultural potential of, 43, 88, 99, 152, 170, 201, 226–35; relatively free of dangerous diseases, 9; scientific interest in, 18, 34, 79, 83–91, 95, 99–104, 109–11, 164–6, 169–78, 181–98, 200, 204–7, 209–10, 212–14, 216, 229, 238–40, 244, 248–52, 255–6, 262, 282–3, 289, 292–3; travel to and in by outsiders, 6, 9, 14, 22, 41–3, 80, 135, 141–2, 161, 164, 171–2, 174, 177, 181, 185, 189, 199, 200–4, 213, 215–6, 226, 250, 253–6, 306m27, 370n7, 372n32. *See also* Columbia District, Mackenzie River District, New Caledonia, Rupert's Land

Hudson's Bay Company–North West Company merger (1821), 27, 38–9, 41, 130, 139–40, 170–71, 209, 226

Hudson's Bay Company–North West Company rivalry, 27, 35, 37–8, 95, 129

Hudson Strait, 52–3, 59, 61, 100, 131, 140

humanitarians, 11–2, 143–5

Humboldt, Alexander von (1769–1859), 194, 211, 259
Hunt, Leigh (1784–1859), 80
Huron. *See* Macdonell, Allan
Hutchins, Thomas (1742–90), 26, 83, 86, 90, 97, 99, 101–2, 124, 278, 296
hydrographers, 71, 109
hydrography, 133–4

ice and icebergs, as subject of inquiry, 54, 62, 88
Icy Cape, 110, 130, **136**
Idotlyazee, 73, 74
Île-à-la-Crosse, 111
Illustrations of Zoology (1831) (Wilson), 178
Industrial Museum of Scotland (Royal Scottish Museum) (Edinburgh Museum of Science and Art), 192–3, 197, 261, 268, 278
insects and spiders, as subjects of inquiry, 88, 141, 144, 160, 167, 195–7, 269. *See also* entomology
Instauratio Magna (1620) (Bacon), 48
Institute of Rupert's Land (1862), 290–1
instruments, scientific, 9–10, 20, 25–6, 34, 36, 51, 61, 62, 77, 83, 85–6, 100, 104–5, 118, 139, 148, 160, 196, 261, 262
International Financial Society (1863), 246
Introduction to the True Astronomy (1721) (Keill), 75
Isbister, Alexander Kennedy (1822–83), 152, 164, 166, 228, 231
Isbister, Thomas (c. 1793–1836), 164, 168
Isham, James (1716–61), 61, 65–71, 73; "Observations on Hudson's Bay," 66

Jacob, Margaret C., 22
Jacobs, Ferdinand (c. 1713–83), 86, 88, 93
James, Thomas (1593–1635), 52
James Bay, 37, 59, 265, 293
Jameson, Robert (1774–1854), 180
Jarvis, Edward (d. c. 1800), 106
Jasper House, 185, 189
Jeffrey, John (1828–54?), 179, 181, 189–90, 192, 198
Jones, Samuel, 67
Jones, Strachan, 271, 284
Journal of the Royal Geographical Society, **149**, 150
Journal of the Society of Arts, 163
Journey from Prince of Wales's Fort in Hudson's Bay to the Northern Ocean (1795) (Hearne), 103–4, 125, 336n31, 337n33, 337n34

Kalm, Peter (Pehr) (1716–79), 89
Kane, Paul (1810–71), 42, 200, **214**, 214–28, 237
Keefer, Thomas (1821–1915), 221
Keill, John (1671–1721), 75–6
Keith, George (1779–1859), 141
Keith, James (1782–1851), 141
Kelsey, Henry (c. 1667–1724), 54, **69**, 104
Kennedy, Robert, 166
Kennicott, Robert (1835–66), 24, 238–9, 243, 250, **251**, 252–3, **254**, 255–6, **257**, 258–66, 269, 271–90
Kensington Museum (South Kensington) (London), 276, 397n208
Kent Peninsula, **136**, 150
King, Richard (1810–76), 142–3, 146, 199, 231
King's College (Aberdeen), 164

Kirkby, William West (1828–1907), 193, 285
Kirtland, Jared Potter (1793–1877), 252–3
Kittson, Norman (1814–88), 207
Knight, James (1640–1720), 67, 131
Knowsley Hall, 184, 186
Kotzebue, Otto von (1787–1846), 133–4

Labrador, 236, 255, 271
Lachlan, Major Robert (1782–1871), 262
Lac La Biche, 116
Lake, Bibye (1684–1744), 60, 67–8
Lake Athabasca, 98, 106, 109–14, 116–17
Lake Manitoba, 106, 117
Lake Winnipeg, 106, 109, 117
Lake Winnipegosis, 106
Lalande, Joseph Jérôme Lefrançois de (1732–1807), 78
Lamb, Charles (1775–1834), 80
Lambert, Johann Heinrich (1728–77), 84–5
Lancaster Sound, 134
La Pierre's House, 152, 254, 256
Latham, John (1740–1837), 90
lay collectors, 22, 27, 36, 263, 266, 278, 280, 289
lay science, 17–20, 28; defined, 18
lay scientists, 20, 27–8, 296
LeConte, John Lawrence (1825–83), 197, 243
Lefroy, John Henry (1817–90), 42, 155, 200, 213–15, **214**, 218, 221, 225–8, 231, 237, 266, 297
Lefroy (Robinson), Emily (d. 1859), 218

leisure, 23–4, 26, 28, 30, 66, 142, 272, 277
Leith, James (1777–1838), 141, 168
Lesser Slave Lake, 116
letters (correspondence): significance of, 22, 27, 29, 31, 58, 135, 152, 184, 188, 190, 194, 216, 218, 225, 235, 250, 261, 262, 263, 292
Levant Company, 6, 91, 296
Lever, Sir Ashton (1729–88), 102
Levere, Trevor, 129
Lewes, John Lee (1792–1872), 213, 216
Lewis and Clark Expedition (1804–6), 121, 171
Liard River, 268
Liard River (River of the Mountains), 168
libraries, 24–6, 98, 172, 215, 221, 249–50, 279–80. *See also* books
Light, Alexander (fl. 1733–43), 61–2, 66, 73
Lincoln, Henry Pelham Fiennes Pelham Clinton, Earl of (later the Duke of Newcastle) (1811–64), 165
Lindheimer, Ferdinand Jacob (1801–79), 187
Lindley, John (1799–1865), 176
Lindsay, Debra, 240, 244, 258, 259, 260, 271, 273, 284, 285, 287
linguistics, 118, 180; vocabularies, 118, 160, 179
Linnaeus, Carolus (1707–78), 82
Linnean Society (1788), 184
lizards, 179
Lockhart, James (b. 1827), 254, 263–5, 269, 271, 275
Logan, William Edmond (1798–1875), 261, 268
London Horticultural Society (Royal Horticultural Society), 41, 171, 174, 176–7, 190

London Journal of Botany, 184, 188
Low, Albert Peter (1861–1942), 293
Lüders, Friedrich G. L. (1818–1904), 187–8
Lyell, Charles (1797–1875), 24

Macdonell, Allan (1808–88), 226–7
Macdonell, Miles (*c.*1767–1828), 37
MacFarlane, Roderick Ross (1833–1920), 193, 249, 254, 260–4, 269, **270**, 272, 275, 278, 280, 289, 296
Mackenzie, Alexander (1764–1820), 112, 116, 119, 121, 136
Mackenzie River, 112, 130, 140, 148
Mackenzie River District, **4**, 148, 191, 215, 232, 238, 240, 249–50, 254–6, 260–1, 265–6, 268, 271, 281, 286–7, 289
MacLeod, Roy, 17–18
Mactavish, William (1815–70), 141, 143, 164, 252, 260–1, 271, 276, 279, 290, 296
mammals (quadrupeds), 54, 87, 90, 92, 141, 150, 167, 168, 191, 205–7, 249, 260, 273, 274
maps, 51, 177, 340n74; by Aaron Arrowsmith, 112–17, 118, 121, 122, 125; by aboriginal people, 67, 73, 116, 158; of the arctic, 130, 135, 138, 140, 143, 149, 159; by David Thompson, 114, 118, 121–3; of Hudson Bay coastline, 62, 73–4, 80, 120; isothermal, 228; by James Clouston, 118; by Joseph Howse, 118; of Pacific Coast of North America, 110–12; by Peter Fidler, 114, 117, 122–3; by Peter Pond, 110–12; by Philip Turnor, 112; of Rupert's Land, 104, 106–7; by Samuel Hearne, 110–11; of the world by Henry Roberts, 109

Marley, John, 109, 338n58
marmots, 346n39
Marten, Humphrey (c. 1729–c. 1790?), 88, 90–1, 97
Martin, Benjamin (1705–82), 26, 77
Martin's Fall, 195–6
Maskelyne, Nevil (1732–1811), 80
Massacre at Bloody Fall, 103, 336n31
Massey, Dr. R. M., 62
Matonabbee, 74, 102, 326n129
McBean, William (b. c 1807), 216
McDonald, Archibald (1790–1853), 24, 141, 188–9
McGill University (College), 197, 210, 384n18
McIntyre, John, 253
McKenzie, Alexander, 271, 275
McKenzie, James, 271
McKenzie (Mackenzie), John George (b. 1808), 191, 265
McKinlay, Archibald, 189
McLean, John (1798–1890), 213
McLeod, John (1788–1849), 204
McLoughlin, John (1784–1857), 25, 39, 170, 172–4, 177, 179, 183, 185, 188–9, 199, 201–4, 296; as Father of Oregon, 173
McMaster, William (1811–87), 221
McMurray, William (c.1820–77), 271
McNab, John (b. 1755), 86, 97, 100
McPherson, Murdoch (c. 1795–1863), 152, 168
McVicar, Robert (c. 1794–1864), 137, 168
Melville Peninsula, 152–3, 158–9
Menzies, Archibald (1754–1842), 171–2, 189
mercury (element), experiments with, 100–1, **101**, 335n16
Merivale, Herman (1806–74), 165

Mersch, Karl Friedrich (1810–88), 187
meteorological records, 26, 28, 30, 34, 80, 85–6, 97–9, 173, 239, 269, 282
meteorology, 20, 32, 36, 50–1, 67, 83, **84**, 85–6, 93, 95, 97–9, 118, 120, 132, 141, 144, 152, 262, 265, 275, 281, 309n45
Methye Portage (Portage La Loche), 111, 117, 250, 255, 258
metropolitan science, 17; defined, 17
Mexican–American War (1846–8), 209
microscopes, 196, 261
Middleton, Christopher (d. 1770), 61–3, 67–8, 70, 73, 124, 131, 158, 199, 296
minerals, 88, 134, 148, 166, 228, **230**, 232, 236, 261
Minnesota, 157, 208, 241, 248, 262, 291
missionaries, 11, 201, 205, 248, 290, 294
Missouri River, 118, 208–9
Mohr, Johan Maurits (c.1716–75), 85
monogenesis, 145, 350n81
Monthly Review, 72
Montreal, 23, 209, 212, 253
Montreal Gazette, 226
Moor (Moore), William (d. 1765), **69**, 70
moose, 168, 275
Moose Factory, 275
Moose Factory (Moose Fort), 58, 100, 155, 166, 191, 254, 265–6, 323n73
Morris, Alexander (1826–89), 221
Morse, Jedidiah (1761–1826), 81
mound builders, 284
mountain goats, 249
Mount Brown, 174
Mount Hooker, 174
Mount St. Helens, 180

Moxon, Joseph (1627–1700), 319n28
Murchison, Sir Roderick Impey (1792–1871), 123
Murray, Andrew Dickson (1812–78), 179, 181, 190, 192–3, 261, 268, 276, 278–9, 297
Murthly Castle, 187, 202
Muscovy Company, 91, 296
Museum of Natural History at Northwestern University, 252
museums, 6, 13, 265, 279, 302n11; George Simpson's private, 253; of the Hudson's Bay Company, 167–8, 171, 253; in Inverness, 179; planned in Scotland, 260; of the Royal Society, 124; in Rupert's Land, 253, 290; Sir Ashton Lever's, 102, 336n22
musk oxen, 205, 207, 249, 273

Napier, Francis (1819–98), 250
Narrative of a Journey Round the World (1847) (Simpson), 15, 388n77
Narrative of a Journey to the Shores of the Arctic Ocean (1836) (King), 144
Narrative of a Journey to the Shores of the Polar Sea (1825) (Franklin), 138–9
Narrative of a Second Expedition to the Shores of the Polar Sea (1828) (Franklin), 141–2
Narrative of the Arctic Land Expedition (1836) (Back), 143–4
Narrative of the Discoveries on the North Coast of America (1843) (Simpson), 151
National Geographic Society (Washington) (1888), 261
National Institute (for the Promotion of Science) (USA (1840), 206
National Museum (USA), 244–5, 253, 274–5
Natural History Society of Montreal (1827), 197, 209–10, 220, 242, 261, 268
natural history, 8, 14–15, 20, 22, 25, 31–33, 36, 42, 50, 54–5, 62, 66–7, 72, 80, 82, 86, 88–9, 91, 93, 95, 100, 103, 118, 120, 132, 134, 139, 143, 148, 159, 167–72, 179–80, 187, 190–1, 193, 197, 210, 239, 244–6, 248–9, 252–3, 255, 259, 261–3, 265, 268–9, 271–2, 275, 281, 282–3, 287–8, 293, 309n45; defined, 32, 313n89
Natural History of Uncommon Birds (1743–51) (Edwards), 62, 66–7
natural philosophy, 299n1; defined, 32, 313n89
natural theology, 169
Navigantium atque Itinerantium Bibliotheca (1748) (Campbell), 73
navigation, 50, 53, 67, 81, 104, 132, 134, 155
Navy (British), 40, 68, 105, 130–4, 146–8, 157–61, 168
Navy (United States), 244, 255
Neile, Sir Paul (1613–86), 47, 55, 56, 318n22
Nelson River, 117
New Atlantis (1627) (Bacon), 32, 48, 51
New Caledonia, 39, 147, 168, 170, 177
New France, 35, 37, 60, 74, 227
Newton, Alfred (1829–1907), 272
New York Historical Society, 268
Nicollet, Joseph Nicholas (1786–1843), 187
Nixon, John (c. 1623–92), 54, 56
Nootka Crisis (1789–90), 111
North Saskatchewan River, 116
North West Company (NWC), 24, 38, 95–6, 98, 106–7, 110, 112, 116, 117,

119, 120–3, 125, 134–5, 137–9, 142, 147, 173, 193, 209, 227, 248, 338n52
Northwest Passage, 33, 41, 47, 53, 62, 67–74, 102–3, 113, 116, 129–34, 140, 142, 150–3, 156, 159, 168–9, 199, 210–11, 320n47, 326n127
Norton, Moses (*c.* 1735–73), 73, 88, 109, 124
Norway House, 164, 196, 213, 253, 255
Novum Organum (1620) (Bacon), 6, 48
Nuttall, Thomas (1786–1859), 42, 186–7, 202, 203, 205, 237–8

octants, 61, 83, 105, 139
Ogden, Peter Skene (*c.* 1790–1854), 24, 29, 204
Ogden Isaac, 110
Oldenburg, Henry (1618–77), 52, 53–4
Oldman River, 113
Oldmixon, John (1673–1742), 60, 74
Onion (Camsell), Julian Stewart (1839–1907), 193, 271
Orcadians, 153
Oregon Botanical Association (British Columbia Botanical Association), 189–90, 192
Oregon Boundary dispute, 40, 118, 164, 170, 176, 201, 205, 208, 231, 316n114
Oregon Country (1818–46). *See* Columbia District (Oregon Country)
Oregon Fever, 205
Oregon Territory (USA) (1846–59), 155, 186. *See also* Columbia District (Oregon Country)
Oregon Treaty (1846), 157, 207
Orkney Islands, 59, 100, 153, 164
ornithology, 168, 279. *See also* birds as subjects of study

Osborn, Henry Fairfield (1857–1935), 263
otters, 274
owls, 66, 150
Owram, Doug, 218, 220
Oxford University, 261

Palliser, John (1817–87), 123, 232, 236
Palliser Expedition (1857–60), 43, **182**, 255
Palliser's Triangle, 232
Parkinson, Sydney (*c.* 1745–77), 82
Parrish, Susan Scott, 20
Parry, William Edward (1790–1855), 133, 140
partridges (ptarmigans), 54, 90
Paxton, Joseph (1803–65), 181
Peel, Robert (1788–1850), 157
Peel River, 152, 164, 165, 256
pelicans, 88
Pelly, John H. (1777–1852), 17, 141, 150–2, 167, 178, 213, 296
Pembina, 208, 241
pemmican, 37, 140
Pennant, Thomas (1726–98), 88, 90, 99, 102–3, 123–4, 167, 297
permafrost (permanently frozen soil), as a subject of study, 152
phenology, 86, 160, 309n45
Philadelphia, 197, 212, 243, 268
Philosophical Transactions of the Royal Society, 6, 33, 51, 61, 63, 67–8, 70, 78, 87–8, 90, 92, 100
Pickstone, John, 7, 296
plants, as subjects of study, 41, 53, 91, 92, 133, 134, 141, 148, 151, 159, 169–91, 196, 202, 236, 246, 261; introduced to Great Britain by David Douglas, 176–7, 181. *See also* botany; trees

Point Barrow, **136**, 141, 148–9
Point Turnagain, **136**, 150
Polk, James K. (1795–1849), 153, 155–6
polygenesis, 242, 384n11
Pond, Peter (1739/40–1807), 110–11, 116–17, 119
Poore, Sir Edward (1826–93), 29
Port Nelson, 58
Pratt, Mary Louise, 44
Prehistoric Man (1862) (Wilson), 222
Present State of Hudson's Bay (1790) (Umfreville), 125
prestige: its connection with science, 3, 15, 17, 29, 33, 78, 96, 131, 133, 169, 244, 246, 280, 294
Prince Rupert (Prince Rupert of the Rhine) (1619–82), 55
Puget Sound, 174
Pullen, William John Samuel (1813–87), 266
purple-flowering currant (*Ribes sanguineum*), 176

quadrants, 61, 83, 105, 139
Quadrupeds of North America (1846–54) (Audubon), 206
Qu'Appelle River, 117

Radisson, Pierre Esprit (*c.* 1640–1710), 52, 58
Rae, John (1813–93), 24, 41, 153–64, **154**, **162**, 253, 266, 271
Rankin, Colin (1827–1921), 271
Red River, 117, 208
Red River Academy, 164
Red River Colony, 22, 37, 147, 156, 174, 196, 207–8, 222, 224, 229, 241–3, 248, 252–3, 258, 284, 290, 306n23
Red River Expedition (1857), 253

Reid, John (*c.* 1826–97), 271, 275
reindeer. *See* caribou
religious faith, and science, 357n2
Repulse Bay, 62, **69**, **136**, **154**, 158–9, 161
Return Reef, **136**, 141, 149
Rich, E. E., 52, 147, 151, 205, 234, 304n17
Richardson, John (1787–1865), 62, 123, 134, 139–41, 143, 152, **154**, 158–60, 164, 167–8, 171–2, 174, 179–80, 196, 231, 249, 268, 297
Roberts, Henry (1757–96), 109
Robertson, Colin (1783–1842), 135
Robinson, John Beverly (1793–1859), 218, 221
Robson, Joseph, 71–4, 125, 165
Roche, Alfred R., 227, 231
Rocky Mountain Fur Company, 187
Rocky Mountains, 118, 141
Romantic science, 103, 138, 151
Rooke, Laurence (1622–62), 51
Ross, Bernard Rogan (1827–74), 30, 191, 193, 214, 242, 249, 250, 254–5, 258, 261, 266, **267**, 269, 271, 273, 277–8, 280, 282–4, 297
Ross, Christina, 269
Ross, Donald (1797–1852), 156, 233, 269
Ross, James Clark (1800–62), 211
Ross, John (1777–1856), 133–4, 142
Ross, Malchom (1754–99), 98, 111
Ross's goose, 273–4, 278, 280
Royal African Company, 60, 296
Royal Botanic Gardens (Edinburgh), 189
Royal Botanical Gardens, Kew (London), 91, 183, 184, 297
Royal Colonial Institute (1868) (Royal Commonwealth Society), 261

Royal Geographical Society (1830), 13, 17, 118, 123, 149–51, 161, 163–4, 232, 261, 268, 307n40
Royal Horticultural Society. *See* London Horticultural Society
Royal Naval Hospital Haslar (Gosport), 180
Royal Observatory, Greenwich, 80–1
Royal Society of Canada, 197
Royal Society of London, 6, 13, 28, 35–6, 47, 49–58, 61, 66, 71, 75, 78–9, 82–92, 97–100, 109, 120, 134, 163, 211, 294, 297
Rumyantsev, Count Nikolai Petrovich (1754–1826), 133
Rupert's Land, 227
Rupert's Land defined, **4**, 300n3, 304n17, 310n58. *See also* Hudson's Bay Company territories
Russia, 132, 139
Russian America, **4**, 15, 40, 140, 152, 177, 179, 212, 237, 256, 258–9, 284–8, 291, 296, 400n279
Russian American Company (RAC), 15, 40, 147, 289, 296, 303n15
Ryerson, Egerton (1803–82), 221

Sabine, Edward (1788–1883), 134, 203, 211–13
Sabine, Joseph (1770–1837), 171
Saskatchewan River system, 106, 113–14, 116, 141, 208, 229, 232, 255
Sault Ste. Marie, 155
Schultz, John Christian (1840–96), 290
science: as apolitical, 3, 7, 11–12, 240–41; defined, 32, 299n1; international, 42, 77–8, 83, 85–6, 200, 210–11, 242; as ostensibly disinterested, 3, 7, 10–12, 33, 41, 53, 110, 129, 131–32, 142, 145, 166, 203, 233, 240, 296, 300n2; state support for, 77, 200, 210; as transcending nationalism, 242, 246
scientific organizations, 8, 13, 14, 28, 36, 40, 42, 44, 49, 56, 169, 198, 248, 263, 268, 289
Scouler, John (1804–71), 168, 172–3, **175**, 179, 201
seals, 150
secrecy: in chartered monopolies, 10, 36; Hudson's Bay Company's, 7, 11, 35, 37, 50, 57–8, 60–1, 72, 75, 79, 91, 97, 102, 104, 107, 125–6, 133, 306n29; James Cook and, 81; North West Company's, 121; in science, 50–1
Seller, John (1630–97), 53
Selwyn, Alfred R. C. (1824–1902), 292
Severn House (Fort Severn), 86, 88–9, 107
Seward, William H. (1801–72), 236–7
sextants, 160
Shaftesbury, Anthony Ashley Cooper, First Earl of (1621–83), 56, 318n22
Shelley, Mary (1797–1851), 132
Sherwood, Morgan, 282
ships: *California*, 70, 124; *Discovery*, 69; *Dobbs Galley*, 70, 124; *Furnace*, 69; *Hannah*, 61; HBC's 306n27, 325n106; *Nonsuch*, 52; *Prince Rupert*, 80; *Prince Rupert* (V), 185; *Unity*, 55; *William and Anne*, 172, 175
Similkameen Valley, 189
Simpson, Alexander (1811–c. 1875), 144, 146
Simpson, George (1787–1860), 15–17, **18**, 27, 40, 135–137, 143–4, 146–8, 151, 153, 155–6, 158, 160, 164, 170, 174,

180, 192, 194–6, 201–2, 204–5, 207, 209, 213–15, 221–26, 231, 233–5, 238, 243, 249–50, 252–3, 255, 260–1, 264, 266, 275, 280, 290, 296
Simpson, Thomas (1808–40), 143, 144, 146, 147, 148, 151, 155, 164
Sitka (Novo-Arkhangel'sk) (Russian America), 212
Sixth Regiment of Foot, 157
Skeena River, 112, 116
skulls, collecting, 173, 180, 222, 360n25, 394n160
Sloane, Hans (1660–1753), 55, 63
Smith, Adam (1723–90), 93, 165
Smith, Edward, 39, 168
Smith, Francis, 70
Smith, William G., 225
Smithsonian Institution, 15, 24, 28, 43, 179, 228, 235, 238–244, 247–49, 251–52, 254, 258, 260, 262, 266, 270, 272, 274–5, 280–1, 284–5, 288–9, 291, 293, 294, 297
Smithsonian Institution Annual Reports, 242, 248–9, 266, 269, 274–5, 278, 284, 286, 288
Smithson, James (*c*.1764–1829), 241, 243
Snake River region, 201
snow buntings, 150
snow geese, 273–4, 278, 280
social intelligence, 21, 44, 80, 124, 143, 169, 198, 262, 329n27
soils, 53, 88, 228, 232, 234
Solander, Daniel Carl (1733–82), 82, 91
solar parallax (and solar distance), 76–9, 81, 327n4, 327n12, 330n35
solar radiation, studies of, 141
Somerset Island, 161
Southern Department, 209
South Saskatchewan River, 116

species: introduced to Europe, 176–8, 181; named after people, 15, 27–8, 203, 278, 364n100; new to science, 25, 66, 67, 89, 134, 274; ranges and distribution of, 159, 282; taxonomy and allotment, 242, 246, 273–4, 278, 372n32
specimens, scientific, 8, 9, 15, 17–18, 21, 23, 28, 31, 54–5, 62, 64, 66–7, 77, 82, 88–92, 100, 102, 124, 133–4, 139, 140–1, 148, 150–1, 158–9, 167–8, 170–9, 183, 185–93, 196–7, 202–10, 222, 236, 238–56, 259–82, 288–9, 297, 333n77, 356n192, 357n192
Spence Bay, **136**, 150
Stanley, Edward Smith, 13th Earl of Derby (1775–1851), 184
Stayner, Thomas (1770–*c*. 1827), 111
steamboats, 208
Stefansson, Vilhjalmur (1879–1962), 161
Stewart, William Drummond (1795–1871), 187, 202
St. Louis (Missouri), 24, 187–8, 202
Stocking, George W., 145
stoneflies, 196
St. Paul (Minnesota), 208, 241, 253, 292
St. Paul Daily Press, 292
Strachan (Robinson), Augusta, 218
Strachan, John (1778–1867), 218
Sullivan, John William, 232
Sumner, Charles (1811–74), 287, 288
surveying, 32, 38, 80, 95–7, 104–6, 108–9, 111, 113–14, 117–18, 120, 148, 153, 155, 158, 176, 214, 245
Sutherland, Donald (1778–1872), 119
Swainson, William John (1789–1855), 168
Swan River, 117

Swan River district, 114
swans, 92–3
Swanston, Thomas, 271

Tahiti, 75, 80, 81, 88
Taylor, James Wickes (1819–93), 291
Taylor, Nicol (b. *c.* 1817), 271
Tennant, James (1808–81), 159
thermometers, 34, 85–6, 100
Thompson, David (1770–1857), **96**, 98, 111, 113, 118–19, 120–23; his "Great Map," 122
tides (ocean), 51, 53
tobacco, as gift, 179
Tocqueville, Alexis de (1805–59), 28
Tolmie, William Fraser (1812–86), 177, 179–80, 186, 203, 248
Tolstoy, Leo (1828–1910), 295
Tomison, William (*c.* 1739–1829), 105, **107**, 113
topography, 141
Toronto, 42, 210, 212, 218, 220–1, 253
Townsend, John Kirk (1809–51), 42, 187, 202–3, 205, 238
Transactions of the American Philosophical Society, 85
Transactions of the Ethnological Society of London, 163
Transactions of the Royal Society of London, 85, 101
Transits of Venus, 36, 75–9, 83–5, 296; in 1639, 75–7; in 1761, 78–9, 85; in 1769, 36, 75, 83, 85, 94, 97, 129; in 1874, 77
trees, 66, 82, 98, 169, 171, 189, 191, 229; conifers, 41, 169, 184; Douglas fir (*Pseudotsuga menziesii*), 178; Garry oak (*Quercus garryana*), 178; grand fir (*Abies grandis*), 178; lodgepole pine (*Pinus contorta*), 178; noble fir (*Abies procera*), 178; oaks, 184; Sitka spruce (*Picea sitchensis*), 178; western hemlock (*Tsuga heterophylla*), 190
trees and forests: as subjects of study, **230**
Trent Affair, 241
tribute: defined, 11, 307n30; examples of, 15, 62, 65–6, 87, 92, 124–5, 138, 141, 143, 167–8, 174, 178, 189, 191–2, 206, 212–3, 223, 235, 265–6, 275; as a motive for scientific activity, 8, 36, 294; significance of, 11–12, 15, 28, 126, 296
Turner, Lucien McShann (1847–1909), 293
Turnor, Philip (*c.* 1751–99 or 1800), 86, 92, **96**, 98, 108–17, 120, 123

Umfreville, Edward (b *c.* 1755), 125, 165
United States expansionism, 156–7, 208, 228–9, 236, 240, 286, 289, 292; manifest destiny, 43, 156, 208, 240, 244, 285, 292
United States Exploring Expedition (1838–42), 204, 245
United States postal service, 208, 248
United States War of Independence (1776–83), 131
universities and colleges, 6, 19, 28, 211
University of Edinburgh, 164, 180, 189, 192
University of Glasgow, 141, 171, 183, 189
University of London, 176
University of Michigan, 253
University of Toronto, 192, 222, 233, 242

Vancouver, George (1757–98), 111, 116, 171, 172
Vancouver Island, 173, 189, 231, 232
VanKoughnet, Philip M. (1822–69), 221, 227, 231
Venus in the Sun (1761) (Martin), 77
Verne, Jules (1828–1905), 256
Victoria, Queen (1819–1901), 151
Victoria Land (Island), 161
Voyage of Discovery, into the South Sea (1821) (Kotzebue), 133
Voyage of Discovery, Made under the Orders of the Admiralty (1819) (Ross), 134
Voyage of Discovery To The North Pacific Ocean (1798) (Vancouver), 172

Wales, William (1734–98), 36, 75, 79–83, 85–6, 88, 92, 97, 103, 105, 108, 297, 329n27
Wallace, Robert (d. 1838), 179, 181, 183, 185, 198, 306n27
Wanderings of an Artist (1859) (Kane), 222, 224, 225, 226
War of 1812, 131, 170
Washington, D.C., 23, 30, 212, 239–40, 243, 246, 252–3, 260, 286
Wealth of Nations (1776) (Smith), 93
Wegg, Samuel (1723–1802), 36, 38, 75, 79, 82, 86–9, 97, 100, 102, 109, 125, 296
Western Union Telegraph, 258, 260, 285–6
Whewell, William (1794–1866), 211
White, Adam (1817–79), 160, 196, 384n18
White, Gilbert (1720–93), 88
Whitway, James (*c.* 1778–*c.* 1838), 122
whooping cranes, 88

Wilkes, Charles (1798–1877), 204
Willamette River, 174
Willamette Valley, 153, 189, 201, 204, 216
Williams, Glyndwr, 57, 60, 67, 71
Williams, William (d. 1837), 135
Wilson, Daniel (1816–92), 222, 290
Wilson, George (1818–59), 192–3, 268, 283
Wilson, James (*c.* 1665–*c.*1730), 178
Wollaston, William Hyde (1766–1828), 117
Wollaston Lake, 117
Wollaston Land (Island), 161
women, as collectors, 269
Wordsworth, William (1770–1850), 103
Work, John (*c.* 1792–1861), 174
Wren, Christopher (1632–1723), 47, 52, 54, 57, 318n22
Wrigley, Joseph (1839–1926), 292
Wyeth, Nathaniel Jarvis (1802–56), 187, 202, 203, 205, 237

XY (New North West) Company, 121, 147

York Factory (York Fort), 24, 58–9, 70, 72, 83, 86, 88, 90, 97–8, 104–5, 107–9, 113–14, 135, 147, 155, 174, 185, 189, 241, 279
Young, James, 54
Yukon River, 256

Zoological Society of London (1826), 168, 184
zoology, 123, 143, 150, 157, 159, 164, 167, 196, 261, 265, 288